Physics of
Highly-Ionized
Atoms

NATO ASI Series

Advanced Science Institutes Series

A series presenting the results of activities sponsored by the NATO Science Committee, which aims at the dissemination of advanced scientific and technological knowledge, with a view to strengthening links between scientific communities.

The series is published by an international board of publishers in conjunction with the NATO Scientific Affairs Division

A	**Life Sciences**	Plenum Publishing Corporation
B	**Physics**	New York and London
C	**Mathematical**	Kluwer Academic Publishers
	and Physical Sciences	Dordrecht, Boston, and London
D	**Behavioral and Social Sciences**	
E	**Applied Sciences**	
F	**Computer and Systems Sciences**	Springer-Verlag
G	**Ecological Sciences**	Berlin, Heidelberg, New York, London,
H	**Cell Biology**	Paris, and Tokyo

Recent Volumes in this Series

Series B: Physics

Physics of Highly-Ionized Atoms

Edited by
Richard Marrus
University of California, Berkeley
Berkeley, California

Plenum Press
New York and London
Published in cooperation with NATO Scientific Affairs Division

Proceedings of a NATO Advanced Study Institute on
Atomic Physics of Highly-Ionized Atoms,
held June 5–15, 1988,
in Cargese, France

Library of Congress Cataloging in Publication Data

NATO Advanced Study Institute on Atomic Physics of Highly Ionized Atoms
(1988: Cargese, France)
    Physics of highly-ionized atoms / edited by Richard Marrus.
       p.   cm.—(NATO ASI series. Series B, Physics; v. 201)
    "Proceedings of a NATO Advanced Study Institute on Atomic Physics of
Highly-Ionized Atoms, held June 5–15, 1988, in Cargese, France"—T.p. verso.
    "Published in cooperation with NATO Scientific Affairs Division."
    Includes bibliographical references.
    ISBN-13: 978-1-4612-8105-4   e-ISBN-13: 978-1-4613-0833-1
    DOI: 10.1007/978-1-4613-0833-1

    1. Ions—Congresses. 2. Atoms—Congresses. 3. Collisions (Nuclear physics)
—Congresses. 4. Quantum electrodynamics—Congresses. I. Marrus, Richard. II.
North Atlantic Treaty Organization. Scientific Affairs Division. III. Title. IV. Series.
QC701.7.N36   1988   89-16347
539.7—dc19   CIP

© 1989 Plenum Press, New York
Softcover reprint of the hardcover 1st edition 1989
A Division of Plenum Publishing Corporation
233 Spring Street, New York, N.Y. 10013

PREFACE

        The progress in the physics of highly-ionized atoms since the
last NATO sponsored ASI on this subject in 1982 has been enormous. New
accelerator facilities capable of extending the range of highly-ionized
ions to very high-Z have come on line or are about to be completed. We
note particularly the GANIL accelerator in Caen, France, the Michigan
State Superconducting Cyclotrons in East Lansing both of which are
currently operating and the SIS Accelerator in Darmstadt, FRG which is
scheduled to accelerate beam in late 1989. Progress in low-energy ion
production has been equally dramatic. The Lawrence Livermore Lab EBIT
device has produced neon-like gold and there has been continued
improvement in ECR and EBIS sources.

        The scientific developments in this field have kept pace with
the technical developments. New theoretical methods for evaluating
relativistic and QED effects have made possible highly-precise calculations of energy levels in one-and two-electron ions at high-Z. The
calculations are based on the MCDF method and the variational method and
will be subject to rigorous experimental tests. On the experimental
side, precision x-ray and UV measurements have probed the Lamb shift in
the one and two electron ions up to Z=36 with increasing precision.
This work will ultimately lead to high precision tests in ions up to
uranium. In another area, the puzzling peaks observed in the positron
spectra resulting from heavy ion collisions in supercritical fields
persist, and attempts to explain them within the confines of conventional
physics have not succeeded. Experiments to test explanations based on
new particles have provided ambiguous or null results.

        The physics of highly-ionized atoms has provided the scientific
base for exciting new developments in the field of x-ray lasers and
fusion diagnostics. Several novel schemes for pumping x-ray lasers have
evolved and led to the coherent generation of very short wavelengths.
Finally, increasing sophistication has evolved in the diagnostics of
fusion plasmas using x-ray and UV spectroscopic techniques.

        The NATO sponsored ASI of June 1988 surveyed all of these
developments in a series of appropriate courses. This book is a
compilation of the resulting lectures. The efforts of many people
contributed to the success of this institute and it is appropriate to
acknowledge them here. Special thanks need to be accorded to my
colleague Jean-Pierre Briand who took much of the burden of the
organizing responsibilities on himself. Thanks are also appropriate for
Marie-France Hanseler who did a superb job of organizing arrangements
at Cargèse.

                                                            R. Marrus

CONTENTS

PARITY AND TIME REVERSAL INVARIANCE IN ATOMS

P.G.H. Sandars

Clarendon Laboratory
Parks Road
Oxford
OX1 3PU
U.K.

INTRODUCTION

In these four lectures we discuss current work on the status of parity (P) and time–reversal invariance (T) in atomic physics. A feature of the treatment will be a unified approach to parity non–conservation (PNC) without associated violation of time–reversal invariance, P not TV, PNC with associated T violation, P and TV, and T violation without PNC, T not PV. The approach will be primarily pedagogical, aiming to give a feel for the subject; completeness will not be attempted; the reader interested in more detail should consult the general reviews to be found in references 1–5.

The structure of the lectures will be as follows:

LECTURE I will give the general background to the subject, starting with the classical definitions of parity and time reversal, together with some simple applications. This is followed by a discussion of the types of experiment and phenomena which test the validity of P and T invariance in atomic physics. We then continue with a brief discussion of P and T in quantum mechanics. To relate this to our present state of knowledge from elementary particle physics, we briefly outline the history of P and T violation and discuss the implications of present theories for atomic physics in the form of additional terms in the atomic Hamiltonian. We then discuss the possible magnitude of these and comment on the choice of atom and experiment which seem to be the most favourable. We follow this with a list of those experiments currently under way throughout the world. Finally we make a few brief remarks about the status of atomic calculations needed to interpret the experiments.

LECTURE II is concerned with P not TV experiments. We deal first with optical rotation in heavy atoms. After outlining the general principles of such experiments and discussing the principles of an ultra–sensitive polarimeter we specialize to bismuth and discuss in detail the recent Oxford experiment[6]. We continue with a brief discussion of the experiment of Wieman[7] the latest in a series of beautiful experiments on electric field induced transitions in Cs initiated by the Bouchiats (see ref. 3). We conclude the lecture with a mention of an ingenious experiment proposed by Bouchiat et al.[8] to look for parity violation using a stimulated emission technique.

LECTURE III reports on experiments to look for atomic P and TV phenomena, particularly electric dipole moments. We first deal with atoms with non–zero electronic angular momentum. After discussing some conventional atomic beam experiments[9], we mention a novel laser experiment under development at Berkeley. We then discuss recent work on the TlF molecule at Yale[10] and conclude with the very beautiful optical pumping experiments at Washington[11,12].

LECTURE IV deals with the atomic theory (see refs 22, 13) needed to interrelate atomic experiments with the underlying physics. After some preliminary generalities, we discuss the simple single particle approach and its limitations. We then discuss in some detail the more complex calculations needed to reach the desired accuracy, concentrating on the many-body perturbation method. We mention the most significant features of the various calculations, though without any detailed discussion. We conclude with a tabulation of the results and a brief conclusion.

LECTURE I

I.1   <u>T and P in classical physics</u>

The classical versions of the two operators P, T are given in table 1.

Table 1   Space and time 'mirror' operations.

Symbol	'mirror' operation
P	x, y, z $\longrightarrow$ -x, -y, -z
T	t $\longrightarrow$ -t

The operation associated with parity is inversion, a sort of three dimensional mirror through the origin; it consists of a mirror image in one plane plus a rotation. Time reversal is actually motion reversal since clearly 'time' cannot be reversed. It can be imitated by running a film or TV recording backwards.

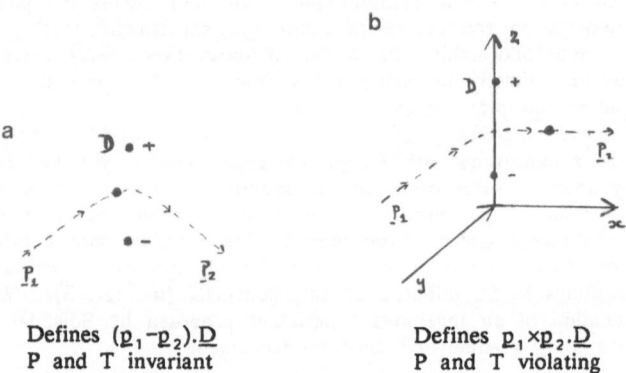

Defines $(\underline{p}_1 - \underline{p}_2) \cdot \underline{D}$
P and T invariant

Defines $\underline{p}_1 \times \underline{p}_2 \cdot \underline{D}$
P and T violating

Figure I.1   In both diagrams the whole motion takes place in the xy plane. In figure 1(a) the charges all lie in this plane while in figure 1(b) they lie on the z axis.

Classical physics is invariant under these operations. This means that if a particular motion is allowed so is its 'mirror' image. Figure 1 gives a pair of possible motions. Figure 1(a) is allowed because its mirror image and time−reverse can be rotated into the original motion; they are therefore P and T invariant. This is not true on the other hand for figure 1(b) where the two mirror images are quite distinct from the original. We see that figure 1(b) could be used to define a handed coordinate system through the triplet of vectors $\underline{p}_1$, $\underline{p}_2$ and $\underline{D}$ the dipole moment of the charge distribution. The direction of time can also be defined through such a motion.

Another very useful way of putting this is to note that figure 1(a) can define the scalar product $(\underline{p}_1 - \underline{p}_2).\underline{D}$ which is even under both P and T whereas figure 1(b) is associated with $(\underline{p}_1 \times \underline{p}_2).\underline{D}$ which is odd under both P and T.

I.2   P and T violating phenomena

It is convenient at this stage to generalize the line of argument made at the end of the previous section to discuss the sort of phenomena and experiments which would constitute evidence for violation of P or T invariance. We can do this in terms of classical concepts since at some level every experiment must ultimately be expressed in terms of macroscopic parameters. Specifically, an experiment to look for or measure P or T violation must involve a combination of quantities which is odd under the appropriate symmetry operation. Since we are interested in atomic and optical physics, we confine our attention to the set of quantities which describe the lowest order interactions of an isolated atom, possibly placed in electric or magnetic fields, with a photon in absorption, emission or fluorescence. The quantities of interest are set out in table 2, together with their transformation properties under P and T.

With these transformation properties it is a simple matter to construct combinations which are odd under the appropriate symmetry operators. The detection of an experimental result which allows the definition of such a signature would constitute evidence for violation of that invariance. Thus detection of an absorption which depends on the handedness $\underline{\sigma}.\underline{k}$ of the photon would indicate parity non−conservation, whereas the

Table 2   Symmetry properties under P and T for some important quantities in photon atom interactions.

Symbol	Property	P	T
J	angular momentum	+	−
k	photon propagation vector	−	−
σ	circular polarization of photon	+	−
ε	linear polarization of photon	−	+
E	electric field	−	+
B	magnetic field	+	−
χ	chirality of handed molecule	−	+

Table 3	Possible atomic signatures for P T violation.

P and T invariant	$\underline{\sigma}.\underline{B}$. $\underline{J}.\underline{B}$, $(\epsilon.E)^2$
P not T violating	$\underline{\sigma}.\underline{k}$, χ , $(\underline{\sigma}_1.\underline{k}_1)(\underline{k}_1 \times \underline{E}.\underline{k}_2)(\underline{\sigma}_2.\underline{k}_2)$
	$\underline{\sigma}.\underline{E} \times \underline{B}$, $\underline{J}.\underline{k}$, $\underline{k}.\underline{B}$
T not P violating	$\underline{k}.\underline{E}$, $\underline{J}_1 \times \underline{J}_2.\underline{B}$, $\underline{\sigma}_1 \times \underline{\sigma}_2.\underline{B}$
P and T violating	$\underline{J}.\underline{E}$, $\underline{\sigma}.\underline{E}$, $\underline{E}.\underline{B}$

detection of an edm interaction $\underline{J}.\underline{E}$ would indicate both P and T violation. We divide our table into the four types of P T violation. We have not attempted to be complete, but have included the majority of terms which are important in practice.

I.3   P, T, C and CPT in quantum mechanics

We now turn to the quantum mechanical description of P and T. In non-relativistic quantum mechanics the inversion operator P remains essentially as in table 1, but in Dirac theory it becomes more complicated because the large and small components of the wavefunction have opposite parity and an additional component of the operator is required to deal with this.

The quantum mechanical time-reversal operator on the other hand requires considerable care since it must correctly reverse such 'motional' operators as momentum $\vec{p}$ and spin $\vec{s}$. The complex form of the former with $\vec{p} = -i\hbar\vec{\nabla}$ suggests that complex conjugation will play an essential part. To also get the correct effect on the spin, an aditional operator on the spin is required. As is explained in detail in reference 14, the time reversal operator must therefore be antiunitary and have the form $\theta = UK$ where U is a suitable unitary operator. The fact that θ is antiunitary whereas P is unitary have very important consequences as we shall see.

The charge conjugation operator is only properly defined in quantum field theory and as we do not need to do so we shall avoid discussion of its detailed form. Suffice to say, one can find a unitary operator C which turns particle into anti-particle. For our purposes, the most important property of C is that it combines with P and T to form the anti-unitary operator CPT. It is one of the most remarkable of theoretical results that in all covariant local field theories CPT is required to be a good symmetry, whereas there are no intrinsic requirements at all about P, T or C separately. The importance of this will be made clear in section I.5.

I.4   Symmetry violation in atoms

Atoms satisfy P and T symmetry to a very high degree, but we are interested in any possible violations and the extent to which they can be detected. We assume therefore that an atom can be described by a Hamiltonian containing a major piece H^0 which is P and T invariant and a smaller piece H^v which is not.

$$H = H^0 + H^v.$$

Using perturbation theory as always we have in zeroth order

$$H^0|\psi_i^0\rangle = W_i^0|\psi_i^0\rangle$$

where $|\psi_i^0\rangle$ has well-defined parity. In first order,

$$|\psi_i^V\rangle \; - \; \sum_J |\psi_j^0\rangle \frac{\langle\psi_j^0|H^V|\psi_i^0\rangle}{W_i^0 - W_j^0}$$

In the case of P violating Hamiltonians H^V, the sum over j spans states opposite in parity to ψ_i^0, whereas if H^V is T <u>not</u> P violating ψ_j^0 spans states of the same parity as ψ_i^0.

Now the majority of the tests for P, T violation in which we are interested involve interaction between an atom and an external electromagnetic field, either static or dynamic. Thus in almost all cases we are interested in elements of the form

$$T_{fi}^V \; - \; \langle\psi_f^0|T|\psi_i^V\rangle + \langle\psi_f^V|T|\psi_i^0\rangle$$

where T is a suitable interaction operator and the two states f and i may or may not be identical.

While the detailed calculation of T_{fi}^V for a particular case may be extremely complex there are certain very general remarks which we can make at this stage concerning the various possibilities for P or T violation. In making this analysis we have assumed that the phases of the states ψ_i^0 have been chosen in the standard way which ensures that the reduced matrix elements of the spherical harmonics are <u>real</u>. With this phase convention we can show for the various choices for H^V:

(i)   P <u>not</u> T violating. The matrix element of H^V between opposite parity states is pure imaginary as therefore are the induced transition elements T_{fi}^V. This means that the diagonal electric dipole moment $E1_{ii}^V$ is zero; as we have already seen, an edm requires P <u>and</u> T violation. Circular polarization on the other hand depends on Imag $E1_{fi}M1_{fi}^*$ which is nonzero when M1 is the normal transition element and E1 is induced by H^V.

(ii)   P <u>and</u> T violating. Here the matrix element of H^V is again between opposite parity states but is now real, as is the induced transition element. A diagonal edm is now possible but the circular polarization will be zero.

(iii)   T <u>not</u> P violating. This case differs from the two above in that H^V acts between states of the same parity, but it is pure imaginary in contrast to the elements of P and T conserving terms. Thus P <u>not</u> T induced elements are pure imaginary relative to the normal ones. One can quite readily show that interference between an M1 induced by H^V and an E1 transition induced by an external electric field leads to an electric field induced transition rate proportional to the T <u>not</u> P violating quantity $\underline{k}.\underline{E}$.

Analysis shows that the rules (i) – (iii) are just those required to produce observable phenomena with the symmetries set out in table 3.

I.5   <u>Historical perspective</u>

In order to put our general observations concerning atomic P and T violation into their elementary particle perspective we inject here a brief and highly selective history of the interaction of particle and atomic physics in this area. An excellent treatment is to be found in the book by Sachs[14].

As the reader will be aware, prior to 1957 it was generally thought that P and T invariance was universal in physics and therefore in atoms too. With the discovery of P violation in β decay there was a flurry of interest in possible atomic implications both P and T. But this died down quite rapidly on the P side because it seemed clear that the weak interactions involved only charged currents and these would not yield direct interactions of order G_F. An additional factor was the realization that the P violation was accompanied by C violation in just such a way as to keep CP an exact symmetry and that through the CPT theorem, T invariance probably remained good. Nonetheless there were a few pioneering experiments by the author and others in this period to look for an atomic electric dipole moment.

The situation changed radically in 1964 with the discovery of the decay of the $K_L^0 \rightarrow 2\pi$ which implied CP violation and hence T violation, and indeed subsequently some indirect evidence was obtained from detailed analysis of the K^0 decay for T violation without the intermediate assumption of CPT. For details, the reader is referred to the book by Sachs[14]. The discovery of CP violation stimulated a range of experiments to look for atomic P and T violation, essentially atomic variants of Ramsey's famous experiment on the neutron edm (see ref. 12 for a review). This activity was not accompanied by any serious work on atomic P not T violation because it was still widely held that only charged currents existed.

The third revolution occurred in 1972 with the discovery of neutral currents in ν scattering at high energies. It was immediately realized by the Bouchiats[15] that atomic P not T experiments were in fact possible at a significant level of precision. After some initial confusion and contradictory experimental results, clear data indicating the existence of P not T violating atomic phenomena of the expected magnitude began to emerge around 1980 – indicating that P violation in the weak interaction does indeed extend right down to the very low energies involved in atoms (see ref. 16 for details).

The present situation is that for P not T violation we have a mature elementary particle theory – that of Weinberg–Salam–Glashow – which is in agreement with all available experimental evidence, including atomic which is now at about the 10% level of precision. On the T violating side, there is now a consensus that the origin may lie in the relative phases of the three generations of quarks, but there is no clearcut experimental evidence (see ref. 17). As a result interest in T violation in atoms has again increased and will be discussed in lecture III.

I.6   The atomic Hamiltonian

In this section we summarise briefly our knowledge of the symmetry violating part of the atomic Hamiltonian in the context of the elementary particle position outlined above. As we shall see, the position differs in the three cases of interest to us and our treatment varies accordingly.

I.6a   P not T violating

Here the Hamiltonian is well-known and can be written in the form:

$$H^{P \text{ not } TV} = \frac{G_F}{2\sqrt{2}} Q_W \gamma^5 \rho_N(r) + \frac{G_F}{2\sqrt{2}} \underline{\mu}_W \cdot \underline{\alpha} \rho_N(r)$$

where G_F is the weak interaction constant, γ^5 and $\underline{\alpha}$ are the usual Dirac operators and $\rho_N(r)$ is a suitable nuclear density. The elementary particle physics is contained in the so-called weak charge Q_W and magnetic moment $\underline{\mu}_W$. Because the weak vector current is conserved Q_W depends directly on electro-weak theory and is free from nuclear structure effects. The value is given in terms of the Weinberg angle Q_W by[18]

$$Q_W = -0.974 \left[N - Z(1-4\sin^2\theta_W) \right]$$

where $\sin^2\theta_W = 0.230 \pm 0.005$. The deviations from integers are electro-weak radiative effects.

$\underline{\mu}_W$ is much more complicated for three reasons:

(i)   The direct electron-nucleon interaction is proportional to $(1 - 4\sin^2\theta_W)$ which is small.

(ii)   It also depends on nuclear structure just as the magnetic moment does.

(iii)   There is an indirect mechanism whereby PNC nuclear forces produce an anapole moment[19] which interacts electromagnetically with the electron giving a term of the same form as the direct interaction. Calculations suggest that the anapole may dominate for heavy atoms.

An important point to note here is that Q_W is independent of nuclear spin and the effect of the nucleons is additive so that it grows roughly proportionally with N (or Z). μ_W, in contrast, depends on nuclear spin and its magnitude remain approximately independent of atomic number.

I.6b   P and T violating

The situation is more complicated here and we do not try to specify the Hamiltonian in complete detail. The main terms are:

$$H^{P \text{ and } TV} = -d_e(\beta-1)\underline{\sigma}\cdot\underline{E} - Q_s\cdot\nabla^2\underline{E}_N - M_e^2\cdot\{\underline{\triangledown}\ \underline{B}\}^2 - Q^3\cdot\{\underline{\triangledown}\ \underline{\triangledown}\ \underline{E}\}^3$$

$$+ \frac{G_F}{2\sqrt{2}}\ C_s\ i\ \beta\ \gamma^5\ \rho_N(r) + \frac{G_F}{2\sqrt{2}}\ C_T\ \underline{I}\cdot\underline{\alpha}\ \rho_N(r)$$

The first term is the electron edm, the next three are the nuclear edm, magnetic quadrupole and electric octupole moments interacting with the fields from the electron. The final two terms are the T violating analogues of the nuclear spin independent and nuclear spin dependent terms in the P not TV interaction. The somewhat unfamiliar forms of the edm interactions are chosen to automatically satisfy Schiff's theorem which states that in the non-relativistic limit the edm on a charged point particle in equilibrium makes no contribution to the energy. The $(\beta-1)$ factor for the electron retains relativistic effects. The use of the so-called Schiff moment Q_S which is proportional to the nuclear radius cubed brings in nuclear size effects to overcome the theorem. The effect of these makes the Schiff edm term of the same order of magnitude as the magnetic quadrupole and octupole terms.

An important difference between $H^{P \text{not} TV}$ and $H^{P \text{and} TV}$ is that we have no accepted theory for T violation and therefore no hard predictions for the magnitudes of the P and TV terms. There are some limits and guesses but experimentally the subject is wide open.

I.6c   T not PV

Very little work has been done on this case and as far as the author is aware no particle physics based atomic Hamiltonian has been proposed. It is in fact not possible to write down a simple local interaction analogous to the P not TV and P and TV terms. An effective Hamiltonian of the right form which is useful for model building is

$$H^{T \text{ not } PV} = C_1\ \underline{s}_1 \times \underline{s}_2\cdot(\underline{\ell}_1 - \underline{\ell}_2) + C_2\ \underline{I}\cdot\underline{s}\times\underline{\ell}$$

Once again the first term is independent of and the second term dependent on the nuclear spin.

I.7   Effective dipole operators

As we pointed out in section I.4 the most important effect of H^v is to admix states with either the wrong symmetry or phase giving rise to a modified transition operator $T\tilde{Y}_i$. An important feature of this is the way in which an induced electric dipole moment depends on the electronic and nuclear angular momenta J and I. This determines selection rules and detailed dependence on hyperfine quantum numbers

P not TV

Here part of the operator is nuclear spin dependent and part not

$$\underline{D}^V = \underline{I}^1_{(e)} + \Sigma_k \left\{ \underline{I}_e^k \, \underline{I} \right\}^1$$

The first term comes directly from the first term of H^{PnotTV} and the second from the second. The different dependence on $\underline{J}$ and $\underline{I}$ mean that in principle these two terms can be unambiguously separated by experiment.

P and TV

Hermiticity and T violation arguments show that the effective edm operator can be written

$$\underline{D}^V = d_{10} \, \underline{J} + d_{01} \, \underline{I} + d_{21} \left\{ \{\underline{J} \ \underline{J}\}^2 \ \underline{I} \right\}^1 + d_{12} \left\{ \underline{J} \ \{\underline{I} \ \underline{I}\}^2 \right\}^1$$

where the parameters are state dependent.

There will in general be a number of independently measurable parameters and these can be related back to particular originating terms. Thus the electron edm will contribute to d_{10}, the nuclear to d_{01}, the magnetic quadrupole to d_{21} etc. It is important to note however that if the hyperfine structure plays an intrinsic role in determining D^V then the above arguments become more complicated. The triple combination of electron edm, magnetic hyperfine interaction and electron electric dipole transition operator can produce a contribution to d_{01}; the hyperfine interaction transfers the PNC effect from electronic to nuclear spin space. This is an important observation which allows experiments on atoms with $J = 0$ to be sensitive to the electron edm.

I.8   Enhancement

The general order of magnitude of the P violating effects which we have been discussing is given by the weak interaction constant $G_F = 2.18 \times 10^{-14}$ atomic units, or probably even smaller in the case of P and T violating effects.

This very small value suggests that we must pick those optimum situations where the effect to be observed is 'amplified' as much as possible. In all cases of experimental interest we make use of state admixtures of the form

$$\text{admixture} = \frac{\text{matrix element}}{\text{energy difference}}$$

It is clear that we need to choose the numerator to be large or the denominator to be small – or both.

(i)   Heavy atoms

In the majority of cases the important atomic matrix element is between $s_{1/2}$ and $p_{1/2}$ states with an operator heavily weighted to the origin.   One finds

$$< s_{1/2} \mid h^{v} \mid p_{1/2} > \quad \propto \quad Z^{3} \text{ or } Z^{2}$$

Z^{3} holds for the spin independent P <u>not</u> TV and P <u>and</u> TV terms and for the electron edm, Z^{2} for the spin–dependent P <u>not</u> TV term and P <u>and</u> TV terms and for the nuclear edm term.   This factor of $10^{4} - 10^{6}$ puts a heavy premium on the heaviest atoms.

(ii)   Degeneracy

It was realized very early in this work that near degeneracy of the opposite parity levels in hydrogen might give major enhancement factors.   This is certainly so, but the smallness of Z and technical problems seem to rule this out as a successful mechanism.

More recently, there has been considerable interest in the rare earths[20] where there are dense overlapping configurations of opposite parity with numerous near coincidences. It remains to be seen whether these are usable in practice.

I.9   <u>Numbers</u>

Let us now consider a few numbers.   For our discussion we take as a typical state admixture $\approx 10^{-10}$.   This will produce in turn an induced moment $\approx 10^{-10}$ ea$_{0}$.   To see whether such an admixed dipole could be measured, we must distinguish P <u>not</u> TV where we are dealing with transitions and P <u>and</u> TV where we have an edm energy to consider.

(a)   P <u>not</u> T

a 10^{-10} fractional effect is clearly very difficult to use but we can improve the situation enormously by using forbidden or highly forbidden transitions with transition dipoles, expressed in the same units of 10^{-3} and 10^{-6} respectively.   It is clear that fractional effects of order 10^{-7} are obtainable and these, as we shall see, are well within the limits of current experiments.

(b)   P <u>and</u> T

an edm of 10^{-10} ea$_{0}$ in an electric field of 10^{5} V/cm gives a frequency shift of order 30 Hz.   Since mHZ sensitivity can readily be obtainable in radiofrequency atomic and molecular beam experiments, our canonical sensitivity can readily be achieved, and as we shall see significantly exceeded.   Unfortunately, the effect of Schiff's theorem is to markedly reduce the sensitivity to nuclear edm effects; the factor involved is of order $10^{-4} - 10^{-6}$.

I.10   <u>Atomic theory</u>

A crucial element if a programme to look for and measure symmetry violating effects in atoms is to be useful is that one has reliable atomic theory to relate atomic experiment to fundamental quantity is of interest (see refs. 13, 22 for details).   Such atomic theory normally goes through three stages:

(i)   Order of magnitude hand waving
(ii)   Central field model, one particle calculations
(iii)   Major calculations

The extent to which one moves down through these stages depends on the situation on the experimental side. Stage (i) is often adequate to indicate whether a project is interesting and viable or not. If it is, stage (ii) must be carried through and is often all that is required until it is clear whether the experiments can produce results of significant accuracy. If they do, then stage (iii) is required. In the P not TV case where accurate experiments are now available a major effort is now under way. This is outlined in lecture IV. In the case of P and TV, where only experimental limits are available, the majority of calculations have been at stage (ii) although some more significant work is under way stimulated partly by the availability of computer codes and expertise from the P not T programme and partly by the worry that the simple minded stage (ii) calculations may possibly be significantly in error.

It is perhaps worth noting in passing that the major calculations require the deployment of a wide range of relativistic, many-body techniques. PNC work now leads non PNC work and suggests that calculations of ordinary properties for heavy atoms can be made at an accuracy which significantly improves on what was previously thought possible.

I.11   Survey of experiments

We bring this birds eye view of the field to its conclusion by listing in tables 4a,b,c without discussion the types of experiment currently underway, together with a few additional suggestions which are being seriously considered. More detailed discussion of a number of the most recent investigations will be given in Lectures II and III.

Table 4a   P not T experiments

Method	Element	Signature	Group/Status
Polarimetry	Bi, Tl Pb	$\underline{\sigma}.\underline{k}$	Seattle[*], Oxford[*], Moscow[*], Novosibirsk[*]
E-field induced fluorescence	Cs	$\underline{\sigma}_i.\underline{k}_1(\underline{k}_1.\underline{E}\times\underline{k}_2)\underline{\sigma}.\underline{k}_2$	Paris
Beam version of above	Cs	$(\underline{\sigma}.\underline{k})(\underline{k}.\underline{E}\times\underline{B})$	Boulder[*]
Stimulated emission	Cs	$(\underline{\epsilon}_1.\underline{\epsilon}_2)(\underline{E}.\underline{\epsilon}_1\times\underline{\epsilon}_2)$	Paris[*]
E-field induced absorption	Tl	$(\underline{\epsilon}.\underline{B})(\underline{\epsilon}.\underline{E}\times\underline{B})$	Berkeley
nmr	Tl mol	$\chi \ \underline{I}.\underline{B}$	Grenoble
laser gyro	?	$\underline{\sigma}.\underline{k}$	idea
molecular energies	?	χ	idea

[*]experiments continuing.

Table 4b   P and TV experiments

Method	Element	Signature	Group/Status
Atomic beam	Cs, Tl, Xe	$\underline{J}.\underline{E}$	Oxford, Berkeley[*]
Molecular beam	TlF	$\underline{I}.\underline{E}$	Oxford, Harvard, Yale[*], Leningrad[*]
Optical pumping	Xe, Hg	$\underline{I}.\underline{E}$	Seattle[*]

Table 4c   P not TV experiments

Method	Element	Signature	Group/Status
Laser gyro	?	$\underline{k}.\underline{E}$	idea

[*]experiments continuing.

11

LECTURE II P not TV experiments

Rather than attempt to review past work which is admirably treated elsewhere[2,3], we describe here the two most recent experiments on Bi, Cs and a novel technique proposed to investigate Cs.

II.1   Optical rotation in bismuth

(i)   General Principles

If parity is not conserved, a pair of levels which have the same basic parity and are joined by a magnetic dipole (M1) transition element can also have a small PNC E1 element admixed.   Because of the phase lag between electric and magnetic fields for circularly polarised radiation absorption and refractive index will now depend on

$$ \left| \left[E1_{PNC} \pm i\ M1 \right]\ \right|^2 \qquad\qquad (II.1) $$

and the interference term implies a difference between right and left polarizations. Simple wave theory then allows one to show that for a single isolated transition plane polarized light suffers a rotation

$$ \varphi_{PNC}(\nu) \ = \ \frac{-4\pi}{\lambda}\ R \int \left[n_{M1}(\nu,\ell) - 1 \right] d\ell \qquad\qquad (II.2) $$

on passing through a region with parity conserving refractive index n_{M1}.   Here

$$ R \ = \ \frac{Im\ E1_{PNC}}{M1} \qquad\qquad (II.3) $$

is a measure of the degree of parity nonconservation and is the experimentally measured quantity.   We see that $\varphi_{PNC}(\nu)$ will have the familiar dispersion shape of refraction index near resonance.   At first sight one can increase its absolute magnitude by increasing the optical path length $\int(n_{M1}(\nu,\ell)-1)d\ell$ but there is a limit to this since in the resonance region where the refractive index varies with wavelength, an essential requirement to separate it from background, there must also be absorption.   It is useful to note that at one absorption length at line centre, $\varphi_{PNC} = \pm R/2$ at its maximum and minimum.   R is therefore a good measure of the magnitude of angle to be measured.

(ii)   The bismuth transitions

There are two allowed M1 transitions in bismuth which are well suited to optical rotation experiments since they fall in a convenient region of the spectrum:

$$ 6p^3\ J = 3/2 \quad \longrightarrow \quad 6p^3\ J' = 3/2 \qquad \lambda\ =\ 876\ nm $$

$$ 6p^3\ J = 3/2 \quad \longrightarrow \quad 6p^3\ J' = 5/2 \qquad \lambda\ =\ 648\ nm $$

We concern ourselves here solely with the 876 nm line[6], a detailed description of work on the 648 nm transition has been given by Taylor et al[1].

This transition is more complicated than indicated in the section above first because bismuth has nuclear spin 9/2 and there is a complex hyperfine structure but also because in addition to the M1 transition amplitude there is also an E2 count interfere with M1 (in the absence of external fields), the E2 does contribute both to absorption and to Faraday effect which we use for calibration and test purposes. Provided that the nuclear spin independent part of the PNC interaction is dominant, equation (II.2) still holds and one can readily compute the spectral form to be expected for the optical rotation hyperfine pattern. The result is illustrated in figure II.1 below. The unknown to be measured is the absolute magnitude which determines R. Calculation suggests that R will be close to 10^{-7} and we therefore use this value for illustrative purposes.

We note in passing that the situation changes markedly if the nuclear spin dependent interaction is significant. This gives rise to an appreciably different hyperfine pattern with the consequential possibility of separating the two effects.

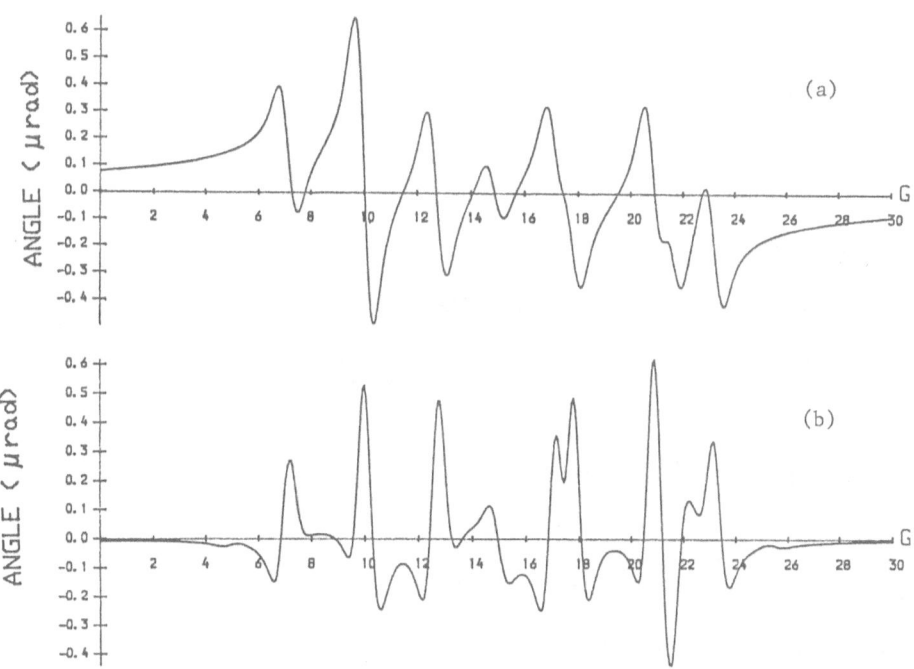

Figure II.1   (a)   Theoretical PNC optical rotation profile φ_{PNC} for $R = 10^{-7}$.

(b)   Theoretical Faraday profile for $B = 1$ mG.

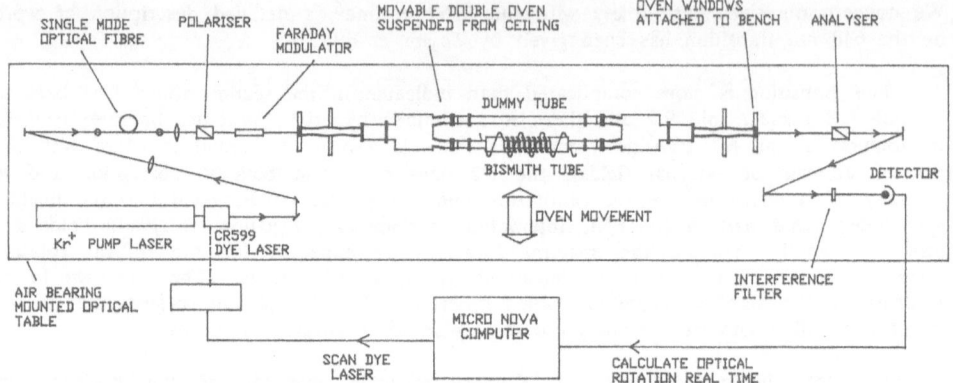

Figure II.2   The Oxford 876 nm optical rotation polarimeter.

(iii)   The apparatus

The apparatus which is sophisticated laser polarimeter is illustrated below

(iv)   Experimental method

One can distinguish four stages in the experiment.

(a) Angle measurement

The basic technique for measuring the optical rotation φ has been described in Tregidgo *et al*[23]. A square wave modulation scheme consisting of angles φ_M = +A, −A, 0 is applied to the Faraday modulator with a period of 1 msec. The corresponding signals I(A), I(−A), I(0) are integrated, normalized for laser intensity, and then used to calculate φ via the square law equation.

$$ I = I_0 \left[\varphi_M + \varphi \right]^2 + background $$

which gives

$$ \varphi = \frac{A}{2} \frac{I(A) - I(-A)}{I(A) + I(-A) - 2I(0)} $$

The denominator is proportional to the transmission of the system and is used to measure the absorption spectrum and hence the optical depth.

(b) Wavelength dependence

In order to utilize the complex signature afforded by the spectrum (figure II.1), the laser is scanned over a significant portion of the hyperfine pattern and the rotation angle at each of a large number (128) of equally spaced frequency points is measured and stored. Simultaneoulsy the absorption is measured at each point as discussed above. Occasionally, a known longitudinal magnetic field is applied and the hyperfine Faraday spectrum is accumulated. This is used to check on our understanding of the spectral lineshape – calculation of the Faraday effect contains many of the same input parameters as does that for the PNC optical rotation. We note at this point that considerable care is taken to reduce the residual magnetic field during the optical rotation runs to a value comparable to the optical rotation – their separation will be discussed below.

(c) Double oven discrimination

A major worry in optical rotation experiments is the possible presence of a wavelength dependent angle of purely optical origin which by chance mimics to some extent the expected PNC spectrum and hence filters through as noise or as an erroneous contribution. A unique feature of the Oxford bismuth experiments is a double oven system which allows a wavelength scan to be made with bismuth vapour present $\varphi_{Bi}(\nu)$ and then a scan without bismuth but with no change in the optical system $\varphi_e(\nu)$. Subtraction of the two should leave only bismuth dependent effects.

(d) Fitting

The rotation spectrum for a complete scan is fitted to a function of the form

$$\varphi_{Bi}(\nu) - \varphi_e(\nu) = \alpha \; \varphi_{PNC}(\nu) + \beta \; \varphi_F(\nu) + Q(\nu)$$

for which the frequency scale, lineshape parameters and average optical density have already been determined. Five parameters are floated: α and β are the scaling factors for the PNC and Faraday rotation theoretical spectra $\varphi_{PNC}(\nu)$ and $\varphi_F(\nu)$, while the other three represent any unsubtracted background rotation as a quadratic infrequency $Q(\nu)$. R is deduced from the fitted value of α. An important feature of the fitting process is the use of a sophisticated weighting procedure to take into account the effect on the signal to noise of the differing absorption throughout the spectrum. Measurements are made at an optical depth which optimizes the overall signal to noise on the final result. This choice has the result that in some regions of the spectrum is almost totally absorbing and the noise is dominated by residual detector background. It is clearly very important to give most weight in the fitting procedure to the regions which have most input on the final answer.

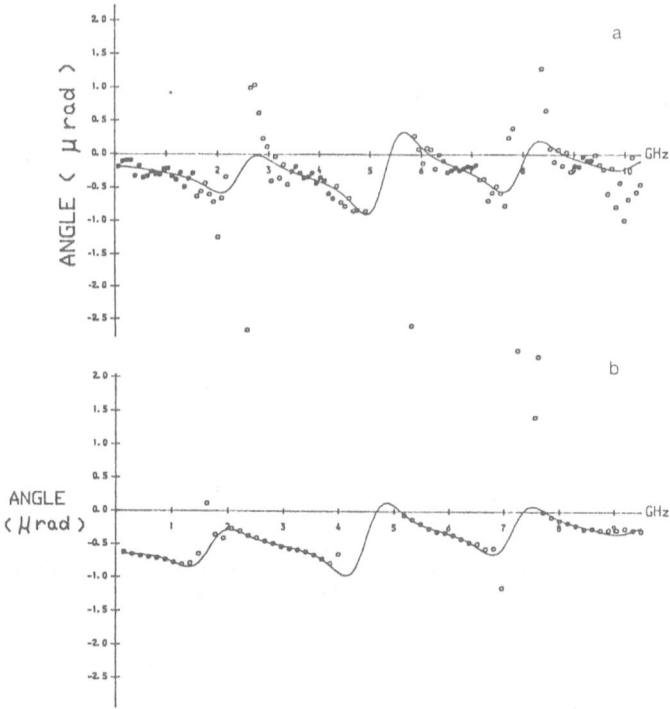

Figure II.3   Optical rotation as a function of frequency (a) single pass, (b) superposition of 58 traces. Open circles represent points of low weight (ref. 6).

(e) Results

Figure II.3(a) shows the results from a single $\varphi_{Bi}-\varphi_e$ run over the red wing of the hyperfine pattern, together with a best fit theoretical curve. Figure II.3(b) shows the superposition of all 58 traces made in the experiment. The frequency dependence is dominated by $\varphi_{PNC}(\nu)$ since the contributions due to $\varphi_F(\nu)$ and $Q(\nu)$ tend to average out.

The final result from the Oxford experiment[6] is

$$R = (-10.0 \pm 1.0)10^{-8}$$

in satisfactory agreement with Fortson's result[24] $R = (-10.4 \pm 1.7)10^{-8}$.

It is interesting to note that the error quoted of $\pm 1 \times 10^{-8}$ is rather conservative because there are one or two unexplained features of the data; the error of measurement is much smaller, about 0.25×10^{-8} and there seems no reason why this sort of overall accuracy cannot be achieved in the near future.

In the slightly longer term, high precision analysis designed to pick out the pattern predicted for the nuclear spin-dependent term may allow the first observation of a nuclear anapole moment. This is expected to occur around the 1% level which is within sight with improved techniques.

II.2 Electric field transitions in cesium

(i) General principles

We discuss here the latest experiment to look for and now measure PNC effects in the highly forbidden M1 transition $6s_{1/2} \rightarrow 7s_{1/2}$ in Cs. This is of course the transition originally proposed by the Bouchiats in their pioneering paper[15] and the main weight of experimental work has come from the Paris group[25].

The relevant Cs energy levels are indicated in figure II.4(c). In the presence of external electric and magnetic fields the transition matrix element between a particular pair of resolved hyperfine states FM → F'M' has three components

Stark: $\qquad A_{ST} = \alpha\, \underline{E}.\underline{\epsilon}_1 + i\beta\, \underline{J}.\underline{E} \times \underline{\epsilon}_1$

Magnetic: $\qquad A_{M1} = -\,M1\, \underline{J}.\underline{k} \times \underline{\epsilon}_1$

PNC E1: $\qquad A_{PNC} = i\, \text{Im}\, E1_{PNC}\, \underline{J}.\underline{\epsilon}$

combining all three transition amplitudes, the transition probability between the particular states is

$$I = |\,A_{ST} + A_{M1} + A_{PNC}\,|^2.$$

The experimental design is chosen to optimise the PNC interference between A_{ST} and A_{PNC} while minimising the PC interference between A_{ST} and A_{M1}. In this arrangement the laser, E and B are mutually perpendicular in the y, x, z directions respectively. With this one obtains a transition probability proportional to:

$$I_{FM}^{F'M'} = \left[\beta^2 E^2 \epsilon_z^2 \pm 2\beta E M1 \left\{ |\epsilon_z^{k+}|^2 - |\epsilon_z^{k-}|^2 \right\} \mp 2\beta E \text{Im} E1_{PNC} \epsilon_z \epsilon_x \right]$$

$$\times \left\{ C_{FM}^{F'M'} \right\}^2 \delta_{m,m' \pm 1}$$

The first term is the pure Stark induced transition which depends on β the 'so-called' vector polarizability. The second term is the interference between the Stark induced and M1 transitions which depends on M the M1 matrix element and as usual with M1 amplitudes reverses with propagetic direction k. The final term is the PNC–Stark interference which is the object of measurement. All terms have a common angular factor $C_{FM}^{F'M'}$ and of course with the geometry used only $\Delta m = \pm 1$ transitions are induced.

We note that the PNC interference term changes sign with (a) the electric field E, (b) the magnetic field B, (c) the handedness of the light ϵ_x / ϵ_z, and (d) the overall sign of m and m', i.e. from one end of the Zeeman pattern to the other. These reversals are essential to separate the PNC interference term from the Stark induced terms which are a factor 10^6 larger. We also note that the Stark–magnetic dipole interference can be suppressed if both propagation directions for the laser $\pm$ k are used simultaneously. It will be clear from the above that the PNC signature for this experiment is $\underline{\sigma}.\underline{k}$ $\underline{k}.E\times B$ where $\underline{\sigma}.\underline{k}$ is the handedness of the laser radiation.

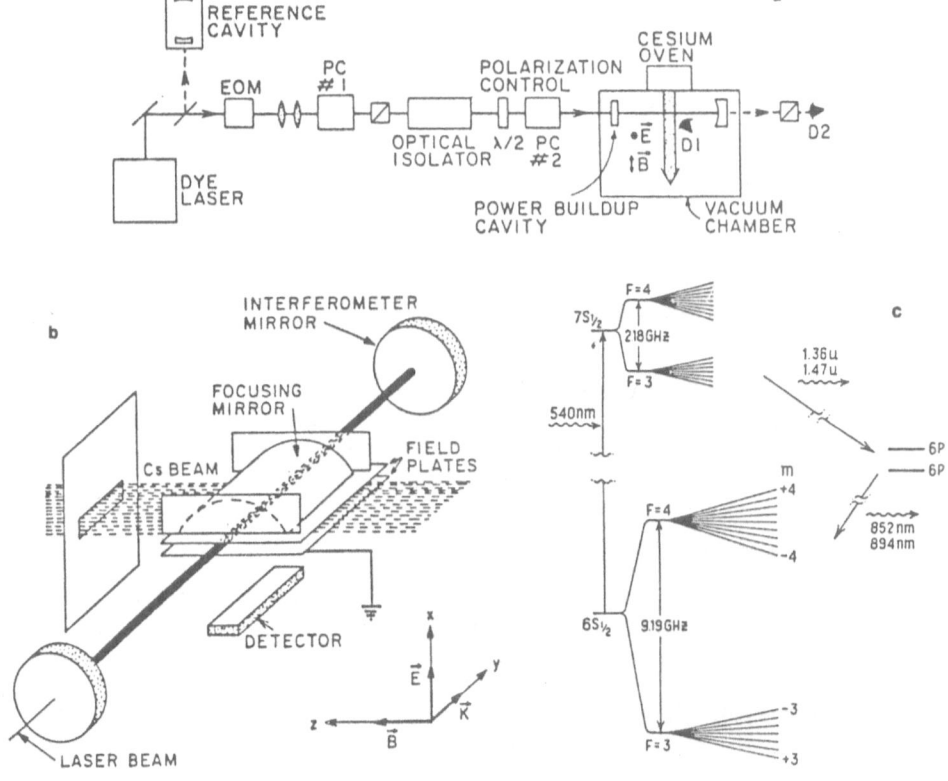

Figure II.4   The Colorado Cs experiment. (a) shows a schematic of the apparatus, (b) indicates the details of the interaction region, and (c) gives the energy levels involved.

(ii)   The apparatus

A schematic of the apparatus is shown in figure II.4(a), with the details of the interaction region in figure II.4b.   The main features to be noted are:

(a)   A frequency stabilized dye laser giving approximately 500 mW of light at 540 nm with linewidth about 100 kHz.

(b)   The polarization control for alternating and controlling the handedness.

(c)   The enhancement cavity which both builds up the useful laser intensity and produces the cancellation of the $A_{ST} \times A_{M1}$ interference.

(d)   A high intensity Cs atomic beam.

(e)   A magnetic field of 70 G which allows the Zeeman splitting to be resolved.

(f)   An electric field of ± 2.5 k V/cm.

(g)   A liquid–nitrogen cooled silicon photo–diode to detect 6s → 7s transition via the 850 and 890 nm light in their subsequent decay cascade.

(iii)   Systematic effects

Systematic effects are very important in this type of experiment because the presence of the external fields allows the possibility of terms in the transition probability which mimic the PNC interference.   The most important of these with their magnitude relative to the PNC term are

(a)   Misalignment of the electric field (E_y) plus a stray field (ΔE_z) (0.01).

(b)   Stray field ΔE_y plus misalignment of the magnetic field (B_x) (0.04).

(c) $A_{ST} - A_{M1}$ interference combined with lack of k reversal and exact photon handedness reversal.   Because the $A_{M1} \approx 10^4 A_{PNC}$ this effect only partially cancels and its magnitude ranged from +0.5 to –0.6.

By intentionally enhancing single features of these effects thay can be investigated and to a considerable extent dealt with.   The experimenters believe that no significant contributions exist which mimic the PNC signal and have not been fully taken into account.

(iv)   Data taking and results

A typical data run consisted of 8 h of data accumulation divided equally between the F = 4→3 and F = 5→4 transitions.   The signal was analysed by subtracting Δ_{PNC} for the low frequency side of the Zeeman multiplet from that on the high frequency.   The result was corrected for systematic effect (c) which was about 50% of the PNC interference in a majority of the runs.

The results obtained were[7]

$$\text{Im E1}_{PNC} / B = -1.51 \pm 0.18 \text{ mV/cm} (F = 4\to3)$$

$$-1.80 \pm 0.19 \text{ mV/cm} (F = 3\to4)$$

$$-1.65 \pm 0.13 \text{ mV/cm} (\text{average})$$

This can be compared with the Bouchiat value $-1.56 \pm 0.17 \pm 0.12$ mV/cm[25,26].

The vector transition polarizability is found by combining several experiments. The result adopted by Wieman et al.[7] is

$$\beta = 27.3 \pm 4 \ a_0{}^3.$$

This yeilds

$$\text{Im E1}_{PNC} = (-0.88 \pm .07) \times 10^{-11} \ ea_0.$$

A comparison of the measurements for the two hyperfine lines can provide information on the nuclear spin dependent term. However, as can be seen this is zero within the limits of error.

(v)

Very recently the Wieman group have completed a similar experiment with improved precision[27]. They obtain

$$\text{Im E1}_{pnc} / \beta = -1.576(34) \ \text{mV/cm}.$$

They also observe a small nuclear spin dependence in agreement with that predicted to arise from a nuclear anapole moment.

II.3   PNC stimulated emission in cesium

(i)   General principles

A major problem in the Cs PNC experiments is the relatively low detection efficiency for the $6s \rightarrow 7s$ transitions. In a very interesting paper Bouchiat et al[8] describe a method which they are developing to overcome this difficulty by means of a technique using stimulated emission.

The energy levels and transitions involved are illustrated in figure II.4(c). A probe laser beam is tuned to the centre of one hyperfine component of the $7s \rightarrow 6p$ transition. Provided its intensity is sufficient for stimulated emission to be more likely than spontaneous emission, any population in the $7s_{1/2}$ state will produce radiation directed in the forward direction. It is essential in such a scheme that the 540 nm laser tuned to the $6s \rightarrow 7s$ and the probe be pulsed. The signal would be observed as a pulse of gain on the probe immediately subsequent to the 540 nm pulse.

An important feature of the scheme is that only one velocity group of atoms is excited and detected so that Doppler broadening is avoided. This allows the various hyperfine splittings to be resolved.

(ii)   The PNC signature

The transition matrix was given previously in section II.2(i). With this one can work out the density matrix of the excited state produced by a particular field configuration. With an electric field collinear with the laser propagation along the z axis and the laser polarized in the x direction, one obtains alignment of the excited state in the y direction from the $\beta\sigma.E_0 \times \epsilon_1$ term (the $\alpha E.\epsilon$ term vanishes for this geometry). However the PNC interference term on the other hand produces alignment at 45° to this direction. This means that overall it produces a small rotation of the axis of alignment around $\underline{E}$. Such a rotation is clearly 'handed'.

This alignment can be probed by a linear polarized second beam and the gain will have a component which depends on the pseudoscalar

$$\beta \ \text{Im} \ E1_{PNC} \ (\underline{\epsilon}_1 \cdot \underline{\epsilon}_2) \ (\underline{E} \cdot \underline{\epsilon}_1 \times \underline{\epsilon}_2)$$

Likewise one can show that if the probe beam is circular polarized the gain will be different for right and left handed polarizations.

(iii)   Technical points

It is not appropriate here to discuss the detailed consideration of gain and signal to noise given by Bouchiat et al[8]. Suffice to highlight two points:

(a)   The parity non-conserving assymmetry can be differentially amplified, by a factor up to 10 in principle.

(b)   The signal to noise should be at least a factor 10 better than in the original Bouchiat experiment, with further factors to come with improvements in laser technology.

This new method is currently being tested by exploring the presence of gain anisotropy resulting from the parity conserving Stark amplitudes and encouraging results have been obtained.

LECTURE III Experiments on P and T violation

III.1   The search for an atomic edm

(i)   Basic principles

If P and T are simultaneously violated then an atom will in general have an electric dipole moment in addition to its magnetic moment. In the absence of complications due to nuclear spin and neglecting the quadratic Stark shift, the Hamiltonian can be written

$$H \ = \ H_0 + \frac{-d_A}{J} \ \underline{J} \cdot \underline{E} + g_J \mu \ \underline{J} \cdot \underline{B}$$

with $\underline{E}$ and $\underline{B}$ parallel the frequency of a transition $M \leftrightarrow M'$ will be

$$h\nu \ = \ \left\{ \frac{-d_A}{J} \ E_z + g_J \mu_B B_z \right\} \left[M - M' \right]$$

The edm term can be separated by looking for a change in frequency

$$h\delta\nu \ = \ \frac{-2d_A}{J} \ E_z \ (M - M')$$

on reversal of the electric field with respect to the magnetic (or vice versa).

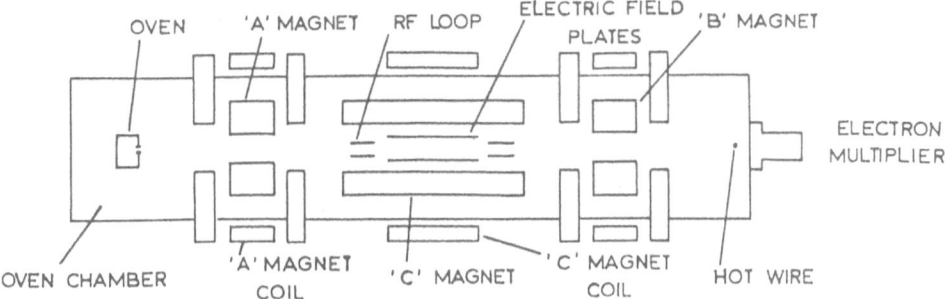

OVEN   'A' MAGNET   RF LOOP   ELECTRIC FIELD   'B' MAGNET
                                      PLATES

                                                         ELECTRON
                                                         MULTIPLIER

OVEN CHAMBER   'A' MAGNET   'C' MAGNET   'C' MAGNET   HOT WIRE
                 COIL   COIL

Figure III.1   Schematic of early atomic beam apparatus to look for
                        atomic edm.   Taken from I. Bellamy, D.Phil. thesis,
                        Oxford, 1966.

It was realized by Ramsey in the case of neutrons and by the author for atoms that this technique was well suited to a beam resonance experiment using separated oscillatory fields as illustrated in figure III.1. Early experiments were carried out in Cs and Tl to give very low limits on the edm of those atoms.

(ii)   <u>The V×E problem</u>

The major limitation in these experiments was the fact that the atoms in moving through the electric field see a motional magnetic field $\underline{V}\times\underline{E}$ which interacts with the magnetic moment of the atom to give a spurious effect which reverses with E. Instead of $\underline{J}.\underline{E}$ which violates P <u>and</u> T we have $\underline{J}.\underline{V}\times\underline{E}$ which violates neither. Of course $\underline{V}\times\underline{E}$ is perpendicular to $\underline{E}$ so that if $\underline{E}$ and $\underline{B}$ were exactly parallel then $\underline{V}\times\underline{E}$ would be perpendicular to the axis of quantization and give no first order effect. But the effect is sufficiently large that alignment to a sufficient degree is very difficult.

An alternative approach adopted by the Oxford group was to use an atom with a very large quadratic Stark effect. The axis of quantization was then determined by the electric field and the frequency given by

$$h\nu \;=\; \left\{-\frac{d_A}{J}\,E + g_J\mu_B B_z\right\}\left[M-M'\right] + \tfrac{1}{2}\alpha_T\left[M^2 - M'^2\right]E^2$$

by inducing a transition $M \to -M$ the frequency remains independent of α_T but the edm can be determined without complications from the $\underline{V}\times\underline{E}$ effect since $\underline{V}\times\underline{E}$ is always perpendicular to $\underline{E}$.

This technique was applied to the $(5p^56s)$ $J=2$ metastable state of Xe with the result[9]

$$d_A \;=\; (0.7 \pm 2.2) \times 10^{-22} \; e \; cm$$

This is one of the most sensitive experiments to an atomic $(J\neq0)$ edm. However, because of the rapid increase in enhancement factor for the electron edm, conventional experiments in Tl remain competitive.

(iii)   Proposed new Tl experiments

An alternative technique for dealing with the $\underline{V} \times \underline{E}$ problem is to vary $\underline{V}$.   An important new atomic beam resonance experiment to look for an edm in the $6p_{1/2}$ in Tl is under construction at Berkeley.   The main features are

(a)   Two beam symmetry so that to a first approximation the $+\underline{V} \times \underline{E}$ will cancel with the $-\underline{V} \times \underline{E}$.

(b)   Velocity selection which allows the magnitude of this term to be altered.

(c)   Optical state selection and resonance detection which both gives high sensitivity and also allows good geometric determination of the beam path since unlike in a conventional atomic beam the state selection is not done by modifying the trajectory through the apparatus.

It is expected that this experiment will comfortably exceed the present Tl limit

III.2   Nuclear spin dependent terms

The sensitivity of experiments such as those on Cs and Tl described above to nuclear effects is limited by the presence of the large V×E effect and other problems resulting from the presence of electronic angular momentum and its associated magnetic moment.   Two quite separate types of experiment have been proposed which are very sensitive to nuclear spin–dependent P and TV terms.   The first of these is to use a polar diatomic molecule and the second is carry out optical pumping in a J = 0 atomic state.   We discuss these in the next two sections.

(i)   The TlF experiments

A number of years ago we pointed out[28] that a heavy polar molecule could be a sensitive system in which to search for P and TV.   The basic idea is that instead of the applied electric field polarizing the atom directly one makes use of the very large internal field in the molecule.   The role of the external field is then simply to align this molecule along with its external field along the desired direction and then to inverse it.   In such a molecule spin $\frac{1}{2}$ nucleus will contain a term in its Hamiltonian of the phenomenological form $H = -d\underline{\sigma}.\underline{\lambda}$ in which d is an effective coupling constant to be measured.   In parallel electric and magnetic fields the normal spin flip resonance frequency will have an addition $\Delta f = -2d\,|\underline{\sigma}.\underline{\lambda}|$ and this has the opposite sign when E and B are antiparallel.   In the field used in the experiments (20 kv/cm) the polarization $|\sigma.\lambda| = -0.46$ so that $d = -1.09\ \Delta f$.

Following the early experiments by the authors group[28,29,30], improved results were obtained by Ramsey[31].   More recently Hinds[10] has reported an even more sensitive result. His apparatus is illustrated in figure III.2.   The first problem to be overcome is that the nuclear spin flip $\Delta m_{Tl} = \pm 1$ is not directly observable since the focussing system discriminates between states of different $|M_J|$ (and J of course).   The solution which is essentially the same as that proposed in ref. 28 is to use a triple resonance technique with subsidiary rf transitions taking place in the state selectors as illustrated.   If no transition takes place in the central region the second state selector undoes the effect of

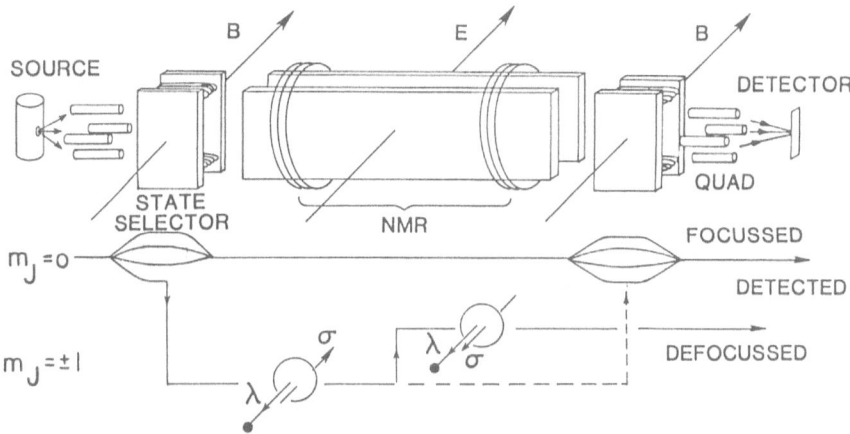

Figure III.2   Schematic of the Yale TlF experiment and the relevant energy levels (from ref. 10).

the first, the molecule reverts to the $M_J = 0$ state and is focussed. If a transition takes place this is no longer the case and the resonance is detected by a drop in signal level at the detector. Because only nuclear moments are involved it is possible to achieve quite narrow resonance lines. The resulting sensitivity to any change of frequency as E is reversed with respect to B is reflected in the very accurate limit on d achieved:

$$\Delta f = \left[-2.2 \pm 2.1 \right] \times 10^{-3} \text{ Hz}.$$

A point to note here is that there is again no $\underline{V} \times \underline{E}$ effect because the axis of quantization lies along the average direction of the internuclear axis which is the same as that for the effective electric field. $\underline{V} \times \underline{E}$ is once again automatically perpendicular to the axis of quantization.

It is expected that the sensitivity of this experiment can be substantially improved by various technical modifications.

(ii)   <u>Optical pumping in xenon and mercury</u>

In the very elegant and sensitive experiment by Fortson and his group[11], the precession of the nuclear spins of xenon atoms in their $^1S_0$ ground state was observed in a magnetic field. Any change in the precession frequency when an applied electric field is reversed relative to the magnetic field would be evidence for P <u>and</u> TV. Using the obvious result that this change of frequency on reversal δf is given by

$$\delta f = 4 d_e \text{ (Xe) E}$$

the Fortson group set the very low limit

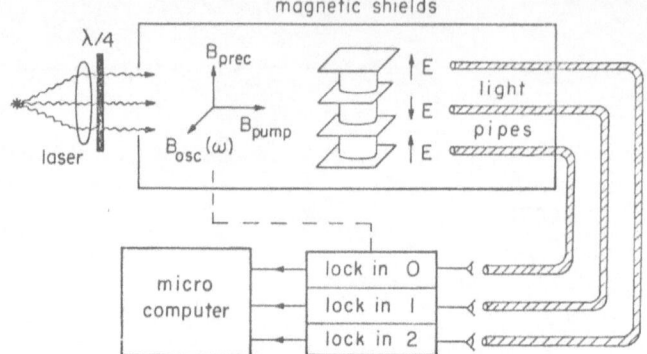

Figure III.3   Schematic of the Washington optical pumping experiment on Xenon.   Taken from reference 11.

$$d_e(Xe) \;=\; \left[\; -0.3 \pm 1.1 \;\right] \times 10^{-26} \text{ excm.}$$

A schematic of their apparatus is given in figure III.3.   The xenon nuclei are polarized and their precession is analysed by spin exchange with optically pumped rubidium vapour.   The precessional relaxation time is very long – of order 500 seconds so that the very great sensitivity is possible provided the adverse effects of residual magnetic fields can be avoided.   This is achieved by using the triple cell illustrated in the figure. It will be seen that by taking the average precession frequencies the two outer cells and subtracting from the centre one, the adverse effects of any residual magnetic field are largely cancelled.

This technique has been extended to Hg and further work is underway[32].   The result

$$d_e(Hg) \;=\; \left[\, 0.7 \pm 1.5 \,\right] \times 10^{-26} \text{ excm}$$

is of similar precision but the higher Z here has given even greater sensitivity to nuclear P _and_ TV effects.   Further appreciable improvements are expected shortly.

LECTURE IV PNC Theory

IV.1   Preliminary remarks

Atomic theory is essential to relate the measurements made by the experimenters to the underlying nuclear or elementary particle properties.   Calculations of some accuracy are clearly necessary in the case of P _not_ TV because the experiments are already at the 10% level and will probably reach 1% precision in the near future.   Considerable care and quite complex calculation is also required in the P _and_ TV case, for a rather different reason. Here, as we have already seen, there is considerable cancellation of the effects and the results can be quite different than one might naively expect. While quite simple techniques sometimes do work well, one needs to go further to ensure that all necessary subtleties have been included.

In all cases of experimental interest, the quantity measured is the E1 matrix element produced by PNC state mixtures. We are therefore interested in the general PNC E1 matrix element

$$E1^{PNC} = \sum_e \frac{<\psi_f \mid \underline{D} \mid \psi_e ><\psi_e \mid H^{PNC} \mid \psi_i >}{W_i - W_e} + \leftrightarrow \qquad IV.1$$

where $\underline{D}$ is an appropriate dipole operator, H^{PNC} is the PNC Hamiltonian, ψ and W are eigenfunctions and eigenvalues of the unperturbed equation

$$H \psi = W \psi$$

$\leftrightarrow$ has an obvious meaning in terms of the interchange of H^{PNC} and $\underline{D}$. When f = i we have an expectation value and an edm; when f $\neq$ i, a transition element.

Before commenting on the detailed calculation of $E1^{PNC}$ there are a number of points which need to be made:

(a)   While H^{PNC} contains a large number of different terms, corresponding to the different possible sources for P not TV and for P and TV, the majority are single particle operators heavily weighted towards the nucleus.

(b)   The dipole operator $\underline{D}$ on the other hand is weighted toward the outer parts of the atom. Thus the calculation of $E1^{PNC}$ requires a subtle interplay of the two regions of the atom. By and large it is the outer part which is the more difficult.

(c)   Nonetheless because we are interested in very heavy atoms and we need good representation of the wavefunctions at the origin it is essential to use Dirac theory.

(d)   It is also essential to include the finite size of the nucleus, though results are in general not very sensitive to the precise model which is adopted.

We conclude these preliminary remarks by some observations concerning the role of empirical input in evaluating $E1^{PNC}$. We first note that such input is peculiarly important in this area because we need not only to calculate the PNC quantity but to have some idea of the reliability of our calculation. Unlike PC effects where we can compare our theories with known results in certain test cases, we must look elsewhere to test for error here. Of course, intercomparison between PNC results for different states and elements can be very valuable, but it is essential to make as much use as possible of related parity conserving data.

The earliest calculations used empirical data as direct input taking the matrix elements of D from lifetimes and transition rates, the energies from observed spectra and by relating the PNC matrix elements to hyperfine interactions. As the calculations become more complex it was realized that the corrections which were needed to this procedure were of the same order of difficulty as an a-priori calculation. Thus the main current approach is to make major a priori calculations of both PC and PNC quantitites on a similar basis. A feel for the accuracy of the PNC effects can be obtained from the reliability of the results for known PC phenomena. However, the need to have the maximum information available on the theoretical side suggests that there is a continuing role for more sophisticated 'semi-empirical' treatments which combine elements of the theory but also contain empirical input.

IV.2   The single particle model and beyond

The central field model of the atom is well known to give a very useful 'first approximation' to the description of a very wide range of atomic phenomena and it is therefore natural to use it for a starting point for calculations of $E1^{PNC}$. In its simplest form equation IV.1 can be written

$$E1^{PNC} = \sum_e \frac{<f\mid \underline{d}\mid e><e\mid h^{PNC}\mid i>}{\epsilon_i - \epsilon_e} + \leftrightarrow \qquad \text{IV.2}$$

where φ_f, φ_e, φ_i are solutions of an appropriate single particle (Dirac) Hamiltonian with eigenvalues ϵ_f etc. It is important to note that in general the sum over e spans all single particle states of opposite parity to i, f, including core states as we shall see.

Two closely related methods have been developed to deal with the infinite sum over states e:

(a) define

$$\mid i^{pnc}> = \sum_e \frac{\mid e><e\mid h^{PNC}\mid i>}{\epsilon_i - \epsilon_e} \qquad \text{IV.3a}$$

then

$$\left\{\epsilon_i - h\right\} = h^{pnc}\mid i> \qquad \text{IV.3b}$$

and

$$E1^{PNC} = <f^{pnc}\mid \underline{d}\mid i> + <f\mid \underline{d}\mid i^{pnc}> \qquad \text{IV.3c}$$

(b) define

$$\mid i^E> = \sum_e \frac{\mid e><e\mid \underline{d}\mid i>}{\epsilon - \epsilon_e - \omega} \qquad \text{IV.4a}$$

$$\left\{\epsilon_i - \omega - h\right\}\mid i^E> = d\mid i> \qquad \text{IV.4b}$$

and

$$E1^{PNC} = <f^E\mid h^{PNC}\mid i> + <f\mid h^{PNC}\mid i^E> \qquad \text{IV.4c}$$

where $\omega = \epsilon_f - \epsilon_i$ is zero when we are dealing with an expectation value. Equations IV.3b and IV.4b are simple inhomogeneous equations which can be reduced to radial form in the usual way and solved numerically for a specified h^{PNC} and specified form of the unperturbed single particle Hamiltonian h. Clearly, IV.3c must give the same answer as IV.4c. We have spelt out these two alternative ways of calculation perturbing either with h^{PNC} or with $\underline{d}$ because this double procedure is mirrored in the more complex calculations to be discussed in the next section.

Before we get into these more complex calculations it is convenient to describe here in physical terms the limitations in this one particle model which have become apparent over the years of work. While not all the difficulties apply in every case, there is sufficient common ground to justify a general comment. The main problems are:

(a) Choice of an adequate single particle Hamiltonian

The main problem is to treat the outer reaches of the atom adequately and to get good eigenvalues. This is essential both to ensure good values for the dipole and matrix elements but also even more vital to get good energy differences between opposite parity states on which the PNC admixtures depend.

(b) PNC core polarization

Equation IV.2 contains only the direct action of h^{PNC}, but there is also an additional indirect effect where a different 'core' orbital is modified and the PNC effect is transmitted onwards with the ordinary electron–electron Coulomb interaction. This PNC core polarization is somewhat similar to that well-known for the magnetic dipole hyperfine interaction. As in this case, the onward transmission depends on exchange interactions.

(c) Dipole shielding

IV.2 assumes that an atomic dipole transition element can be represented adequately by the usual dipole operator. In reality, one knows that there are very strong collective shielding phenomena always present. This follows from the observation that when an extrenal field E^{ext} is applied to an atom it must rearrange itself so that

(i) the field at the nucleus is zero $(\underline{E}^{ext} + \underline{E}^{int})_{r=0} = 0$.

(ii) the average field on the electrons is zero $<\sum_i (\underline{E}_i^{int} + \underline{E}_i^{ext})>$.

(iii) the average field on the electron edm is zero $<\psi | \sum_i \underline{\sigma}_i \cdot (\underline{E}_i^{ext} + \underline{E}_i^{int}) | \psi> = 0$.

(d) Near degeneracy

In this section we have assumed that it is possible to relate the many–electron calculation IV.1 to a single one particle approximation IV.2. This is often not true, particularly where open shells with more than a single electron are involved. This is the case in a number of the atoms of interest particularly bismuth where the breakdown of jj coupling is very important and large mixing of states of the same parity has to be included. An even more extreme case occurs in the rare earths where two nearly lying states of opposite parity are extremely complex and it is not possible at all to reduce IV.1 to IV.2.

IV.3   Many body perturbation theory (MBPT)

Because of the importance of calculations of PNC induced E1 transition elements and electric dipole moments, a number of different theoretical techniques have been brought to bear on the problem. But by far the most common is MBPT which because of its clear diagrammatic representation and the ease with which different types of effect can be separated and treated by the most appropriate method has become the basis for the majority of the calculations made recently. Since our aim is to pass on a general idea of what is involved in this work we confine attention to this method, details of which are to be found in the text by Lingren and Morrison[33].

(i)   Lowest order

MBPT divides up the many particle PC Hamiltonian into a single particle part and a residual perturbation.

$$H^0 = \sum_i \left[\beta_i mc^2 + \underline{\alpha}_i \cdot c \underline{p}_i - \frac{Ze^2}{4\pi\epsilon_0 r_i} - eU_i(r_i) \right] \qquad \text{IV.5a}$$

$$H^1 = \tfrac{1}{2} \sum_{i \neq j} \frac{e^2}{4\pi\epsilon_0 r_{ij}} + \sum_i eU(r_i) \qquad \text{IV.5b}$$

to which must be added H^{PNC} and H^E the PNC Hamiltonian and the interaction with an external field, either static for the edm or time varying for the transition elements.

The solutions to H^0 are of course determinants and the combined perturbation of $H^1 + H^{PNC} + H^E$ is expressed as matrix elements between them. The essence of MBPT is to reduce these many electron matrix elements to one and two particle form represented by simple diagrams. Thus figure IV.1a,b represents equation IV.2. We note that the excitation out of the closed shell (b) into the open shells must be added to excitation out of the open shell (a) to an unoccupied one to produce IV.2, a point which we have already mentioned.

In figure IV.1a,b the two open shell states can either represent the same state for an edm or different states for a transition. While this figure represents the lowest order for an open shell atom, figure IV.1c represents a transition from a closed shell to an excited

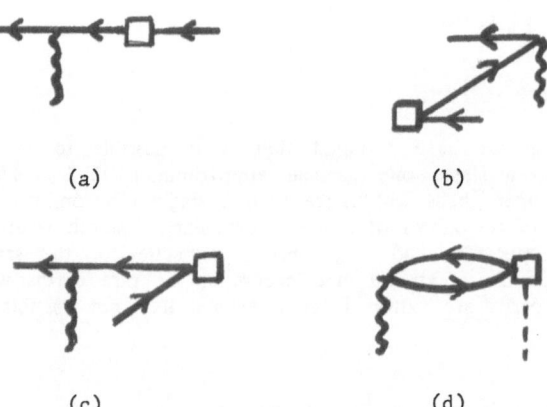

(a)   (b)

(c)   (d)

Figure IV.1   Some lowest order PNC diagrams. (a) and (b) represent the left hand term in equations IV.2. (c) represents a transition in which a core electron is excited. (d) represents part of the expectation value in a closed shell in which the nuclear spin dependent PNC Hamiltonian prodces an atomic electric dipole moment.

state and figure IV.1d an edm proportional to $\underline{I}.\underline{E}$ induced by a nuclear spin–dependent part of H^{PNC}.

(ii)   PNC Hartree–Fock

The next set of diagrams which we encounter are illustrated in figure IV.2. In these the PNC interaction acts on a core state exciting it to an unoccupied orbital which then interacts onwards via the Coulomb electron–electron interaction. An important feature to note is that the 'direct' process, at the bottom left of figure IV.2, is forbidden for all P not T interactions and for P and T scalar interactions because of a combination of parity and time–reversal arguments. Thus except in the case of the nuclear spin–dependent edm only the exchange diagrams need to be included.

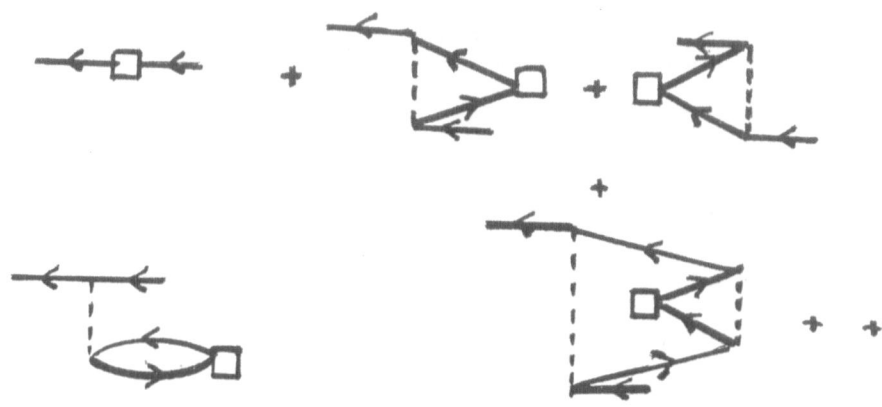

Figure IV.2   The effect of H^{PNC} and its iterations leading to PNC Hartree–Fock. The bottom left hand diagram is zero since only exchange interactions contribute.

A number of years ago, we pointed out[34] that these diagrams simply consituted a set of PNC Hartree–Fock diagrams in which each orbital was written as a sum of PC and PNC parts.

$$\tilde{\varphi}_i = \varphi_i + \varphi_i^{PNC} \qquad\qquad IV.6$$

in which case the usual HF equations break up into two (the first order in PNC):

$$\left\{\epsilon_i - h - V_{HF}\right\} |i> = 0 \qquad\qquad IV.7a$$

$$\left\{\epsilon_i - h - V_{HF}\right\} |i^{PNC}> = \left\{h^{PNC} + V_{HF}^{PNC}\right\} |i> \qquad\qquad IV.7b$$

Here V_{HF}^{PNC} is defined by a logical extension of the usual definition:

$$V_{HF}^{PNC} = \sum_{core} < i^{PNC} | \frac{(1-P_{12})e^2}{4\pi\epsilon_0 r_{12}} | i > + < i | \frac{(1-P_{12})e^2}{4\pi\epsilon_0 r_{12}} | i^{PNC}>$$

$$IV.8$$

h is simply a shorthand for the first three terms in H^0 (IV.5a).

29

Application of equations IV.7a,b and IV.8 is for the most part as straightforward as solving the PC Hartree–Fock equations. The usual problems occur in deciding the most appropriate definition for V_{HF}^{PNC} when open shells are involved, but these are equally present for V_{HF}. A more complicated situation occurs when one has both a nuclear spin–dependent PNC term and an open shell. Care must be taken to keep track of perturbation terms which have been omitted from the particular set of equations solved.

(iii)   Self consistent electric field treatment

Figure IV.3 shows a number of diagrams in which the electric field perturbs the states and the perturbation is then propagated as an effective interaction via the normal electron–electron forces. The diagrams look very similar to those in figure IV.2e and this is not surprising since both are single particle perturbations but here both direct plus exchange are non–zero. But there is a major difference when $\underline{E}$ is a transition field since we are evaluating a matrix element not an expectation value. However even this case can be incorporated into the Hartree–Fock mechanism by using so–called TDHF (time–dependent Hartree Fock) techniques.

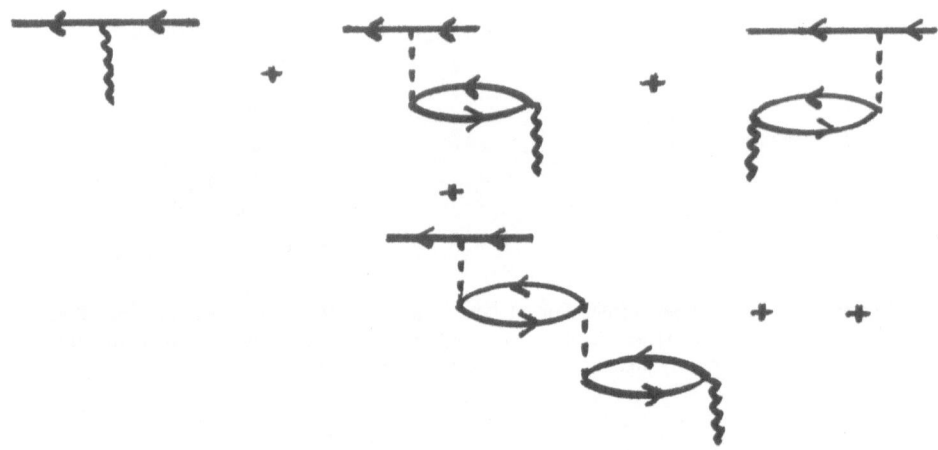

Figure IV.3   The effect of an external electric field and its iterations leading to an appropriate self–consistent dipole field

The upshot is that the figures IV.3 can be treated by replacing each orbital φ_i, by φ_i^+, φ_i^- where the $\varphi_i^{\pm}$ satisfy[4]

$$\left\{ \epsilon_i \pm \omega - h - V_{HF} \right\} i^{\pm} > \; = \; \underline{d} \, | i > + V_{HF}^{\pm} \, | i > - \text{orthogonality} \qquad IV.9$$

where

$$V_{HF}^{\pm} \; = \; \sum_i < c^{\mp} \, | \; \frac{(1-P_{12})e^2}{4\pi\epsilon_0 r_{12}} \; | \, e > + < e \, | \; \frac{(1-P_{12})e^2}{4\pi\epsilon_0 r_{12}} \; | \, c^{\pm} >$$

$$IV.10$$

The orthogonality terms ensure that $\varphi_i^{\pm}$ are orthogonal to all the occupied statese. ω is essentially the transition frequency. For an edm we put $\omega = 0$ and $\varphi^{\pm}$ become identical but equations IV.9 and IV.10 are still valid.

These equations too are straightforward to solve although again one has some subtleties involved in the treatment of open shells since the perturbation is automatically a vector operator.   The resultant excitation of the core yields a vector for $V_{HF}^{\pm}$ as for the nuclear spin dependent term it does not automatically do so for open shell excitation.

(iv)   PNC plus electric field

It should not surprise the reader that diagrams such as figure IV.4 in which both PNC and electric field effects are involved can also be dealt with by equations of the general form of IV.7 and IV.9 since both involve one particle operators.   However, considerable care is now needed because we are interested in terms of first order in PNC and first order in E the combination can produce a cross-term which has even parity and which therefore modifies the PC equation.

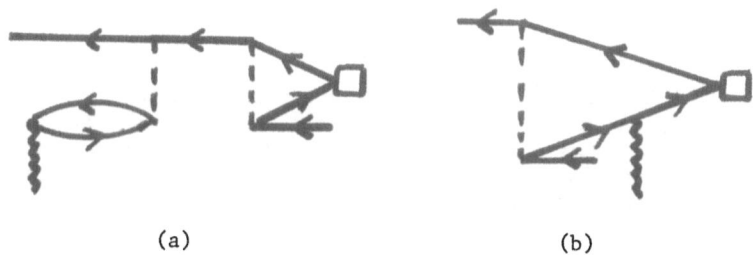

(a)   (b)

Figure IV.4   Diagrams involving both PNC and electric field.   (a) is separable but in (b) both perturbations are in the same excitation.

(v)   Higher order diagrams

While the processes described by the diagrams in figures IV.2, IV.3 and IV.4 contain many of the most important physical effects, calculations to the accuracy now necessary requires additional diagrams characterized by having two simultaneous excited states; these involve genuine electron correlations which can't be treated by one particle methods.   An example is figure IV.5a which represents a modification to the potential in which the electron moves.   More complicated diagrams in which PNC and E are involved with the correlation are illustrated in figures IV.5b and IV.5c.

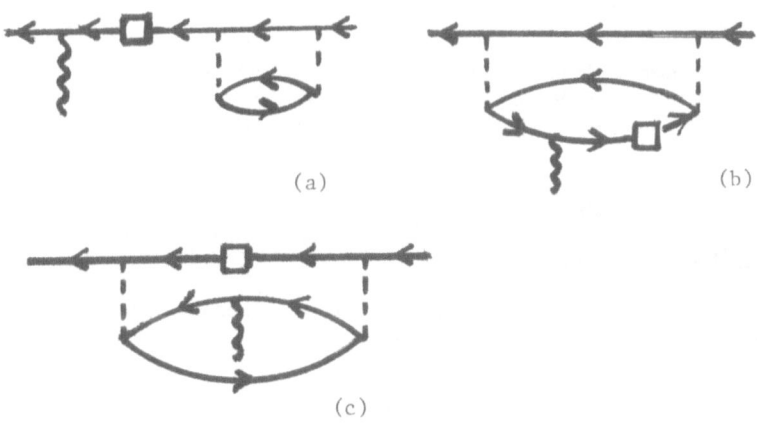

(a)   (b)

(c)

Figure IV.5

31

IV.4   Calculational points

A vast range of PNC calculations have been carried out over the years and space does not allow detailed discussion: the interested reader is referred to reference 35.   Here we shall simply highlight some major points and difficulties in various calculations.

(i)   Computational methods

To the extent that the calculation consists of a series – perhaps self–consistent – of single particle equations as discussed in sections IV.3(i) to (iv) it is now essentially straightforward with well established and powerful codes.   Occassionally some considerable skill is required to achieve convergence.   The order of difficulty changes completely when we move on to genuine correlation diagrams which involve a simultaneous double sum over excited orbitals.   Three approaches have been adopted:

(a) Double summation

The use of a number of 'tricks' to interpolate and extrapolate series of matrix elements and also to reduce to manageable numbers the continuum contributions means that it is possible to carry out a relatively straightforward evaluation of a double summation.   The Novosibirsk group have exploited this in a very successful series of calculations, both transition E1 and edm in the heavy elements Cs^{36}, Tl^{18}, and recently Bi^{37}.

(b) Splines

The Notre Dame group have developed a powerful technique[38,39] in which the atomic wavefunctions are expanded in a 'spline' basis set which has a number of very convenient features for subsequent evaluation.   The double summation is now reduced to a matrix manipulation for which very fast and powerful algorithms are available.   A number of calculations on both E1 and edm in Cs, Tl have been reported.   A particular feature has been the complete treatment of the correlation to lowest order.

(c) Two particle operators

A technique which looks very promising for PNC calculations although no calculation has yet been reported has been developed by the Goteborg group.   This involves turning the double summation over excited states into a two particle differential equation.   Such pair equations are relatively straightfoward in the non–relativistic limit and have been much used in calculations of the correlation energy.   The relativistic approach which is needed for PNC problems has a number of conceptual difficulties to do with negative energy states.   These are now well understood and PNC calculations can be expected in the near future.

By whatever means these calculations are carried out, the computational load is a major limiting factor.   It is to be expected that the new generation of very high performance parallel processors will enable more complex calculations to be carried out.

(ii)   Features of particular calculations

(a) Cs 6s → 7s and edm in 6s

Cs seems a very straightfoward element for calculation.   The one particle part contains the major part of the final answer and the calculations seem stable and convergent.   One initial problem was the difficulty in obtaining one particle energies for some of the core states that were sufficiently realistic to be a basis for perturbation theory.   The inclusion of correlation contributions to these energies has largely solved these problems.   An accuracy of order 1% seems quite feasible.

(b)   Tl: $6p_{1/2} \rightarrow 7p_{1/2}$, $6p_{1/2} \rightarrow 6p_{3/2}$ and edm in $6p_{1/2}$

In principle, this is a one particle problem like Cs.   However, in practice a major contribution to all the PNC effects comes from excitation of the $6s_{1/2}$ into the 6p shell. The energies of the various possible $6s6p^2$ levels are not well described by a single particle model and this means that a conventional perturbation approach does not readily converge.   In fact a semiempirical approach turns out to have some merit.   In spite of the difficulty E1 calculations of order a few percent have been reported – though there are still some difficulties with the edm.

(c)   Pb: J = 0 → 1

This has some advantages for calculation because of the J = 0 starting state. However no correlation calculations have yet been reported, presumably because the only experiment is of limited accuracy.   A potential difficulty is the breakdown of jj coupling within the $6p^2$ state.

(d)   Bi $6p^3$ J = 3/2 → 3/2', 5/2

A major problem here is the complexity due to the presence of three open shell electrons.   A complete calculation to lowest order in the residual Coulomb interaction has been made and a more detailed calculation including some correlation has been reported recently.   One can be optimistic and predict that in due course calculations in this element will approach the present limits of accuracy for say Tl.

(e)   Xe, Hg, $^1S_0$

No particular problem is found in treating the effects of a direct nuclear spin dependent interaction.   Since we have a J = 0 starting point, symmetry requirements are automatically satisfied.

A much more complicated situation arises when one attempts to calculate the effect of an electron edm.   This can only appear through a combined effect the magnetic dipole field from the nucleus.   This can either be 'direct' in which the electron edm interacts with the $\underline{V} \times \underline{B}$ notional electric field, or indirect when the magnetic hyperfine interaction H_{hfs} has to be combined with the edm interaction H_{edm} and the interaction with an external field H_{dip}.   A problem of this sort with three perturbations is manageable in the independent particle approximation, but rapidly beocmes very complicated when the two particle Coulomb interaction is included.   This is nonetheless a very important calculation since the measurements described in lecture III have enormous potential accuracy and may be the best source of limit on the electron edm.

(f)   TlF

This calculation[43] in a diatomic molecule is very different from all the others reported because the 'electric field' is built into the molecule along the internuclear axis. The problem is complicated by the need to use relativistic functions close to the nucleus when only non–relativistic functions are available for the molecule.   The matching is not unique nor is it quite clear that the requirement that the field vanishes at each nucleus is properly satisfied.   A much improved calculation is urgently required.

(h)   Nearly degenerate levels in the rare earths

A quite different type of calculational problem occurs in the rare–earths where there are pairs of opposite parity states which are very close together[20].   In such a case one must revert back to the full expression IV.1 but with just a single term in the expansion over excited states.   Unfortunately, this enormous simplification is more than compensated by the complexity of these rare–earth states.   Calculations are urgently required.

(i)   P not TV

The present world situation for P not TV is summarized in table IV.1. Where appropriate, the lecturer has arbitrarily combined perhaps incompatible experimental data and has picked a highly subjective 'best' theoretical value. In order to avoid continual change of the theoretical value with particle physics input, it has become conventional to express the theory in terms of $Q_W = -N$ and then to incorporate the small difference through the scaling quantity η. Because the ratio of protons to neutrons is accurately the same for all the elements considered here η is constant for all transitions of interest. In column 3 we view this as the quantity to be measured. The combined result

$$\eta = 0.907 \pm 0.038$$

compares well with the standard model including radiative corrections:

$$\eta = 0.974 \left\{ 1 - \frac{Z}{N} (1 - 4 \sin^2\theta_W) \right\}$$

with $\sin^2\theta_W = 0.230$ giving $\eta = 0.920$. It is interesting to more that the reduction in error by recent developments in Cs has meant that we have perhaps seen the PNC contributions of the proton for the first time.

Only one limit has been given for the nuclear spin dependent term. For Cs, Wieman[27] has obtained the dimensionless constant defined by Khriplovich and co-workers[19]

$$K_a = 0.72 \pm 0.39$$

Table IV.1   World situation for $\eta = Q_W/-N$

	Experiment	Atomic Theory ($Q_W = -N$)	η measured
Bismuth 876 nm R × 10^8	-10.1 ± 0.9	-10.4 ± 1.0	0.975 ± 0.13
Bismuth 648 nm R × 10^8	-10.48 ± 1.0	-12 ± 3	0.865 ± 0.29
Lead 1.28 μm R × 10^8	-9.9 ± 2.5	-10.4 ± 0.8	0.95 ± 0.25
Cesium 539 nm iE1 × 10^{11}	0.848 ± 0.067	0.91 ± 0.04	0.92 ± 0.08
Cesium 539 nm*			0.89 ± 0.05
Thallium 293 nm iE1 × 10^{10}	0.76 ± 0.14	0.93 ± 0.08	0.82 ± 0.17
		Combined atomic result:	$\eta = 0.91 \pm 0.04$

*Intermediate details are not given in in ref. 27.

Table IV.2   Limits on P and T violating atomic quantities.

Quantity	System	Limit	Reference
Electron edm	Tl	$1 - 2 \times 10^{-24}$ excm	12
Nuclear edm	TlF	2×10^{-21} excm	43
Schiff moment of nucleus	Hg	4×10^{-10} exfm^3	12
Magnetic quadrupole moment	Cs	10^{-4} $\mu_N R_N$	44
C_S	Hg	2×10^{-5} G_F	12
C_T	Hg	2.5×10^{-7} G_F	12

which is consistent with the predicted magnitude expected from a nuclear anapole moment[27].

We conclude, first that the agreement at the 40% level above confirms that the electroweak interactions are indeed present as expected at the very low energies involved in atoms. Second, that an increase in precision to the 1% mark, if this proves possible, will yield information on very interesting electro-weak radiative corrections. Finally, turning the argument around, we have a new and fascinating field of electro-weak atomic physics which may prove very fruitful to our understanding of atomic phenomena.

(ii)   P and TV

We collect together in table IV.2 the best available limits for the various P and TV atomic quantities. These limits should be taken as order of magnitude limits only pending a more thorough understanding of the various calculations.

References

1. P.G.H. Sandars, Phys. Scripta 36 904-10 (1987).

2. E.D. Commins, Phys. Scripta 36 468-475 (1987).

3. M.A. Bouchiat and L. Pottier in Atomic Physics 9, ed. R.S. van Dyck Jr and E.N. Fortson, World Scientific (1984) pp.246-271.

4. P.G.H. Sandars in Atomic Physics 9, ed. R.S. van Dyck Jr and E.N. Fortson, World Scientific (1984) pp.225-245.

5. F.J. Raab in Atomic Physics 9, ed. R.S. van Dyck Jr and E.N. Fortson, World Scientific (1984) pp.272-284.

6. M.J. Macpherson, D.N. Stacey, P.E.G. Baird, J.P. Hoare, P.G.H. Sandars, K.M.J. Tregidgo and Wang Guowen, Europhysics Lett. 4 811-816 (1987).

7. S.L. Gilbert and C.E. Wieman, Phys. Rev. A34 792-803 (1986).

8. M.A. Bouchiat, J. Guena, P.H. Jaquier, M. Lintz and L. Pottier, Opt. Commun. 56, 100–106 (1985).

9. M.A. Player and P.G.H. Sandars, J. Phys. B3 1620 (1970).

10. S., Schropp, D. Cho, T. Vold and E.A. Hinds, Phys. Rev. Lett. 59 991–4 (1987).

11. T.G. Vold, F.J. Raab, B. Heckel and E.N. Fortson, Phys. Rev. Lett. 52 2229 (1984).

12. B. Heckel in Atomic Physics 9, ed. R.S. van Dyck Jr and E.N. Fortson, World Scientific (1984) pp.285–291.

13. C. Bouchiat in Atomic Physics 7, ed. D. Kleppner and F.M. Pipkin, Plenum (1981) pp.

14. R.G. Sachs, 'The Physics of Time Reversal', The University of Chicago Press (1987).

15. M.A. Bouchiat and C.C. Bouchiat, Phys. Lett. 48B 111–4 (1974).

16. E.N. Fortson and L. Wilets, Advances in atomic and molecular physics 16 319–73 (1980).

17. B.W. Lynn in Atomic Physics 9, ed. R.S. van Dyck Jr and E.N. Fortson, World Scientific (1984) pp.212–224.

18. V.A. Dzuba, V.V. Flambaum, P.G. Silvestrov and O.P. Sushkov, J. Phys. B20 3297–311 (1987).

19. V.V. Flambaum, I.B. Khriplovich and O.P. Sushkov, Phys. Lett. 146B 376–9 (1984).

20. V.A. Dzuba, V.V. Flambaum and I.B. Khriplovich, Z. Phys. D1 243–5 (1986).

21. J.D. Taylor, P.E.G. Baird, R.G. Hunt, M.J.D. Macpherson, G. Nowicki, P.G.H. Sandars and D.N. Stacey, J. Phys. B20 5423–5442 (1987).

22. P.G.H. Sandars, Physica Scripta 21 284–93 (1980).

23. K.M.J. Tregidgo, P.E.G. Baird, M.J.D. Macpherson, C.W.P. Palmer, P.G.H. Sandars, D.N. Stacey and R.C. Thompson, J. Phys. B19 1143–52 (1986).

24. J. Hollister, G.R. Apperson, L.L. Lewis and E.N. Fortson, Phys. Rev. Lett. 46 643–6 (1981).

25. M.A. Bouchiat, J. Guena, L. Hunter and L. Pottier, Phys. Lett. 117B, 358–64; 134B 493 (1984).

26. M.A. Bouciat, J. Guena, L. Pottier and L. Hunter, J. Physique 47 1709–30 (1986).

27. M.C. Noecker, P.B.P. Masterson and C.E. Wieman, Private communication (1988).

28. P.G.H. Sandars, Phys. Rev. Lett. 19 1396–8 (1967).

29. G.E. Harrison, P.G.H. Sandars and S.J. Wright, Phys. Rev. Lett. 22 1263–5 (1969).

30. E.A. Hinds and P.G.H. Sandars, Phys. Rev. A21 471 (1980).

31. D.A. Wilkening, N.F. Ramsey and D.J. Larson, Phys. Rev. A29 425 (1984).

32. S.K. Lamoreaux, J.P. Jacobs, B.R. Heckel, F.J. Raab and N. Fortson, Phys. Rev. Lett. 59 2275–8 (1987).

33. I. Lindgren and J. Morrison, Atomic and many-body theory, Springer-Verlag (1982).

34. P.G.H. Sandars, J. Phys. B$\underline{10}$ 2983-95 (1977).

35. Relativistic Many-Body Problems, ed. I. Lindgren, Phys. Scripta RS$\underline{7}$ (1987).

36. V.A. Dzuba, V.V. Flambaum, P.G. Silvestrov and O.P. Sushkov, J. Phys. B$\underline{18}$ 597-613 (1985).

37. V.A. Dzuba, V.V. Flambaum, P.G. Silvstrov and O.P. Sushkov, preprint NBI-88-15 (1988).

38. W.R. Johnson, D.S. Guo, M. Idress, J. Sapirstein, Phys. Rev. A$\underline{34}$ 1043-57 (1986).

39. W.R. Johnson, S.A. Blundell, Z.W. Lie, J. Sapirstein, preprint (1988).

40. E. Lindroth, Phys. Scripta $\underline{36}$ 485-492 (1987).

41. V.V. Flambaum and I.B. Khriplovich, J.E.T.P. $\underline{62}$ 872 (1985).

42. A.-M. Martensson-Pendrill and P. Oster, Phys. Scripta $\underline{36}$ 444-452 (1987).

43. P.V. Coveney and P.G.H. Sandars, J. Phys. B$\underline{16}$ 3727-3740 (1983).

44. I.B. Khriplovich, J.E.T.P. $\underline{44}$ 25 (1976).

QUANTUM ELECTRODYNAMICS (QED) IN STRONG COULOMB FIELDS:

CHARGED VACUUM, ATOMIC CLOCK, AND NARROW POSITRON LINES

Berndt Müller

Institut für Theoretische Physik
J. W. Goethe Universität
Postfach 11 19 32
D-6000 Frankfurt am Main, West Germany

INTRODUCTION

Slow collisions of two heavy atoms or ions offer a unique laboratory for the study of the motion of electrons, and the behaviour of the QED vacuum, in very strong electric fields. It is for this reason that the theoretical and experimental aspects of such collisions have been studied very intensely during the last two decades. Early on, theory focussed on the change of the vacuum state from a neutral to a charged realization at a critical nuclear charge $Z_c = 173$, and the associated spontaneous emission of positrons of well-defined energy. Later, the view broadened into the quest of gaining an understanding of dynamic phenomena caused by the time-dependence of the collision process, e.g. dynamically induced pair production, K-shell ionization, and delta-electron production. This will be discussed in the first part of my lectures, however rather concisely, because an extensive treatise of this subject exists [1].

In recent years two further developments have attracted widespread interest: the "atomic clock" phenomenon, and the narrow electron-positron coincidence lines detected at GSI. The first of these, which allows for a fairly precise, model-independent determination of nuclear reaction times, is by now well established and forms the subject of the second part of these lectures. The narrow $e^+ - e^-$ lines, on the other hand, have been a surprising experimental discovery that is still controversial and unexplained. After a (theorist's) discussion of the present status of the experiments, I will present the pro-s and (mostly) con-s of various mechanisms that have been suggested as eplanation of the data. Much of this discussion will be concerned with the hypothesis that the lines are the result of the $e^+ - e^-$ decay of new neutral particles. The lectures conclude with some exciting aspects that will open up when new accelerator facilities become available in the near future.

THE ELECTRON-POSITRON VACUUM IN STRONG ELECTRIC FIELDS

Definition of the Vacuum State

Let me begin with a simple, nonrigorous discussion of the concepts underlying the definition of the vacuum state in QED of strong fields. In the presence of an external Coulomb source j_{ext}^ν the equations of motion for the electron-positron field

$\hat{\Psi}(x)$ and the electromagnetic field $\hat{A}_\mu(x)$ read:

$$\left(\gamma^\mu(i\partial_\mu - e\hat{A}_\mu) - m_e\right)\hat{\Psi} = 0, \tag{1}$$

$$\partial_\mu \hat{F}^{\mu\nu} = e\hat{\Psi}^\dagger \gamma^0 \gamma^\nu \hat{\Psi} + j^\nu_{ext}. \tag{2}$$

Here and in the following, the 'hat' symbol indicates second quantized fields. In the Heisenberg picture, which is most useful for our purposes, these equations specify the time-evolution of the system, while the state vector remains fixed. Whenever the external source is very strong, as it is the case in the situations of interest to us here, the dynamical source may be neglected to lowest order on the right-hand side of eq.(2), and the resulting solution, A^{ext}_μ, may be inserted in the Dirac equation, eq.(1).

If used as the basis for a systematic iteration procedure this leads to a perturbation expansion in terms of the electric charge e (but not in the external charge Ze !), known as Furry picture or *bound state QED*, which is the subject of Peter Mohr's lectures [2]. The interested reader is referred to these lectures for more details. Note that, in the presence of a large number of electrons, it is useful to replace the Coulomb field of the nuclear charge, A^{ext}_μ, by a screened Coulomb potential that accounts for the average screening experienced by a given electron or positron due to the presence of all others. Higher order corrections then describe the effect of fluctuations of the electromagnetic field around the mean potential.

With this simplification the evolution of the second quantized electron field operator can be solved exactly by an eigenfunction expansion

$$\hat{\Psi}(x) = \sum_n \hat{b}_n \phi_n(x), \tag{3}$$

with help of a complete set of time-dependent single particle solutions of the Dirac equation:

$$\left(\gamma^\mu(i\partial_\mu - eA^{ext}_\mu) - m_e\right)\phi_n(x) = 0. \tag{4}$$

It is easy to see that the single particle operators $\hat{b}_n$ are time- independent, even if the external potential varies with time. In a static field the time-evolution of the Dirac eigenfunctions is given by an energy phase, and the expansion for the field operator becomes:

$$\hat{\Psi}(x) = \sum_n \hat{b}_n \phi_n(\vec{x}) \exp(-iE_n t). \tag{5}$$

From the equal-time anticommutation relations for the Dirac field operator one deduces the well-known relations for the single-fermion operators

$$[\hat{b}_m, \hat{b}^\dagger_n]_+ = \delta_{mn}, \qquad \text{etc.,} \tag{6}$$

in terms of which the second quantized Hamiltonian takes an especially simple form, if the external field is static:

$$\hat{H} = \frac{1}{2}\int d^3x \left[\hat{\Psi}^\dagger(x), (-i\vec{\alpha}\nabla - e\vec{\alpha}\vec{A} + eA^0 + \beta m_e)\hat{\Psi}(x)\right]_- = \sum_n E_n(\hat{b}^\dagger_n \hat{b}_n - \frac{1}{2}). \tag{7}$$

The contribution of each mode is expressed in terms of the single-particle energy E_n and the electron number operator $\hat{b}^\dagger_n \hat{b}_n$, which has eigenvalues 0 and 1. If we tentatively define the vacuum state $|0\rangle$ as the state of lowest energy in the given external field, we must choose the eigenvalue 0 for modes with positive energy E_n, and the eigenvalue 1 for modes with negative energy, i.e.

$$\begin{aligned}\hat{b}^\dagger_n \hat{b}_n|0\rangle &= 0 \qquad \text{for } E_n > 0, \\ \hat{b}_n \hat{b}^\dagger_n|0\rangle &= 0 \qquad \text{for } E_n < 0.\end{aligned} \tag{8}$$

Introducing positron (or "hole") creation operators $\hat{d}_n^\dagger$ for the electron annihilation operators $\hat{b}_n$ of negative energy modes, the Hamiltonian assumes the normal-ordered form

$$\hat{H} = \sum_{E_n>0} E_n \hat{b}_n^\dagger \hat{b}_n - \sum_{E_n<0} E_n \hat{d}_n \hat{d}_n^\dagger + E_{vac}, \tag{9}$$

where the energy of the vacuum state is given by a sum over all modes:

$$E_{vac} = \langle 0|\hat{H}|0\rangle = -\frac{1}{2}\sum_n |E_n|. \tag{10}$$

This expression for the vacuum energy is, of course, divergent and must be regularized and renormalized properly, but this need not concern us here.

The main point to be noted is that the above definition of the vacuum state as the state of minimal energy does not make much sense, because this state sometimes is not accessible physically. The reason for this is that electric charge is conserved exactly, so that only states with the same total charge can communicate. In order to account for charge conservation, one adds the charge operator $\hat{Q}$ with a Lagrange multiplier to the Hamiltonian, and looks for the state that minimizes $(\hat{H} - \mu\hat{Q}/e)$. Here μ is the energy required to supply a unit charge to the system from outside. This is $\mu = -m_e$, because the system may emit a positron, carrying with it the rest energy m_e. One can now go through the same steps as before, except that now the discrimination between electron and positron states occurs at $E = -m_e$ instead of $E = 0$.

More generally, the boundary between particle and antiparticle (hole) states is called the Fermi energy E_F, and the definition of the vacuum state corresponds to the choice $E_F = -m_e$. Other choices $E_F > -m_e$ represent configurations where a finite number of electrons are present in states above the Dirac sea; in particular, $E_F = +m_e$ describes a neutral atom. Of course, any choice between two discrete bound states is equivalent. In this way bound state QED can be formulated for many-electron atoms in a unified way.

Properties of the Vacuum State

A mathematically very convenient method [3] to describe properties of the vacuum state is based on the Feynman propagator, which is defined as

$$\begin{aligned}iS_F(x,x') &= \sum_n \phi_n(x)\bar{\phi}_n(x')\Big(\theta(t-t')\theta(E_n-E_F) - \theta(t'-t)\theta(E_F-E_n)\Big)\\ &= \langle 0|T\left(\hat{\Psi}(x)\hat{\Psi}^\dagger(x')\gamma^0\right)|0\rangle. \end{aligned} \tag{11}$$

where T denotes the time-ordered product of operators and $E_F = -m_e$. In the presence of an external field S_F is sufficient to completely specify the vacuum state. (If one wants to include the effect of interactions among the electrons, all n-particle propagators must be known.) For example, the charge density contained in the vacuum,

$$\rho_{vac}(x) = \langle 0|\hat{j}^0(x)|0\rangle = \frac{e}{2}\sum_{E_n<E_F}\phi_n^\dagger(x)\phi_n(x) - \frac{e}{2}\sum_{E_n>E_F}\phi_n^\dagger(x)\phi_n(x), \tag{12}$$

can be expressed in terms of the Feynman propagator as

$$\rho_{vac}(x) = -ie\,\lim_{x'\to x} tr\left(\gamma^0 S_F(x,x')\right). \tag{13}$$

Formally, the limit here must be taken symmetrically from the left and the right. A short-hand notation, now for the full current four-vector is:

$$j^\mu_{vac}(x) = -ie\, tr\left(\gamma^\mu S_F(x,x)\right).$$ (14)

Similarly, with help of the propagator the vacuum energy can be written in the form

$$E_{vac} = -i \int d^3x \lim_{x' \to x} tr\left(\gamma^0 H_D(x) S_F(x,x')\right),$$ (15)

where $H_D(x)$ is the single-particle Dirac Hamiltonian in the external field that was explicitly written in eq.(7). It is useful to consider the change of the vacuum energy under a change in the external field by introducing a strength parameter λ that runs from 0 to 1. Subtracting the trivial, and here uninteresting, vacuum energy in the absence of an external field, we may write:

$$E'_{vac}[A_\mu] = E_{vac}[A_\mu] - E_{vac}[A_\mu = 0] = \int_0^1 d\lambda \frac{d}{d\lambda} E_{vac}[\lambda A_\mu].$$ (16)

Inserting the previous equation and making use of the normalization of the single-particle eigenfunctions $\phi_n(x)$, one can show that the vacuum energy can be expressed in the following form:

$$E_{vac} = E_F \sum_n \left(\theta(E_F - E_n[A_\mu]) - \theta(E_F - E_n[0])\right) + \int_0^1 d\lambda \int d^3x\, A_\mu j^\mu_{vac}[\lambda A_\mu].$$ (17)

The first term simply counts all electron states that have passed through the Fermi surface when the external potential was switched on. Such states are usually called *supercritical* bound states. The second term describes the interaction energy of the vacuum charge density with the external potential. The number N_{sc} of supercritical states also occurs in the analogous expression for the total vacuum charge:

$$Q_{vac} = e \sum_n \left(\theta(E_F - E_n[A_\mu]) - \theta(E_F - E_n[0])\right) = eN_{sc},$$ (18)

where e denotes the electron charge. In other words: the total charge of the vacuum is equal to the number of states that lie below the Fermi surface $E_F = -m_e$, i.e. of states bound by more than $2m_e$.

Supercritical Atoms

For atomic nuclei of normal nuclear density, the first bound state - the 1s-state - is predicted to acquire a binding energy of $2m_e$ and become supercritical at a critical nuclear charge $Z_c = 173$. The next state, the $2p_{1/2}$-state would 'dive' into the Dirac sea at about $Z = 190$, and so on. This is illustrated in Fig. 1, which shows the energy eigenvalues of atomic inner-shell states as function of nuclear charge Z [4,5]. Note that s- and $p_{1/2}$-states exist for $Z > 137$ only when the finite nuclear size is taken into account.

What happens when a bound states reaches the top of the negative energy continuum at $E = -m_e$, is best described in terms of the formalism normally employed for the treatment of autoionizing states, i.e. bound states imbedded in a continuum [4]. Let us denote the just critically bound state by ϕ, and the negative continuum states by $\psi_E, E < -m_e$. We write symbolically:

$$H_e\phi = -m_e\phi, \qquad H_e\psi_E = E\psi_E \quad (E < -m_e),$$ (19)

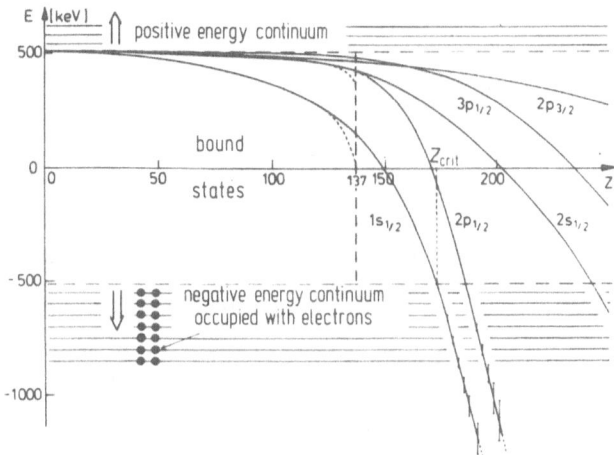

Figure 1. Relativistic energy eigenvalues of atomic inner-shell states. The calculations account for nuclear finite size and electron screening. The 1s-state become supercritical at nuclear charge $Z = 173$.

where H_c stands for the Dirac Hamiltonian of the atom with $Z = Z_c$. Now we add a few, $Z' \ll Z_c$, protons to the nucleus and seek for the negative energy continuum states of the supercritical atom with nuclear charge $(Z_c + Z')$:

$$(H_c + Z'U(r))\Psi_E = E\Psi_E \quad (E < -m_e). \tag{20}$$

$U(r)$ here stands for the Coulomb potential of a finite size nucleus, with $U(r) = -1/r$ for $r > R$, R being the nuclear radius. Eq. (20) can be solved analytically, when the matrix elements of the additional interaction between continuum states are neglected, by expanding:

$$\Psi_E = a(E)\phi + \int_{-\infty}^{-m_e} dE' b_{E'}(E)\psi_{E'}. \tag{21}$$

The result for the admixture of the critical bound state to a supercritical continuum state of energy E is:

$$|a(E)|^2 = \frac{\Gamma_E/2\pi}{(E + m_e - \Delta E)^2 + \frac{1}{4}\Gamma_E^2}, \tag{22}$$

where

$$
\begin{aligned}
\Delta E &= \langle\phi|Z'U(r)|\phi\rangle + \frac{1}{2\pi}P\int_{-\infty}^{-m_e} dE'\frac{\Gamma'_E}{E - E'}, \\
\Gamma_E &= 2\pi|\langle\psi_E|Z'U(r)|\phi\rangle|^2.
\end{aligned}
\tag{23}
$$

Eq. (22) express the fact that the bound state is drawn into the continuum by an amount ΔE and its strength is spread over a range of continuum states of width $\Gamma_r = \Gamma_{-m_e+\Delta E}$. In other words, the bound state becomes a resonance. The inverse width of this resonance gives the lifetime of the vacant supercritical state against spontaneous creation of an electron- positron pair. Typical numerical values are $\Gamma_r = 1 - 5$ keV, i.e. $\Gamma_r^{-1} = 10^{-19} - 10^{-18}$ s.

In order to make quantitative predictions for the physical processes that occur when the bound 1s-state becomes supercritical, we must specify the precise way in which the potential changes from subcritical to supercritical and, possibly, back to subcritical strength. We have to distinguish three general cases:

1. If the bound state is initially occupied by an electron, nothing observable happens at all. The additional state joining the Dirac sea increases the vacuum charge by one unit, but at the same time the discrete atomic bound state, that carried a unit charge, disappears.

2. The bound 1s-state is initially vacant, and the additional potential is switched on suddenly and remains permanently. In this case a pair is created, the electron occupies the supercritical state, and the positron is emitted with a narrow energy distribution given by $|a(E)|^2$, see eq. (22). The narrow line at kinetic energy $E_r = |E_B(1s)| - 2m_e$, where $E_B(1s)$ is the binding energy of the 1s-state, is the signature of spontaneous pair production. As Fig. 1 shows, E_B and therefore E_r is a strong function of Z, changing by about 30 keV per unit of nuclear charge.

3. The bound 1s-state is initially vacant, and the additional potential is switched on for a finite length of time T. Unless $T \gg \Gamma_r^{-1}$, the resulting spectrum of created positrons will depend on the precise value of T, according to:

$$\bar{N}_E(T) = |a(E)|^2 \left(1 - 2e^{-\Gamma_r T/2} \cos(E - E_r)T + e^{-\Gamma_r T}\right). \tag{24}$$

For $T \ll \Gamma_r^{-1}$ the width of this distribution is given by the lifetime of the supercritical potential T^{-1}, which expresses the uncertainty principle, whereas for $T \gg \Gamma_r^{-1}$ the width is just the spontaneous decay width Γ_r. The decay probability is $[1 - \exp(-\Gamma_r T)]$, approaching 1 in the limit of large T. Thus, the signature of spontaneous pair creation, viz. the narrow positron line, emerges only when the supercritical field exists for a period of time that is at least comparable to Γ_r^{-1}.

One may, finally, ask whether higher order corrections of QED, such as virtual vacuum polarization or electron self-energy effects, can modify this picture substantially, e.g. prevent the "diving" into the Dirac sea. The answer was given in a number of elaborate calculations [6,7,8] which showed that the corrections to first order in the coupling constant e^2 remain small at Z_c. Moreover, the vacuum polarization correction for the critical 1s-state, -10.7 keV, and the self-energy correction, +11.0 keV, cancel almost completely. The resulting shift in the binding energy of the 1s-state at Z_c is less than 1 keV, to be compared with a total binding of 1022 keV. There is no reason to believe that higher order corrections would change this picture.

Strong Electric Fields in Heavy Ion Collisions

Atomic nuclei with charge $Z > 170$ do not exist, but electric fields of equal strength can be generated transiently in collisions of two very heavy atoms or ions, e.g. uranium and uranium with a combined nuclear charge $Z_u = (Z_1 + Z_2) = 184$. Here the parameter that controls the intensity of the electric field is the internuclear separation R. Of greatest interest for possible tests of QED of strong fields are collisions at energies around the Coulomb barrier, i.e. where the beam energy is just sufficient to bring the nuclei into contact·in a head-on collision. At these energies, typically about 6 MeV per nucleon (MeV/u), the nuclei move with velocity $v/c \approx 0.05$ in the centre-of-mass frame, thus being much slower than the motion of inner-shell electrons which move almost with the speed of light c. The adiabatic picture, where one views the collision as a succession of snapshots with electrons moving in the Coulomb field of two frozen nuclei, should therefore be an excellent starting point for a more complete description of the collision process.

This concept leads to the Born-Oppenheimer approximation, which defines *quasimolecular* states $\varphi_i(r, R)$ as the eigenstates of the instantaneous two-centre Hamiltonian

$$H_{TC}[R] = -i\vec{\alpha}\nabla + \beta m_e + V_1(r, R) + V_2(r, R), \tag{25}$$

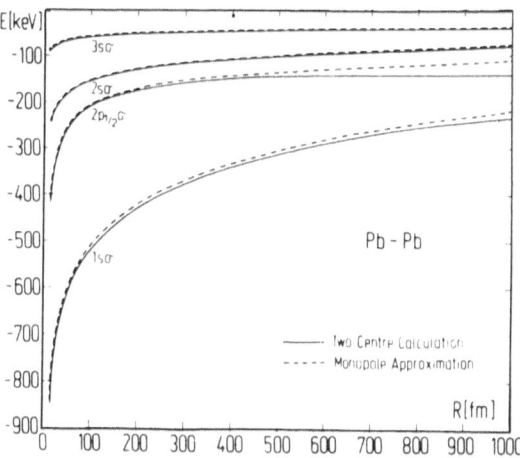

Figure 2. Adiabatic level diagram of the four innermost states of the (Pb+Pb) quasi-molecule ($Z_u = 164$). Solid lines: full two-centre potential; dashed lines: monopole approximation.

with eigenvalue $E_i(R)$. Here V_1 and V_2 are the screened Coulomb potentials of the two nuclei. In general, the energies $E_i(R)$ decrease when the nuclei approach each other, i.e. the binding energy of the quasi- molecular states grows with shrinking nuclear distance R, as shown in Fig. 2 for the collision system Pb + Pb. The value of R where the lowest electron state becomes bound by more than $2m_e$ is called the *critical distance* R_c. After taking into account the finite nuclear radius and electron screening the critical distance is 27 fm in U + U and 35 fm in U + Cm collisions, respectively.

The effect of the collision dynamics on the single particle wavefunctions is obtained by solving the time-dependent Dirac equation

$$i\frac{\partial}{\partial t}\phi_i(t) = H_{TC}[R(t)]\phi_i(t) \tag{26}$$

with a prescribed nuclear trajectory $R(t)$ and the initial condition that the wave-function enters the collision in some specified adiabatic state at $R = \infty$, i.e. in a certain eigenstate of the individual atoms: $\phi_i(t = -\infty) = \varphi_i(R = \infty)$. The practical solution of eq. (26) under the stated conditions is best achieved by expanding the wavefunction $\phi_i(t)$ in terms of the adiabatic quasimolecular states:

$$\phi_i(r,t) = \sum_k a_{ik}(t)\varphi_k(r, R(t))e^{-i\chi_k(t)}, \tag{27}$$

where $\dot{\chi}_k(t) = E_k[R(t)]$ is the adiabatic phase of the state k. The single-particle occupation amplitudes $a_{ik}(t)$ satisfy a set of coupled differential equations ("coupled-channel" equations) of the form

$$\dot{a}_{ik} = -\sum_{j\neq k} a_{ij}\langle\varphi_k|\frac{\partial}{\partial t}|\varphi_j\rangle \exp(i\chi_k - i\chi_j). \tag{28}$$

The partial differential operator $\partial/\partial t$ acts on the parametric dependence of the basis states φ_i on the nuclear relative distance vector $\vec{R}$.

One distinguishes a *radial* motion, i.e. a change in the length $|\vec{R}|$, and a *rotational* motion, i.e. a change in the orientation of $\vec{R}$. Extensive investigations [9] have shown that the rotational motion is of only minor importance in the processes leading to ionization of strongly bound states and to pair production in collision systems with $Z_u > 137$. The reason is that these processes involve almost exclusive quasimolecular states with angular momentum $j = \frac{1}{2}$ in the united atom limit $R \rightarrow 0$, which have a spherically symmetric density distribution in this limit. On the other hand, numerical calculations have revealed that a very large number of such states must be retained in the coupled equations 28 in order to achieve convergence [10]. It has, therefore, been found to be quite appropriate to restrict the adiabatic basis to states with $j = \frac{1}{2}$, not only in the united atom limit but for all values of R. This procedure, which is known as the *monopole approximation*, has proved to exceedingly successful in describing inner-shell excitation processes and pair-production in systems with $Z_u > 137$. As shown in Fig. 2, the binding energies of the innermost quasimolecular states are well reproduced in this approximation up to rather large distances R comparable to the K-shell radius of the separated atoms. By contrast, inner-shell excitation and pair production amplitudes are essentially determined at distances below a few 100 fm.

In the monopole approximation, the two-centre potential $V(\vec{r}, \vec{R})$ is replaced by its spherically symmetric part $V(r, R)$ where r and R are now scalars. As the wavefunctions φ_i now also depend on the distance R alone, the partial differential operator in eq. (28) can be replaced by the operator $\dot{R}\frac{\partial}{\partial R}$. The matrix elements can be evaluated conveniently with the help of the Hellmann–Feynman theorem, and are then tabulated for a large number of grid points in the variable R. Inclusion of electron screening is very important to obtain quantitative agreement with experimental data. In the calculations of our group screening is routinely described in the adiabatic Hartree- Fock- Slater approximation for discrete as well as continuum states [11].

As long as the electron-electron interaction is described by an average screening potential, i.e. neglecting correlation effects, all observables can be expressed in terms of the single-particle occupation amplitudes $a_i(t \rightarrow \infty)$ [12]. For example, the number of electrons excited into an initially vacant state i is given by

$$N_i = \sum_{E_k < E_F} |a_{ki}|^2 \qquad (E_i > E_F), \qquad (29)$$

while the number of vacancies in a state below the initial Fermi surface E_F is represented by

$$\bar{N}_i = \sum_{E_k > E_F} |a_{ki}|^2 \qquad (E_i < E_F). \qquad (30)$$

In our calculations the Fermi level E_F is usually chosen just above the $3s_{1/2}$ and $4p_{1/2}$ states, respectively, because the higher states with $j = \frac{1}{2}$ have an orbital velocity that is smaller than the ion velocity and are therefore ionized early in the collision with very large probability. Eq. (30) also describes positron production, since a positron can be viewed as a vacancy in the negative energy states forming the Dirac sea. The expression for the number of particle- hole pairs, e.g. electron positron pairs, contains two terms:

$$N_{i,j} = N_i \bar{N}_j - \sum_{E_k < E_F} \sum_{E_n > E_F} a_{ki}^* a_{kj} a_{nj}^* a_{ni} \qquad (E_i > E_F, E_j < E_F). \qquad (31)$$

The first term obviously described accidental coincidences between a particle and a hole, whereas the second term describes correlated particle- hole excitations, as can be shown in perturbation theory. In the case of correlated emission of a free electron-positron pair one can ask for the possibility of an angular correlation between the two particles. The relevant formula is:

$$N_{i,j}(\theta) = N_i \bar{N}_j + \frac{1}{2} \sum_{\sigma=\pm} |\sum_k a_{ki}^{(\sigma)} a_{kj}^{(\sigma)*}|^2$$

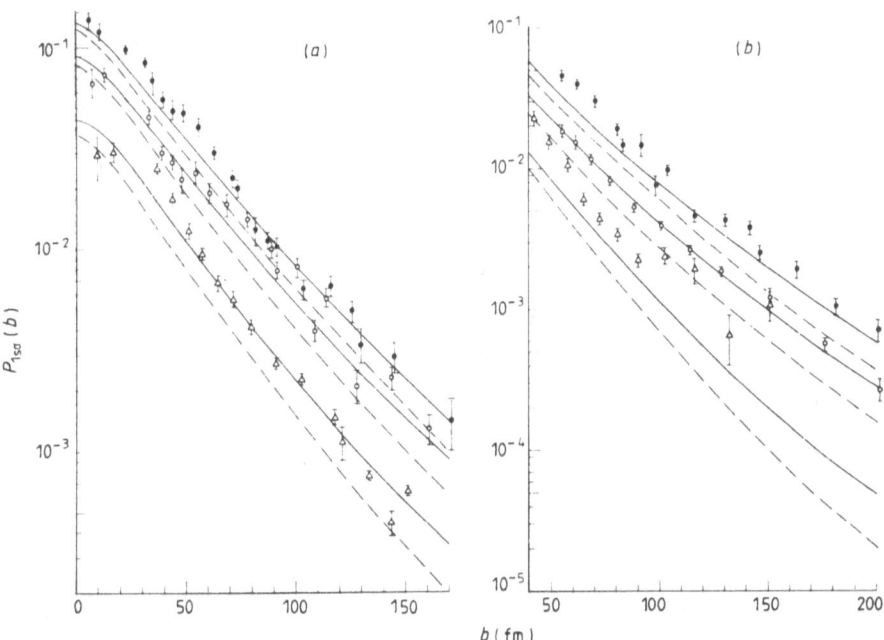

Figure 3. Impact parameter dependence of (left) Pb K-shell ionization by Sm ions for three bombarding energies and (right) Cm K-shell ionization by Pb ions at 5.9 MeV/u.

$$+ \cos\theta \; \Re[e^{i\Delta}(\sum_k a_{ki}^{(+)} a_{kj}^{(+)*})(\sum_n a_{ni}^{(-)} a_{nj}^{(-)*})]. \tag{32}$$

Here θ is the angle between the direction of emission of the electron and that of the positron, and the superscript $(\pm)$ denotes the parity of the $j = \frac{1}{2}$ state, i.e. $(+)$ stands for s-states and $(-)$ for $p_{1/2}$ states. Δ denotes the asymptotic phase differnce between states with opposite parity, which turns out to be π. Clearly, a nonvanishing angular correlation requires pair production in s- as well as $p_{1/2}$-states, and modifications can be expected when the balance between the two contributions is changed.

The time-dependent coupled channel formalism based on the quasimolecular picture, in combination with the monopole approximation, has proved to be extremely successful in describing experimental data on inner-shell ionization, electron and positron emission in heavy ion collisions with $Z_u > 137$. We cannot here give a full account of this success, and only a few examples must suffice. Fig. 3, which shows the impact parameter dependence of Pb K-shell ionization in collisions with Sm nuclei ($Z_u = 82 + 62 = 144$) at three different energies, and of Cm K-shell ionization by Pb projectiles ($Z_u = 96 + 82 = 178$) at 5.9 Mev/u [13] clearly proves the importance of including screening effects in the calculation for quantitative agreement with experiment [11].

The ^{208}Pb nucleus is particularly suited for such experiments since it has no low-lying excited states that could strongly contribute to K-vacancy production by internal conversion. This fact has been exploited in many experiments, e.g. in the one reproduced in Fig. 4 showing δ-electron emission in the collision system I + Pb ($Z_u = 135$). The comparison with theory is made for δ-electrons emitted in coincidence with a Pb K X-ray, signalling the simultaneous presence of a K-vacancy in Pb [14].

The last figure shows a comparison of theory and experiment for positron emission in collision systems ranging from $Z_u = 163$ up to $Z_u = 188$, at energies around

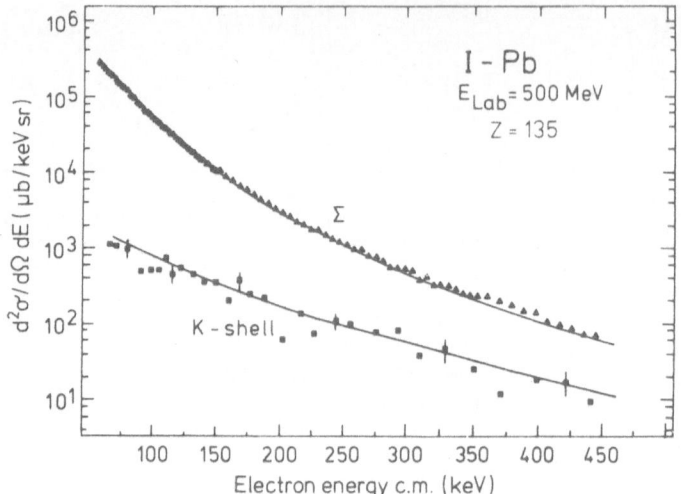

Figure 4. Spectrum of δ-electrons emitted in I + Pb collisions at 500 MeV beam energy. Upper data: all δ-electrons; lower data: electrons in coincidence with a K X-ray in Pb.

the Coulomb barrier. One sees how the contribution from strong field QED pair production (dashed lines) increases rapidly in comparison with the nuclear positron background (dotted lines). The sum of both contributions (solid lines) is in excellent agreement with the measurements, without need foran overall normalization factor.

Supercritical Heavy Ion Collisions

As it has been formulated in the previous section the bound state interaction picture is not directly applicable to supercritical collision systems, i.e. those systems where the combined charge of both nuclei is sufficiently large to let the quasimolecular 1s-state enter the Dirac sea at a critical distance R_c. The difficulty resides in the presence of the supercritical 1s-state as a very narrow resonance ($\Gamma \sim 1$ keV) moving rapidly through the negative energy continuum ($dE/dR \sim 10$ keV/fm). In order to obtain numerically reliable results for the evolution of the occupation amplitudes in thsee continuum states it would be necessary to include continuum states spaced by much less than 1 keV at several points per 1 fm on the nuclear distance grid. This is far beyond numerical possibilities. Besides, such a treatment would obscure the nature of the physical process: a 1s-electron would be represented by the coherent superposition of amplitudes belonging to several continuum states and not by an amplitude of its own, and the mechanism of spontaneous positron emission would be virtually impossible to disentangle from other processes.

The difficulty can be avoided by employing an improved version of the auto-ionization picture discussed previously. One artificially constructs a normalizable resonance wavefunction $\tilde{\varphi}_r$ for the supercritical 1s-state, e.g. by cutting off a continuum wavefunction in the centre of the resonance at its first zero (more sophisticated procedures have been devised and are routinely used [12]). In the next step a set of orthogonal states $\tilde{\varphi}_E$ are constructed in the negative energy continuum with the help of a projection technique. Those states are solutions of the projected Dirac equation

$$(H_{TC} - E)\tilde{\varphi}_E = \langle \tilde{\varphi}_r | H_{TC} | \tilde{\varphi}_E \rangle \, \tilde{\varphi}_r \qquad (E < -m_e). \qquad (33)$$

Since $\tilde{\varphi}_r$ and $\tilde{\varphi}_E$ do not diagonalize the two-centre Hemiltonian, there exists a non-vanishing static coupling between the truncated 1s-resonance state and the modified

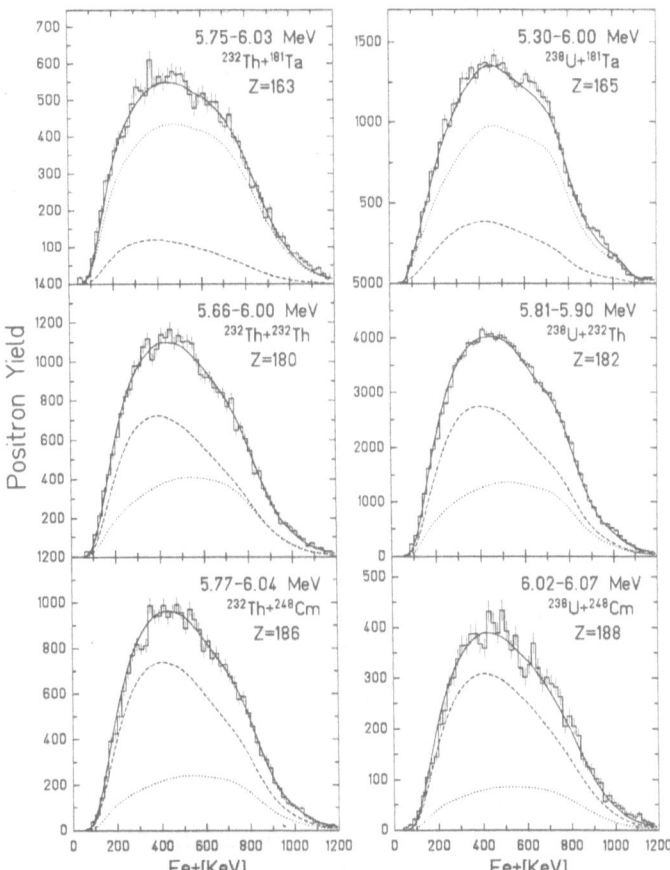

Figure 5. Total positron spectra for various collision systems. Data: EPOS collaboration; dashed lines: QED pair production; dotted lines: nuclear pair conversion; solid lines: sum of both.

negative energy continuum which describes the spontaneous decay of a vacancy in the supercritical 1s-state. As is eq. (23), the spontaneous decay width is given by the expression

$$\Gamma_E = 2\pi |\tilde{V}_E|^2, \qquad \tilde{V}_E = \langle \tilde{\varphi}_E | H_{TC} | \tilde{\varphi}_r \rangle. \qquad (34)$$

Similarly the coupled differential equations for the amplitudes $a_{ik}(t)$ are amended by a "spontaneous" matrix element that does not vanish in the limit where the nuclei do not move:

$$\dot{a}_{ik} = -\sum_{j \neq k} a_{ij} \left(\langle \tilde{\varphi}_k | \frac{\partial}{\partial t} | \tilde{\varphi}_j \rangle + i \langle \tilde{\varphi}_k | H_{TC} | \tilde{\varphi}_j \rangle \right) \exp(i\chi_k - i\chi_j). \qquad (35)$$

Careful investigations have shown that the asymptotic amplitudes $a_{ik}(\infty)$ are insensitive to the precise way of constucting the supercritical 1s-state, although the individual matrix elements may differ somewhat for the various procedures. This means that the concept of "spontaneous" pair production has no unique definition in the dynamical environment of a heavy ion collision, except in the limiting case when the nuclei fuse (for some time) into a single compound nucleus. For collisions without

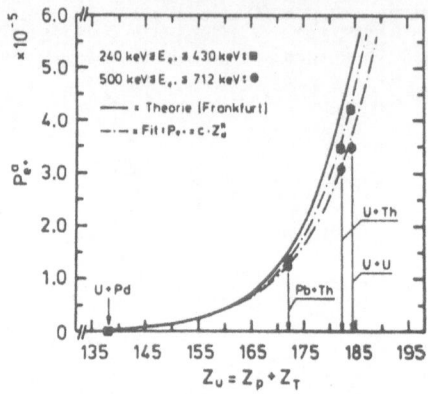

Figure 6. Total positron yield as function of combined nuclear charge $Z_u = Z_1 + Z_2$.

nuclear contact, when the two nuclei move on hyperbolic Rutherford trajectories, the contribution from the "spontaneous" coupling constitutes only a small fraction, in any event. Accordingly, the calculations do not yield any perceptible change in the predicted positron spectra for such collisions when one goes from subcritical to supercritical systems. The total positron yield at fixed beam energy per nucleon for $Z_u > 137$ is predicted to grow at a very rapid rate that can be parametrized by the effective power law

$$\sigma_p(Z) \propto Z^n, \qquad n \approx 20, \tag{36}$$

showing no discontinuity at the transition to supercriticality. The large value of the exponent demonstrates the entirely nonperturbative nature of pair production in collisions of heavy ions, indicating the coherent participation of an average number of 10 virtual photons in the process. The fact that this prediction has been verified in the experiments at GSI is a major confirmation of our ability to accurately treat quantum electrodynamics in strong Coulomb fields by the theoretical methods described above, based on the adiabatic quasimolecular basis and the monopole approximation.

Besides being a source of strong electric fields, colliding heavy nuclei carry also strong magnetic fields. The magnetic field strength can exceed 10^{16} gauss, probably the highest value achievable anywhere in nature. The presence of this magnetic dipole field leads to a splitting between the quasimolecular states with opposite spin projection perpendicular to the scattering plane, which reaches about 30 fm in grazing collisions of the heaviest nuclei [15]. As a result of this Zeeman splitting the two spin states are ionized with different intensity, so that a net polarization of the emitted δ-electrons of the order of ten percent has been predicted. Unfortunately, this effects seems to be hardly detectable in an experiment.

Recently, it was suggested by Scharf and Twerenbold [16] that the magnetic dipole coupling could be the cause of oscillatory structures in the positron spectrum, presumably as a result of interference of the magnetic dipole amplitude for pair production with the dominant amplitude from the Coulomb monopole field. The origin of such an effect is not easy to understand, because it would imply the existence of two different time scales for the magnetic and the electric field. Indeed, a subsequent calculation by Soff and Reinhardt [17], using superior numerical methods, has not confirmed the results of ref. [16]. Even when the magnetic dipole coupling is included, the total positron spectra appear completely structureless.

The "Atomic Clock" Phenomenon

When in the course of a heavy ion collision the two nuclei come into contact, a nuclear reaction occurs that lasts a certain time T. The length of this contact or *delay time* depends on the nuclei involved in the reaction and on the beam energy. For light and medium heavy nuclei the nuclear attraction is greater than, or comparable with, the repulsive Coulomb force, thus allowing for rather long reaction times of the order of 10^{-20}s or even more. For very heavy nuclei, in or beyond the Pb region, however, the Coulomb interaction is by far the dominant force between the nuclei, so that delay times are typically much shorter and probably do not exceed $1 - 2 \times 10^{-21}$s. Nuclear reaction models predict that the delay time increases with the violence of the collision, as measured by inelasticity (negative Q-value), and mass or angular momentum transfer. [Some model calculations have hinted at the possibility of much longer reaction times in collisions of strongly deformed nuclei right at the Coulomb barrier [18]. More detailed investigations [19] and experimental studies have not substantiated this conjecture, so far.]

That a delay in the collision due to a nuclear reaction can lead to observable modifications in atomic excitation processes was recognized more than twenty years ago (for the history and an overview of the field see [20]). Early investigations were primarily concerned with light ion reactions, because extremely long delay times can occur there in the case of narrow resonance scattering. However, in these cases the method is not of practical interest, because the delay time can be determined much more conveniently from nuclear scattering data. As pointed out in 1979 by Anholt and by Soff et al., this is different for heavy ion reactions above the Coulomb barrier (deep-inelastic reactions), since there no other model-independent way of measuring reaction times exists. The two main observable effects in such collisions are: (a) interference patterns in the spectrum of δ-electrons [21], and (b) a change in the probability for K-vacancy formation [22]. These effects have become known as *atomic clock* for deep-inelastic nuclear reactions.

The origin of the atomic clock effect is most easily understood in a semiclassical model for the nuclear motion, where the nuclear trajectory is described by the classical function $R(t)$ and the only effect of the nuclear reaction is to introduce a time delay T between approach and separation of the nuclei, i.e. $\dot{R}(t) = 0$ for $0 \leq t \leq T$. The consequences of this assumption can be analyzed either in the atomic picture or in the quasimolecular picture. As the latter is more appropriate for heavy projectile nuclei, we shall base our discussion on it. To retain lucidity of the argument we further, for an exception, make use of first-order perturbation theory for the excitation amplitudes a_{ik}:

$$a_{ik}(\infty) = - \int_{-\infty}^{\infty} dt\, \dot{R}(t) \langle \varphi_k | \frac{\partial}{\partial R} | \varphi_i \rangle \exp[i \int_0^t dt' (E_k - E_i)]. \tag{37}$$

The range of the main integral splits into three parts: (a) $t < 0$, (b) $0 \leq t \leq T$, and (c) $t > T$. Because $\dot{R}$ enters as a factor in eq. (37), the median part does not contribute. For the last part one can rewrite $t \rightarrow t + T$ so that the integral runs from 0 to ∞. Because this exit part of the nuclear trajectory is just the time-reverse of the entrance part, i.e. $\dot{R}(-t) = -\dot{R}(t)$, the amplitude from part (c) can be expressed as the complex conjugate of the amplitude $a_{ik}(0)$ from part (a) of the integral, except for a phase factor resulting from the variable substitution in the phase integral in eq. (37). Thus we find the relation:

$$a_{ik}^T(\infty) = a_{ik}(0) - a_{ik}^*(0) \exp[iT(E_k - E_i)], \tag{38}$$

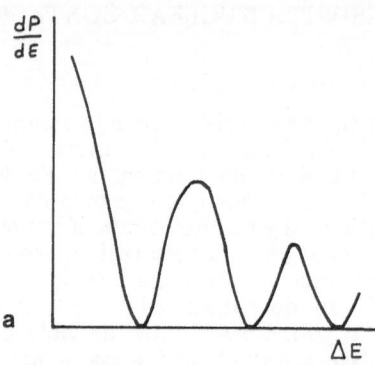
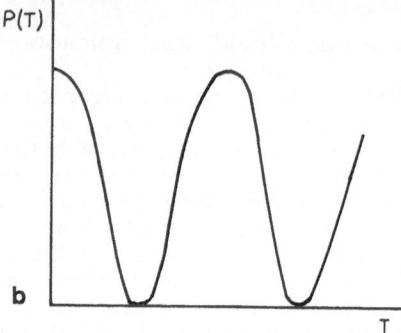

Figure 7. Atomic clock phenomena: (a) oscillations for fixed ΔE; (b) osillations for fixed T as function of transition energy.

where the energies have to be taken at the distance of nuclear contact. If we write $a_{ik}(0)$ in the symbolic form $a_0 e^{i\alpha}$, the final probability for excitation between states i and k becomes

$$P_{ik}(T) = |a_{ik}^T(\infty)|^2 = 4a_0^2 \sin^2(\frac{1}{2}T\Delta E - \alpha), \qquad (39)$$

where $\Delta E = (E_k - E_i)$ is the transition energy. The excitation probability is obviously an oscillating function, either of T for a given transition $i \to k$, or a function of transition energy ΔE for fixed delay time T, as sketched in Fig. 7. The first case applies, e.g., to the probability of K-vacancy production, which is dominated by transitions from the quasiatomic $2p_{1/2}$-state into threshold continuum states. The second case occurs when δ-electron spectra are measured in coincidence with a nuclear reaction.

In reality, of course, things are more complicated. Except in truly elastic collisions the outgoing trajectory is not a precise mirror image of the approaching trajectory. Furthermore, multi-step excitations play an important role in very heavy systems, as discussed previously. The total excitation amplitude therefore contains a mixture of contributions from different intermediate states, see eq. (28). The simple expression, eq. (38), must then be replaced by the formula

$$a_{ik}^T(\infty) = \sum_j a_{ij}^{in} e^{-iE_j T} a_{jk}^{out}. \qquad (40)$$

Finally, the nuclear delay time T is usually not sharply defined, so that an average over a distribution $f(T)$ of delay times has to be taken in eq. (39). The common result of these refinements is a dilution of the interference patterns, i.e. the oscillations become less pronounced. For short delay times and a large uncertainty of T all that remains is a partially destructive interference between the incoming and outgoing branches of the trajectory, observable as a decrease in the K-vacancy yield or a steepening of the slope of the low-energy part of the δ-electron spectrum.

A more serious objection appears to be that a nuclear reaction is a quantum mechanical process that is not really adequately described by a classical variable such as T. However, it can be shown that no new effects are introduced by the transition to a quantal description of the nuclear reaction. As first derived by Tomoda [23], the quantum mechanical analogue of eq. (40) is:

$$a_{ik}^T(\infty) = \sum_j a_{ij}^{in} S(E + E_i - E_j) a_{jk}^{out}, \qquad (41)$$

where $S(E)$ denotes the energy-dependent nuclear reaction S-matrix. The argument of $S(E)$ accounts for the fact that the energy at which the nuclear reaction occurs depends on the energy absorbed by the intermediate electronic state. The total measurable transition probability $P_{ik}(T)$ is obtained by averaging the square of eq. (41) over the incident beam energy E. It can be shown [24] that this procedure is tantamount to taking an average of the classical expression, eq. (40), over the delay time T with the distribution function

$$f(T) = \int_{-\infty}^{\infty} dE' e^{iE'T} \langle S^*(E)S(E - E') \rangle_{av.}.$$ (42)

Quite remarkably this is the same expression for the distribution of nuclear reaction times as occurs in the theory of crystal blocking. In some sense, therefore, the atomic clock phenomenon can be understood as a microscopic version of the blocking effect.

Experimental Results

Over the last few years several groups have carried out experiments aimed at the measurement of nuclear reaction times in deep-inelastic collisions with the help of the atomic clock effect. Meyerhof and collaborators have concentrated on K-vacancy production in U+U collisions, as determined from the K X-ray yield from uranium-like fragments. The main experimental problem here is that highly excited uranium nuclei have a strong tendency to fission, yielding no uranium-like K X-ray in such a case. Somewhat surprisingly, Meyerhof's group has succeeded in identifying unfissioned uranium nuclei even for reactions with a Q-value of -200 MeV, although only one nucleus in 10^4 survives at this inelasticity. Another difficulty is that the nuclei partly de-excite by internal conversion processes, e.g. ejecting a K-shell electron. This contribution must be subtracted from the measured K X-ray yield.

The left part of Fig. 8 shows the reduction of the experimentally determined K-vacancy yield P_K as function of the reaction Q-value (note that P_K is normalized to the value 4). A comparison with the also reproduced theoretical relation between P_K and T indicates the steady increase of the delay time with increasingly negative Q-value, up to about 10^{-21}s for the most violent collisions.

The implications of these results are best analyzed in terms of semiclassical statistical models, in which nucleon transfer between the two nuclei is treated in a space of a few selected collective parameters, one of them being the nuclear separation R, another the angle Θ of the major axis of inertia to the beam direction. Two of the most successful models of this type are those of Wolschin et al. [26,27], and that of Feldmeier [28]. In Wolschin's model the evolution of the nuclear system along the collective coordinates is described by Newtonian equations of motion:

$$\mu\ddot{R} = -\frac{\partial}{\partial R}(V_C + V_N) + \mu R\dot{\Theta}^2 - \gamma_R f(R)\dot{R}$$
$$\mu R^2 \ddot{\Theta} = -\gamma_\Theta f(R)R^2\dot{\Theta},$$ (43)

where $\gamma_R = (0.16 \text{ MeV/fm})^{-1}$ and $\gamma_\Theta = (40 \text{ MeV/fm})^{-1}$ are the radial and tangential friction coefficients, respectively. V_C and V_N denote the Coulomb and nuclear potential, and the form factor of the friction force is given by $f(R) = (\partial_N/\partial R)^2$. The nuclear interaction potential is taken as a proximity potential with a correction term for dynamic deformations. The friction parameters have been adjusted to fit the measured kinetic energy and angular momentum dissipation into internal nuclear degrees of freedom.

Due to the deterministic nature of eqs. (43) the nuclear reaction time T and the energy loss $(-Q)$ are unique functions of the impact parameter b. The other models

53

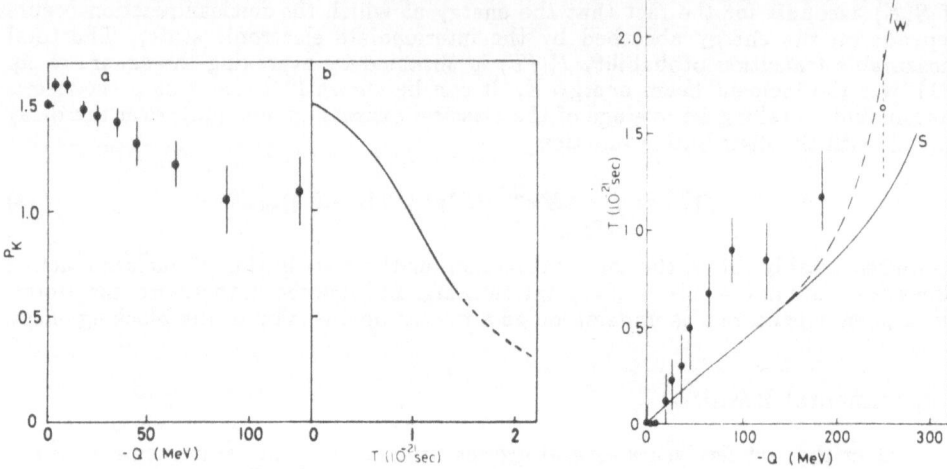

Figure 8. Left: (a) Measured K-vacancy production as function of Q-value; (b) theoretical decrease with nuclear delay time T. Right: Correlation between T and $(-Q)$, in comparison with theoretical models.

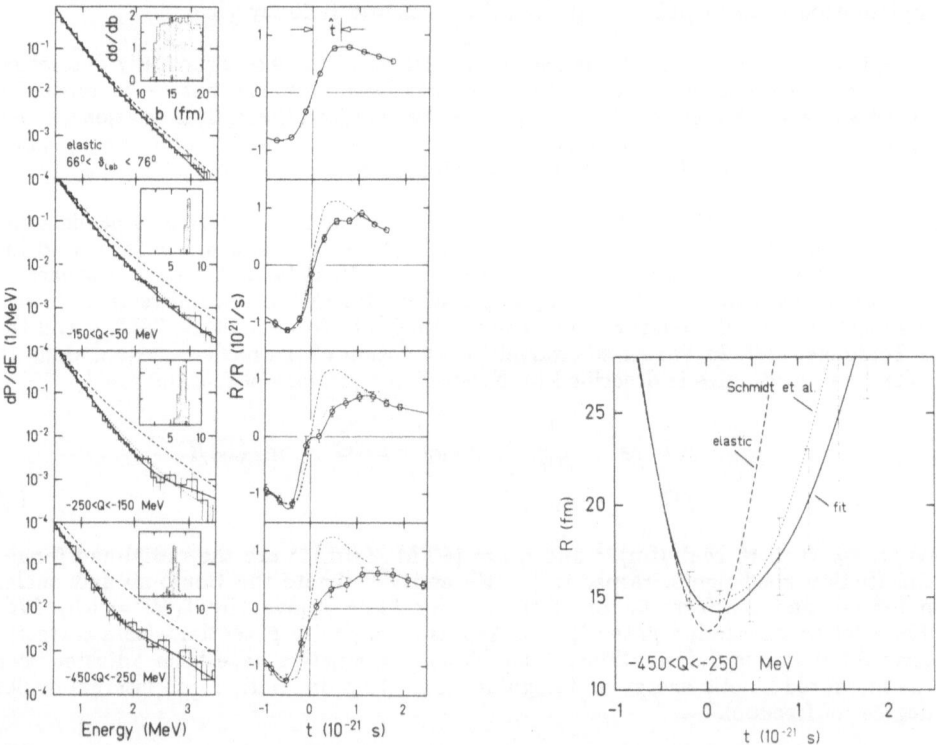

Figure 9. Left: δ-electron spectra measured in Pb+Pb collisions at 8.6 MeV/u and fits of the function $\dot{R}/R$. Right: Trajectories $R(t)$. Predictions of a friction model are indicated by dotted lines.

differ only slightly, except for the fact that in Feldmeier's reaction model a fluctuating Langevin force $\delta \vec{F}(t)$ is added on the right-hand side of eqs. (43). In this way also the mean fluctuations in the collective variables for fixed b are determined. This feature has so far not yet been fully exploited in the analysis of δ-electron spectra, although first attempts to deduce the spread in the time delay distribution were made by the group of Backe, Senger et al. [31] (see below).

The right part of Fig. 8 shows how the correlation between Q-value and delay time T, as extracted from the experiment of Meyerhof et al., compares with two versions of the Wolschin model. The solid line, denotes by (S), represents the model of ref. [27], whereas the dashed line, labeled by (W), contains the prediction following ref. [26]. Clearly, the trend of an almost linear increase of T versus $(-Q)$ is reproduced, but the data indicate somewhat larger time delays. However, given the large correction to the measured K X-ray yield for internal conversion, it is not clear whether this discrepancy is significant. Another source of caution is the selection of a very small fraction of all events by requiring unfissioned uranium-like nuclei. It is not at all obvious that the reactions in this subset of all events should follow the average trajectory attributed to all events.

The necessity of selecting a small subset of all events to avoid fission does not exist in the second class of experiments, i.e. those measuring δ-electron spectra in coincidence with a deep-inelastic nuclear reaction. This circumstance, when combined with the fact that one does not measure only a single number, renders this type of experiment superior to the X-ray experiment. Two groups have measured δ-electron production in deep-inelastic collisions at GSI over the past few years. The TORI group (TH Darmstadt) uses a toroidal magnetic transport system that permits the simultaneous measurement of electron and positron spectra, in coincidence with mainly binary nuclear final states detected in parallel plate counters [29,30]. The other group, a Mainz-GSI-Heidelberg-Frankfurt collaboration, has measured electron spectra by means of an orange-shaped magnetic spectrometer that is inserted into a large scattering chamber. The chamber also contains large-area particle counters capable of determining the nuclear kinetic energy loss with high accuracy also in fission events [31].

The TORI group has analyzed the electron spectra measured in deep-inelastic Pb+Pb collisions at 8.6 MeV/u in terms of a nuclear trajectory model involving several free parameters. To wit, the function $R(t)$ was discretized inside the nuclear interaction range, and the about ten discrete values $\dot{R}(t_k)/R(t_k)$ were fitted such that the spectrum calculated from the perturbative expression, eq. (37), was is agreement with the measured spectrum, apart from an overall normalization factor. These fits are shown in Fig. 9 for a series of Q-value windows and for elastic collisions. The spectra and the fitted values $\dot{R}/R$ are also compared with the predictions of the friction model of Schmidt et al. [27] (dotted lines). The dashed lines show the (unrealistic) results that would be obtained on the basis of Rutherford trajectories. Also shown in the inserts is the range of impact parameters associated with the respective Q-value window according to ref. [27].

The right-hand part of Fig. 9, shows the fitted trajectory function $R(t)$ for the highest Q-value window 250 MeV $< -Q <$ 450 MeV. Again, the dotted line indicates the prediction of the friction model. As in the X-ray experiment discussed before, the nuclear time delay extracted from the data is somewhat larger than that predicted by theory. However, the discrepancy is just at the edge of experimental uncertainties, and the general agreement must be regarded as quite spectacular. Our group has solved the full set of coupled equations for the excitation amplitudes for the Schmidt friction model, in order to see whether the agreement might be an artefact of the perturbative treatment. Indeed, the deviations from the measured spectra are somewhat larger, but more extensive studies are needed before final conclusions can be drawn.

The experimental set-up pf the second group, around P. Senger and H. Backe [31], is most suited to measure electron spectra in coincidence with scattering events

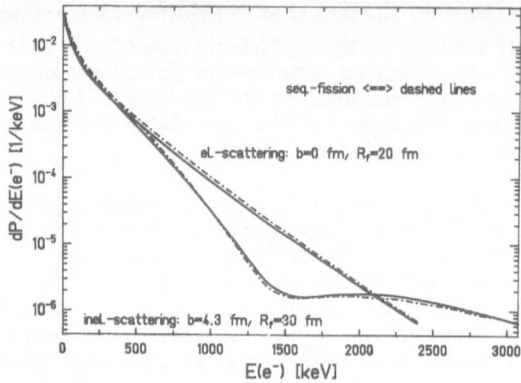

Figure 10. δ-electron spectra calculated for Pb+U collisions with binary (solid lines) and ternary exit channel (dash-dotted lines).

leading to fission of at least one nucleus. They have recently studied the system U+Au at a beam energy of 8.6 MeV/u, where the uranium-like nuclear fragment fissions in most cases for negative Q-values beyond about 100 MeV. Concentrating on these ternary events one can determine the nuclear time delay for the dominant reaction channel.

In this case theory has to check whether the fission process, which occurs some (unknown) time after the deep-inelastic reaction, has any influence on the predicted δ-electron spectrum. This question was studied by S. Graf [32], who extended the coupled-channel code for the quasi-molecular occupation amplitudes a_{ik} to allow for the presence of three nuclei in the exit channel. This does not pose any new problems of principle in the context of the monopole approximation, but the complexity of the calculations is increased considerably, because the basis states φ_i now depend on three parameters corresponding to the three distances R_{12}, R_{13}, and R_{23} between the three nuclei. Limitations of data storage do not permit to tabulate the transition matrix elements at a sufficient number of points, and therefore the matrix elements must be recalculated for every single exit channel trajectory.

Fortunately, the calculations have shown conclusively that the influence of the fission process on the electron spectra is negligible, as illustrated in Fig. 10. The calculation was performed for the system Pb+U at 8.4 MeV/u for the inelastic collision, and at 6 MeV/u for the quasi-elastic collision. Assuming the same friction trajectory in either case, the results for a binary exit channel (solid lines) are compared with those for a ternary exit channel (dash-dotted lines). Here fission of the uranium nucleus was assumed to occur at a distance $R_f = 20$ fm for the quasi-elastic collision and at $R_f = 30$ fm for the deep-inelastic collisions. No significant difference was found when the distance of fission was varied. The conclusion is that the δ-electron spectra are sensitive to the evolution of the di-nuclear system during the deep-inelastic reaction, but not to processes occurring after the end of the reaction, such as (sequential) fission.

Fig. 11 shows the measured δ-electron spectra in ternary U+Au collisions for

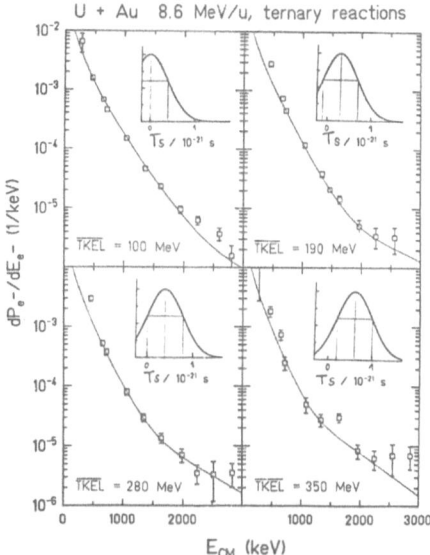

Figure 11. δ-electron spectra measured in 8.6 MeV/u U+Au collisions. The lines and inserts represent fits to the data with a gaussian distribution of reaction times.

four different Q-value windows. The change in shape with increasing inelasticity is clearly visible. The flattening of the spectra at high electron energy appears to be the remnant of an interference minimum at about 2 MeV, corresponding to a nuclear delay time in the range of 10^{-21}s. The lines drawn through the data points represent an analysis in terms of first order perturbation theory with a gaussian distribution of nuclear reaction times T_s, where the centre and the width of the gaussian have been fitted to the data. The time distributions, reproduced in the inserts, show a systematic shift towards larger reaction times for increasing $(-Q)$-value, but little change in width.

Since the trajectories used in the analysis include also small transitory regions required to smoothly join the velocity to Coulomb trajectories, the total time delay T is somewhat larger than the parameter T_s. Its value as function of negative Q-value (or: total kinetic energy loss = TKEL) is shown in Fig. 12 (left part). The values compare well with those found in the other experiments discussed above. The right part of Fig. 12 contains a comparison of the fitted trajectories with those predicted by Feldmeier's friction model (dotted lines). The two solid lines in the exit channel indicate the width of the gaussian distribution of reaction times. For $(-\bar{Q}) = 190$ MeV the agreement is perfect, but for $(-\bar{Q}) = 350$ MeV the friction model predicts a larger delay time. A more detailed analysis of the data, e.g. as function of the mass transfer between the nuclei, is possible and will provide a significant test of the various nuclear reaction models.

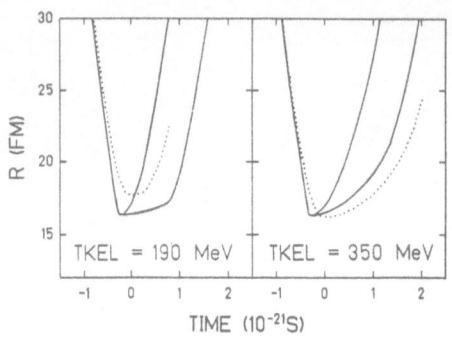
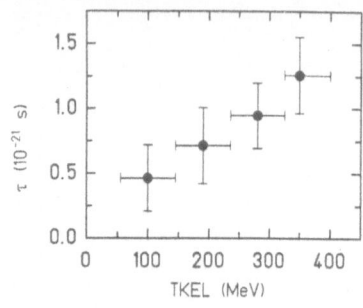

Figure 12. Left part: Nuclear delay times in deep-inelastic U+Au collisions. Right part: Comparison of the fitted trajectories with a friction model (dotted line).

Positron Production in Delayed Collisions

In principle, the positron spectra contain the same information about nuclear time delay as the electron spectra. However, because of their low emission probabilty, positrons are not as useful from a practical point of view, at least in subcritical collision systems. Nevertheless, positron spectra emerging from deep-inelastic heavy ion collisions have been measured [29], and the first experimental observation of the atomic clock phenomenon in heavy ion collisions came in fact from positrons [33]. The yield argument does not necessarily apply to supercritical collision systems for two reasons. Firstly, a reaction-induced nuclear time delay may allow for the detection of spontaneous pair-creation in these systems, as will be discussed below. Secondly, a tiny component of very long reaction times ($T > 10^{20}$s) might become visible in the positron spectrum, because the spontaneous emission mechanism acts as a kind of "magnifying lens" for long delay times [34].

In order to see why this is so, we return to eq. (35) for the amplitudes a_{ik} in a supercritical system, which contained the additional time-independent couplings $\tilde{V}_E$ between the resonant bound state and the (modified) positron continuum states. The presence of this coupling has the effect that the contribution to the integral in eq. (37) from the central time interval $0 \leq t \leq T$ does not vanish any longer. The total amplitude a_E^T for emission of a positron from the supercritical bound state contains therefore an additional term compared with eq. (38):

$$a_E^T(\infty) = a_E(0) - a_E^*(0)\exp[iT(E - E_r)] - \tilde{V}_E \frac{e^{iT(E-E_r)} - 1}{E - E_r}, \qquad (44)$$

where E_r is the energy of the supercritical state when the nuclei are in contact. For the U+U system this about $E_r = -700$ keV, and for the U+Cm system one has $E_r = -830$ keV, depending somewhat on the degree of ionization. For $T = 0$ the new term vanishes, but grows rapidly with increasing T. For delay times considerably greater than 10^{-21}s the additional term in eq. (44) begins to dominate over the first two terms , causing the emergence of a peak in the positron spectrum at the energy of the supercritical bound state:

$$|a_E^T(\infty)|^2 = \frac{\Gamma_E}{2\pi}T^2 \frac{\sin^2[(E - E_r)T/2]}{[(E - E_r)T/2]^2}, \qquad (T \ll \Gamma_E^{-1}). \qquad (45)$$

The energy distribution has a width $\Gamma(T) = 2\pi/T$ as would be expected on grounds of the uncertainty relation, and the total probability for positron emission grows

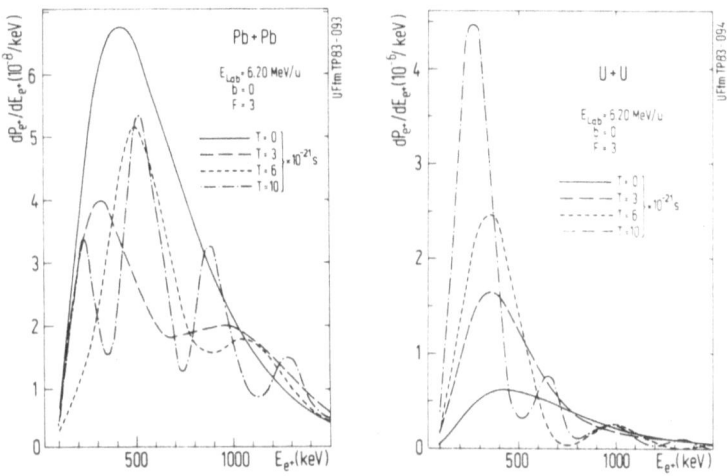

Figure 13. Effect of a nuclear time delay on the positron spectrum. Left: subcritical system. Right: supercritical system.

proportional to T. For very, very long delay times, $T > \Gamma_E^{-1}$, perturbation theory no longer applies because the integrated emission probability approaches unity. One can show that the energy distribution then goes over into a Breit-Wigner curve centred at E_r with width equal to the spontaneous decay width Γ_E.

The emergence of a peak in the positron spectrum for strongly delayed supercritical collisions is strikingly demonstrated in Fig. 13, where the effect of a nuclear time delay is compared for a subcritical system (Pb+Pb, $Z_u = 164$) and a supercritical system (U+U, $Z_u = 184$). In the subcritical case the delay causes interference patterns like those already known from electron spectra, effectively reducing the positron yield. In the supercritical case, on the other hand, the positron yield increases dramatically when the delay time exceeds about 3×10^{-21}s. Unfortunately, this is beyond the range accessible for the average time delay in deep-inelastic reactions where not much more than 10^{-21}s has been observed. Still, the situation may not be entirely hopeless, because the intensity of the line structure grows with T and simultaneously becomes more localized at the resonance energy. In principle, even a very small tail of the delay time distribution $f(T)$ could acquire sufficient weight to be visible in the positron spectrum. Model calculations [35] have yielded that a fraction of 5×10^{-3} of all events with a delay time of the order of $T = 4 \times 10^{-20}$s would clearly show up as a peak on top of the dynamical positron "background".

As briefly mentioned in the introduction such long delay times could only occur if an attractive pocket is present in the internuclear potential for supercritical collision systems, for which no conclusive theoretical or experimental evidence exists at present. But even if a pocket were there, the existence of a sufficiently large tail of long delay times in the distribution $f(T)$ is not ensured. Studies of a schematic model by Heinz et al. and Pinkston [36] have not provided reason to be optimistic. Although very narrow positron peaks were obtained for beam energies in a small window around the Coulomb barrier, their intensity was much too low to allow for observation. But again it must be emphasized that these models are too simple to permit definite conclusions for the realistic case. These can come from experiments alone. So far, no evidence has been found.

THE GSI POSITRON PEAKS: STATUS OF EXPERIMENTAL RESULTS

Starting in 1980 line-structures were observed in the positron spectra from very heavy collision systems by two experimental groups working at GSI: The ORANGE collaboration (München-GSI) has made use of an orange-type magnetic spectrometer in combination with a direction sensitive solid state detector, whereas the EPOS collaboration (GSI-Yale-Fankfurt-Heidelberg-Mainz) has combined a solenoidal magnetic transport system with high resolution solid state counters.

A few years later, in 1985, it was discovered that these lines are correlated with similar structures in the coincident electron spectra. The most fascinating aspects of this discovery were that the peak found in the sum energy spectrum of these correlated electron-positron lines was much narrower than the lines in the singles spectra, and that there is indirect evidence for back-to-back emission of the two leptons.

Although much has been said, written, and speculated about their unexpected experimental results, the record of published papers in regular journals is rather thin [37,38,39,40,41,42,43,54]. More extensive material is contained in some conference reports and review articles [44,45,46,47,48,49], which the interested reader may wish to consult. For obvious reasons, the present discussion is from a theorist's standpoint, and thus naturally biased and incomplete.

Structures in the Positron Singles Spectrum

When the line structures in the positron spectra were first detected at GSI, they were associated with the spontaneous positron emission line that was predicted by theory for supercritical collision systems with long nuclear time delay. This was quite natural, because that had been the aim and inspiration of the experiments from the beginning. For the first two systems that were investigated, U+Cm and U+U [37,38], this explanation worked quite nicely; the position of the line agreed rather well with the expected spontaneous emission peak. The measured spectra could be described in the framework of schematic models involving the intermediate formation of a long-lived ($T \approx 5 \times 10^{-20}$s) "giant" di-nuclear system [35,38,50].

Of course, the experimentalists were very cautious to make sure that the lines would not be a trivial artefact caused by pair decay of some excited nuclear state. This can be checked experimentally, since a pair-decaying nuclear state can always decay in another way, either by photon emission (if the transition multipolarity is not $L = 0$) or by internal conversion to a K-shell electron. The latter process works for any multipolarity. The branching ratios for the various decays can be calculated essentially model independently, because the nucleus is small compared to the wavelength of the emitted real or virtual photon. The photon and electron spectra were measured simultaneously in the relevant energy range (beyond 1 MeV), but no associated structure was found [37,38].

The observed line width of 70-80 keV provided a second argument against nuclear pair decay. If the structures were emitted from the scattered nuclei they would have to be Doppler broadened due to the motion of the source. At 45° (lab) scattering angle the broadening would amount to 100 keV, i.e. more than the observed line width even for an intrinsically monochromatic structure. However, the positron spectrum emerging from a normal nuclear pair decay is not monochromatic at all! The energy of the transition is shared between the electron and the positron, and a broad peak develops at the upper end of the positron spectrum only for heavy nuclei due to Coulomb effects. Normal pair conversion could thus be ruled out by line width arguments as well [37].

An intrinsically monochromatic positron line could, in principle, be caused by a process called monoenergetic pair conversion, which can occur if an inner atomic shell is not fully occupied. The electron can then be captured in this bound state, and the positron carries away the full remaining energy. The sharply defined energy

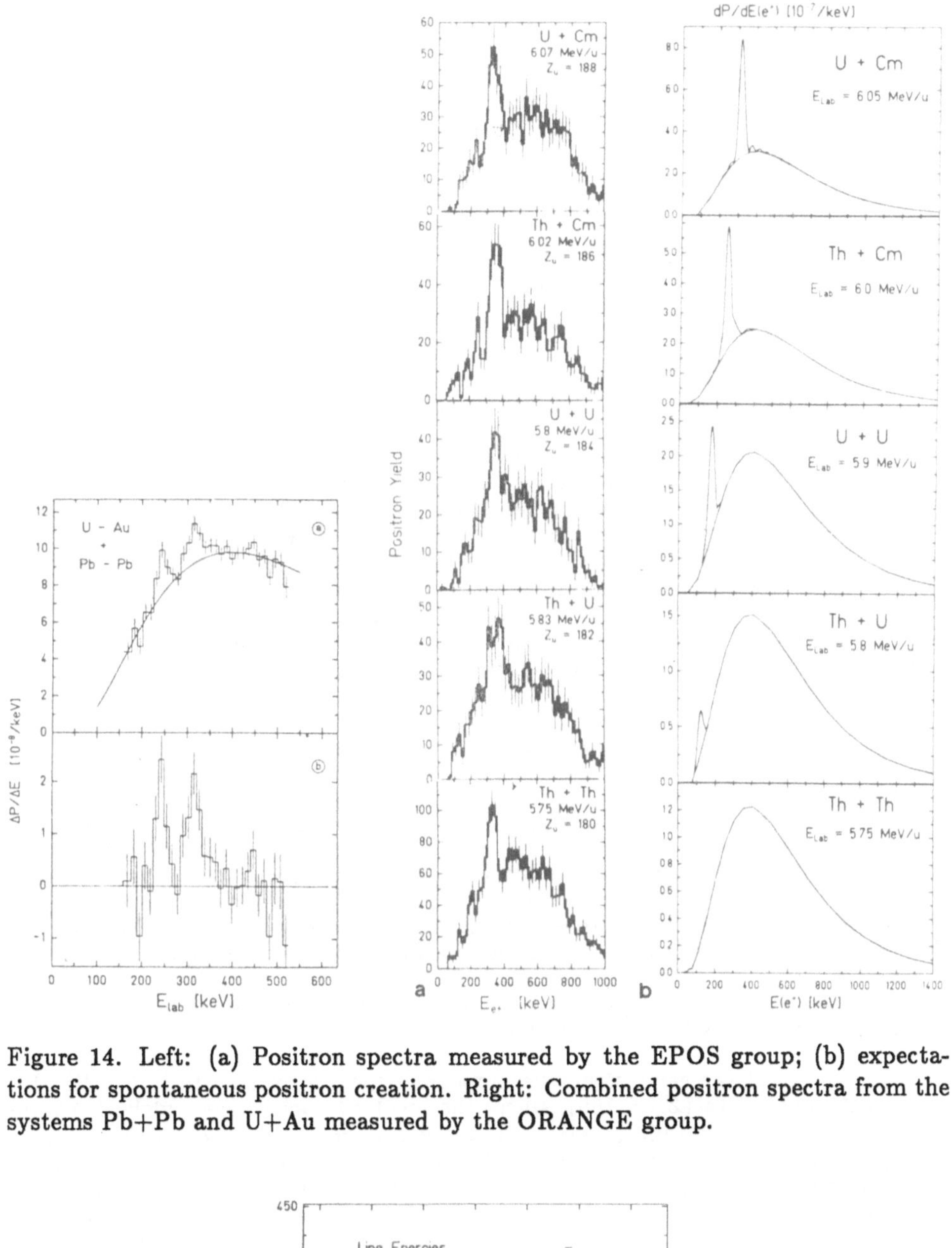

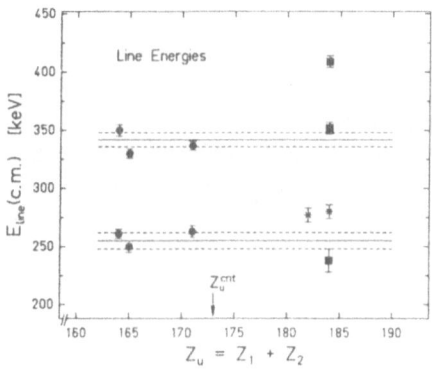

Figure 14. Left: (a) Positron spectra measured by the EPOS group; (b) expectations for spontaneous positron creation. Right: Combined positron spectra from the systems Pb+Pb and U+Au measured by the ORANGE group.

Figure 15. Peak positions measured as function of Z_u by the ORANGE group.

is characteristic of a two-body decay $A^* \rightarrow (A+e^-) + e^+$, whereas the normal pair decay into a free electron-positron pair is a three-body decay $A^* \rightarrow A + e^- + e^+$. Although a large number of inner-shell vacancies are created in the heavy ion collision, monoenergetic pair conversion is expected to be strongly suppressed, because the vacancies are filled by transitions from outer shells within about 10^{-17}s. This filling time is at least two orders of magnitude shorter than the lifetime of nuclear excited states. Therefore, a possible origin of the line structures by monoenergetic pair conversion was ruled out, too, on experimental [37] and theoretical grounds [51].

The dependence of the line structures as function of combined nuclear charge $Z_u = Z_1 + Z_2$ afforded a crucial test of the hypothesis that they could be attributed to spontaneous positron production. The line must then occur at the positron kinetic energy corresponding to the energy E_{1s} of the 1s-resonance that is imbedded in the Dirac sea: $E_{peak} = |E_{1s}(Z_u)| - m_e$. The surprising result of such a study by the EPOS collaboration [39] is shown in the left-hand part of Fig. 14: The position of the peak was always in the range 350 ± 30 keV essentially independent of Z_u! (It seems now that the remaining fluctuations in the peak position are caused by the presence of a multiplicity of lines.) For comparison, the expectation for a peak caused by spontaneous pair creation in the strong Coulomb field is also shown in part (b) of the figure. Starting at about 320 keV in the U+Cm system the line should move to lower energies and decrease in intensity, assuming similar nuclear delay times for all systems. For Th+Th ($Z_u = 180$) the structure should be gone, unless a spherical compound nucleus $^{464}_{180}$A is formed - a scenario that is beyond imagination in the framework of standard nuclear physics, because the Coulomb energy of such a configuration would be much too high. The independence from Z_u was also confirmed by an experiment with Th+Ta ($Z_u = 163$), where the same positron line appeared [55].

Positron experiments by the ORANGE collaboration [42] for subcritical collision systems, i.e. for $Z_u < 172$, supported these findings and conclusively ruled out spontaneous pair creation as origin of the observed line structures. Besides being present in spectra from subcritical systems, the lines appeared in multiple structures, at energies so much independent of Z_u that their statistical significance could be improved by adding spectra from systems with different Z_u [41]. This is illustrated in Fig. 14 (right part) showing the added positron spectra from the systems Pb+Pb and U+Au. Fig. 15 demonstrates that the observed structures fall into three groups, at the positron energies 250, 330, and 400 keV, respectively, with an uncertainty of about 20 keV. (Note that these are not the measured laboratory energies, but energies corrected for Doppler shift, assuming that the source moves with the velocity of the centre of mass.) When the EPOS data are added, more evidence for the two highest energies, i.e. 330 and 400 keV, is accumulated.

The fact that the same structures appear in spectra from systems like Pb+Pb ($Z_u = 164$) and U+Ta ($Z_u = 165$), where the involved nuclei have widely disparate properties - ^{208}Pb is doubly magic with a first excited state at 2.6 MeV, whereas ^{238}U and ^{181}Ta are both strongly deformed with dense excitation spectra -, makes any explanation in terms of nuclear physics rather unlikely. [In principle, fission could lead to the appearence of the same daughter nuclei in both cases, but the fission mode is experimentally ruled out by the coincident detection of both scattered nuclei.] A second rather general argument against any nuclear origin of the positron peaks is provided by the variation of the line intensities with nuclear charge. This was found by the ORANGE group to be indistinguishable from the Z_u-dependence of the dynamic QED background, i.e. $(d\sigma/d\Omega)_{peak} \propto Z_u^{20}$, as shown in Fig. 16. This would be natural if the process responsible for the appearance of the line structures somehow involves the strong combined Coulomb field of both nuclei, but it would be hard to understand for a pure nuclear physics effect.

We conclude this section by summarizing the experimental results for the line structures in positron singles spectra:

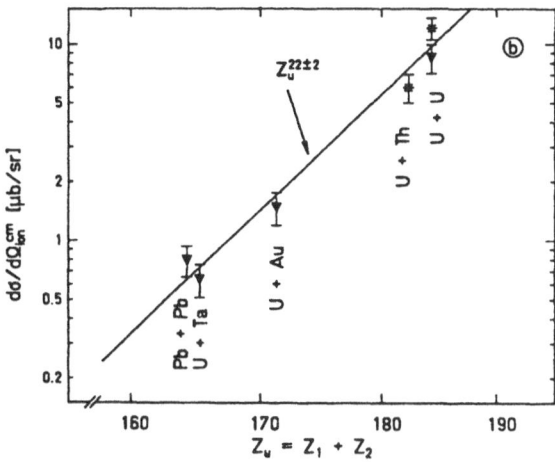

Figure 16. Z_u-dependence of the intensity of the positron lines (left part) and of the QED background (right part).

- Lines have been observed for a large variety of collision systems, ranging from $Z_u = 163$ (Th+Ta) up to $Z_u = 188$ (U+Cm) and involving nuclei with widely different structure.

- The line positions appear to fall into several groups between 250 and 400 keV; their width is about 70 keV, if all positron emission angles are covered. This value corresponds to the Doppler width of a sharp line emitted by a source moving with centre-of-mass velocity.

- A number of lines are common to different collision systems and to both experiments (ORANGE and EPOS).

- Nuclear pair conversion processes ($A^* \rightarrow A e^- e^+$) appear to be excluded from γ-ray and electron spectra, linewidth, and A-independence.

Finally, it should be mentioned that the line intensity may depend very sensitively, almost erratically, on some yet unknown parameter, possibly the beam energy or the target quality. This circumstance has been a constant source of worry for the experimental groups, and it may well convey some important clue toward the origin of the positron peaks (for a painstaking investigation of target effects, see ref. [53]).

Correlated Electron-Positron Lines

The A- and Z-invariance of the line energies strongly hint at a common source that in itself is not related to the nuclei or the strong electric field, although the strong Coulomb field might play a role in the production of this source. Since no Z-dependence is seen, the most natural candidate for such a source would be some (neutral) object that moves with the velocity of the centre of mass and eventually decays into a positron and a single other particle. (A two-body decay must be invoked to explain the narrow linewidth, as mentioned before.) Could the second decay product simply be a second electron, i.e. could it be that one sees the pair decay of a neutral particle, $X^0 \rightarrow e^+ e^-$, with a mass somewhat below 2 MeV?

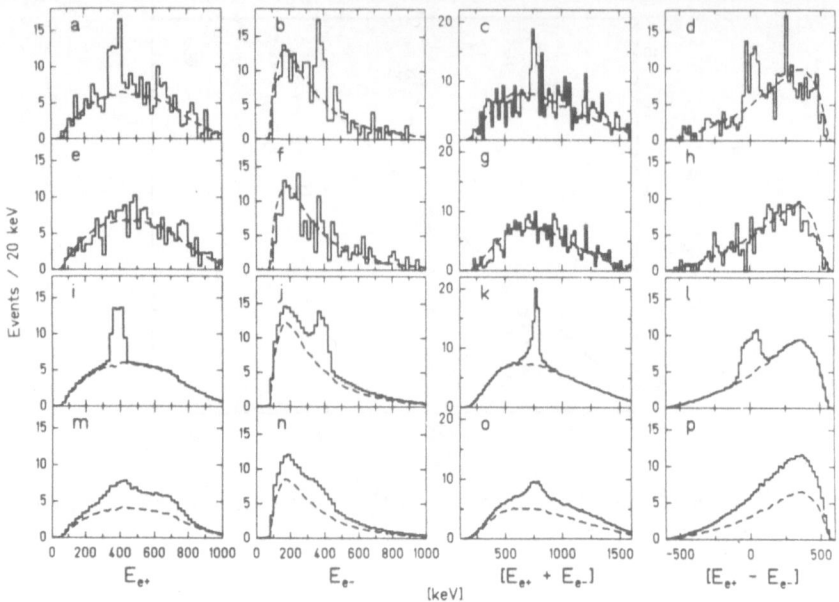

Figure 17. Spectra taken from e⁺e⁻ coincidence events in U+Th collisions at 5.83 MeV/u (EPOS data), compared with computer simulations of pair decay of a neutral particle and of a nucleus.

When the EPOS collaboration, inspired by this hypothesis, added an electron counter opposite of the positron detector to their device in order to investigate this question, they indeed found a correlated line structure in the electron spectrum at precisely the same energy of 380 keV in the U+Th system [54]. Fig. 17 shows the published results, which contain a comparison of the measured spectra with Monte Carlo simulations of the decay of a slowly moving neutral particle and of the pair decay of one of the nuclei. An event required the coincident detection of an electron, a positron, and two elastically scattered nuclei. Part (a) shows the positron spectrum in coincidence with an electron in the energy range 340-420 keV; part (b) shows the electron spectrum in coincidence with a positron in that energy range; while (c) shows the sum energy spectrum under the condition that electron and positron energy are the same within the limits provided by Doppler broadening from c.m. motion. Finally, part (d) shows the difference energy spectrum for all events where the sum of electron and positron energy is 760 ± 40 keV. Parts (e) - (h) show the same spectra, but for adjacent energy windows. Evidently a clear peak is visible in any of the top four spectra, but in none of the four reference spectra. The peak intensity is just what would be expected if every positron in the singles peak shown in Fig. 14 is accompanied by an electron of the same energy.

The computer simulations reproduced in the two bottom rows are quite conclusive: Pair decay of a slowly moving neutral particle X⁰ of mass 1.78 MeV, parts (i) - (l), nicely fits all the spectra of the top row, whereas nuclear pair decay, parts (m) - (p), does not. In particular, the appearence of a peak in the difference spectrum (last column) cannot be understood except for a two-body decay. That nuclear pair decay does not even correctly reproduce the narrow sum energy line is due to the fact that the nuclear pair is not emitted back-to-back, so that the linear Doppler shift from the nuclear motion does not cancel. A second argument against nuclear E0-pair decay comes from the coincidence yield, which saturates the positron singles line if back-to-back emission is assumed. For the $(1+\cos\theta_{ee})$ distribution of nuclear E0-pairs the observed intensity of e⁺e⁻ coincidences would be too high by a large factor [54].

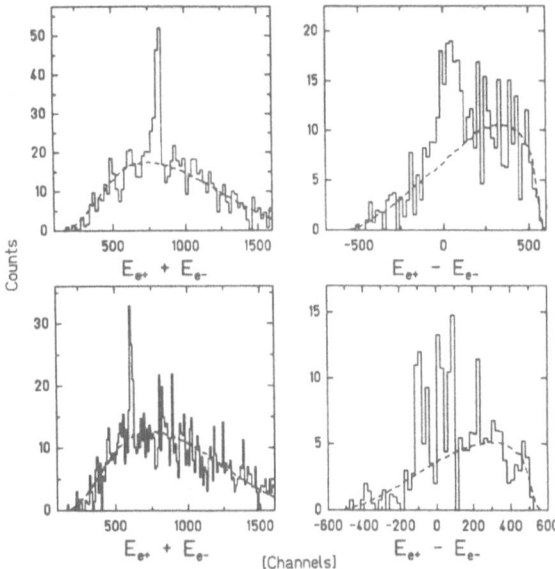

Figure 18. Sum and difference energy e^+e^- spectra in U+Th collisions at 5.82-5.87 MeV/u (EPOS data), for prompt (top row) and delayed (bottom row) coincidences .

When the same system was remeasured in 1986, at 5.82-5.87 MeV/u beam energy, again narrow sum energy lines were found, but now at different energies [52,55] (see Fig. 18). The top two subfigures show the sum and difference spectra obtained requiring a prompt coincidence between the two leptons. A narrow line now emerges at the sum energy $E_1 = 809 \pm 8$ keV with a width $\Gamma_1 = 41 \pm 5$ keV. The two lower subfigures show similar spectra, now taken with the condition of a slightly delayed coincidence, with the positron arriving about 5 ns after the electron. Such a delay would not necessarily indicate that the positron was emitted after the electron, but could simply mean that the positron followed a longer trajectory before it arrived at the detector. This would occur if the positron is emitted at a large angle with respect to the axis of the solenoidal transport system, being forced to move on a large helical orbit. Fig. 18 shows that an even narrower sum peak appears for such events, but now at the sum energy $E_2 = 608 \pm 8$ keV and with width $\Gamma_2 = 23 \pm 3$ keV. In both cases there is also a broad peak in the difference energy spectrum, again around equal energies of electron and positron.

For the next run, in 1987, the EPOS group modified their lepton detectors so that electrons and positrons emitted forward from, or in front of, the target could be distinguished from those emitted backwards. The two parts of the detectors will be denoted by "F" and "B". The aim was to find out whether the leptons were, indeed, emitted back to back as required by the particle decay hypothesis, or with uncorrelated angles, or even preferentially in the same direction as in nuclear pair decay. This could be done by comparing the combinations (FB+BF) and (FF+BB), where the first letter indicates the part of the electron counter, and the second letter that of the positron counter. Tests with a radioactive ^{90}Sr source, which emits pairs from a 1.77 MeV transition in ^{90}Zr, proved that the discrimination could be achieved if the pair is produced within a few mm from the target [56].

The measurements were this time performed with the collision system U+Ta for reasons of target quality, and a broad range of beam energies around the Coulomb barrier was investigated [56]. In the energy range 5.93-6.16 MeV/u a sum energy peak was found for delayed positrons at $E_3 = 748 \pm 8$ keV with width $\Gamma_3 = 26 \pm 5$ keV. In the higher energy range 6.24-6.38 MeV/u two peaks were observed, which appeared to

Table 1. Sum energies and widths of correlated e^+e^- lines observed at GSI

System	Expt.	Beam energy (MeV/u)	condition	Energy (keV)	width (keV)
U + Th	EPOS	5.82-5.87	delayed	608 ± 8	23 ± 3
U + Ta	EPOS	6.24-6.38	delayed	620 ± 8	20 ± 3
U + Th	EPOS	5.83		760 ± 20	80 ± 20
U + Ta	EPOS	5.93-6.16	delayed	748 ± 8	26 ± 5
U + Th	EPOS	5.82-5.87	prompt	809 ± 8	41 ± 5
U + Ta	EPOS	6.24-6.38	prompt	805 ± 8	30 ± 4
U + U	ORANGE	5.9	$(180 ± 18)°$	815 ± 8	40 ± 15
U + Pb	ORANGE	5.9	$(180 ± 18)°$	802 ± 8	32 ± 15

be the same as those discovered in 1986 in the U+Th system. The peaks could again be isolated by selecting prompt or delayed coincidences. All the measured positions and widths of peaks in the sum energy spectrum are listed in Table 1.

The two lines observed at the higher beam energy, i.e. the lines at 620 and 805 keV, were only visible in the spectrum from the detector combination (FB+BF), indicating a back-to-back correlation of the lepton pair. This was more or less the expected result, because trajectory simulations for the lines seen in the 1986 run with U+Th had yielded that the relative angle of emission had to be at least 150° to explain the narrow width of the sum energy line. However, the 748 keV line seen at the lower beam energies showed up preferentially in the FF detector. Moreover, the difference energy spectrum in this case does not show a peak at equal energies of electron and positron. Note that 748 keV is precisely the sum energy associated with the 1.77 MeV transition in ^{90}Zr discussed above in connection with the radioactive source test. The long lifetime of the ^{90}Zr state (62 ns) allows to rule this explanation out on the basis of timing measurements. Also, no line was seen at 1.77 MeV in the γ-ray spectrum, but a transition of multipolarity E0 has not been ruled out [56].

Intrigued by this discovery, the ORANGE group has added a second orange-type magnetic spectrometer to their set-up, permitting simultaneous measurement of the electron (forward) and positron (backward) spectrum. Since both lepton detectors are subdivided into six azimuthal angular ranges of 60° ("pagoda" counters), spectra for specific angular correlation, e.g. 60° or 180°, can be selected. In 1987 a first run with U+U at 5.9 MeV/u was carried out [57]. The sum energy spectra of e^+-e^- coincidence events obtained in this run is shown in Fig. 19. The energies of the most prominent line detected in the 180° correlated spectrum coincides almost exactly with one of the two lines that appeared to be back-to-back in the latest EPOS measurement: 815 keV. As the figure shows, this line structure appears to be absent in the spectrum taken for an angular correlation in the range 40° − 170°, confirming the EPOS results. A third line appears at about 630 keV sum energy, but no line was seen that could be identified with the third EPOS energy (748 keV). The 810 keV line was also observed in the system U+Pb.

Most recently, also the TORI group has measured e^+-e^- coincidences in the U+Th system at 5.85 MeV/u with kinematical conditions corresponding to a large relative angle between electron and positron (70° - 180°) or to a small opening angle (0° - 110°), respectively. As listed in Table 1, two line structures were found in the backward correlated events at roughly the same energies as in the 1986 data of the EPOS group. However, structures were also observed in the spectrum of forward

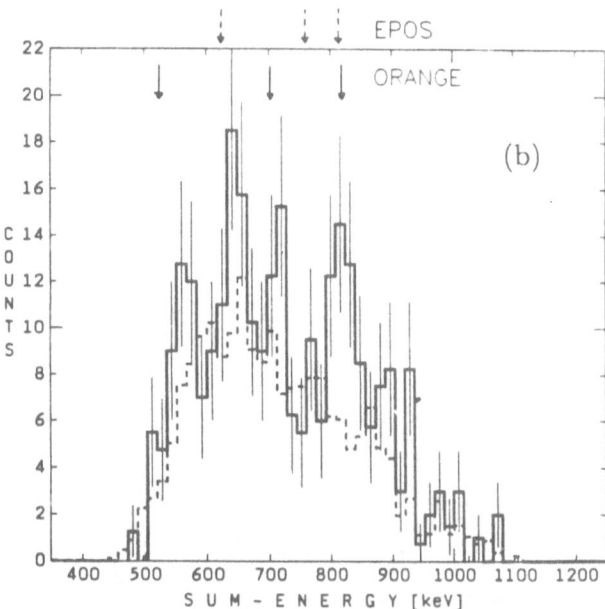

Figure 19. Sum energy e^+e^- spectra measured in U+U collisions at 5.9 MeV/u with the double-orange spectrometer. Solid line: $180° \pm 18°$ correlations; dotted line: correlations with $40° - 170°$ opening angle (scaled).

correlated events, and no firm conclusion about the opening angle distribution has been reached so far [58].

We conclude the section by summarizing the experimental results for the correlated line structures in electron-positron coincidence spectra:

- Lines at 620 and 810 keV sum energy have been observed by the EPOS and the ORANGE collaboration. A third line at 750 keV was only seen by the EPOS group. The peaks seem to occur at the same positions in various systems, e.g. U+Ta ($Z_u = 165$), U+Th ($Z_u = 182$) and U+U ($Z_u = 184$).

- The width of the sum energy peaks lie in the range 20-40 keV; they are much narrower than the positron singles peaks. The source must move slowly ($\beta_s \leq 0.05$); if it is not at rest, the data imply an opening angle close to 180° between the two leptons.

- The 620 and 810 keV lines appear to be caused by back-to-back emission. The 750 keV line in U+Ta could be forward correlated.

- The difference-energy spectra exhibit a broad peak near zero energy, indicating that the lepton pair is not produced inside the strong Coulomb field, or in the vicinity of a third body. Again, the 750 keV line in U+Ta appears to be an exception to this rule.

- The e^+e^- coincidence line intensity exhausts the full strength of the positron singles line, when the experimental efficiency is taken into account ($d\sigma/d\Omega \approx 10\mu b/sr$). [The line intensity appears to be smaller in the data taken by the double-orange spectrometer.]

- The different spectra obtained for prompt and delayed coincidences hint at non-isotropic emission of the lepton pair. This may also explain the lack of intensity exhibited by the double-orange data.

Thus almost everything appears to be compatible with the assumption that one observes the pair decay of at least three neutral particle states in the mass range between 1 and 2 MeV. These states must have a lifetime of more than 10^{-19}s (because of the narrow linewidth) and less than about 10^{-9}s (because the vertex of the lepton pair is within 1 cm of the target). Surely, a few pieces of data do not really fit into this picture, e.g. the characteristics of the 748 keV line observed in U+Ta. But it is by no means clear at the present time, whether these features provide conclusive evidence against the particle hypothesis, especially if one is not dealing with simple, elementary particle states.

Nevertheless, the very idea that a whole family of neutral particle states in the MeV mass range should have remained undetected through more than 50 years of nuclear physics research is hard to accept for the conservative mind. Most physicists, when first confronted with the GSI data, have therefore tried to explain the data in terms of known nuclear or atomic physics. As mentioned before, nuclear pair decay would be the most natural explanation. In fact, it is rather easy to invent some scenario that would give rise to nuclear pairs of the correct sum energy, e.g. due to target impurities, deexcitation of fission products, neutron activation of the target frame, and so on. However, none of those scenarios has yet stood up against a detailed comparison with the experiments. As long as this is so, nuclear pair decay cannot be considered as a viable alternative to the particle hypothesis. Similar remarks apply to attempts to explain the GSI peaks in terms of atomic physics. None of the ideas that were studied quantitatively have been successful, even if they were based on plausible, but unfounded, ad hoc assumptions. We will discuss some of these attempts later, after a review of the neutral particle models.

THE CASE AGAINST NEW ELEMENTARY PARTICLES

In this chapter we will review the experimental arguments against the existence of neutral *elementary* particles with mass below 2 MeV that can decay into an e^+-e^- pair. We will see that precision experiments set severe limits on the permissible coupling strength of such particles to the electron-positron field, but do not rule out the full range of lifetimes relevant to the GSI peaks. Beyond that, dedicated searches have now eliminated the possibility of an axion in the relevant mass range. Beam dump experiments have closed the remaining lifetime gap for weakly interacting point-like particles, but have left room for extended particle states. Bhabha scattering below 2 MeV centre-of-mass energy, which is the unique model-independent probe for the hypothetical particle states, has just begun to set new limits but is still far from the sensitivity required to reject the particle hypothesis once and for all.

Limits on Light Neutral Bosons from Precision Experiments

Even when the hypothesis of a new neutral particle was first seriously discussed [59,60], it was recognized that the precision experiments of quantum electrodynamics provide stringent limits on the coupling of such light particles to the electron-positron field and to the electromagnetic field. The strength of this argument lies in the fact that any particle X^0 which decays into an e^+-e^- pair must couple to the electron-positron field. At least in the low-energy limit, the coupling can be expressed by an effective interaction of the form:

$$L_X = g_i(\bar{\psi}\Gamma_i\psi)\phi, \qquad (46)$$

where ψ denotes the electron-positron field, ϕ the X^0 field, and Γ_i with $i =$ S,P,V,A stands for the vertex operator associated with the various possible values of spin and parity of the X^0 particle. Given the interaction Lagrangian (46) one can calculate the lifetime of the X^0 particle against pair decay as well as the contributions to QED processes by virtual exchange of an X^0. The most sensitive of these is the anomalous magnetic moment, because of the high experimental accuracy [61]:

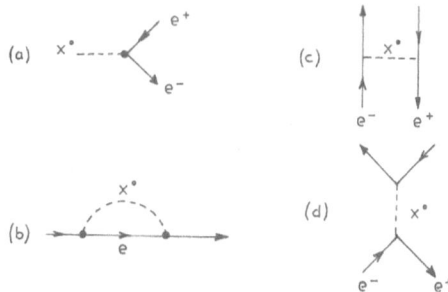

Figure 20. Feynman diagrams for (a) pair decay of X^0, (b) contribution to the electron anomalous magnetic moment, (c,d) positronium hyperfine splitting.

$$a_e^{exp} = (g_e - 2)/2 = (1159652193 \pm 4) \times 10^{-12} \ ,$$

and its excellent agreement with the value predicted by QED [62]:

$$a_e^{QED} = (1159652230 \pm 52 \pm 41) \times 10^{-12} \ .$$

where the first error is due to the uncertainty in the electromagnetic coupling constant $\alpha = 0.00729735308 \pm 33$, and the second error contains the uncertainties in the evaluation of the third and fourth order Feynman diagrams. A reasonable estimate for the 95% confidence limit is $\Delta a_e < 2 \times 10^{-10}$.

As illustrated in Fig. 20, the contribution of a hypothetical X^0 particle to the value of a_e involves two vertices between an electron or positron and an X^0, and are thus proportional to the effective coupling constant $\alpha_{Xe} = g_i^2/4\pi$. The same applies to the decay rate τ_X^{-1}, which involves the square of an amplitude with a single vertex, and to the contribution of an X^0 particle to the hyperfine splitting of the positronium ground states. (The last process does not yield a limit rivalling that derived from a_e in accuracy, but permits to eliminate possible cancellations between contributions from different particles to the anomalous magnetic moment [63,64].)

The limits derived from these considerations [63] on the X^0-coupling constant and its lifetime are listed in Table 2. Particles with lifetime $\tau_X > 10^{-13}$s cannot be ruled out by this argument. Considering that the experimental conditions only require a lifetime below about 1 ns, there remains an unexplored range of four orders of magnitude in τ_X.

Similar upper limits can be derived for the coupling of an X^0 boson to other known particles [63]. That for the coupling to the muon, $\alpha_{X\mu}$, is about one order of magnitude weaker because of the lower accuracy in the value of a_μ. A limit on the product of the coupling constants to the electron and to nucleons is obtained from the Lamb shift in hydrogen and from the K-shell binding energy in heavy elements, one finds $\alpha_{Xe}\alpha_{XN} < 10^{-14}$. For scalar particles an extremely stringent bound on the coupling to nucleons can be derived from low-energy neutron scattering: $\alpha_{XN} < 10^{-9}$ [65]. Special bounds for vector (gauge) bosons were considered by Zee [66]. Finally, measurements of nuclear Delbrück scattering yield an upper limit on the coupling of

Table 2. Limits on the coupling constant, lifetime, and pair decay width of neutral bosons with mass $M_X = 1.8$ MeV derived from the anomalous magnetic moment of the electron.

Particle type	Spin J^π	Vertex Γ_i	Max. coupling $\alpha_{Xe} = g_i^2/4\pi$	Min. lifetime τ_X (s)	Max. width Γ_X^{ee}(meV)
Scalar (S)	0^+	1	7×10^{-9}	2×10^{-13}	3.0
Pseudoscalar (P)	0^-	$i\gamma_5$	1×10^{-8}	1×10^{-13}	6.8
Vector (V)	1^-	γ_μ	3×10^{-8}	4×10^{-14}	16
Axial vector (A)	1^+	$\gamma_\mu\gamma_5$	5×10^{-9}	5×10^{-13}	1.4

a spinless X^0 boson to the electromagnetic field through an effective interaction of the type

$$L_{X\gamma\gamma} = g_S(E^2 - H^2)\phi_X \quad \text{(scalar)}$$

$$L_{X\gamma\gamma} = g_P(\vec{E} \cdot \vec{H})\phi_X \quad \text{(pseudoscalar)}. \tag{47}$$

The limits are: $g_S < 0.02$ GeV^{-1} and $g_P < 0.5$ GeV^{-1}. They provide lower limits for the lifetime against decay into two photons: $\tau_{\gamma\gamma}(X^0) > 6 \times 10^{-11}$s for a scalar particle and $\tau_{\gamma\gamma}(X^0) > 4 \times 10^{-13}$s for a pseudoscalar particle [67].

Inadequacy of Perturbative Production Mechanisms

One consequence of these results is that the particle hypothesis cannot be rejected off-hand. On the other hand, the condition that the coupling constant between the hypothetical X^0 boson and the particles involved in the heavy ion collision, i.e. electrons and nucleons, must be very small creates severe problems for any attempt to explain the measured production cross section of about 100 μb by a perturbative interaction of the type shown in eq. (46) [59,60,63,68]. (The treatment of ref. [69] suffers from an incorrect treatment of nuclear, as opposed to nucleon, spin.) Also the cross section for production by the strong electromagnetic fields present in the heavy ion collision falls short by several orders of magnitude, if it is based on the Lagrangian (47) or similar perturbative interactions [70,71,72,73,75,76,78].

A second serious difficulty with the interactions (46) and (47) is that they favour the production of particles with high momenta due to phase space enhancement. For collisions with nuclei moving on Rutherford trajectories the calculated spectra typically are very broad, peaking at velocities $\beta_X > 0.5$. As discussed in the previous chapter, the experiments would require an average particle velocity $\beta_X < 0.05$!

Both these problems could, in principle, be circumvented by the assumption that a very long-lived, excited giant compound nucleus is formed [70,76], but only at the price of violating other boundary conditions set by the experimental data, e.g. the absence of a much larger peak in the positron spectrum caused by spontaneous pair production [77]. A further counterargument is that the emission of slow particles is tremendously suppressed by phase space factors [78]. One might also consider the possibility that the X^0 particles are somehow slowed down after production, but this cannot be achieved with the interactions discussed above.

Two mechanisms remain, which can conceivably ensure the survival of the particle hypothesis: (1) a form factor that cuts off production at large momenta [74];

and (2) a non-perturbative production mechanism, e.g. production in a bound state around the two nuclei [79,80,87]. Both mechanisms require particles with internal structure. An additional advantage of a composite particle is that it would be expected to have a number of excited states, providing a natural explanation for the fact that several narrow lines in the sum energy spectrum are observed at GSI. Models for such particles are discussed in the next chapter.

Axion Searches

At first the axion, i.e. the light pseudoscalar Goldstone boson associated with the breaking of the Peccei-Quinn symmetry required to inforce time-reversal invariance in quantum chromodynamics, seemed like a plausible candidate for the suspected X^0 boson. The interest in an axion was revived when it was realized that there was, indeed, a gap left by previous axion search experiments for a short-lived axion in the mass range around 1 MeV [81]. However, new experimental studies of J/Ψ and Υ decays [82,83,84] quickly ruled out the standard axion [85].

The reason, why the heavy quarkonium states provide the best test for the Peccei-Quinn axion, is that it couples to all quarks according to their mass, except for a parameter x that determines the ratio of the coupling constants to quarks with weak isospin $+1/2$ (u,c,t) and $-1/2$ (d,s,b). It is possible to modify the axion model in such a way that the axion coupling to the light quarks (u,d) is essentially independent of the coupling to heavy quarks [86,87]. However, even in these variant axion models the coupling constants to the light quarks, g_u and g_d, remain predictable numbers, if the axion mass is known. In terms of the quark masses m_u, m_d, the pion mass and decay constant m_π, f_π, and a basis symmetry breaking scale $f \leq 250$ GeV, on has:

$$g_u = \frac{m_u x}{f}, \qquad g_d = \frac{m_d}{x f}, \qquad m_a = \frac{m_\pi f_\pi \sqrt{2 m_u m_d}}{2 f (m_u + m_d)} \left(x + \frac{1}{x} \right), \qquad (48)$$

and a similar coupling to the electron, g_e, which is proportional to the electron mass m_e. The existence of a variant axion with mass above 1 MeV can therefore be ruled out by experiments on $e^+ e^-$-decays of excited hadronic states involving light quarks, e.g. pion decays and decays of excited nuclear states. There are also strong limits from electromagnetic decays of strange hadrons, such as $\Sigma^+ \to p e^+ e^-$ and $K^+ \to \pi^+ e^+ e^-$ [88,89,90,91], and from neutral pion decay [92].

The rare radiative pion decay $(\pi^+ \to e^+ \nu e^+ e^-)$ was studied at SIN in order to search for a decay branch $(\pi^+ \to e^+ \nu \phi)$ where the axion ϕ decays later into an electron-positron pair [93]. No such events were found, limiting the branching ratio of this decay mode to less than 10^{-10}. Variant axion models typically predict a branching ratio of order 10^{-6} [94]. In order to rigorously eliminate all variant axions one must turn to nuclear decays. Hallin et al. [95] analyzed a new experimental study of the pair decay of the 9.17 MeV 2^+-state in ^{14}N [96] and older experiments on ^{10}B and ^{6}Li, and showed that these combined results ruled out all axions that decay into an electron-positron pair. Similar analyses by Bardeen et al. [97] and by Krauss and Zeller [64] reached the same conclusion. (However, a variant axion below 1 MeV is not yet completely ruled out, see ref. [98].)

Further experiments on the 3.59 MeV 2^+-state in ^{10}B [99], the 3.68 MeV $\frac{3}{2}^-$-state in ^{13}C [100], the 15.1 MeV state in ^{12}C [101], the 18.15 and 17.6 MeV 1^+-states in ^{8}Be [104], and the 1.115 MeV state in ^{65}Cu [103] have also not shown any indication for the presence of a short-lived, pair-decaying axion. Searches for a light, pair-decaying scalar or vector particle emitted in the decay of the 6.05 MeV 0^+-state in ^{16}O [101,104] have also been negative, yielding an upper limit of 2×10^{-9} for the coupling constant α_{XN} of such a particle to nucleons. Of course, no useful limits are provided for particles that interact only with leptons, not with quarks, except electromagnetically. (An experiment that does not rely on the coupling of the axion to quarks was suggested in ref. [105].)

Beam Dump experiments

Beam dump experiments, in particular those with a high-energy electron beam, are an excellent source of rather model-independent bounds on the properties of hypothetical light neutral particles. The idea behind these experiments is rather simple: When an electron enters a solid piece of material, e.g. a lead block, it is slowed down and deflected by collisions with the target electrons and nuclei. In particular, the electron can be scattered off mass shell in the Coulomb field of a target nucleus, and return to the mass shell by emission of a real photon. This process is called bremsstrahlung. Of course, instead of radiating a photon, the electron can get rid of its excess energy by emitting some other light neutral particle X^0, if any exists. Except for effects from the particle mass and spin, the expected cross section is given by the cross section for photon radiation, multiplied by the ratio of the coupling constant of the emitted particle to the electron and the electromagnetic coupling constant:

$$\frac{d\sigma_X}{d\Omega dE} = \frac{\alpha_{Xe}}{\alpha} \frac{d\sigma_\gamma}{d\Omega dE}. \tag{49}$$

An upper limit for the measured cross section for the X^0-particle cross section hence yields an upper limit for the coupling constant α_{Xe}. A simple formula for the bremsstrahlung cross section has been given by Tsai [106] (see also ref. [107]).

However, for every beam dump experiment there is not only a lower bound for the range of excluded coupling constants but also on upper bound, for the following reason. The lifetime of the hypothetical particle against pair decay is inversely proportional to the α_{Xe}. For sufficiently large values of the coupling constant almost all produced particles therefore decay inside the beam dump, and the e^+e^- pair produced in the decay is absorbed in the target. It is clear that a good value for this other limit requires a short beam dump, whereas high cross-section and low background require a thick target.

Hence, the result of a beam dump experiment is a region of excluded values of α_{Xe}, i.e. the coupling cannot be in the range $\alpha_{Xe}^{min} < \alpha_{Xe} < \alpha_{Xe}^{max}$. In the analysis one assumes that the neutral particles interact so weakly that they pass essentially undisturbed through the target. For elementary particles this assumption is not critical: for the SLAC experiment discussed below (ref. [110]) it has been shown that even a nuclear absorption cross section as large as 50 mb per nucleon (this is more than the total nucleon-nucleon cross section at high energy!) would not seriously affect the limits derived from the data.

Three such beam dump experiments with high-energy electrons have been performed recently, at KEK [108], Orsay [109], and at SLAC [110], and there exists a relevant older experiment at much lower energy [111]. The conditions and results of these experiments are listed in Table 3, where the excluded ranges of the coupling constant are given for pseudoscalar particles of mass 1.8 MeV. For a scalar particle the bounds would be similar, but for spin-one particles about one order of magnitude better lower limits would be obtained. Also listed in Table 3 is a proton beam dump experiment performed at Fermilab. Due to the production of secondary electrons and positrons in the target, a limit is obtained also for the coupling to electrons.

Together, the experiments exclude the range of coupling constants α_{Xe} between 10^{-14} and 10^{-7}, corresponding to lifetimes against pair decay in the range 10^{-14}s $< \tau_X < 10^{-7}$s. When combined with the bounds derived from the electron anomalous magnetic moment a_e and by experimental conditions, the beam dump results conclusively rule out any elementary neutral particle as source of the GSI e^+e^- events. To

Table 3. Excluded ranges of the coupling constant α_{Xe} of a pseudoscalar particle of mass 1.8 MeV derived from beam dump experiments.

Experiment	Beam	Target	α_{Xe}^{min}	α_{Xe}^{max}
Konaka et al. (KEK) [108]	e^- (2.5 GeV)	W + Fe(2m)	10^{-14}	4×10^{-8}
Davier et al. (Orsay) [109]	e^- (1.5 GeV)	W (10cm)	10^{-11}	10^{-8}
Riordan et al. (SLAC) [110]	e^- (9.0 GeV)	W (10-12cm)	10^{-12}	10^{-7}
Bechis et al. [111]	e^- (45 MeV)		10^{-13}	10^{-10}
Brown et al. (FNAL) [112]	p (800 GeV)	Cu (5.5m)	10^{-10}	10^{-7}

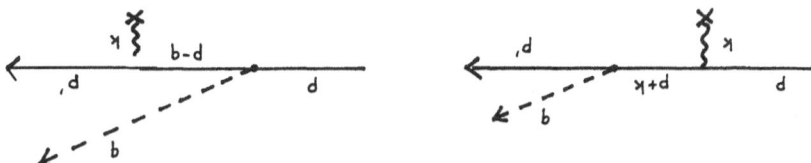

Figure 21. Feynman diagrams for (a) radiation of a neutral boson by an electron scattering in the Coulomb field of a nucleus.

wit, the bound from a_e gave $\alpha_{Xe} < 10^{-8}$ (see Table 2), and the experimental conditions required a mean lifetime below 10^{-9}s, or $\alpha_{Xe} > 10^{-12}$. The remaining interval is covered by the beam dump results.

Does this rule out the particle scenario altogether? As we shall see in a moment, the answer is, no! (Whether that is good news or bad news, of course, is a question of taste!) That there is still a "loop-hole" for neutral particles left was revealed by a recent analysis by A. Schäfer, who calculated the bremsstrahlung production cross section for extended particles [113]. He showed that a finite form factor can invalidate the experimental bounds, if the emitted particle has a radius of more than about 100 fm (10^{-11}cm). Let us try to understand why this is so.

In lowest order of perturbation theory the radiation emitted by a Coulomb scattered electron is given by the sum of the two Feynman graphs depicted in Fig. 21. In order for the electron to be able to emit an X^0-boson, e.g., in the left-hand diagram it must be off mass shell. More precisely, its invariant mass must exceed the sum of the rest masses of the two particles in the final state:

$$(p+k)^2 = (p^0)^2 - (\vec{p} + \vec{k})^2 \approx m_e^2 - 2\vec{p} \cdot \vec{k} > (m_e + m_X)^2. \tag{50}$$

This means that the longitudinal momentum of the virtual photon must be larger than a minimal value $|k_\parallel| > m_X(m_e + m_X/2)/E$, where E is the incident energy of the electron. Since the Coulomb scattering cross section falls as $|\vec{k}|^{-4}$, the bremsstrahlung cross section is dominated by events where the invariant mass $(p+k)^2$ just barely exceeds the critical threshold. In other words: as seen from the rest frame of the emitted boson, the electron is slow before and after the emission process. The width of the momentum distribution of the electron in the X^0 rest frame, measured by the Lorentz invariant variable $\tilde{p} = [(p+k) \cdot q]/m_X$, is of order m_X.

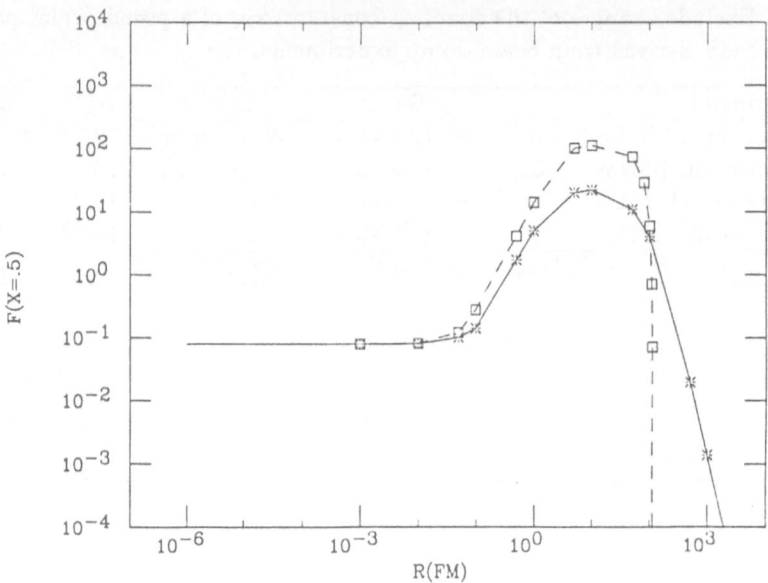

Figure 22. Bremsstrahlung strength function $F(x)$ at $x = \frac{1}{2}$ versus particle size. Solid line: pseudoscalar boson; dashed line: vector bosons.

We are now in a position to discuss the effect of a finite form factor of the X^0-boson on the bremsstrahlung cross section. According to our considerations, the characteristic de Broglie wavelength of the electron in the rest frame of the X^0-boson is given by $\tilde{\lambda}/2\pi = \langle \tilde{p}^{-1} \rangle \approx m_X^{-1}$. If this is much larger than the size R_X of the radiated particle, the emission will remain unaffected, otherwise the emission of an X^0 will be suppressed. The amount of suppression is expressed by the form factor $G_X(\tilde{p}R_X)$, which has the limits $G_X(0) = 1$ and $G_X(x) \ll 1$ for $x \gg 1$. The bremsstrahlung cross section will, therefore, be strongly suppressed if $R_X \gg m_X^{-1} \approx 100$ fm.

This qualitative consideration is confirmed by the results of the complete calculation [113], which are represented in Fig. 22. The cross section is here expressed in terms of the dimensionless strength function $F(x)$, defined by

$$\frac{d\sigma_X}{dx} = \frac{2(Z\alpha)^2 \alpha_X}{m_e^2} \cdot F(x), \tag{51}$$

where $x = |\vec{q}|/|\vec{p}|$ is the fraction of the initial electron momentum carried away by the X^0-boson. The X^0 form factor has been assumed to be of monopole form

$$G_X(\tilde{p}R_X) = \frac{1}{1 + (\tilde{p}R_X)^2}. \tag{52}$$

In Fig. 22 the value of $F(x = \frac{1}{2})$ is plotted as function of R_X for emission of a pseudoscalar or vector boson. Clearly, for $R_X > 1000$ fm the bremsstrahlung cross section is so much reduced that the limits derived from the beam dump experiments discussed above are no longer relevant.

The curious fact that F initially rises rapidly is explained as follows. The contributions from the two diagrams in Fig. 21 almost cancel for point particles, because

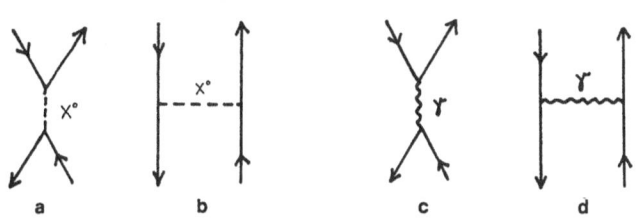

Figure 23. Feynman diagrams contributing to Bhabha scattering. Only the s-channel diagram (a) is resonant and can compete with the QED process represented by the diagrams (c) and (d).

the exchange of the order of the vertex operators γ^0 and γ^μ or γ_5 changes the sign of the scattering amplitude, leaving the magnitude almost unchanged. This cancellation is gradually alleviated in the presence of a form factor and the cross section rises, until the cut-off due to the finite size sets in. A consequence of this effect is that the beam dump limits would actually become considerably sharper, not weaker, for an X^0 boson with a radius in the range between 1 and 10 fm! (This remark bears upon the "micro-positronium" states discussed in the next section.)

Bhabha Scattering at MeV Energies

All the limits on the possible existence of a light neutral X^0-boson discussed so far were derived assuming that the particle has no internal structure. When one allows for a particle with finite size, they become model dependent, as was demonstrated at the example of the limits derived from beam dump experiments. Similar considerations apply to the bounds from the anomalous magnetic moment, Delbrück scattering, positronium hyperfine structure, and so on. The reason for this model dependence lies in the fact that all these processes involve particles off their mass shell, either the electron or the X^0-boson, whereas all particles are on mass shell in the pair decay $X^0 \to e^+e^-$. A form factor, therefore, enters in different ways into these processes.

In order to obtain model-independent bounds it is necessary to consider the process, in which the boson is produced on shell by electrons and positrons that are also on mass shell. This process is called Bhabha scattering on resonance, and represented by the Feynman diagram (a) in Fig. 23. Averaging over electron and positron spin the cross section from this diagram alone is narrowly peaked around the beam energy E_R corresponding in the centre-of-mass system to the rest mass of the X^0-boson:

$$\bar{\sigma}_X(E) = \frac{\pi \alpha_{X_e}^2 f_{J^\pi}(m_X/m_e)}{4(E - E_R)^2 + (m_X \Gamma_X/m_e)^2} \tag{53}$$

where $E_R = (m_X^2/2m_e) - 2m_e$, and $f_{J^\pi}(x)$ is a dimensionless function of order unity that depends on spin and parity of the X^0-boson [63]. Right on resonance, i.e. for $E = E_R$, the cross section exhausts the unitarity limit for a single partial wave

$$\bar{\sigma}_X(E_R) = \frac{\pi \alpha_{X_e}^2 m_e^2 f_{J^\pi}}{m_X^2 \Gamma_X^2} \approx \frac{\pi}{m_X^2} \cdot \frac{\Gamma_X^{ee}}{\Gamma_X}, \tag{54}$$

unless other final states, such as $\gamma\gamma$, $\gamma\gamma\gamma$ or $\bar{\nu}\nu$, contribute significantly so that the partial e^+e^- width Γ_X^{ee} is not equal to the total width Γ_X. On the other hand, the

Table 4. Upper limits for the resonance signal of a particle of mass 1.8 MeV in Bhabha scattering at $\theta_{Lab} = 29°$. The energy resolution was assumed to be 20 keV.

S (0^+)	P (0^-)	V (1^-)	A (1^+)
0.7 %	1.3 %	8.0 %	0.7 %

QED Bhabha scattering cross section, described by the Feynman diagrams (c) and (d) in Fig. 23, is of the order

$$\bar{\sigma}_{QED}(E_R) \approx \frac{\alpha^2}{m_X^2} \approx 10^{-4} \bar{\sigma}_X(E_R). \tag{55}$$

At first glance, therefore, it appears as if the resonance caused by the new particle would give a tremendous signal in Bhabha scattering. Unfortunately, this is only so in the ideal world where the experimenter can measure an excitation function with infinite energy resolution. (Here "infinite" means at least comparable to the decay width Γ_X, which is of order meV, according to Table 3.) In reality, the cross section must be averaged over a finite energy interval ΔE, which depends on the experimental conditions. The QED cross section (55) must then be compared with the energy averaged resonance cross section

$$\frac{1}{\Delta E} \int dE \bar{\sigma}_X(E) = \frac{2\pi^2 (2J+1)}{m_X^2 - 4m_e^2} \cdot \frac{\Gamma_X^{ee}}{\Delta E}. \tag{56}$$

In practice, the energy resolution is not determined by the uncertainty in the positron beam energy, but by the Fermi motion of the electrons in the target. If k_F denotes the Fermi momentum of the target electrons, the energy resolution is given by

$$\Delta E = 2k_F \sqrt{2E_R/m_e} = 2k_F \sqrt{\frac{m_X^2}{m_e^2} - 4}. \tag{57}$$

At fixed beam intensity this value can be reduced only at the expense of scattering rate, because the Fermi momentum is related to the electron density n_e in the target, viz. $k_F = (3\pi n_e)^{1/3}$.

Most experiments have used low-Z targets, such as beryllium or hydrocarbon foils. The energy spread for such targets can be calculated precisely, if the Compton profile of the electron momentum distribution is known [114]. Such calculations give a typical value ΔE of about 20 keV, reducing the signal-to-background ratio by more than 10^{-6}. One therefore expects, at best, a resonance signal of about 1% of the QED background. The maximal signal compatible with the limits from Table 3 is listed in Table 4 for the various spin-parity assignments. With a reasonably intense positron beam it is not difficult to obtain a statistical error of this size. The major difficulty lies in the elimination of systematic errors, e.g. uncertainties in the determination of the beam intensity, which is an especially serious source of error when the excitation function is measured point by point.

The first Bhabha scattering experiment [115] was performed on a Th target ($Z = 90$) because it was thought that the strong Coulomb field of the heavy nucleus could play a role in the phenomena observed at GSI. Note that, according to our above considerations, thorium with its high electron density forms the worst possible target! Nevertheless, the energy spectra of scattered electrons and positrons were

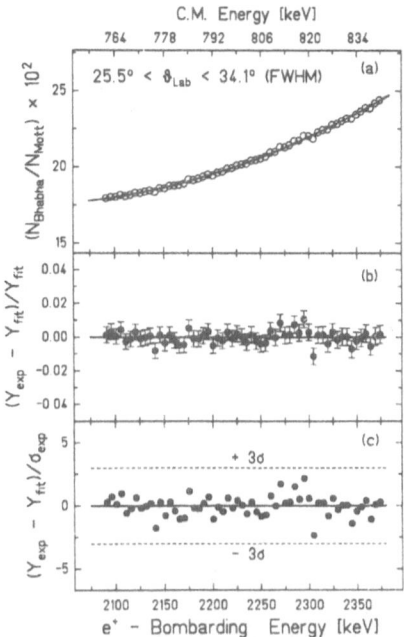

Figure 24. Energy dependence of Bhabha scattering in the invariant mass region between 1.80 and 1.86 MeV, measured at Grenoble. No structure is seen at an error level of 0.25%.

measured, and a line was observed in the coincidence spectrum at 340 keV. The experiment was repeated and confirmed by two independent goups [116,117]. However, subsequent experiments by other groups [118,119] showed that this effect was caused by contamination due to Compton scattering of 511 keV annihilation γ-rays in the detectors.

Over the past two years, several groups have performed Bhabha scattering experiments with low-Z targets. The experimental techniques that have been used differ widely, as do the positron sources. Some groups have made use of the broad band spectrum emitted by radioactive sources [121,122,123,124,128], others have used positrons produced by a reactor [125,129], an intense electron beam [126], or have accelerated slow positrons with an electrostatic accelerator [120,127,130]. While some of the experiments have reported the observation of structures in the Bhabha scattering excitation function [121,127], most have not seen an effect outside the experimental error bars.

Presently, the highest sensitivity has been achieved by the group working at Grenoble with reactor positrons scattered on a beryllium foil. The statistical and systematic errors are below 0.5% [129]. The energy dependence, as measured in this experiment, is flat and shows no sign of a resonance in the invariant mass range 1.797 MeV $< m_X <$ 1.862 MeV (see Fig. 24). The upper limit for the energy integrated resonance cross section is reported as 0.5 b·eV/sr, corresponding to the following values for the e^+e^- width and lifetime of a hypothetical spinless X^0-boson:

$$\Gamma_X^{ee} < 1.9\,\text{meV}, \qquad \tau_X < 3.5 \times 10^{-13}\text{s}. \tag{58}$$

All structures reported by other experiments are clearly excluded by this limit. We can conclude that the Bhabha scattering experiments have now reached a sensitivity that allows to establish limits on new particles which are comparable to those derived from other QED precision experiments. The crucial advantage of the new limits is that they are measured on mass shell and, therefore, independent of assumptions about the structure of the X^0-boson.

The present limit (58) is still several orders of magnitude away from the bound set by the GSI experiments, i.e. $\tau_X < 10^{-9}$s. With the presently used experimental techniques the limit certainly cannot be improved by more than a factor of ten, so that the search for lifetimes above 10^{-12}s requires new methods. One such technique may be to make use of the smaller energy loss as a trigger on those events where the e^+-e^- pair was travelling as a neutral boson during part of its trip through the target [124]. Another way to improve the sensitivity is to search for inelastic scattering events. Here the QED background is reduced because the bremsstrahlung process contains one more order of α. A positive signal would correspond to (radiative) transitions between excited states of an intermediate particle, which is formed in a higher state but decays back into e^+e^- in a state of lower energy. The Grenoble experiment has reported a limit of 0.15 b·eV/sr for such events.

However, one has to bear in mind that such improvements in method usually rely on rather specific model assumptions about the structure and the properties of the sought particle. E.g., in the case of the energy loss technique one implicitly assumes that the particle's size is sufficiently small so that electromagnetic polarization effects in the target may be neglected. This condition is not satisfied for some of the models discussed in the next section. It is not even clear that the X^0-boson would escape undestroyed from a thick target, or even a beam dump, in some specific models. Nevertheless, such improved experiments would be very useful in eliminating large classes of extended particle models, and serve to further tighten the already severe conditions on viable models.

Correlated Two-Photon Lines

Any pair-decaying boson with spin different from one can also decay into two photons. This prompted Meyerhof and collaborators to search for correlated narrow two-photon lines emitted in U+Th collisions at the Super-HILAC accelerator in Berkeley. Unfortunately, the branching ratios for the decay modes e^+e^- and $\gamma\gamma$ are strongly model-dependent; their relative intensity can vary anywhere between 10^{-5} and 10^5 [131]. Therefore, the absence of a signal does not allow to draw any conclusions. The experimental set-up consists of a large number of high resolution γ-ray counters arranged in pairs opposite to each other at an angle close to 180°. Beam and target conditions were chosen as in the 1985 experiment of the EPOS group, but the scattered nuclei were *not* detected in coincidence.

A first experimental run did not reveal any narrow structure at a sum energy in the range corresponding to the events observed at GSI, and an upper limit of about 100 μb was established [131]. A second, longer experiment again showed no structure in the mass region between 1.6 and 1.9 MeV, but a very narrow (3 keV) sum energy peak was found at 1062 keV [132]. Because of its narrow width, the peak was originally not associated with a cascade of nuclear γ-ray transitions (but see below!). This result initially obtained some further support by the fact that experiments concerning pair production by γ-rays on atoms close to threshold have consistently yielded results that were higher than QED predictions [133]. This excess could, in principle, be explained by the production of an (extended) neutral particle with the mass of the observed two-photon line [134]. Unfortunately, a recent repetition of the Berkeley experiment by the same group has shown that the two-photon line must be attributed to a nuclear γ-ray cascade: The $32^+ \rightarrow 30^+$ transition (543 keV) and the $30^+ \rightarrow 28^+$ transition (519 keV) in ^{238}U combine to give a 1062 keV sum line. For the given detector arrangement the line can appear very narrow due to kinematic restrictions on the excitation of the nuclear high-spin states [135].

A particle that can decay both into an e^+e^- pair and into two photons will also show up as a narrow resonance in the excitation function of electron-positron

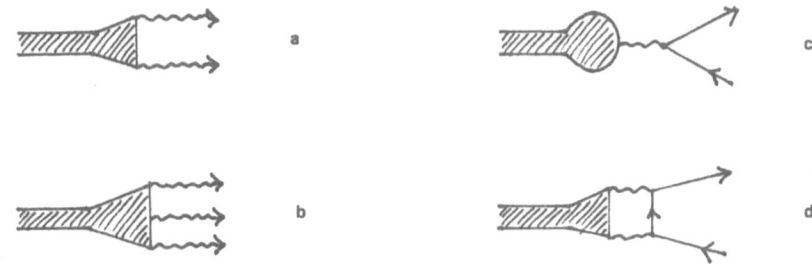

Figure 25. Diagrams for electromagnetic decay modes of a neutral boson: (a) two-photon decay, (b) three-photon decay, (c) pair decay ($J^\pi = 1^-$), (d) pair decay (other J^π).

annihilation in flight. Two such experiments, which are complementary to Bhabha scattering, have been reported to date. The first experiment, performed at Johannesburg, did not show the presence of a resonance structure in the invariant mass range 1.43 - 2.02 MeV [136]. This result allowed to set a limit of about 300 meV on the width of resonances in this mass range. The second experiment, carried out at the University of Tokyo, aimed at the invariant mass region 1.065 ±0.02 MeV. Also here no indication for a resonance structure was found, and a very tight limit $\Gamma_{\gamma\gamma}\Gamma_{ee}/\Gamma < 0.11$ meV was established [137].

MODELS OF NEW EXTENDED NEUTRAL PARTICLES

General Considerations

The postulate of new neutral particles with finite size, or substructure, can simultaneously solve several general difficulties of any explanation of the GSI data in terms of particle decay. These are:

- The fact that several line structures have been seen is naturally explained as the decay of internally excited states of the same particle.

- The small velocity of the pair-decaying source may be explained in two ways: either as a high-momentum cut-off due to the X^0 form factor, if $R_X > 20m_X^{-1} \approx$ 2000 fm; or by production of the X^0-boson in a bound state around both nuclei.

- A composite particle with electrically charged constituents could be efficiently produced by some non-perturbative mechanism that requires the presence of strong Coulomb fields.

- As already argued in the previous section, a general bonus is that all experimental limits are rendered irrelevant for a sufficiently large radius R_X, with the exception of those derived from resonant Bhabha scattering.

Moreover, a general conclusion can be drawn with respect to the competition between two-photon and pair decay. Unless the particle is a bound state of electron-positron pairs, or has a fundamental coupling to the electron field (as the axion would!), the photon decay dominates for all states except those with spin one and negative parity. The argument goes as follows: One may first rule out pair decay due to the weak neutral current, because this would yield a lifetime much longer than 1 ns. For electromagnetic decays, two-photon decay and pair decay via a virtual photon (diagrams (a) and (c) in Fig. 25) are both of order α^2, three-photon decay (diagram (b)) is of order α^3 and pair decay via two virtual photons (diagram (d)) is of order α^4. Decay into two photons is possible for all particles except those with spin one, for these three photons are needed in the final state. On the other hand, pair decay via a single virtual photon requires that the particle carries the quantum numbers of the photon, i.e. $J^\pi = 1^-$. Thus pair decay dominates for a vector boson, three-photon decay is the main decay mode for an axial vector boson, and all other particles would predominantly decay into two photons. (Again it must be emphasized that this conclusion does not apply if the particle is already composed of one or several electron-positron pairs, which may be emitted without interaction with an intermediate virtual photon.)

Two general routes can be taken by the theorist who wants to construct a model of extended particles in the mass range between 1 and 2 MeV:

- One can speculate that there exists an undiscovered, "hidden" sector of low-energy phenomena within the framework of the standard model of particle physics, i.e. within the $SU(3) \times SU(2) \times U(1)$ gauge theory. This might be a non-perturbative, strongly coupled phase of quantum electrodynamics, low-energy phenomena associated with the Higgs sector of the Glashow-Salam-Weinberg model, or some unknown long-range properties of QCD. It has even been speculated that the standard electromagnetic interaction between charged particles with spin behaves quite differently at short distances than normally assumed in perturbation theory.

- One can invoke new interactions which, for some reason that remains to be explained, do not normally show up in experiments. Examples are many-body forces between electrons and positrons that do not contribute in positronium, or new light fermions that are confined by equally new, medium ranged interaction.

Both roads have been extensively explored during the past three years, overall with little success. One must be aware that any attempt to fit a scheme of new low-energy composite particles into the standard model faces awesome obstables, viz. the wealth of experimental data and precision measurements accumulated over fifty or more years. For the first class of models, i.e. those models based on some obscure aspect of the standard model itself, the problem is that there is essentially no free parameter. Every conjectured phenomenon can be calculated reliably, at least in principle. Many theorists, who took to this road, have (therefore?) speculated about states which are so complicated in nature that their calculation is, at present, beyond technical means. Nevertheless, a rather comprehensive no-go theorem for a class of e^+-e^- states has been established, and calculations within lattice gauge theory have started to reject speculations about new phases of QED.

Although free parameters can be introduced in abundance, the second route is no less treacherous. A new force active in the MeV energy range can potentially show up in every atomic, nuclear, or particle physics experiment. This has led, for instance, to the rejection of speculations about many-body forces between electrons. On the other hand, attempts to fit new interactions with consequences at low energies into the standard model are not entirely hopeless. If some yet unknown interaction at the TeV range could be responsible for the observed mass spectrum of leptons and quarks, as technicolour models suggest, why should it not be possible to construct new low-mass particles in a similar manner?

The scope of these lectures, clearly, does not permit to cover the full range of composite particle models that have been proposed to explain the GSI data (for more details see refs. [138,139]). Maybe, the wide range of ideas that have been put forth is best indicated by quoting a few titles of relevant articles. The list begins with "Is there a tightly bound poly-positronium state?" [140], "Exotic states in QED?" [141] and "Abnormal QED atoms" [142], then proceeds over such daring speculations as "The production of a light new neutral particle in high-energy anomalon collisions" [143] or "Are tachyons present?" [144] (remember: the big problem was to explain why the source of the e^+e^- peaks appears to move so *slowly!*), and ends with the almost mystic titles "Anomalous positron line and fractal time" [145] and "Quadronium: Rosetta stone for the e^+e^- puzzle" [146]. (I apologize to anyone whose title is missing from this brief list, but would deserve to be in it.)

In the following I will try to discuss the variety of models from rather general viewpoints in an attempt to emphasize the generic merits or problems of these models, independent of their specific details. I will begin with the postulated highly relativistic "magnetic" bound states of an electron-positron pair, which I will call "micro-positronium" because of their suggested small size. General arguments have been derived why such states with a dominant component in the single e^+e^- pair channel do not exist within the framework of QED. I then discuss the hypothesis of strongly bound states of several e^+e^- pairs under the keyword "poly-positronium". Next, we will turn to speculations about a new ("abnormal") confining phase of QED. Finally I will discuss some general merits and difficulties of composite particle models in the context of a schematic model of meson-like bound states with a long-ranged confining force.

Micro-positronium and the Virial Theorem

The idea that leptons can form (quasi-)bound states on a distance scale of about 10^{-13} cm due to the interaction of their magnetic moments has been entertained by Barut and his collaborators for many years [147]. They were first associated with the GSI data by C.Y. Wong and Becker [148] on the basis of a two-body equation that treated the magnetic moments of the electron and positron as those of classical, spinning point particles. This is clearly inappropriate because it does not take into account the "Zitterbewegung" of relativistic leptons. Similar problems exist in Barut's approach, where the binding force originates with the anomalous magnetic moment which, as a radiative effect, is distributed over a region of the size of the Compton wavelength (386 fm) around the bare electron, according to perturbative QED [149]. Arguments that the form factor could shrink to dimensions of 1 fm due to nonperturbative effects in the micro-positronium bound state [150] have never been substantiated.

Moreover, it was shown by Geiger et al. [151] that Barut's equation fails to reproduce the spectrum of (normal) positronium due to an incorrect treatment of retardation effects. It therefore certainly cannot be relied upon to describe the highly relativistic, hypothetical micro-positronium states. Also, such states could not even be found as solutions of Barut's equation for a highly localized magnetic form factor [151]. A resonance at 1.42 MeV obtained by Dehnen and Shahin on the basis of an approximate treatment of the Bethe-Salpeter equation [152] was also incorrect.

Recently, Wong and collaborators have made an attempt to derive strongly localized, quasi-bound micropositronium states in the framework of the quasi-potential approach to the Bethe-Salpeter equation [153,154]. In principle, although not in practice, the two-body problem in QED can be treated exactly by this approach, and the spectrum of positronium can be described correctly. The authors of ref. [153] found that their interaction leads to singular behaviour in the $^3P_0$ channel which required the introduction of annihilation terms into their equation, i.e. naturally leads beyond the two-body problem. At present, the method has not been sufficiently far developed to allow a final verdict.

However, a very general argument on the basis of the relativistic virial theorem has been presented by Grabiak et al. [155,156] against the existence within QED of quasi-bound states that are mainly composed of a single electron-positron pair. Because it is quite illuminating, let us investigate the basic idea behind this argument in some detail. We proceed in a number of steps:

(a) Let us start with an electron or positron interacting with a static potential $V(\vec{r})$. The motion of the particle is then described by the Dirac Hamiltonian

$$H_D = -i\vec{\alpha} \cdot \vec{\nabla} + \beta m_e + V(\vec{r}). \tag{59}$$

For a normalizable eigenstate Ψ_0 with energy E_0 one has

$$\langle \Psi_0 | i\vec{\alpha} \cdot \vec{\nabla} + \vec{r} \cdot \vec{\nabla} V | \Psi_0 \rangle = \langle \Psi_0 | [\vec{r} \cdot \vec{\nabla}, H_D] | \Psi_0 \rangle = (E_0 - E_0)\langle \Psi_0 | \vec{r} \cdot \vec{\nabla} | \Psi_0 \rangle = 0. \tag{60}$$

For a Coulomb potential $V(r) = \pm e^2/r$ one finds $\vec{r} \cdot \vec{\nabla} V = -V(r)$, and therefore

$$0 = \langle \Psi_0 | i\vec{\alpha} \cdot \vec{\nabla} - V(r) | \Psi_0 \rangle = \langle \Psi_0 | - H_D + \beta m_e | \Psi_0 \rangle = -E_0 + m_e \langle \Psi_0 | \beta | \Psi_0 \rangle. \tag{61}$$

Now it is easy to see that the expectation value of the Dirac matrix β is always less than unity. One thus obtains the general inequality $|E_0| < m_e$, i.e. that all normalizable eigenstates must lie below the continuum threshold.

(b) In order to see how this relation can be generalized to quasi-bound states, i.e. resonance states, one must study the derivation of the virial theorem by means of scale transformations. This is implicit in the above equations, because the operator $\vec{r} \cdot \vec{\nabla}$ is just the generator of such transformations. However, whereas a normalizable eigenstate does not vary under infinitesimal changes of scale, a continuum solution does so. It is possible to show that the magnitude of this variation is inversely proportional to the energy derivative of the continuum phase shift, and hence proportional to the resonance width in the case of a quasi-bound state [156]. The relation derived above is therefore generalized to resonance states in the form

$$|E_{res}| < m_e + g\Gamma_{res}, \tag{62}$$

where g is a numerical factor of order one.

(c) The argument can be extended to arbitrary electromagnetic fields generated by point particles, where the interaction Hamiltonian is $V(x) = e\gamma^0\gamma^\mu A_\mu$. The electromagnetic potential of a point charge e satisfies the wave equation

$$\partial^\nu \partial_\nu A^\mu = e \int d\tau u^\mu(\tau) \delta^4[x - z(\tau)], \tag{63}$$

where $z^\mu(\tau)$ is the world line of the point charge and u^μ its velocity four-vector. The relativistic scaling operator $(x \cdot \partial) = x^\nu \partial_\nu$ yields the following relations:

$$(x \cdot \partial)\partial^\nu \partial_\nu A^\mu = \partial^\nu \partial_\nu [(x \cdot \partial) A^\mu] - 2\partial^\nu \partial_\nu A^\mu$$

$$(x \cdot \partial)\delta^4(x) = -4\delta^4(x)$$

$$(x \cdot \partial)x^\mu = x^\mu, \tag{64}$$

and therefore one finds that $(x \cdot \partial) A^\mu$ also satisfies a wave equation like eq. (63), but with the opposite sign of the charge, i.e. with $(-e)$. Hence, $(x \cdot \partial) A^\mu = -A^\mu$, and the remainder of the argument given under (a) and (b) goes through in the same way. For N pairs, i.e. $2N$ particles, the energy bound becomes:

$$|E^{res}_{(e^+e^-)^N}| < 2Nm_e + g\Gamma_{res}. \tag{65}$$

(d) The core of this argument based on the virial theorem is that the scale in electrodynamics is set solely by the particle mass, here the electron rest mass m_e. This statement must be somewhat modified in the context of *quantum* electrodynamics, because here a second scale, the four-momentum cut-off Λ, has to be introduced in the renormalization procedure. A detailed analysis [156] shows that the cut-off scale can be absorbed in the renormalized mass of the electron, with the exception of correction terms due to radiative processes, which are of order α. The final bound on the energy of a resonance due to a quasibound state of N electron-positron pairs then takes the following form:

$$|E^{res}_{(e^+e^-)^N}| < 2N(1 + c\alpha)m_e + g\Gamma_{res}, \qquad (66)$$

where c is a numerical constant of order one, but depending (logarithmically) on the spatial extension of the quasibound state.

From these considerations one can conclude that *a narrow resonance that is mainly composed of a single e^+e^--pair cannot occur at 1.6 or 1.8 MeV*. This eliminates the idea originally conceived by Wong and Becker to explain the GSI events as the result of the decay of magnetically bound positronium states.

Poly-positronium

The argument presented in the previous section cannot be used to rule out states that are predominantly built up from two or more e^+e^--pairs. However, no mechanism is known within the framework of QED for the strong binding required to bring such states far below the threshold of at least $4m_e$. This constitutes the main objection against recent claims that strongly bound states of two electron-positron pairs ("quadronium") hold the key to the solution of the GSI positron puzzle [146]. An equation of Bethe-Salpeter type has been derived for this system [157], but no indication for a strongly bound state has been found.

(The $(e^+e^-)^2$ system is known to have a very weakly bound state with binding energy of a few eV [158], which has the structure of an ordinary positronium molecule. The analogous system $(e^+e^+e^-)$ was originally proposed as explanation for the GSI events by Wong [159] when only the positron singles peaks were known. However, the branching ratios for the interesting two-body decay modes $(e^+e^+e^-) \rightarrow e^+\gamma$ and $(e^+e^-)^2 \rightarrow e^+e^-$ are minute [160,161], eliminating any such possibility for reasons of intensity. An attempt by Wong to circumvent this difficulty [162] turned out to be based on an incorrect interpretation of the Dirac equation [163].)

Thus, strongly bound $(e^+e^-)^n$ states, the so-called "poly-positronium" states, appear to require the assumption that some new, non-QED force exists between electrons and positrons [140]. On this basis, a rather satisfactory phenomenological explanation of the GSI events could be constructed, if the poly-positronium system would have a size of several 100 fm. The widths for decay into a single e^+e^--pair or into two photons would then be roughly in agreement with the bounds derived from QED. The states would be expected to be produced in the heavy ion collision by the action of the strong electric fields with a cross section and kinematic charcteristics similar to that of the QED pairs[140,164]. (The behaviour of micro-positronium in strong electric fields was studied in refs. [165,166]. The latter of these seems to suffer from an incorrect treatment of gauge invariance.)

Is the required new interaction between electrons and positrons compatible with our knowledge of e^+e^- physics? E.g., one might postulate the existence of a short range attractive many-body force that does not act between a single e^+e^--pair, thus avoiding problems in electron-positron scattering at high energy and in the normal positronium system that is well described by QED. The question was systematically studied by Ionescu et al. [167], who considered the limits set by spectroscopic data from heavy atoms on nonlinear interactions of the form

$$L_{int} = \lambda(\bar{\psi}\psi)^n, \qquad (67)$$

where n is some integer greater than one. Such forces would contribute measurably to the K-shell binding energy in heavy atoms, if the effective coupling constant λ is too large. The following limits were obtained in this way: $\lambda(n = 2) < 5 \times 10^{-4}$ and $\lambda(n = 3) < 2 \times 10^{-3}$. On the other hand, the values of λ required to support a poly-positronium bound state are at least $8(n = 2)$ or $130(n = 3)$, respectively [167]. Higher exponents n or even nonpolynomial interactions do not give more favourable results. Thus, *poly-positronium states based on a new e^+e^--interaction of type (67) can be excluded.*

Abnormal QED Vacuum

In spite of the fact that it is not tenable for the reason explained above, the idea that the GSI events could be attributed to the decay of tightly bound e^+e^- states has many attractive aspects. If a new interaction among electrons and positrons is incompatible with atomic physics, and if the standard forces of QED do not appear to support such states, one may ask whether the QED force between electrons and positrons could not be modified by the action of very strong electric fields such as those present in heavy ion collisions.

The first considerations in the context of pure QED were concerned with the possible existence of collective modes in the electron-positron vacuum in intense external fields [168,169]. A high vacuum polarization density could indeed support plasma oscillations of virtual e^+e^--pairs with a plasma frequency between 1.5 and 2 MeV, which would predominantly decay into an electron-positron pair [169]. However, if such collective excitations exist at all, they appear to be much too broad to be associated with the GSI events [168]. Furthermore, the strong fields are present only for a very brief time, and it is not at all clear how these modes could survive sufficiently long to give rise to narrow e^+e^- lines, simply on the basis of the uncertainty relation $\Delta E \cdot \Delta t \geq \hbar$.

A more radical approach is based on the hypothesis, first put forward by Celenza et al. [141,172], that QED may possess an alternative vacuum state resembling in its properties the normal vacuum of QCD (quantum chromodynamics), and that this new vacuum may be formed in heavy ion collisions. Indeed, it has been known for some time that the U(1) lattice gauge theory has a strong coupling phase, which confines electric charges due to the formation of a condensate of magnetic monopoles [170,171]. If a similar phase transition also occurs in the (non-compact) continuum gauge theory with dynamical fermions, i.e. in the QED of real physics, then the GSI events might be explained crudely as follows [141,173,174] (see Fig. 26): A region of the confining, strongly coupled phase of QED (shaded in the figure) is formed in the vicinity of the colliding heavy ions due to the action of their strong electric fields. An electron-positron pair may be captured in this region, where it becomes permanently bound because the force between the two leptons is now of a long-ranged, confining nature. The meson-like structure survives for some time after the collision of the nuclei and finally decays into a free e^+e^--pair with simultaneous disappearance of the "bag" of abnormal QED vacuum.

If one is willing to follow this speculation for the moment being, the multiplicity of e^+e^- peaks observed at GSI may be naturally interpreted as decays of various "abnormal QED mesons" [141,173,174,175,176]. Depending on the specific details of the slightly different models, the energy splittings between the various GSI peaks can be interpreted as mass splittings due to internal (radial or rotational) excitations or different spin couplings of the QED mesons. Typical values required to fit the data are $\kappa = (166 \text{ keV})^2$ for the slope of the linear confinement potential [174], or $B = (250 \text{ keV})^4$ for the QED bag constant [175], while the size of these states lies in the range between 1000 and 3000 fm.

As alluring as the speculation of abnormal, confined e^+e^- states may be, one

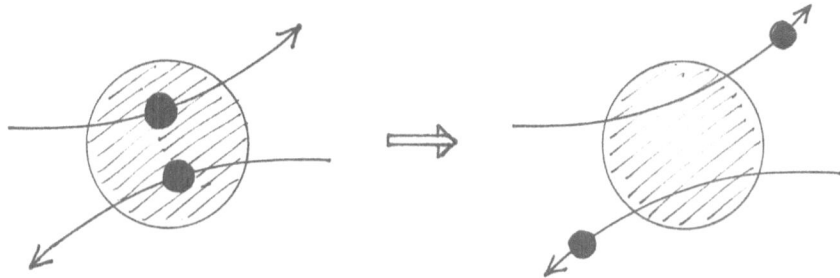

Figure 26. Schematic illustration of the "abnormal phase" interpretation of the GSI events. See text for detailed explanation.

must be aware that it relies on an unproven hypothesis, viz. the existence of a second, strongly coupled QED vacuum state. It is now quite clear that QED has a phase in which chiral symmetry is broken, if the intrinsic coupling constant α is sufficiently large (probably for $\alpha > \alpha_c = \pi/3$) [177,178]. However, we all know that in the real world we have $\alpha \approx 1/137 \ll \alpha_c$, which is why the normal, chirally symmetric QED vacuum state is realized. The question is then, whether the strong electric fields present in a heavy ion collision can somehow catalyze a transition to the abnormal vacuum state, i.e. effectively increase the electromagnetic coupling constant beyond the critical value.

This question was recently studied by Dagotto and Wyld [179], who investigated lattice QED in the presence of an electric background field. They found that the critical coupling constant α_c actually *increases* if an external electric field is present. This behaviour is not really surprising, because it has an analogue in superconductivity. There the superconducting phase, which exhibits confinement of magnetic fields (the Meissner effect), is based on the condensation of Cooper pairs, i.e. of electric charges. This condensate is perturbed, and finally disrupted, by a strong external magnetic field. Similarly, the magnetic monopole condensate of strongly coupled QED is weakened by an external electric field, and a higher value of α is required to keep it in existence. Dagotto and Wyld found that the chirally broken, abnormal phase altogether ceases to exist beyond a certain strength of the electric background field, where it goes over to the charged QED vacuum already studied in Chapter 1 in connection with supercritical nuclear charge Z. (This charged vacuum is also present in the lattice gauge theory.) The phase diagram of QED in the presence of an external charge Z as function of the inverse (!) coupling constant α^{-1} is shown schematically in Fig. 27. Thus, at present, the results of lattice gauge theory do not at all support the hypothesis of the formation of abnormal QED meson states in heavy ion collisions.

The question whether strong external electric fields may facilitate the transition into a new, abnormal phase of QED, was studied by Peccei et al. from a more phenomenological point of view [180]. Their analysis of nonlinear effects in QED showed that these are generally suppressed by a factor of order α, compared to possible naive expectations relying on the large value of $Z\alpha$. Another interesting, non-perturbative approach to QED in strong background fields indicated the possible presence of thresholds in the production of virtual e^+e^--pairs as function of the external field strength [181]. However, this treatment suffers from severe analytic approximations whose validity in the real physical situation is difficult to assess. Finally, it should be mentioned that the chirally broken, strongly coupled phase of QED may be stabilized by the presence of a magnetic background field [182]. While this is not implausible, it

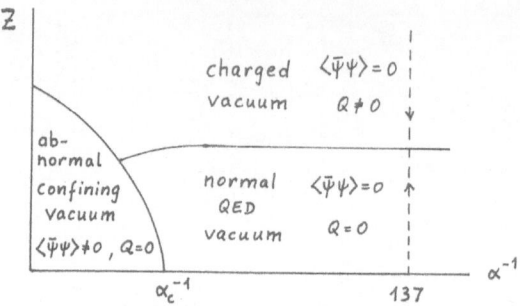

Figure 27. Schematic representation of the phase diagram of QED. In heavy ion collisions the diagram is explored along the dashed line.

is difficult to see how it would apply to heavy ion collisions at the Coulomb barrier, where the magnetic fields are characteristically much weaker than the electric fields ($H/E < 0.1$).

In conclusion, let me discuss some other problematic aspects of this speculative explanation of the GSI peaks:

- Would the region of space filled with abnormal QED vacuum and the e^+e^--pair confined in it detach itself from the heavy ions after the collision and somehow lead a "life of its own"? This would require that the abnormal QED phase can be metastable even in the absence of a strong background field. No clear answer to this question has been provided.

- The precise mechanism by which the QED meson states would decay into a free electron-positron pair has not been described in the literature. Would the decay occur still in the presence of the strong background field, or in a more or less field-free region of space after the QED meson has left the interaction region? It appears that the latter alternative must be realized, because a background field would act as a third body that can absorb momentum, thereby destroying the two-body characteristics of the e^+e^- decay. This, as we remember, was the basis for the particle hypothesis in the first place.

- Would the QED meson states show up as resonances in Bhabha scattering? It has been argued that they would not, because the strong electric field is instrumental in their creation [174]. However, if the decay can occur in a field-free region, the creation should also be possible in the absence of an external field, although the cross section for this process might be very small. Thus, the states should be visible in e^+-e^- scattering unless time reversal invariance is broken, for which there is no indication.

Other Models

A fair number of models for extended particles outside the scope of QED but more or less within the standard model of elementary interactions, $SU(3) \times SU(2) \times U(1)$, has been proposed. In fact, the earliest theoretical speculation about particle creation and decay in association with the GSI positron peaks was based on the unified

SU(2)×U(1) gauge theory of electroweak interactions [183]. The idea started from the observation that this theory contains a doublet of charged Higgs bosons. Thus one might suspect that the vacuum state of the Higgs field can be influenced by sufficiently strong electric fields. (The same model was later re-invented in ref. [184].) However, a careful analysis of the field equations of the electroweak gauge theory reveals that any low-energy distortion of the Higgs vacuum can be absorbed in a gauge transformation, i.e. it does not correspond to a physical change in the vacuum state [185]. This mechanism therefore does not work.

One might think of introducing a more complicated Higgs structure, where some low-energy modes of the Higgs field survive as physical degrees of freedom, which may be influenced by a strong electric background field [186]. A general problem of this class of models is that the Coulomb interaction between the nuclei would be substantially modified at the same time, with drastic consequences for the Rutherford cross section that have not been observed in the experiments. (This cannot be associated with the slight deviations from elastic scattering observed by the EPOS collaboration.) Alternatively one may think of introducing a Higgs potential with two almost degenerate minima [187]. This can occur naturally in the Coleman-Weinberg scheme of symmetry breaking via radiative corrections. However, this scheme is obstructed by the existence of large surface energy at the phase boundary caused by the high barrier between the minima.

Neutral particles of low mass might also occur in the colour-SU(3) sector of the standard model, if the SU(3) gauge theory were broken on a length scale much beyond 1 fm [78,188]. In these models the extended particles are regarded as some nontrivial excitation of the electrically neutral glue field. The strong electric field would therefore not be expected to be the cause of the production of the particles, but perhaps the large number of baryons (almost 500) contained in the colliding nuclei might play a major role. The broken colour models have not been studied in sufficient quantitative detail to be put to a meaningful test against experimental facts.

The possibility that strong electromagnetic fields might influence the vacuum structure of QCD has also been investigated. It has been shown that an effective CP-violating "θ-term" of the form occurs when electric and magnetic fields are simultaneously present [78,189]. Such a term would yield corrections to the coupling between the electromagnetic field and an axion. Since the existence of this particle has been ruled out, the effect does not seem to be of direct interest. The θ-term might cause nontrivial effects in the renormalization scheme, leading to the appearance of a second mass scale in QCD [187], but this path has not been explored quantitatively. The influence of electrically induced isospin breaking on the hadronic mass spectrum also appears to be negligible [77]. A paper [190], which made the opposite claim in connection with the Skyrme model, has not been substantiated.

Two experimental papers have claimed that neutral bosons in the MeV mass range had been observed in nuclear reactions at high energy in emulsions [143,191]. The validity of these claims is hard to assess, although "anomalous" events in emulsion experiments have a long, though not well-reputed, history. For the sake of completeness, we finally mention by reference only a number of rather esoteric suggestions for the solution of the GSI positron puzzle, such as tachyons [144], QED strings [192], and a fractal time dimension $\frac{3}{2}$ [145].

A Schematic Model of New Extended Particles

In view of the fact that none of the specific models discussed above appears to be a good candidate for the particle interpretation of the GSI events, it may be worthwhile to study a schematic model of extended particles that has no direct connection with known physics. Such a model, which resembles those studied in the context of an abnormal QED vacuum state [141,173,174,175,176] was recently

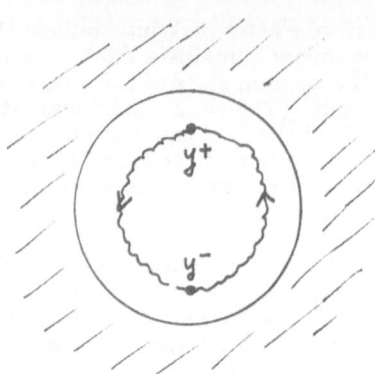

Figure 28. Bag containing two charged fermionic constituents Y^+ and Y^-, as schematic model for an extended X^0-boson.

suggested by Schramm et al. [193]. Here one postulates the existence of a *new* kind of electrically charged light fermions, called Y^+ and Y^-, that interact among each other by an also *new*, confining interaction. This interaction is *not* assumed to be of electrodynamic origin, although it might be described by a strongly coupled U(1) gauge group, and the Y-particles are not assumed to have any relation to electrons.

Following the experience with QCD, the bound states of a Y^+Y^--pair were described in terms of the MIT-bag model but, of course, with a value B_X for the bag constant that has no relation at all with the bag constant of QCD. (It will turn out that the motion of the Y-constituents in the bag is non-relativistic for the preferred values of the parameters; a confining potential model would, therefore, furnish a more appropriate description. Such a model yields only minor quantitative modifications [194] which are not essential in the present context.) Denoting the constituent mass by m_y, the total energy of a Y-bag of radius R_X, representing the X^0-boson, is:

$$M_X = \frac{4\pi}{3} R_X^3 B_X + 2 \frac{\xi(m_y R_X)}{R_X}, \tag{68}$$

where ξ is the dimensionless eigenvalue of the Dirac equation with bag boundary condition. ξ depends on $m_y R_X$; for $m_y = 0$ the lowest eigenvalue is $\xi = 2.04$. The expression (68) must be minimized with respect to the bag radius R_X, which is thus determined as function of the two parameters m_y and B_X of the model. If the total bag mass M_X is fixed, e.g. to the energy of one of the e^+e^- lines seen at GSI, a one parameter family of solutions is obtained. The solutions are shown in Fig. 29 versus m_y as independent parameter. Taking $M_X = 1.8$ MeV, one finds $B_X^{1/4} \approx 250$ keV for $m_y = 0$, and the bag radius is about 600 fm. For larger values of m_y one finds that B_X decreases while R_X increases, which is desirable as we shall see. (Of course, m_y must remain smaller than $\frac{1}{2}M_X = 900$ keV.) A reasonable solution is, e.g., $m_y = 880$ keV with $R_X = 4200$ fm [193].

In order to be compatible with the GSI data, the model must satisfy several minimal requirements:

- The spectrum of excited states must be dense, but not too dense. In our schematic bag model, excited states are separated by $\Delta E \approx n\pi^2/R_X^2 m_y$, i.e. about 40 keV for $R_X = 4200$ fm. This is not incompatible with the splitting between the lines observed at GSI.

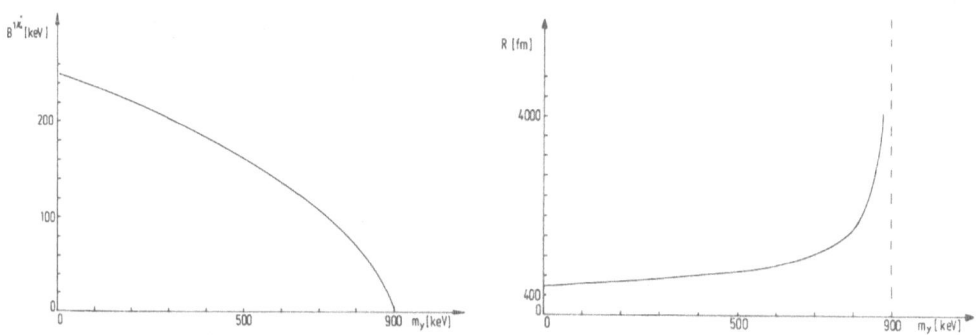

Figure 29. Bag constant (left axis) and bag radius (right axis) as function of the constituent mass for $M_X = 1.8$ MeV.

- The size of the particles should be sufficiently large to evade the limits posed by the beam-dump experiments, i.e. R_X should not be smaller than about 1000 fm.

- The lifetime against electromagnetic decay must be longer than about 10^{-13}s, because of the bounds from QED and from Bhabha scattering. Since the decay rates are proportional to R_X^{-3}, large radii are preferred.

- An efficient production mechanism requires that the strong electric fields present in the heavy ion collision can strongly interact with the overall neutral X^0 boson. This interaction is proportional to the electric polarizability of the particle, which grows as the volume, R_X^3.

The rates for decay into an e^+e^--pair or into two photons according to the Feynman diagrams in Fig. 25 are of order

$$\Gamma(y^+y^- \to e^+e^-), \Gamma(y^+y^- \to \gamma\gamma) \sim \left(\frac{\alpha}{M_X}\right)^2 |\psi_{yy}(0)|^2 \sim \frac{\alpha^2 M_X}{(M_X R_X)^3}, \qquad (69)$$

where ψ_{yy} is the relative wavefunction of the constituents. For $M_X = 1.8$ MeV and $R_X = 4200$ fm one finds lifetimes between 10^{-14}s and 10^{-13}s, but the precise value will depend on details of the relative wavefunction. Even larger radii are preferable from this point of view.

The interaction energy between our composite X^0-boson and the electric field of the heavy ions cannot be calculated perturbatively, because the available field energy is much larger than the particle rest mass. To wit, the total available field energy inside the bag volume is

$$E_{int} \sim (Z\alpha)^2 \int_{R_N}^{R_X} \frac{1}{r^4} r^2 dr \approx \frac{(Z\alpha)^2}{R_N} \approx 20\text{MeV}, \qquad (70)$$

where $R_N \approx 15$ fm is the radius of the nuclear charge distribution. Schramm et al. [193] calculated the interaction energy by solving the Dirac equation for the $y^\pm$ with the two nuclei at the centre of the bag. This leads to a large reduction of the effective mass of the bag state in the presence of the two heavy ions, e.g. two uranium nuclei, from 1.8 MeV down to a mere 46 keV in the case of the special parameter set ($m_y = 880$ keV).

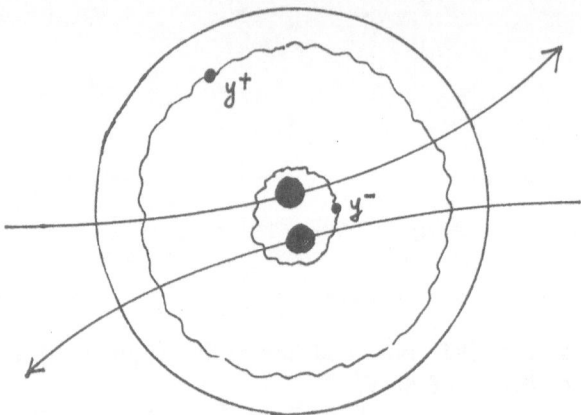

Figure 30. Schematic illustration of the strong binding of a composite X^0-boson in the field of two heavy ions.

The reason for this effect is just the same as that for the strong binding of K-shell electrons in the field of two heavy nuclei, which was discussed in detail in the first section of this report. The negatively charged constituent y^- becomes bound in a tight orbit around the two nuclei, while the positively charged constituent y^+ is repelled toward the edge of the bag, as illustrated in Fig. 30. The net result of this rearrangement is a stronly attractive interaction between the X^0-boson and the two nuclei.

This has two major advantages: First, the lowering of the effective mass due to binding will increase the production cross section tremendously. This was already pointed out in ref. [72], where it was estimated that the electromagnetic production cross section could approach the desired value of about 100 μb, if the effective mass is reduced below 300 keV. Another way to estimate the production cross section in a heavy ion collision is to neglect the presence of the bag and to calculate the total cross section for production of y^+y^--pairs in the same way as for e^+e^--pairs in section 1. In this way one obtains similar values [194].

Secondly, the production of the X^0-boson occurs in a bound state that moves with the centre of mass of the heavy ions. When the nuclei separate, the binding is strongly reduced, and the X^0-boson may be released almost at rest in the centre of mass, as desired according to the experimental results. Unfortunately, for the specific set of parameters discussed here in detail the bag remains bound even to the individual nuclei [193]. A certain fraction of X^0-bosons may be stripped off by the action of the collision dynamics, but these will not be slowly moving in the c.m. frame. The particles that remain bound would decay in the presence of a third body, hence its decay products (e^+ and e^-) would not exhibit the desired two-body characteristics. This difficulty has not yet been satisfactorily resolved.

Is this schematic model compatible with what we know about QED, atomic and high energy physics? Although we have constructed the model in such a way that the e^+e^--width of any single particle state does not violate the bounds imposed by the precision experiments, the answer is not simple, because there exists an infinite number of excited states that can contribute to certain physical processes. Let us discuss two specific examples. In high energy e^+-e^- collisions a pair of constituents y^+y^- could be created with high relative momentum. Because there is a confining force between the y^+ and y^- one would expect that a force string extends between the two particles that finally breaks into a multitude of y^+y^--pairs that eventually form X^0-bosons, just as the glue string between a high energy quark-antiquark pair fragments into hadrons (see Fig. 31).

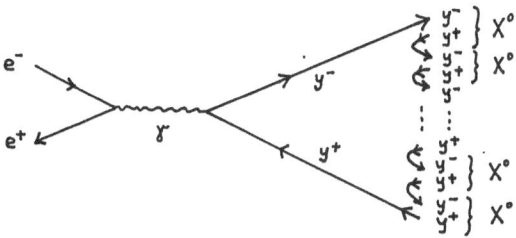

Figure 31. Formation of a jet of y^+y^--pairs in electron-positron collisions at high energy.

However, if one estimates the breaking rate of the y^+-y^- string according to the standard Schwinger formula [195]

$$\frac{dP}{d^4x} = \frac{\kappa^2}{4\pi^3} \sum_{n=1}^{\infty} \frac{1}{n^2} \exp(-\frac{n\pi m_y^2}{\kappa}), \tag{71}$$

where $\kappa = (32\alpha_y B_X/3)^{1/2}$ is the string constant, one encounters a surprise. For $m_y = 880$ keV one has $B_X^{1/4} \approx 25$ keV and $\kappa \approx (45$ keV$)^2 = 1$ MeV/Å (with $\alpha_y = 1$). Multiplying the expression (71) by the volume $V = E_{yy}/B_X$ when the string is fully extended, one finds that the breaking rate is negligible by all comparisons:

$$\frac{dP}{dt} = \frac{8\alpha_y}{3\pi^3} E_{yy} \exp(-\frac{\pi m_y^2}{\kappa}) = \frac{1}{12} E_{yy} \exp(-1200). \tag{72}$$

Clearly, formation of a high energy y^+y^- jet does not occur. But then what are the observable final states predicted by this model? At the moment we simply do not know the answer to this question.

The second kind of process, where the large number of internally excited states becomes relevant, is the exchange of virtual X^0-bosons, e.g. in the vacuum polarization correction to energy levels in muonic atoms (see Fig. 32). The interesting muonic states have an orbital radius of 100 fm or less, i.e. they are much smaller than the X^0-boson states in our schematic model. It appears therefore justified to neglect the confinement force between the virtual y^+y^--pair, so that the total contribution of all excited X^0-boson states to the vacuum polarization can be estimated by the Feynman diagram shown in the left-hand part of Fig. 32 with free $y^\pm$ propagators. Because m_y is not much different from the electron mass m_e, this diagram yields a contribution of almost the same size as the standard electronic vacuum polarization, and thus is in flagrant contradiction to experiment. Is there a way out of this difficulty? One mechanism that could possibly save our model is a short-range repulsive force between the y^+y^--pair, so that the total y-y potential looks as illustrated in the right-hand part of Fig. 32. If the repulsion is sufficiently strong it would probably reduce the contribution of the virtual y^+y^--loop. However, one should not be too optimistic about the prospects of saving the schematic extended particle model with the help of such "tricks".

In conclusion, it seems fair to say that no convincing candidate for a model of extended neutral particles exists at present that could provide an explanation for the

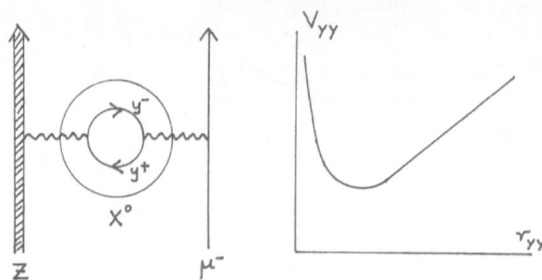

Figure 32. Left: Contribution of virtual X^0-boson exchange to vacuum polarization. Right: y-y potential with short-range repulsion and long-range confining attraction.

e^+e^- peaks observed at GSI. The speculations about an "abnormal", confining phase of QED have not been supported by results from lattice gauge theory. A schematic model of composite X^0-bosons runs into serious difficulties when the contribution to vacuum polarization in muonic atoms is computed. If the GSI events are really caused by the two-body decay of a neutral particle, its nature remains a mystery.

NON-PARTICLE MODELS

General Considerations

The difficulty of accomodating new neutral particles in the framework of established physics and the lack of success of all attempts to find their trace in other phenomena beside the GSI electron-positron peaks has motivated the search for explanations of the GSI events that do not involve new particles. Because these models must involve phenomena which almost look like particle decay, but not quite so, it is useful to remind oneself of what constitutes a particle and what not. A "particle" is a quantum mechanical state which has a definite value of energy, momentum, invariant mass, spin, and parity, or expressed in somewhat more formal terms, which transforms under an irreducible representation of the Poincaré group. In the eyes of an experimentalist, any system that decays strictly back-to-back into an electron-positron pair of equal energy in some reference frame is a particle, and everything else is not a particle. It is with phenomena of the latter kind that we are concerned in this section.

Broadly speaking, three types of models not involving new particles have been proposed: (1) Atomic physics models, invoking some complex process that occurs during the collision of two heavy ions, but has not been taken into account in the simplified description presented in section 1; (2) models which are based on some specific nonlinear property of QED in strong fields; and (3) models which exploit some scheme involving pair conversion processes of nuclear excited states. Some aspects of these models have been briefly touched upon in the previous sections, but we will here discuss all non-particle models in connection.

Models involving Atomic Physics

The possibility that complex atomic excitation mechanisms can produce structures in the electron and positron spectra observed in heavy ion collisions was emphasized by Lichten et al. [196,198]. Indeed, it is well known that the matrix elements

$\langle \varphi_k | \partial/\partial t | \varphi_j \rangle$ in the coupled channel equations (28) can have pronounced structures at some internuclear distance R_0, if one goes beyond the monopole approximation. A typical example is provided by the rotational coupling between the $2p_{1/2}\sigma$ and the $2p_{/2}\sigma$ states, which has a large, broad peak at $R_0 \approx 500$ fm. Assuming an analogous coupling also to states of the negative energy continuum, structures emerged in the positron spectrum [196]. However, a dynamical calculation by Reinhardt et al. [197] showed that this phenomenon is absent if one uses realistic single-particle matrix elements in the coupled channel equations.

It was argued by Lichten that this absence of dynamic structures might be an artefact of the single-particle approximation, and that many-body correlation effects might possibly yield highly structured matrix elements at some particular value of R [198]. No indication was given as to the specific nature of these correlations, and it was not explained why they could be as narrow as the line structures seen in the experiment.

A specific mechanism for narrow structures was considered by de Reus et al. [199], who took up an old suggestion that sudden rearrangements could occur in the ionic charge clouds when the inner electron shells begin to penetrate each other [200]. In principle, a sudden change in the screening potential could lead to rather sharp structures in *all* matrix elements in the coupled equations, at a characteristic two-centre distance $R_0 \approx 2000$ fm. The contribution to the pair creation amplitude from this structure would interfere with the main contribution created at very small distances $R < 50$ fm by the strong electric fields. Depending on the value of the energy of the electron or positron this interference would be constructive or destructive, resulting in interference patterns similar to those predicted in connection with the "atomic clock" effect.

Further study of this model revealed that such a general structure in the dynamic matrix elements would ruin the good agreement of the calculations with experimental data on the impact parameter dependence of K-shell ionization [201]. This problem could be avoided, if the structure in the matrix elements at R_0 was assumed to be of a particular nature, i.e. extremely narrow ($\Delta R \approx 50$ fm) and oscillating with vanishing total integral. Then the structure does not contribute to excitation processes with low average energy transfer, such as ionization, but it is still actively contributing to pair creation. Unfortunately, the effect of the structures was much reduced, even in the singles spectra [201]. The effect essentially disappeared in the sum-energy coincidence spectrum, when the same wedge-shaped cut was taken as in the EPOS experiment, as shown in the right-hand part of Fig. 33. Besides the fact that the model does not seem to work, it is not clear at present whether the assumed sudden rearrangements actually exist. Rapid charge oscillations have been found to occur in time dependent Hartree-Fock calculations [202], but their effect on pair production has not been studied so far.

Several authors have claimed that the continuum matrix elements of the two-centre Dirac equation could contain resonance structures even in the adiabatic approximation, because charged particle may be captured along the molecular axis between the two nuclei [203,204,205]. In one space dimension this is indeed correct, but the phenomenon does not seem to occur for the real three-dimensional two-centre potential. A phase shift analysis of the positron continuum of the two-centre Dirac equation did not reveal any resonance structure except the one caused by the super-critical bound state [207]. But even if such resonances would occur, their specific location would vary with nuclear distance R, so that it is hard to understand how they could result in narrow spectral structures. (The mechanism suggested in ref. [206] appears to be completely at fault, because the vacuum energy cannot be tapped to create real pairs, unless the field is supercritical.)

Two mechanisms for the formation of structures in positron and electron spectra

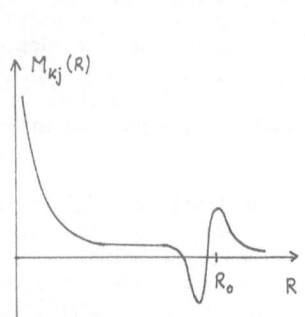

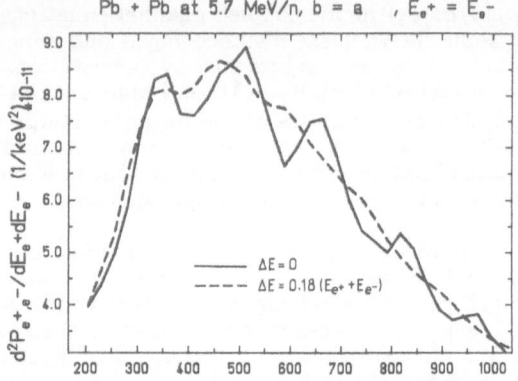

Pb + Pb at 5.7 MeV/n, b = a , $E_{e^+} = E_{e^-}$

Figure 33. Left: Assumed structure in the matrix elements for dynamical pair production, attributed to sudden change in the screening potential. Right: Small structures in the sum-energy spectrum disappear after averaging over experimental cuts.

have been proposed which are based on the presence of strong magnetic fields in the heavy ion collision. The one that was based on the interference between the pair creation amplitudes due to electric (monopole) and magnetic (dipole) coupling has already been discussed at the end of the first section, where it was argued that it does not work [16,17]. The second mechanism was based on the assumption that the magnetic field produces Landau resonances in the positron or electron continuum [208]. However, the numerical solution of the two-centre continuum states of the Dirac equation in the presence of a megnetic dipole field showed that such Landau orbits are absent [209]. Therefore, magnetic effects seem to be excluded as source of the e^+e^--lines.

Models involving Nonperturbative QED

Phenomena associated with QED of strong fields are natural candidates for the explanation of the GSI events. After all, the experiments were designed to study this previously untested area of QED, especially spontaneous pair creation in strong fields. As already argued in section 3, this particular process is excluded because it would not give rise to a free electron. This remark also applies to the suggestion of Scharf [210], whose criterion for enhanced positron production turned out to be just spontaneous positron creation from a vacant supercritical bound state in a new disguise [211].

The ideas of Dietz and Römer [212] appear to be misled by an incorrect interpretation of the solutions of the single particle Dirac equation. Their second-order wave equation corresponds to the Dirac equations with both signs of the mass term $[i\gamma \cdot (\partial - eA) \pm m]\psi = 0$, so that half of their solutions are redundant. The failure to eliminate these by some projection method leads to an apparent coexistence of bound electron and positron states, which cannot be correct. Moreover, no plausible argument is given why electron and positron lines should occur at some definite enrgy in the continuum.

Suggestions of Lemmer and Greiner [168] and of Midorikawa and Yamaguchi [169] the GSI lines could be caused by the decay of collective modes in the highly polarized QED vacuum around the two nuclei have already been discussed in section 5. As argued in ref. [169] the polarized vacuum in the vicinity of the two nuclei may be regarded as an electron-positron plasma with temperature $T \approx 0.5$ MeV and chemical potential $\mu \approx 1$ MeV. Unfortunately, the mass of the plasmon mode for these parameters is quite small ($m_P < 0.1$ MeV), so that further, handwaving arguments must be introduced to get into the relevant mass region 1-2 MeV. In any case, the assumption of a homogeneous QED plasma over distances corresponding to the plasmon wavelength appears to be questionable. The work of ref. [168] treats the QED plasma problem around a high-Z nucleus in RPA approximation, correctly taking into account the inhomogeneity of the Coulomb field. A broad peak at about 2 MeV is found only for $Z = 169$, but not for $Z = 160$. Besides the fact that the peak is very much too broad, the strong Z-dependence makes this result useless as a possible explanation for the GSI peaks.

A completely different approach to the strong field QED problem around a high-Z nucleus was proposed by Hirata and Minakata [213], who applied fermion bosonization techniques to QED in the monopole approximation. In principle, this technique should yield the same results as calculations within the bound state interaction picture (see section 1), except that the interaction with the vacuum polarization is treated self-consistently. Since the first order vacuum polarization correction comes out very small in the Furry picture (1% energy change) this should not lead to major differences, at least in the static case. The calculations of ref. [213] yielded a value $Z_c \approx 150$ for the critical nuclear charge instead of $Z_c \approx 170$. The difference might be caused by an incorrect treatment of the asymptotic form of the vacuum charge distribution. Notwithstanding this discrepancy, the problem of neutral oscillations around the ground state was investigated by two groups [214,215], who found broad structures in the energy range between 1 and 2 MeV. (In ref. [215] it was claimed without proof that the modes have a width of 1 keV.) The location of the modes depends too strongly on Z to explain the GSI data. The relation to the results of ref. [168] is presently not clear.

Models involving Nuclear Pair Conversion

Pair decay of nuclear excited states is a well known source of e^+e^--pairs in heavy ion collisions. As explained in section 3, the experiments were designed to eliminate this background, and many experimental checks have been made to rule this effect out as source of the narrow e^+e^--lines. Although the possibility that these events are caused by nuclear pair decay has been considered again and again, no working scenario has been found up to now. For details the reader is referred to the experimental literature, in particular in refs. [44,45], and to ref. [51].

More recently, it was speculated that very narrow e^+e^--lines could be produced by conversion processes, if long-lived compound nuclei with $Z > Z_c$ ("giant nuclei") were formed [216,217]. In their supercritical fields most of the conversion strength of the positron continuum is localized in the imbedded 1s-resonance. For a given nuclear transition energy then also the electron energy becomes well defined (but the emission is not back to back!). More complicated versions of this scenario assume cascade transitions involving intermediate bound states [216,218]. The cascade process could also work in producing narrow e^+e^--lines in subcritical cases, e.g. when the first nuclear transition ejects an electron from the K-shell, and this vacancy serves to capture the electron from a second nuclear transition, the positron escaping with sharp energy. The difficulties of this mechanism are the large Doppler broadening, the lack of back-to-back correlation, Z- and A-dependence, the smallness of the relevant conversion coefficients, and the short lifetime of a K-vacancy.

CONCLUSIONS AND FUTURE PROSPECTS

Before discussing some possible future prospects of the study of QED of strong fields with heavy ions, let us briefly summarize the results of the previous sections:

- The narrow positron peaks and the even narrower e^+e^- coincidence lines appear to be well established experimentally. The different experiments give a consistent picture overall, but a few aspects, e.g. the beam energy dependence, require further investigation. It has also not been clarified, precisely under what conditions the various lines appear and disappear, and why.

- From an experimental point of view the three most important questions that remain are:

    1. Are the electron and positron really emitted back to back, i.e. do the coincident pairs have a definite invariant mass?
    2. What is the lowest nuclear charge Z for which the lines can be observed?
    3. Is nuclear pair decay really excluded?

- On the side of theory, there is still some room left for the "new particle" interpretation, but it is shrinking rapidly. Elementary particles have been conclusively ruled out, and the remaining alternatives are rather exotic. It is hard to believe that any of them could be true.

- Other explanations have also been unsuccessful, and the "principle of the most conservative assumption" is the only serious argument presently in favour of nuclear pair decay.

- Bhabha scattering is the unique method to look for neutral particles without specific model-dependent assumptions, but probably the required sensitivity can only be reached with an e^+-e^- collider. It is not yet clear whether such an experiment is feasible.

Finally, it may be good to come back to the original aim of the QED experiments with heavy ions, i.e. detecting spontaneous pair creation in supercritical electric fields! Two routes are open to attack this problem. First, a distinct line is predicted to appear in the positron spectrum for sufficiently long nuclear time delay, as discussed at the end of section 2. Such long reaction times have not yet been observed for systems with $Z_u > 170$, but they are known to occur for lighter collision systems. Maybe they can be found and put to use in the positron experiments with improved methods of selecting nuclear final states, e.g. collisions with large mass transfer, large negative Q-value, etc.

The second route becomes viable with the advent of fully stripped beams of very heavy ions, now at Berkeley, and in the near future at GSI. Bringing the K-vacancies into the collision increases the positron yield almost by two orders of magnitude [219]. Since this is true for spontaneously emitted positrons as well as for the dynamic background, the use of fully stripped beams alone is not enough. However, one could make use of the fact that there is no background in the electron spectrum due to ionized atomic electrons if the projectile and target nuclei are both fully stripped. Each emitted electron is then associated with a partner positron. It was recently demonstrated by O. Graf et al. [220] that, under such conditions, delayed collisions would give a unique signal for spontaneous pair creation due to a change in the angular correlation between electron and positron that occurs only in supercritical systems: For nuclear delay times of 2×10^{-21}s the normal forward correlation turns into a backward correlation.

The reason for this behaviour becomes apparent from eq. (32), which shows that the size of the angular correlation term depends on the amplitudes for pair creation in the two partial waves $p_{1/2}$ and $s_{1/2}$ together. In supercritical fields, however, their reaction on a nuclear time delay T is quite different. The supercritical $s_{1/2}$-wave amplitude grows rapidly with increasing T, whereas the subcritical $p_{1/2}$-wave amplitude actually shrinks. (This is analogous to the behaviour of the total positron spectrum in subcritical and supercritical systems, compared in Fig. 13.) Until this effect shows up in the total positron spectrum T has to become quite large, but the effect on the angular correlation is immediate, so that even moderately large delay times are sufficient.

Of course, the feasibility of such experiments depends on many things, e.g. the magnitude of the background from nuclear pair decay, but the situation does not appear hopeless. With further studies of the yet unexplained narrow e^+e^- coincidence lines, and with the prospects of colliding fully stripped uranium beams, the future promises to remain exciting.

ACKNOWLEDGEMENTS

I would like to take this opportunity to thank all my friends and collaborators: J. Augustin, K. Geiger, M. Grabiak, O. Graf, S. Graf, C. Ionescu, U. Müller, G. Plunien, J. Rafelski, J. Reinhardt, T. de Reus, K. Rumrich, A. Schäfer, A. Scherdin, S. Schramm, G. Soff, and especially W. Greiner, without whose enthusiasm this field may not have developed to its present state. I am particularly grateful to J. Reinhardt for the permission to use his list of references. I also thank all members of the experimental groups for their patience in explaining their results, and especially H. Bokemeyer, J.S. Greenberg, and W. König, for careful reading of section 3 of this manuscript. Last, but not least, I would like to thank the organizers of this Cargèse school, R. Marrus and J.P. Briand, for providing this unique opportunity in so pleasant an atmosphere.

References

[1] W. Greiner, B. Müller, and J. Rafelski, "Quantum Electrodynamics of Strong Fields", Springer, Berlin-Heidelberg (1985).

[2] P. Mohr, Lectures at this NATO Institute.

[3] G.Plunien, B. Müller, and W. Greiner, The Casimir effect, Phys. Rep. 134:62 (1986).

[4] B. Müller, J. Rafelski, and W. Greiner, Electron shells in overcritical external fields, Z. Phys. 257:62 (1972); Autoionization of positrons in heavy ion collisions, Z. Phys. 257:183 (1972).

[5] Ya. B. Zel'dovich and V. S. Popov, Electronic Structure of Superheavy Atoms, Sov. Phys. Usp. 14:673 (1972).

[6] M. Gyulassy, Higher order vacuum polarization for finite radius nuclei, Nucl. Phys. A 244:497 (1975).

[7] G. A. Rinker and L. Wilets, Vacuum polarization in strong realistic electric fields, Phys. Rev. A 12:748 (1975).

[8] G. Soff, P. Schlüter, B. Müller, and W. Greiner, The self-energy of electrons in critical fields, Phys. Rev. Lett. 48:1465 (1982)

[9] G. Soff, W. Greiner, W.Betz, and B. Müller, Electrons in superheavy quasi-molecules, Phys. Rev. A 20:169 (1979).

[10] J. Reinhardt, B. Müller, W. Greiner, and G. Soff, Role of multi-step processes in heavy ion inner-shell excitations, Phys. Rev. Lett. 43:1307 (1979).

[11] T.H.J. de Reus, J. Reinhardt, B. Müller, W. Greiner, G. Soff, and U.Müller, The influence of electron-electron interactions on inner-shell processes in heavy ion collisions, J. Phys. B 17:615 (1984).

[12] J. Reinhardt, B. Müller, and W. Greiner, Theory of positron production in heavy ion collisions, Phys. Rev. A 24:103 (1981).

[13] P.H. Mokler and D. Liesen, X-rays from superheavy collision systems, in: "Progress in Atomic Spectroscopy C", p.321, H.J. Beyer and H. Kleinpoppen, ed., Plenum, New York (1984).

[14] G. Mehler, T. de Reus, J. Reinhardt, G. Soff, and U. Müller, Delta electron emission in superheavy quasiatoms with $Z \geq 137$, Z. Phys. A 320:355(1985).

[15] J. Rafelski and B. Müller, Magnetic splitting of quasimolecular electronic states in strong fields, Phys. Rev. Lett. 36:517 (1976).

[16] G. Scharf and D. Twerenbold, The origin of positron lines in heavy-ion collisions, Phys. Lett. B 198:389 (1987).

[17] G. Soff and J. Reinhardt, Positron creation in heavy ion collisions: The influence of the magnetic field, Phys. Lett. B in print.

[18] M. Seiwert, W. Greiner, and W.T. Pinkston, Do heavy-ion potentials have pockets?, J. Phys. G 11:L21 (1985); M. Ismail, M. Rashdan, A. Faessler, M. Trefz, and H.M.M. Mansour, The effect of deformation on the nucleus-nucleus optical model potential and how it produces pockets, Z. Phys. A 323:399 (1986).

[19] V.E. Oberacker, M.W. Katoot, and W.T. Pinkston, Theories of heavy-ion interaction potentials for giant dinuclear systems, in: "Physics of Strong Fields", p. 511, W. Greiner, ed., Plenum, New York (1987); M.W. Katoot and V.E. Oberacker, Microscopic theory of heavy-ion interaction potentials, preprint, Vanderbilt University (1988).

[20] U. Heinz, Interplay of nuclear and atomic physics in ion-atom collisions, Rep. Prog. Phys. 50:145 (1987); W.E. Meyerhof and J.-F. Chemin, Nuclear reaction effects on atomic inner-shell ionization, Adv. At. Mol. Phys. 20:173 (1985).

[21] G. Soff, J. Reinhardt, B. Müller, and W. Greiner, Delta-electron emission in deep-inelastic heavy ion collisions, Phys. Rev. Lett. 43:1981 (1979).

[22] R. Anholt, Electronic $1s\sigma$ vacancy production in deep-inelastic nuclear reactions, Phys. Lett. B 88:262 (1979).

[23] T. Tomoda, Semiclassical approach to the theory of atomic excitation processes associated with heavy-ion collisions, Phys. Rev. A 29:536 (1984).

[24] J. Reinhardt, B. Müller, W. Greiner, and U. Müller, Description of atomic excitations in heavy ion reactions, Phys. Rev. A 28:2558 (1983).

[25] Ch. Stoller, M. Nessi, E. Morenzoni, W. Wölfli, W.E. Meyerhof, J.D. Molitoris, E. Grosse, and Ch. Michel, Nuclear reaction times in the deep-inelastic U+U collision deduced from K-shell ionization probabilities, Phys. Rev. Lett. 53:1329 (1984); M. Nessi, Ch. Stoller, E. Morenzoni, W. Wölfli, W.E. Meyerhof, J.D. Molitoris, E. Grosse, and Ch. Michel, Phys. Rev. C 36:143 (1987).

[26] G. Wolschin, Relaxation phenomena in deeply inelastic heavy ion collisions, Nukleonika 22:1165 (1977);

[27] R. Schmidt, V.D. Toneev, and G. Wolschin, Mass transport and dynamics of the relative motion in deeply inelastic heavy-ion collisions, Nucl. Phys. A 311:247 (1978).

[28] H. Feldmeier, Dynamics of dissipative heavy-ion reactions, in: "Nuclear Sructure and Heavy-Ion Dynamics", p. 274, L. Moretto and R. Ricci, eds., North-Holland, Amsterdam (1984).

[29] R. Krieg, E. Bozek, U. Gollerthan, E. Kankeleit, G. Klotz-Engmann, M. Krämer, U. Meyer, H. Oeschler, and P. Senger, Reaction dynamics studied via positron and electron spectroscopy, Phys. Rev. C 34:562 (1986).

[30] M. Krämer, B. Blank, E. Bozek, E. Kankeleit, G. Klotz-Engmann, C. Müntz, H. Oeschler, and M. Rhein, Collision dynamics between heavy ions determined from δ-electron spectroscopy, Phys. Lett. B 201:215 (1988).

[31] J. Stroth, H. Backe, M. Begemann-Blaich, H. Bokemeyer, P. Glässel, D. Harrach, W. Konen, P.Kosmadakis, S. Mojumder, P. Senger, and K. Stiebing, Nuclear contact times in dissipative heavy ion collisions measured via δ-ray spectroscopy, in: Proceed. XXVI Intern. Winter Meeting on Nuclear Physics, Bormio/Italy 1988, p. 659, I. Iori, ed., Ric. Sci. Educ. Perm. Suppl. 63:659 (1988).

[32] S. Graf, J. Reinhardt, U. Müller, B. Müller, W. Greiner, T. de Reus, and G. Soff, Electronic excitations in heavy ion collisions with sequential fission, Z. Phys. A 329:365 (1988).

[33] H. Backe, P. Senger, W. Bonin, E. Kankeleit, M. Krämer, R. Krieg, V. Metag, N. Trautmann, and J.B. Wilhelmy, Estimates of the nuclear time delay in dissipative U+U and U+Cm collisions derived from the shape of positron and δ-ray spectra, Phys. Rev. Lett. 50:1838 (1983).

[34] J. Rafelski, B. Müller, and W. Greiner, Spontaneous decay of supercritical nuclear composites, Z. Phys. A 285:49 (1978).

[35] J. Reinhardt, U. Müller, B. Müller, and W. Greiner, The decay of the vacuum in the field of superheavy nuclear systems, Z. Phys. A 303:173 (1981).

[36] U. Heinz, U. Müller, J. Reinhardt, B. Müller, and W. Greiner, Time structure and atomic excitation spectra in heavy-ion collisions with nuclear contact, Ann. Phys. 158:476 (1984); W.T. Pinkston, Can potential pockets explain "nuclear sticking"?, J. Phys. G 11:L169 (1985).

[37] J. Schweppe, A. Gruppe, K. Bethge, H. Bokemeyer, T. Cowan, H. Folger, J.S. Greenberg, H. Grein, S. Ito, R. Schule, D. Schwalm, K.E. Stiebing, N. Trautmann, P. Vincent, and M. Waldschmidt, Observation of a peak structure in positron spectra from U+Cm collisions, Phys. Rev. Lett. 51:2261 (1983).

[38] M. Clemente, E. Berdermann, P. Kienle, H. Tsertos, W. Wagner, C. Kozhuharov, F. Bosch, and W. Koenig, Narrow positron lines from U-U and U-Th collisions, Phys. Lett. B 137:41 (1984).

[39] T. Cowan, H. Backe, M. Begemann, K. Bethge, H. Bokemeyer, H. Folger, J.S. Greenberg, H. Grein, A. Gruppe, Y. Kido, M. Klüver, D. Schwalm, J. Schweppe, K.E. Stiebing, N. Trautmann, and P. Vincent, Anomalous positron peaks from supercritical collision systems, Phys. Rev. Lett. 54:1761 (1985).

[40] H. Tsertos, E. Berdermann, F. Bosch, M. Clemente, P. Kienle, W. Koenig,

C. Kozhuharov, and W. Wagner, On the scattering angle dependence of the monochromatic positron emission from U+U and U+Th collisions, Phys. Lett. B 162:273 (1985).

[41] H. Tsertos, F. Bosch, P. Kienle, W. Koenig, C. Kozuharov, E. Berdermann, S. Huchler, and W. Wagner, Multiple positron structures observed from U-U collisions at 5.6 MeV/u and 5.9 MeV/u, Z. Phys. A 326:235 (1987).

[42] W. Koenig, F. Bosch, P. Kienle, C. Kozhuharov, H. Tsertos, E. Berdermann, S. Huchler, and W. Wagner, Positron lines from subcritical heavy-ion collision systems, Z. Phys. A 328:129 (1987).

[43] H. Tsertos, F. Bosch, P. Kienle, W. Koenig, C. Kozuharov, E. Berdermann, M. Clemente, and W. Wagner, On the production mechanism of the narrow positron lines observed in heavy-ion collisions, Z. Phys. A 328:499 (1987).

[44] "Quantum Electrodynamics of Strong Fields", W. Greiner, ed., NATO Advanced Study Institute Series B, vol. 80, Plenum, New York (1981).

[45] "Physics of Strong Fields", W. Greiner, ed., NATO Advanced Study Institute Series B, vol. 153, Plenum, New York (1986).

[46] H. Backe and B. Müller, Positron production in heavy ion collisions, in: "Atomic Inner-Shell Physics", p. 627, B.Crasemann, ed., Plenum Press, New York (1985).

[47] F. Bosch and B. Müller, Positron production in heavy-ion collisions, Progr. Part. Nucl. Science 16:195 (1985).

[48] J.S. Greenberg and P. Vincent, Heavy ion atomic physics - experimental, in: "Treatise on Heavy Ion Science", vol. 5, p. 141, D.A. Bromley, ed., Plenum Press, New York (1985).

[49] P. Kienle, Positron production from heavy ion collisions, Ann. Rev. Nucl. Part. Science 36:605 (1986).

[50] U. Müller, G. Soff, T. de Reus, J. Reinhardt, B. Müller, and W. Greiner, Positrons from supercritical fields of giant nuclear systems, Z. Phys. A 313:263 (1983).

[51] P. Schlüter, T. de Reus, J. Reinhardt, B. Müller, and G. Soff, Can the observed structures in positron spectra be caused by internal conversion?, Z. Phys. A 314:297 (1983).

[52] H. Bokemeyer, H. Folger, T. Cowan, J.S. Greenberg, J. Schweppe, K. Bethge, K. Sakaguchi, P. Salabura, K.E. Stiebing, D. Schwalm, P. Vincent, and H. Backe, Positron electron coincidence spectroscopy in very heavy ion collisions, in: "GSI Scientific Report 1986", p.167, publication GSI-87-1 (1987).

[53] K. E. Stiebing, Special aspects of the EPOS experiments on positron emission from superheavy collision systems, in: ref. [45], p. 253.

[54] T. Cowan, H. Backe, K. Bethge, H. Bokemeyer, H. Folger, J.S. Greenberg, K. Sakaguchi, D. Schwalm, J. Schweppe, K.E. Stiebing, P. Vincent, Observation of correlated narrow peak structures in positron and electron spectra from superheavy collision systems, Phys. Rev. Lett. 56:444 (1986).

[55] T.E. Cowan and J.S. Greenberg (EPOS collaboration), Narrow correlated positron-electron peaks from superheavy collision systems, in: ref. [45], p. 111.

[56] H. Bokemeyer, H. Folger, T. Cowan, J.S. Greenberg, J. Schweppe, K. Bethge, K. Sakaguchi, P. Salabura, K.E. Stiebing, D. Schwalm, and H. Backe, Multiple line structures in positron-electron sum energy spectra observed in heavy ion collisions, in: "GSI Scientific Report 1987", p.173, publication GSI-88-1 (1988).

[57] E. Berdermann, F. Bosch, P. Kienle, W. Koenig, C. Kozhuharov, H. Tsertos, S. Schuhbeck, S. Huchler, J. Kemmer, and A. Schröter, Monoenergetic (e^+e^-) pairs from heavy-ion collisions, preprint GSI-88-35 (1988).

[58] B. Blank, E. Bozek, E. Ditzel, H. Friedemann, H. Jäger, E. Kankeleit, G. Klotzmann-Engmann, M. Krämer, V. Lips, C. Müntz, H. Oeschler, A. Piechaczek, M. Rhein, I. Schall, Sha Yin, and C. Wille, Angular correlation study of electron-positron pairs emitted in heavy ion collisions, in: "GSI Scientific Report 1987", p.175, publication GSI-88-1 (1988).

[59] A. Schäfer, J. Reinhardt, B. Müller, W. Greiner, and G. Soff, Is there evidence for the production of a new particle in heavy-ion collisions?, J. Phys. G 11:L69 (1985).

[60] A.B. Balantekin, C. Bottcher, M. Strayer, and S.J. Lee, Production of a new particle in heavy-ion collisions, Phys. Rev. Lett. 55:461 (1985).

[61] R.S. Van Dyck, P.B. Schwinberg, and H.G. Dehmelt, The electron and positron geonium experiments, in: "Atomic Physics 9", p. 53, ed. R.S. Van Dyck and E.N. Fortson, World Scientific, Singapore (1984).

[62] P. Mohr, Status of precision QED in light and heavy atoms, in: ref. [45], p.17. [The value of α was replaced by the new recommended value of the 1986 fit of fundamental physical constants, see Physics Today, August 1987.]

[63] J. Reinhardt, A. Schäfer, B. Müller, and W. Greiner, Phenomenological consequences of a hypothetical light neutral particle in heavy ion collisions, Phys. Rev. C 33:194 (1986).

[64] L.M. Krauss and M. Zeller, e^+e^- peaks at 1.8 MeV: Phenomenological constraints on nuclear transition axions and particle interpretation, Phys. Rev. D 34:3385 (1986).

[65] R. Barbieri and T.E.O. Ericson, Evidence against the existence of a low mass scalar boson from neutron-nucleus scattering, Phys. Lett. B 57:270 (1975); U.E. Schröder, Evidence against a hypothetical light scalar particle produced in heavy-ion collision systems, Mod. Phys. Lett. A 1:157 (1986).

[66] A. Zee, Phenomenological and theoretical bounds on a low mass gauge boson, Phys. Lett. B 172:377 (1986).

[67] A. Schäfer, J. Reinhardt, W. Greiner, and B. Müller, Elementary light neutral bosons: New limits from precision experiments, Mod. Phys. Lett. A 1:1 (1986).

[68] B.A. Li and H.T. Nieh, A neutral scalar meson at 1.8 MeV?, preprint ITP-SB-86-23, Stony Brook (1986): W.S. Hou and R.S. Willey, Comments on the particle interpretation of the narrow positron and electron peaks observed in heavy ion collisions, preprint PT-8605-86, Pittsburgh (1986).

[69] D.Y. Kim and M.S. Zahir, On a new low mass pseudoscalar boson, Phys. Rev. D 35:886 (1987).

[70] A. Chodos and L.C.R. Wijewardhana, Production mechanisms for a new neutral particle below 2 MeV, Phys. Rev. Lett. 56:302 (1986).

[71] K. Lane, Difficulties for the particle interpretation of positron anomalies in high-Z collisions, Phys. Lett. B 169:97 (1986); M. Suzuki, Arguments against particle physics interpretation of anomalous positrons in heavy-ion collisions, preprint UCB-PTH 85/54, Berkeley (1985).

[72] B. Müller and J. Rafelski, Production of light pseudoscalar particles in heavy-ion collisions, Phys. Rev. D 34:2896 (1986).

[73] Y. Yamaguchi and H. Sato, Positron and electron peaks produced by sub-Coulomb barrier heavy ion collisions, Phys. Rev. C 35:2156 (1987).

[74] S. Barshay, Possibility of a light spin-one boson produced electromagnetically in heavy-ion collisions, Mod. Phys. Lett. A 1:653 (1986).

[75] A. Schäfer, B. Müller, and J. Reinhardt, Consequences of a light vector boson in nuclear decays, Mod. Phys. Lett. A 2:159 (1987).

[76] D. Carrier, A. Chodos, and L.C.R. Wijewardhana, Electromagnetic production of spinless neutral particles in heavy-ion collisions, Phys. Rev. D 34:1332 (1986).

[77] B. Müller and J. Reinhardt, Comment on "Production mechanisms for a new neutral particle below 2 MeV", Phys. Rev. Lett. 56:2108 (1986).

[78] A. Schäfer, J. Reinhardt, B. Müller, and W. Greiner, Models for new-particle production in nuclear collisions, Z. Phys. A 324:243 (1986).

[79] B. Müller, Positron production in heavy ion collisions - A puzzle for physicists, in: "Intersections between Particle and Nuclear Physics", D.F. Geesaman, ed., AIP Conf. Proceed. 150:827 (1986); B. Müller and J. Rafelski, Spontaneous production of light neutral particles in heavy-ion collisions, preprint UFTP-170/86, Frankfurt (1986).

[80] S. Brodsky and M. Karliner, private communication.

[81] N.C. Mukhopadhyay and A. Zehnder, Are there "Visible" Axions?, Phys. Rev. Lett. 56:206 (1986).

[82] G. Mageras, P. Franzini, P.M. Tuts, S. Youssef, T. Zhao, J. Lee-Franzini, and R.D. Schamberger, Search for light short-lived particles in radiative upsilon decays, Phys. Rev. Lett. 56:2672 (1986).

[83] T. Bowcock et al. (CLEO collaboration), Upper limits for the production of light short-lived neutral particles in radiative Υ decay, Phys. Rev. Lett. 56:2676 (1986).

[84] H. Albrecht et al. (ARGUS collaboration), Search for exotic decay modes of the $\Upsilon(1s)$, Phys. Lett. B 179:403 (1986).

[85] R.D. Peccei and H.R. Quinn, Constraints imposed by CP conservation in the presence of pseudoparticles, Phys. Rev. D 16:1791 (1977); S. Weinberg, A new light boson?, Phys. Rev. Lett. 40:223 (1978); F. Wilczek, Problem of strong P and T invariance in the precence of instantons, Phys. Rev. Lett. 40:279 (1978).

[86] R.D. Peccei, T.T. Wu, and Y. Yanagida, A viable axion model, Phys. Lett. B 172:435 (1986).

[87] L.M. Krauss and F. Wilczek, A short-lived axion variant, Phys. Lett. B 173:189 (1986).

[88] E. Ma, Consequences of possible new light pseudoscalar particle, Phys. Rev. D 34:293 (1986); E. Ma, Naturally light pseudoscalar boson in an extended model of quantum chromodynamics, preprint, UC Davis (1986).

[89] M. Suzuki, Constraints on couplings of a fast decaying axion, Phys. Lett. B 175:364 (1986).

[90] C.M. Hofman, Short lived axions and kaon decay, Phys. Rev. D 34:217 (1986).

[91] N.J. Baker, H.A. Gordon, D.M. Lazarus, V.A. Polychronakos, P. Rehak, M.J. Tannenbaum, J. Egger, W.D. Herold, H. Kaspar, V. Chaloupka, E.A. Jagel, H.J. Lubatti, C. Alliegro, C. Campagnari, P.S. Cooper, N.J. Hadley, A.M. Lee, and M.E. Zeller, Search for short-lived neutral particles emitted in K^+ decay, Phys. Rev. Lett. 59:2828 (1987).

[92] E. Massó, Constraints on axions from the $\pi^0 \rightarrow e^+e^-$ decay, Phys. Lett. B 181:388 (1986).

[93] R. Eichler, L. Felawka, N. Kraus, C. Niebuhr, H.K. Walter, S. Egli, R. Engfer, Ch. Grab, E.A. Hermes, H.S. Pruys, A. van der Schaaf, W. Bertl, N. Lordong, U. Bellgardt, G. Otter, T. Kozlowski, and J. Martino, Limits for short lived neutral particles emitted in μ^+ or π^+ decay, Phys. Lett. B 175:101 (1986).

[94] L.M. Krauss and M.B. Wise, Constraints on short-lived axions from the decay $\pi^+ \rightarrow e^+e^-e^+\nu$, preprint YTP86-13/ CALT-68-1356.

[95] A.L. Hallin, F.P. CalaPrice, R.W. Dunford, and A.B. McDonald, Restrictions on a 1.7 MeV axion from nuclear pair transitions, Phys. Rev. Lett. 57:2105 (1986).

[96] M.J. Savage, R.D. McKeown, B.W. Filippone, and L.W. Mitchell, Search for a short-lived neutral particle produced in nuclear decay, Phys. Rev. Lett. 57:178 (1986).

[97] W.A. Bardeen, R.D. Peccei, and T. Yanagida, Constraints on variant axion models, Nucl. Phys. B 279:401 (1987).

[98] L.M. Krauss and D.J. Nash, A viable weak interaction axion?, Phys. Lett. B 202:560 (1988).

[99] F.W.N. deBoer, K. Abrahams, A. Balanda, H. Bokemeyer, R. van Dantzig, J.F.W. Jansen, B. Kotlinski, M.J.A. de Voigt, and J. van Klinken, Search for short-lived axions in a nuclear isoscalar transition, Phys. Lett. B 180:4 (1986).

[100] C.V.K. Baba, D. Indumathi, A. Roy, and S.C. Vaidya, Search for a light neutral particle in the decay of the 3.68 MeV state in ^{13}C, Phys. Lett. B 180:406 (1986).

[101] C.V.M. Datar, S. Fortier, S. Gales, E Hourani, H. Langevin, J.M. Maison, and C.P. Massolo, Search for short lived neutral particle in the 15.1 MeV isovector transition of ^{12}C, Phys. Rev. C 37:250 (1988).

[102] F.W.N. deBoer, J. Deutsch, J. Lehmann, R. Prieels, and J. Steyaert, Search for elusive neutral particles in nuclear decay, J. Phys. G 14:L131 (1988).

[103] F.T. Avignone III, C. Baktash, W.C. Barker, F.P. CalaPrice, R.W. Dunford, W.C. Haxton, D. Kahana, R.T. Konzes, H.S. Miley, D.M. Moltz, Search for axions from the 1115-keV transition of ^{65}Cu, Phys. Rev. D 37:618 (1988).

[104] M.J. Savage, B.W. Filippone, and L.W. Mitchell, New limits on light scalar and pseudoscalar particle production in nuclear decay, Phys. Rev. D 37:1134 (1988).

[105] S.J. Brodsky, E. Mottola, I.J. Muzinich, and M. Soldate, Laser produced axion photoproduction, Phys. Rev. Lett. 56:1763 (1986).

[106] Y.S. Tsai, Axion bremsstrahlung by an electron beam, Phys. Rev. D 34:1326 (1986).

[107] H.A. Olsen, Axion creation in high-energy scattering of electrons on atoms, Phys. Rev. D 36:959 (1987).

[108] A. Konaka, K. Imai, H. Kobayashi, A. Masaike, K. Miyake, T. Nakamura, N. Nagamine, N. Sasao, A. Enomoto, Y. Kukushima, E. Kikutani, H. Koiso, H. Matsumoto, K. Nakahara, S. Ohsawa, T. Taniguchi, I. Sato, and J. Urakawa, Search for neutral particles in electron beam-dump experiment, Phys. Rev. Lett. 57:659 (1986).

[109] M. Davier, J. Jeanjean, and H. Nguyen Ngoc, Search for axion-like particles in electron bremsstrahlung, Phys. Lett. B 180:295 (1986).

[110] E.M. Riordan, M.W. Krasny, K. Lang, P. de Barbaro, A. Bodek, S. Dasu, N. Varelas, X. Wang, R, Arnold, D. Benton, P.Bosted, L. Clogher, A. Lung, S. Rock, Z. Szalata, B.W. Filippone, R.C. Walker, J.D. Bjorken, M. Crisler, A. Para, J. Lambert, J. Button-Shafer, B. Debebe, M. Frodyma, R.S. Hicks, G.A. Peterson, and R. Gearhart, An electron beam dump search for light, short-lived particles, Phys. Rev. Lett. 59:755 (1987).

[111] D.J. Bechis, T.W. Dombeck, R.W. Ellsworth, E.V. Sager, P.H. Steinberg, L.J. Tieg, J.K. Joh, and R.L. Weitz, Search for axion production in low-energy electron bremsstrahlung, Phys. Rev. Lett. 42:1511 (1979).

[112] C.N. Brown, W.E. Cooper, D.A. Finley,A.M. Jonckheere, H. Jostlein, D.M. Kaplan, L.M. Lederman, S.R. Smith, K.B. Luk, R. Gray, R.E. Plaag, J.P. Rutherford, P.B. Straub, K.K. Young, Y. Hemmi, K. Imai, K. Miyake, Y. Sasao, N. Tamura, T. Yoshida, A. Maki, J.A. Crittenden, Y.B. Hsiung, M.R. Adams, H.D. Glass, D.E. Jaffe, R.L. McCarthy, J.R. Hubbard, and Ph. Mangeot, New limits on axion production in 800 GeV hadronic showers, Phys. Rev. Lett. 57:2101 (1986).

[113] A. Schäfer, Bremsstrahlung production of light extended particles, preprint, Caltech (1988).

[114] J. Reinhardt, A. Scherdin, B. Müller, and W. Greiner, Resonant Bhabha scattering at MeV energies, Z. Phys. A 327:367 (1987).

[115] K.A. Erb, I.Y. Lee, and W.T. Milner, Evidence for peak structures in $e^+ +$Th interactions, Phys. Lett. B 181:52 (1986).

[116] Ch. Bargholz, L. Holmberg, K.E. Johansson, D. Liljequist, P.-E. Tegner, L. Bergström, and H. Rubinstein, Investigations of anomalous spectral structure in low-energy positron scattering, J. Phys. G 13:L265 (1987).

[117] M. Sakai, Y. Fujita, M. Inamura, K. Omata, S. Ohya, T. Miura, Studies of electron peaks in $e^+ +$Th, U and Ta interactions, INS-Reports 632 and 651.

[118] R. Peckhaus, Th. W. Elze, Th. Happ, and Th. Dresel, Search for peak structure in $e^+ +$Th collisions, Phys. Rev. C 36:83 (1987).

[119] T.F. Wang, I. Ahmad, S.J. Feedman, R.V.F. Janssens, and J.P. Schiffer, Search for sharp lines in $e^+ - e^-$ coincidences from positrons on Th, Phys. Rev. C 36:2136 (1987).

[120] K. Maier, W. Bauer, J. Briggmann, H.D. Carstanjen, W. Decker, J. Diehl, V. Heinemann, J. Major, H.E. Schaefer, A. Seeger, H. Stoll, P. Wesolowski, E. Widmann, F. Bosch, and W. Koenig, Experimental limits for narrow lines in the excitation function of positron-electron scattering around $E^* = 620$ keV and $E^* = 810$ keV, Z. Phys. A 326:527 (1987).

[121] U. von Wimmersperg, S.H. Connell, R.F.A. Hoernlé, and E. Sideras-Haddad, Observation of Bhabha scattering in the center-of-mass kinetic energy range 342 to 845 keV, Phys. Rev. Lett. 59:266 (1987).

[122] J. van Klinken, Comment on "Observation of Bhabha scattering in the center-of-mass kinetic energy range 342 to 845 keV", Phys. Rev. Lett. 60:2442 (1988).

[123] A.P. Mills and J. Levy, Search for a Bhabha-scattering resonance near 1.8 MeV/c^2, Phys. Rev. D 36:707 (1987).

[124] J. van Klinken, W.J. Meiring, F.W.M. deBoer, S.J. Schaafsma, V.A. Wichers, S.Y. van der Werf, G.C.Th. Wierda, H.W. Wilschut, and H. Bokemeyer, Search for resonant Bhabha scattering around an invariant mass of 1.8 MeV, Phys. Lett. B 205:223 (1988).

[125] H. Tsertos, C. Kozhuharov, P. Armbruster, P. Kienle, B. Krusche, and K. Schreckenbach, Sensitive search for neutral resonances in Bhabha scattering around 1.8 MeV/c^2, Phys. Lett. B 207:273 (1988).

[126] U. Kneissl, C. Kozhuharov, et al. Gießen, experiment in progress.

[127] K. Maier, E. Widmann, W. Bauer, F. Bosch J. Briggmann, H.-D. Carstanjen, W. Decker, J. Diehl, R. Feldmann, B. Keyerleber, D. Maden, J. Major, H.E. Schaefer, A. Seeger, and H. Stoll, Evidence for a resonance in positron-electron scattering at 810 keV centre-of-mass energy, Z. Phys. A 330:173 (1988).

[128] E. Lorenz, G. Mageras, U. Stiegler, and I. Hszár, Search for narrow resonance production in Bhabha scattering at center-of-mass energies near 1.8 MeV, preprint, Munich (1988).

[129] H. Tsertos, C. Kozhuharov, P. Armbruster, P. Kienle, B. Krusche, and K. Schreckenbach, New limits for resonant Bhabha scattering around the invariant mass of 1.8 MeV/c^2, preprint GSI-88-32.

[130] J.S. Greenberg, K. Lynn, et al., Brookhaven, experiment in progress.

[131] W.E. Meyerhof, J. Molitoris, K. Danzmann, D. Spooner, F.S. Stephens, R.M. Diamond, E.M. Beck, A. Schäfer, and B. Müller, Search for correlated narrow-peak structures in the two photon spectrum from 6 MeV/N U+Th collisions, Phys. Rev. Lett. 57:2139 (1986).

[132] K. Danzmann, W.E. Meyerhof, E.C. Montenegro, X.-Y. Xu, E. Dillard, H.P. Hülskotter, F.S. Stephens, R.M.Diamond, M.A. DelePLanque, A.O. Macchiavelli, J. Schweppe, R.J. McDonald, B.S. Rude, and J.D. Molitoris, 180^0 correlated equal-energy photons from 5.9 MeV/N U+Th collisions, Phys. Rev. Lett. 59:1885 (1987).

[133] F.T. Avignone, W.C. Barker, H.S. Miley, H.A. O'Brien, F.J. Steinkruger, and P.M. Wanek, Near-threshold behaviour of pair production cross sections in a

lead target, Phys. Rev. A 32:2622 (1985); A. Coquette, Méthode de mesure absolue de la section efficace total de création de paires, près du seuil, au moyen d'une source d'énergie variable, Nucl. Instr. Meth. 164:337 (1979).

[134] J. Reinhardt, B. Müller, and W. Greiner, Resonant two-photon positron annihilation and pair production, in: "GSI Scientific Report 1987", p.196, publication GSI-88-1 (1988).

[135] W.E. Meyerhof, private communication.

[136] S.H. Connell, R.W. Fearick, RFA Hoernlé, E. Sideras-Haddad, and J.P.F. Sellschop, Search for low energy resonances in the electron-positron annihilation in-flight photon spectrum, Phys. Rev. Lett. 60:2242 (1988).

[137] M. Minowa, S. Orito, M. Tsuchiaki, and T. Tsukamoto, Search for resonances in $e^+e^- \rightarrow \gamma\gamma$ process in the mass region around 1.062 MeV, preprint UT-ICEPP-88-05, Tokyo (1988).

[138] A comprehensive review of this field is presently being written by A. Schäfer, J. Reinhardt, and G. Soff(to be published).

[139] A. Chodos, Narrow e^+e^- peaks in heavy-ion collisions: Fact and fancy, Comm. Nucl. Part. Phys. 17:211 (1987).

[140] B. Müller, J. Reinhardt, W. Greiner, and A. Schäfer, Is there a tightly bound poly-positronium state ?, J. Phys. G 12:L109 (1986); ibid., 12:477 (Erratum).

[141] L.S. Celenza, V.K. Mishra, C.M. Shakin, and K.F. Liu, Exotic states in QED, Phys. Rev. Lett. 57:55 (1986).

[142] C.W. Wong, Abnormal QED Atoms, Phys. Lett. B 205:115 (1988).

[143] M. El-Nadi and O.E. Badawy, The production of a new light neutral boson in high energy anomalon collisions, preprint, Kairo (1987).

[144] J.J. Steyaert, Are tachyons present?, preprint, Louvain-la-Neuve (1987).

[145] L. Nottale, On time, mass, and energy, preprint, Meudon (1987); L.Nottale, Anomalous positron line and fractal time, preprint, Meudon (1988).

[146] J.J. Griffin, Quadronium: Rosetta stone for the e^+e^- puzzle, preprint PP#88-245, Maryland (1988); J.J. Griffin, Quadronium: Key to the e^+e^- puzzle, preprint, GSI (1988).

[147] A.O. Barut and J. Kraus, Resonances in e^+-e^- system due to anomalous magnetic moment interactions, Phys. Lett. B 59:175 (1975); A.O. Barut, Nonperturbative treatment of magnetic interactions at short distances and a simple magnetic model of matter, in: Ref. [44], p. 755.

[148] C.Y. Wong and R.L. Becker, Scalar magnetic e^+e^-- resonance as possible source of anomalous e^+ peak in heavy ion collisions, Phys. Lett. B 182:251 (1986).

[149] B. Lautrup, The short distance behaviour of the anomalous magnetic moment of the electron, Phys. Lett. B 62:103 (1976).

[150] A.O. Barut and J. Kraus, Form-factor corrections to superpositronium and short-distance behaviour of the magnetic moment of the electron, Phys. Rev. D 16:161 (1977).

[151] K. Geiger, J. Reinhardt, B. Müller, and W. Greiner, Magnetic moment inter-

actions in the $e^+ - e^-$ system, Z. Phys. A 329:77 (1988).

[152] H. Dehnen and M. Shahin, Magnetically bounded superpositronium states, preprint, Konstanz (1987).

[153] H.W. Crater, C.Y. Wong, R.L. Becker, and P. VanAlstine, Non-perturbative covariant treatment of the e^+e^- system using two-body Dirac equation from constraint dynamics, prepint, Oak Ridge (1987).

[154] C.Y. Wong, On the possibility of QED (e^+e^-) resonances at 1.6-1.8 MeV, in: "Windsurfing the Fermi Sea", Vol.2, T.T.S. Kuo and J. Speth, Eds., Elsevier, Amsterdam (1987), p. 296.

[155] M. Grabiak, B. Müller, and W. Greiner, Exclusion of Quasi-Bound e^+e^- state at MeV energies within the framework of QED, Ann. Phys. in print.

[156] M. Grabiak, Dissertation, unpublished, Frankfurt (1988).

[157] S.K. Kim, B. Müller, and W. Greiner, Bethe-Salpeter equation for a four-fermion system, Mod. Phys. Lett. in print (1988).

[158] E.A. Hylleraas and A. Ore, Binding energy of the positronium molecule, Phys. Rev. 71:493 (1947); Y.K. Ho, Binding energy of positronium molecules, Phys. Rev. A 33:3584 (1986).

[159] C.Y. Wong, Polyelectron P^{++-} production in heavy ion collisions, preprint, Oak Ridge (1985).

[160] K.G. Lynn, D.N.Lowy, and I.K. Mackenzie, Search for positron annihilation with a single photon emission, J. Phys. C 13:919 (1980).

[161] M.C. Chu and V. Pönisch, Calculation of the one-photon decay rate of the polyelectron P^{++-}, Phys. Rev. C 33:2222 (1986).

[162] C.Y. Wong, Anomalous positron peak in heavy-ion collisions, Phys. Rev. Lett. 56:1047 (1986).

[163] H. Lipkin, Anomalous positron peaks and the Dirac equation (Comment), Phys. Rev. Lett. 58:425 (1987).

[164] J.M. Bang, J.M. Hansteen, and L. Kocbach, Anomalous positron production and its dependece on the heavy ion scattering angle, J. Phys. G 13:L281 (1987).

[165] D.H. Jakubassa-Amundsen, Comment on magnetic e^+e^- resonances in a central field, Phys. Lett. A 120:407 (1986).

[166] C.Y. Wong, Condensation of (e^+e^-) due to short-range non-central attractive forces, preprint, Oak Ridge (1988). C.Y. Wong, Interaction of a neutral composite particle with a strong Coulomb field, preprint, Oak Ridge and Tokyo (1988).

[167] D.C. Ionescu, J. Reinhardt, B. Müller, W. Greiner, and G. Soff, Nonlinear interactions and poly-positronium bound states, J. Phys. G 14:L143 (1988).

[168] R.H.Lemmer and W. Greiner, Enhancement of positron photoproduction in nearly critical strong fields, Phys. Lett. B 162:247 (1985).

[169] S. Midorikawa and Y. Yamaguchi, Screening effect in relativistic QED plasma, preprint INS-567, Tokyo (1986).

[170] M. Creutz, L. Jacobs, and C. Rebbi, Monte Carlo study of abelian lattice gauge theories, Phys. Rev. D 20:1915 (1979).

[171] A.H. Guth, Existence proof of a nonconfining phase in four-dimensional U(1) lattice gauge theory, Phys. Rev. D 21:2291 (1980); J. Fröhlich and T. Spencer, Massless phases and symmetry restoration in abelian gauge theories and spin systems, Comm. Math. Phys. 83:411 (1982).

[172] L.S. Celenza, C.R. Ji, and C.M. Shakin, Nontopological solitons in strongly coupled QED, Phys. Rev. D 36:2144 (1987).

[173] D.G. Caldi and A. Chodos, Narrow e^+e^- peaks in heavy ion collisions and a possible new phase of QED, Phys. Rev. D 36:2876 (1987).

[174] Y.J. Ng and Y. Kikuchi, Narrow e^+e^- peaks in heavy-ion collisions as possible evidence of a confining phase of QED, Phys. Rev. D 36:2880 (1987).

[175] C.W. Wong, Electrons and photons trapped in bags of abnormal QED vacuum, Phys. Rev. D 37:3206 (1988).

[176] C.W. Wong, Generations of abnormal QED states, UCLA-preprint (1987).

[177] J. Kogut, E. Dagotto, and A. Kocić, A new phase of quantum electrodynamics: A non-perturbative fixed point in four dimensions, Phys. Rev. Lett. 60:772 (1988).

[178] P. Fomin, V. Gusynin, V. Miransky, and Yu. Sitenko, Dynamical symmetry breaking and particle mass generation in gauge field theories, Riv. Nuovo Cim. 6:1 (1983); V. Miransky, Dynamics of spontaneous chiral symmetry breaking and the continuum limit in quantum electrodynamics, Nuovo Cim. A 90:149 (1985).

[179] E. Dagotto and H.W. Wyld, Nonperturbative study of QED in a strong Coulomb field, Phys. Lett. B 205:73 (1988).

[180] R.D. Peccei, J. Solà, and C. Wetterich, A new phase of QED?, Phys. Rev. D 37:2492 (1988).

[181] H.M. Fried and H.T. Cho, On strongly coupled $(QED)_4$ and the emission of e^+e^- pairs in heavy ion collisions, preprint HET#631.

[182] Y. Kikuchi and Y.J. Ng, Non-perturbative quantum electrodynamics in a photon-condensate background field, preprint IFP-314-UNC, CTP-TAMU-14/88 (1988).

[183] A. Schäfer, B. Müller, and W. Greiner, Spontaneous symmetry breaking in the presence of strong fields, Phys. Lett. B 149:455 (1984).

[184] L. S. Celenza, A. Pantziris, C. M. Shakin, and H.W. Wang, Excitation of charged Higgs-field condensate in large Z heavy-ion collisions, preprint BCCNT-88/101/174, Brooklyn (1988).

[185] M. Grabiak, G. Staadt, B. Müller, W. Greiner, and A. Schäfer, Spontaneously broken gauge theory in strong external fields, Phys. Lett. B 183:259 (1987).

[186] L. S. Celenza, A. Pantziris, C. M. Shakin, and H.W. Wang, Confined phase of electroweak theory and the e^+e^- peaks observed in large Z heavy-ion collisions, preprint BCCNT-88/031/175, Brooklyn (1988).

[187] A. Schäfer, B. Müller, and W. Greiner, Local vacuum excitations in strong electromagnetic fields, Int. J. Mod. Phys. A 3:1751 (1988).

[188] G.L. Shaw, Spectrum of narrow states produced in low-energy heavy-ion collisions, Phys. Lett. B 199:560 (1987).

[189] B. Müller and J. Rafelski, Vacuum structure and non-relativistic heavy-ion collisions, preprint, Cape Town (1985).

[190] J. Dey, M. Dey, and P. Ghose, Has a skyrmion been observed at 1.7 MeV in heavy-ion collisions at GSI?, preprint, Calcutta (1986).

[191] F.W.M. de Boer and R. van Dantzig, On the possible observation of light neutral bosons in nuclear emulsions, NIKHEF preprint, Amsterdam (1988).

[192] E. Vatai, The interpretation of non-separability by means of connective strings and its experimental evidence, preprint, Debrecen (1988).

[193] S. Schramm, B. Müller, J. Reinhardt, and W. Greiner, Decay of a composite particle as a hypothetical explanation of e^+e^- coincidences observed at GSI, Mod. Phys. Lett. A 3:783 (1988).

[194] S. Graf, Frankfurt, to be published.

[195] N.K. Glendenning and T. Matsui, Creation of $q\bar{q}$ pairs in a chromoelectric flux tube, Phys. Rev. D 28:2890 (1983).

[196] W. Lichten and A. Robatino, New atomic mechanism for positron production in heavy-ion collisions, Phys. Rev. Lett. 54:781 (1985).

[197] J. Reinhardt, B. Müller, and W. Greiner, Comment on "New atomic mechanism for positron production in heavy-ion collisions", Phys. Rev. Lett. 55:134 (1985).

[198] W. Lichten, What causes the sharp positron spectrum in heavy atom collisions? The atomic hypothesis, in: "Relativistic and QED Effects in Heavy Atoms", AIP Conf. Proceed. 136:319 (1985). W. Lichten, On an alternative to the X^0 particle, preprint, Yale (1986)..

[199] T. de Reus, G. Soff, O. Graf, and W. Greiner, The consequences of sudden rearrangements of electronic shells, J. Phys. G 12:L303 (1986).

[200] R.K. Smith, B. Müller, W. Greiner, J.S. Greenberg, and C.K. Davis, Sudden rearrangements in intermediate molecular systems, Phys. Rev. Lett. 34:117 (1975).

[201] T. de Reus, U. Müller-Nehler, G. Soff, O. Graf, B. Müller, W. Greiner, Can multiple structures in positron spectra be caused by atomic effects?, Z. Phys. D 8:305 (1988).

[202] K.R. Sandhya Devi and J.D. Garcia, He^{2+}+He collisions in time-dependent Hartree-Fock theory, J. Phys. B 16:2837 (1983); W. Stich, H.J. Lüdde, and R.M. Dreizler, TDHF calculations for two-electron systems, J. Phys. B 18:1195 (1985).

[203] Yu.N. Demkov and S.Yu. Ovchinnikov, Nature of the narrow energy peak of positrons produced in heavy-ion collision, Pis'ma ZhETF 46:14 (1987).

[204] M. O'Connor, Continuum structure in strong electromagnetic fields, M. Sc. thesis, Cape Town, unpublished (1987).

[205] L.G.Ixaru and D. Pantea, Periodicity effects in the two-center Dirac equation, preprint FT-323-1987, Bucharest.

[206] L. G. Ixaru, e^+e^- pair resonant production in heavy-ion collisions: A theoretical scenario, preprint FT-329-1988, Bucharest.

[207] K. Rumrich, W. Greiner, G. Soff, K.H. Wietschorke, and P. Schlüter, Continuum states of the two-centre Dirac equation, Phys. Lett. B in print (1988).

[208] K. Bethge, Frankfurt, private communication (1986).

[209] K. Rumrich, W. Greiner, and G. Soff, The influence of strong magnetic fields on positron production in heavy-ion collisions, Phys. Lett. A 125:394 (1987).

[210] G. Scharf, Plain mechanism for positron resonances in heavy ion collisions, Phys. Lett. B 177:429 (1986).

[211] S. Schramm, J. Reinhardt, B. Müller, and W. Greiner, Electron-positron scattering matrix in supercritical heavy ion collisions, Phys. Lett. B 198:136 (1987).

[212] K. Dietz and H. Römer, The role of the η-invariant for the description of electrons in a strong Coulomb field, Nucl. Phys. B 300:313 (1988); K. Dietz and H. Römer, Avoided level crossings and pair production in strong Coulomb fields, preprint, Bonn (1988).

[213] Y. Hirata and H. Minakata, Quantum field theories around a large-Z nucleus, Phys. Rev. D 34:2493 (1986); Y. Hirata and H. Minakata, Hydrogenlike atom in bosonized QED, Phys. Rev. D 35:2619 (1987).

[214] Y. Hirata and H. Minakata, Soliton-antisoliton pair creation in strong external fields, Phys. Rev. D 36:652 (1987).

[215] A. Iwazaki and S. Kumano, Oscillations of the polarized vacuum around a large Z 'nucleus', preprint ILL-(TH)-88#8, Urbana (1988).

[216] P. Schlüter, U. Müller, G. Soff, T. de Reus, J. Reinhardt, and W. Greiner, Conversion processes in superheavy and giant atoms, Z. Phys. A 323:139 (1986).

[217] D. Carrier and L.M. Krauss, Spontaneous e^+e^- pair creation in heavy ion collisions, Phys. Lett. B 194:141 (1987).

[218] D. Carrier and L.M. Krauss, Higher order pair conversion peaks in heavy-ion collisions, preprint YTP 87-31, Yale (1987).

[219] U. Müller, T. de Reus, J. Reinhardt, B. Müller, W. Greiner, and G. Soff, Positron production in crossed beams of bare uranium nuclei, Phys. Rev. A 37:1149 (1988).

[220] O. Graf, J. Reinhardt, B. Müller, W. Greiner, and G. Soff, Angular correlations of coincident electron-positron pairs produced in heavy ion collisions with nuclear time delay, preprint GSI-88-22, Darmstadt (1988).

110

QUANTUM ELECTRODYNAMICS OF HIGH-Z FEW-ELECTRON ATOMS

Peter J. Mohr

Institute for Theoretical Physics
University of California
Santa Barbara, California 93106
and
National Bureau of Standards
Gaithersburg, Maryland 20899

INTRODUCTION

Atomic theory makes predictions for the properties and interactions of atoms. Properties of interest include energy levels, lifetimes, transition probabilities, cross sections, *etc.* This paper is concerned with the calculation of energy levels. A correct method of making predictions must take into account relativistic effects, quantum electrodynamic (QED) effects, and many-body aspects of atoms. In addition, the method of calculation must be manageable so that numerical results can be obtained. It is a current challenge to develop an approach that will meet these requirements at a level of precision comparable to that of experiments. Among the available methods are the Schrödinger equation, many-body perturbation theory, the Dirac equation, Hartree-Fock methods, bound-state QED, and the Bethe-Salpeter equation.

Increasing precision in the theory is made through the following steps: The nonrelativistic approximation, given by the Schrödinger equation, gives the gross structure of atoms for the case of one or many electrons. At the next level, the Dirac equation gives relativistic corrections in the one-electron case, but an exact many-electron analog is not known. Finally, bound-state QED yields QED corrections that presumably give the most precise predictions. Calculation of these corrections for atoms with more than one electron is a task that requires more work.

To be more precise about what is meant by QED effects, we refer to corrections that require renormalization such as the self-energy and vacuum polarization that make up a large part of the Lamb shift. Self-energy corrections arise from the process in which an atom emits and reabsorbs a virtual photon. Coulomb interactions of an electron and the nucleus are modified by vacuum polarization effects in which the exchanged photon excites a virtual electron-positron pair from the vacuum.

The structure of conventional theory for one-electron atoms in the external-field approximation is shown in Fig. 1. External-field approximation refers to the fact that the methods are based on the assumption that the nucleus is a fixed source of an external potential. The starting point is shown on the top line. The basic ingredients are the Dirac equation for an electron in the external potential of the nucleus and QED. These can be combined to formulate bound-state QED which then makes presumably complete predictions for energy levels in hydrogen-like atoms. The other routes to predictions from the Dirac equation can be via nonrelativistic approximations to obtain less precise predictions for properties of hydrogen.

Conventional theory employed for two-electron atoms is shown in Fig. 2. The starting point is again the one-electron external-field Dirac equation plus QED. There

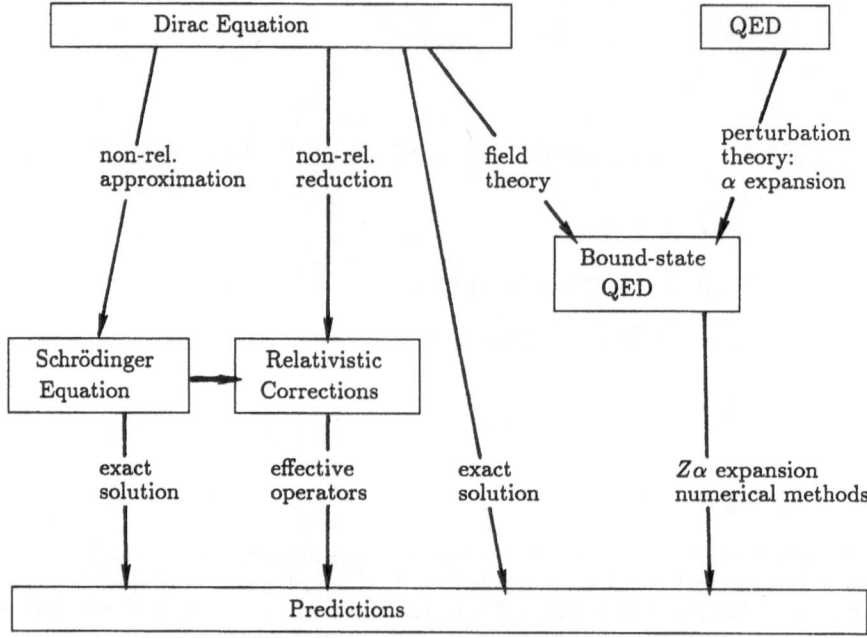

Figure 1. One-electron atom theory

is no exact two-electron relativistic equation analogous to the Dirac equation, but the Dirac equation and QED can be combined to produce the external-field Bethe-Salpeter equation, which is then the starting point for various approximate calculations. It is also possible to write bound-state QED for two electrons. In this case, the perturbation expansion in interactions of the electron with the photon field produces a series in $1/Z$ that is the relativistic analog of the conventional $1/Z$ expansion of the Schrödinger equation.

For many-electron atoms, as shown in Fig. 3, the theoretical framework is similar to the two-electron case, with the exception that there is no well-developed many-electron analog of the external-field Bethe-Salpeter equation. Another difference is the fact that for many electrons, the $1/Z$ expansion is effectively an N/Z expansion where N is the number of electrons in the atom. This has the consequence that an effective- potential many-body approach is more likely to be a useful method for obtaining predictions from the theory than a straightforward perturbation expansion.

This paper will focus on the path through bound-state QED and numerical methods.

Conventional Theory Methods
(External-Field Approximation)

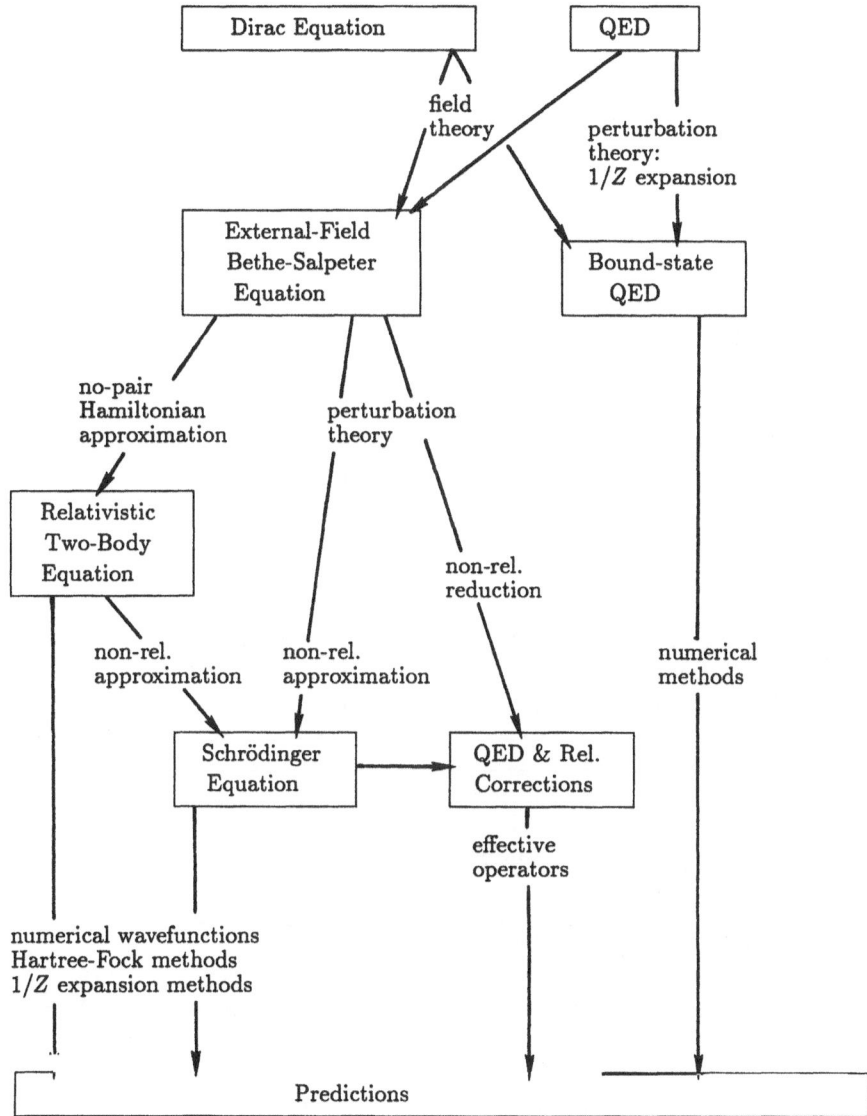

Figure 2. Two-electron atom theory

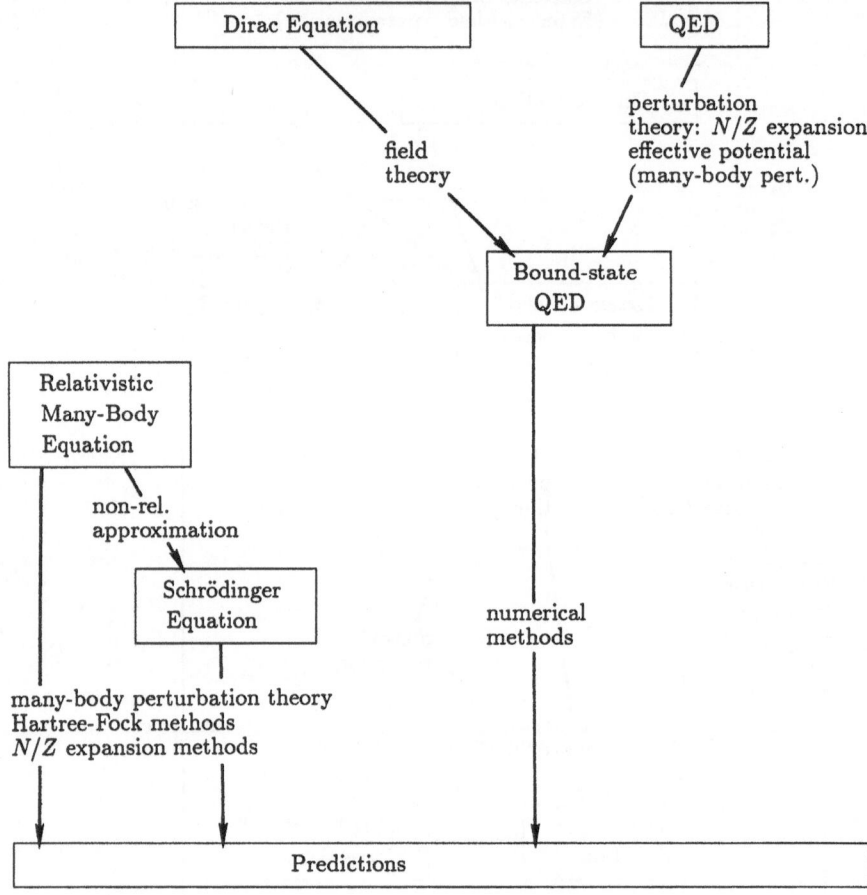

Figure 3. Many-electron atom theory

BOUND-STATE QED (FURRY PICTURE)

Bound interaction QED is formulated by starting with solutions of the Dirac equation (with units chosen so that $\hbar = c = m_e = 1$)[1]

$$\left[-i\vec{\alpha} \cdot \vec{\nabla} + V(\vec{x}) + \beta - E_n \right] \phi_n(\vec{x}) = 0 \tag{1}$$

with corresponding time-dependent solutions given by

$$\phi_n(x) = \phi_n(\vec{x}) e^{-iE_n t} \tag{2}$$

The potential in the Dirac equation is assumed to be the nuclear Coulomb field, although the case where it is an effective potential for a many-electron atom will be discussed briefly in the last section. The electron-positron field operator is expanded in terms of electron and positron annihilation and creation operators as

$$\psi(x) = \sum_{E_n > 0} a_n \phi_n(x) + \sum_{E_m < 0} b_m^\dagger \phi_m(x) \tag{3}$$

where a_n is the electron annihilation operator for an electron in state n $(E_n > 0)$, and $b_m^\dagger$ is the positron creation operator for a positron in state m $(E_m < 0)$ with the conventional anticommutation relations

$$\{a_n, a_m^\dagger\} = \delta_{nm} \ ; \qquad \{a_n, a_m\} = 0 \ ; \qquad etc. \tag{4}$$

Unperturbed states

The unperturbed state vectors are of the form

$$|n\rangle = a_n^\dagger |0\rangle \tag{5}$$

for one-electron states, or

$$|n\rangle = \sum_{ij} c_{ij} a_i^\dagger a_j^\dagger |0\rangle \tag{6}$$

for two-electron states.

For these states, the unperturbed energy operator is given by

$$H_0 = \int d\vec{x} : \psi^\dagger(x) H_D \psi(x) := \sum_{E_n > 0} E_n a_n^\dagger a_n - \sum_{E_m < 0} E_m b_m^\dagger b_m \tag{7}$$

where H_D is the Dirac Hamiltonian

$$H_D = -i\vec{\alpha} \cdot \vec{\nabla} + V(\vec{x}) + \beta \tag{8}$$

Interaction

Interaction between the electron-positron field and the photon field is provided by the interaction Hamiltonian

$$H_I(x) = j^\mu(x) A_\mu(x) - \delta M(x) \tag{9}$$

where j is the current operator

$$j^\mu(x) = -\frac{e}{2} \ [\bar{\psi}(x)\gamma^\mu, \psi(x)] \tag{10}$$

$A_\mu(x)$ is the quantized photon field operator, and δM is the mass renormalization counterterm

$$\delta M(x) = \frac{\delta m}{2} \ [\bar{\psi}(x), \psi(x)] \tag{11}$$

The magnitude of the mass counterterm coefficient δm is fixed by the condition

$$\lim_{Z\alpha \to 0} \Delta E_n(Z\alpha) \to 0 \tag{12}$$

where ΔE_n is the level shift defined below. Commutators in j and δM refer to operators only, *i.e.*,

$$[\bar{\psi}(x), \psi(x)] \equiv \sum_{E_m>0,E_n>0} \bar{\phi}_m(x)\phi_n(x) \left[a_m^\dagger, a_n\right] + \ldots \tag{13}$$

and not Dirac matrices.

Level shifts

It is convenient to introduce the adiabatically damped interaction Hamiltonian

$$H_I^\epsilon(t) = e^{-\epsilon|t|} H_I(t) \tag{14}$$

with

$$H_I(t) = \int d\vec{x}\, H_I(x) \tag{15}$$

as the basic interaction for the level shift calculations. The damping factor makes terms in the perturbation expansion that are infinite in the limit $\epsilon \to 0$ finite. In this way, singularities that appear in separate terms of the level shift expression can be seen to cancel if appropriate diagrams are included.

Time development of the fields is governed by

$$i\, \frac{\partial}{\partial t}\, U_{\epsilon,\lambda}(t, t_0) = \lambda H_I^\epsilon(t) U_{\epsilon,\lambda}(t, t_0) \tag{16}$$

with the well-known solution

$$U_{\epsilon,\lambda}(t, t_0) = \sum_{j=0}^{\infty} (-i\lambda)^j \int_{t_0}^{t} dt_j \int_{t_0}^{t_j} dt_{j-1} \ldots \int_{t_0}^{t_2} dt_1\, [H_I^\epsilon(t_j) \ldots H_I^\epsilon(t_1)] \tag{17}$$

or, with the aid of the time ordering symbol T,

$$U_{\epsilon,\lambda}(t, t_0) = \sum_{j=0}^{\infty} \frac{(-i\lambda)^j}{j!} \int_{t_0}^{t} dt_j \int_{t_0}^{t} dt_{j-1} \ldots \int_{t_0}^{t} dt_1\, T\left[H_I^\epsilon(t_j) \ldots H_I^\epsilon(t_1)\right] \tag{18}$$

The S-matrix is defined by

$$S_{\epsilon,\lambda} = U_{\epsilon,\lambda}(\infty, -\infty) \tag{19}$$

with the perturbation expansion

$$S_{\epsilon,\lambda} = 1 + \sum_{j=1}^{\infty} S_{\epsilon,\lambda}^{(j)} \tag{20}$$

where

$$S_{\epsilon,\lambda}^{(j)} = \frac{(-i\lambda)^j}{j!} \int d^4 x_j \ldots \int d^4 x_1\, e^{-\epsilon|t_j|} \ldots e^{-\epsilon|t_1|}$$
$$\times\, T\left[H_I(x_j) \ldots H_I(x_1)\right] \tag{21}$$

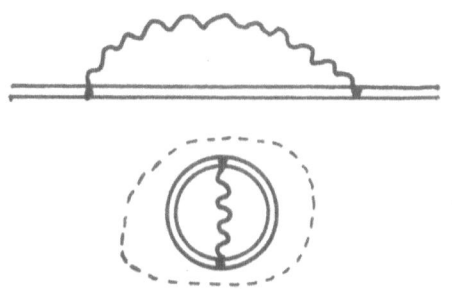

Figure 4. Example of a disconnected graph

A level shift ΔE_n may be defined by

$$\Delta E_n = \lim_{\substack{\epsilon \to 0 \\ \lambda \to 1}} \frac{\langle n|U_{\epsilon,\lambda}(\infty,0)[H_0 + \lambda H_I^\epsilon(0) - E_n]U_{\epsilon,\lambda}(0,-\infty)|n\rangle}{\langle n|U_{\epsilon,\lambda}(\infty,-\infty)|n\rangle} \qquad (22)$$

Gell-Mann, Low[2] and Sucher[3] show that

$$\langle n|U_{\epsilon,\lambda}(\infty,0)[H_0 + \lambda H_I^\epsilon(0) - E_n]U_{\epsilon,\lambda}(0,-\infty)|n\rangle$$
$$= \frac{i\epsilon\lambda}{2} \frac{\partial}{\partial\lambda}\langle n|S_{\epsilon,\lambda}|n\rangle \qquad (23)$$

which leads to

$$\Delta E_n = \lim_{\substack{\epsilon \to 0 \\ \lambda \to 1}} \frac{i\epsilon\lambda}{2} \frac{\frac{\partial}{\partial\lambda}\langle n|S_{\epsilon,\lambda}|n\rangle}{\langle n|S_{\epsilon,\lambda}|n\rangle} \qquad (24)$$

for the level shift. In order to eliminate disconnected graphs, such as the one circled with a dotted line in Fig. 4, the S-matrix element is factorized by taking advantage of the fact that

$$\langle n|S_{\epsilon,\lambda}|n\rangle = \langle n|S_{\epsilon,\lambda}|n\rangle_c \langle 0|S_{\epsilon,\lambda}|0\rangle \qquad (25)$$

where the subscript c means that only connected graphs are included; a connected graph is one in which no subgraph that has no external legs can be circled without crossing any lines in the graph.

Differentiation of the product form of the matrix element yields

$$\Delta E_n = \lim_{\substack{\epsilon \to 0 \\ \lambda \to 1}} \frac{i\epsilon\lambda}{2} \frac{\left[\frac{\partial}{\partial\lambda}\langle n|S_{\epsilon,\lambda}|n\rangle_c\right]\langle 0|S_{\epsilon,\lambda}|0\rangle + \langle n|S_{\epsilon,\lambda}|n\rangle_c\left[\frac{\partial}{\partial\lambda}\langle 0|S_{\epsilon,\lambda}|0\rangle\right]}{\langle n|S_{\epsilon,\lambda}|n\rangle_c\langle 0|S_{\epsilon,\lambda}|0\rangle} \qquad (26)$$

or

$$\Delta E_n = \lim_{\substack{\epsilon \to 0 \\ \lambda \to 1}} \frac{i\epsilon}{2} \frac{\frac{\partial}{\partial\lambda}\langle n|S_{\epsilon,\lambda}|n\rangle_c}{\langle n|S_{\epsilon,\lambda}|n\rangle_c} + \text{constant} \qquad (27)$$

where the constant term in (27) is the same for all levels and does not contribute to transition energies. For future reference, the perturbation expansion of the ratio in Eq. (27) is

Figure 5. Feynman diagram for atomic
excitation followed by photon emission

$$\lambda \frac{\partial}{\partial \lambda} \frac{\langle S_{\epsilon,\lambda} \rangle_c}{\langle S_{\epsilon,\lambda} \rangle_c} \bigg|_{\lambda=1} = \frac{\langle S_{\epsilon}^{(1)} \rangle_c + 2\langle S_{\epsilon}^{(2)} \rangle_c + 3\langle S_{\epsilon}^{(3)} \rangle_c + \cdots}{1 + \langle S_{\epsilon}^{(1)} \rangle_c + \langle S_{\epsilon}^{(2)} \rangle_c + \langle S_{\epsilon}^{(3)} \rangle_c + \cdots}$$

$$= \langle S_{\epsilon}^{(1)} \rangle_c + 2\langle S_{\epsilon}^{(2)} \rangle_c - \langle S_{\epsilon}^{(1)} \rangle_c^2$$

$$+ 3\langle S_{\epsilon}^{(3)} \rangle_c - 3\langle S_{\epsilon}^{(1)} \rangle_c \langle S_{\epsilon}^{(2)} \rangle_c + \langle S_{\epsilon}^{(1)} \rangle_c^3 \qquad (28)$$

$$+ 4\langle S_{\epsilon}^{(4)} \rangle_c - 4\langle S_{\epsilon}^{(1)} \rangle_c \langle S_{\epsilon}^{(3)} \rangle_c - 2\langle S_{\epsilon}^{(2)} \rangle_c^2$$

$$+ 4\langle S_{\epsilon}^{(1)} \rangle_c^2 \langle S_{\epsilon}^{(2)} \rangle_c - \langle S_{\epsilon}^{(1)} \rangle_c^4 + \cdots$$

where

$$\langle S_{\epsilon,\lambda} \rangle_c = \langle n | S_{\epsilon,\lambda} | n \rangle_c$$
$$\langle S_{\epsilon}^{(j)} \rangle_c = \langle n | S_{\epsilon,1}^{(j)} | n \rangle_c \qquad (29)$$

Although the definition for the level shift is quite plausible, the question remains as to how this definition relates to the physical line center displacement observed in an experiment. This issue has been examined by Low in the context of QED by examining the scattering process depicted in Fig. 5 which represents a hydrogen atom excited by an external potential, denoted by an ×, interacting with the virtual radiation field in the intermediate state, and decaying by photon emission.[4] The triple line between the vertices represents the full electron propagation function that takes into account the resonant level shift and width that result from interactions with virtual photons. Low demonstrated that the shift in the resonance center is equal to high accuracy to the radiative level shift of the bound-state energy, which in turn is given to high accuracy by the Gell-Mann, Low expression, as shown by Sucher.[3]

Wick's theorem

To evaluate the time-ordered products of Fermion operators in the expressions for the S-matrix, defined by

$$T[A(t_A)B(t_B)\ldots] = (-1)^{\pi} B(t_B)\ldots A(t_A)\ldots \qquad t_B > \ldots > t_A > \ldots \qquad (30)$$

where π is the number of nearest-neighbor exchanges to bring the order of the operators in brackets to the order on the right-hand-side of the equation, it is convenient to employ Wick's theorem[5] in the form:

$$T[ABCD\ldots] = :ABCD\ldots : + :A\underset{\sqcup}{B} CD\ldots : + \cdots$$

$$+ :\underset{\sqcup}{AB} \underset{\sqcup}{CD}\ldots : + \text{ all possible contractions} \qquad (31)$$

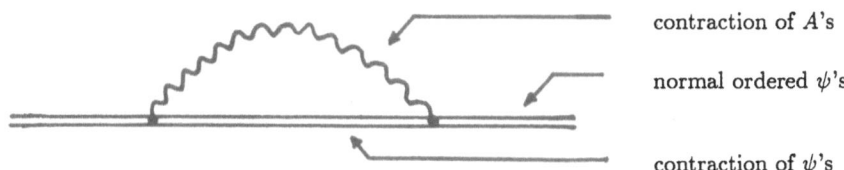

contraction of A's

normal ordered ψ's

contraction of ψ's

Figure 6. Example of a second-order
one-electron Feynman diagram

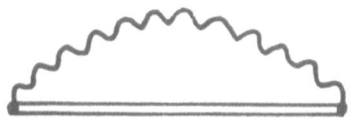

Figure 7. Example of an operator Feynman diagram

The contraction symbol is defined by

$$AB = \langle 0|T[AB]|0\rangle \tag{32}$$

For two equal-time fermion operators, as in $j^\mu(x)$, the time-ordering symbol is defined
as

$$T[A(t)B(t)] = \tfrac{1}{2}A(t)B(t) - \tfrac{1}{2}B(t)A(t) \tag{33}$$

With this definition

$$T\left[AB\tfrac{1}{2}(CD - DC)EF\ldots\right] = T[ABCDEF\ldots] \tag{34}$$

for equal times in C and D. As a consequence, the current may be written in a simple
product form in T rather than as a commutator. In order to correctly generate the
vacuum polarization diagrams, contractions are made between all operators, including
those with equal-time arguments.

Feynman diagrams

This procedure leads to expressions that may be represented by Feynman diagrams or Feynman-diagram-like operator diagrams. For example, for one electron
in second-order, the diagram in Fig. 6 appears. In this figure, the external double
lines correspond to normal-ordered ψ's, the internal double line corresponds to a contraction of ψ's, and the wavy line corresponds to a contraction of A's. If the initial
and final electron states are left arbitrary, the expression may be represented by an
operator Feynman diagram with no external legs as in Fig. 7. This diagram can be
evaluated for any set of external legs, such as a two-electron atom as shown in Fig.
8.

Propagation functions

The contraction of electron-positron field operators, denoted by the double line
in the Feynman diagrams, corresponds to the bound-electron propagation function

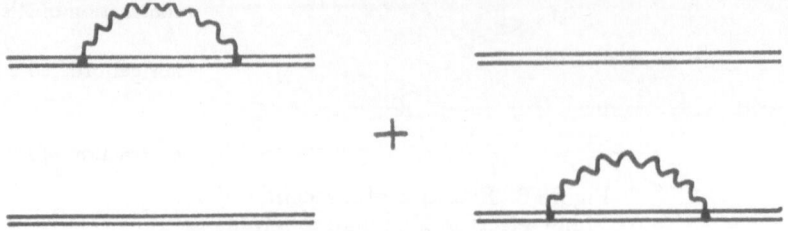

Figure 8. Example of the operator in
Fig. 7 evaluated in a two-electron state

$$S_F(x_2, x_1) = \langle 0 | T[\psi(x_2)\bar{\psi}(x_1)] | 0 \rangle$$

$$= \begin{cases} \displaystyle\sum_{E_n > 0} \phi_n(x_2)\bar{\phi}_n(x_1) & t_2 > t_1 \\[2ex] -\displaystyle\sum_{E_n < 0} \phi_n(x_2)\bar{\phi}_n(x_1) & t_2 < t_1 \end{cases} \tag{35}$$

$$= \frac{1}{2\pi i} \int_{-\infty}^{\infty} dz \sum_n \frac{\phi_n(\vec{x}_2)\bar{\phi}_n(\vec{x}_1)}{E_n - z(1 + i\delta)} e^{-iz(t_2 - t_1)}$$

$$= \frac{1}{2\pi i} \int_{-\infty}^{\infty} dz \, G(\vec{x}_2, \vec{x}_1, z(1 + i\delta))\gamma^0 e^{-iz(t_2 - t_1)}$$

In (35), $G(\vec{x}_2, \vec{x}_1, z)$ is the Dirac Green's function that satisfies the inhomogeneous Dirac equation

$$\left[-i\vec{\alpha} \cdot \vec{\nabla}_2 + V(\vec{x}_2) + \beta - z \right] G(\vec{x}_2, \vec{x}_1, z) = \delta(\vec{x}_2 - \vec{x}_1) \tag{36}$$

For the purpose of calculations, it is useful to take advantage of the analyticity properties of the Green's function in the complex z plane. The Green's function is an analytic function of z, except for the bound-state poles and branch cuts corresponding to the condition $\text{Re}((1 - z^2)^{\frac{1}{2}}) > 0$, as indicated in Fig. 9. In that figure the poles are denoted by filled circles and the cuts are denoted by heavy lines.

The photon propagation function, depicted as a wavy line in the Feynman diagrams, corresponds to the contraction of two photon-field vector potential operators as in free-particle QED. It is given by

$$\langle 0 | T[A_\mu(x_2)A_\nu(x_1)] | 0 \rangle = g_{\mu\nu} D_F(x_2 - x_1) \tag{37}$$

where

$$D_F(x_2 - x_1) = -\frac{i}{(2\pi)^4} \int d^4q \, \frac{e^{-iq \cdot (x_2 - x_1)}}{q^2 + i\delta}$$

$$= \frac{1}{2\pi i} \int_{-\infty}^{\infty} dq_0 \, H(\vec{x}_2 - \vec{x}_1, q_0) e^{-iq_0(t_2 - t_1)} \tag{38}$$

In (38), the photon Green's function $H(\vec{x}_2 - \vec{x}_1, q_0)$, that is the analog of the Dirac Green's function in the expression for the electron-positron propagator, is given by

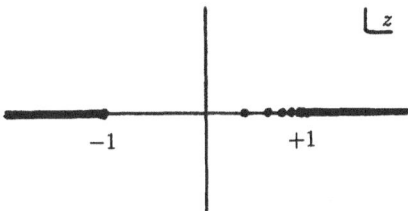

Figure 9. Singularities of the Dirac Green's function

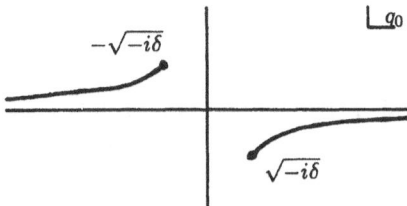

Figure 10. Singularities of the photon Green's function

$$H(\vec{x}_2 - \vec{x}_1, q_0) = -\frac{e^{-bx_{21}}}{4\pi x_{21}}$$

$$x_{21} = |\vec{x}_2 - \vec{x}_1| \; ; \; b = -i(q_0^2 + i\delta)^{\frac{1}{2}} \; , \; \text{Re}(b) > 0$$

(39)

In this case, it is an elementary function calculated by evaluating the three-vector integration in the first line of Eq. (38). The analytic properties of the photon Green's function are depicted in Fig. 10. It is an analytic function of q_0 except for the branch points of b defined in (39). The curved lines are branch cuts that insure that the condition $\text{Re}(b) > 0$ is valid in the cut q_0 plane. The curve is the boundary $\text{Re}(b) = 0$ or $\text{Re}(q_0)\text{Im}(q_0) = -\delta/2$.

APPLICATIONS OF BOUND STATE QED

The formalism presented so far determines energy levels for the general case. In the following, specific applications are examined.

Simple example: atom in an external potential

The interaction of an atom with an external potential is formulated by replacing the vector potential of the quantized electromagnetic field by the external-field vector potential in the interaction Hamiltonian. This example is sufficiently transparent

121

Figure 11. Lowest-order Feynman diagrams for the
interaction of a bound electron with an external field

that all of the steps in the evaluation of the perturbation theory expression can be displayed explicitly.

In terms of the electrostatic field $\phi(\vec{x})$, the vector potential is

$$A^e_\mu(x) = \delta_{\mu 0}\phi(\vec{x}) \tag{40}$$

This gives the interaction Hamiltonian

$$H_I(x) = -\frac{e}{2}\left[\bar{\psi}(x)\gamma^0, \psi(x)\right]\phi(\vec{x}) = \tfrac{1}{2}V(\vec{x})\left[\bar{\psi}(x)\gamma^0, \psi(x)\right] \tag{41}$$

where $V(\vec{x})$ is the potential energy of the electron. The first order S-matrix is

$$S^{(1)}_\epsilon = -i\int d^4x\, e^{-\epsilon|t|}T\left[H_I(x)\right] \tag{42}$$

An elementary application of Wick's theorem yields

$$
\begin{aligned}
T\left[H_I(x)\right] &= V(\vec{x})T\left[\bar{\psi}(x)\gamma^0\psi(x)\right] \\
&= V(\vec{x})\left[:\bar{\psi}(x)\gamma^0\psi(x): + \bar{\psi}(x)\gamma^0\psi(x)\right]
\end{aligned} \tag{43}
$$

The Feynman diagrams corresponding to the terms in (43) are shown in Fig. 11 in which the $\times$ denotes interaction with the external field. The connected part $\langle S^{(1)}_\epsilon\rangle_c$ contains only the first term in (43). In the following, the expressions will be specialized to the case in which the bound state is comprised only of electrons, which leads to a considerable saving in writing. The evaluation proceeds with an elementary integration over the time t

$$
\begin{aligned}
\langle S^{(1)}_\epsilon\rangle_c &= -i\int d^4x\, e^{-\epsilon|t|}V(\vec{x})\left\langle \sum_{nm} a^\dagger_n\phi^\dagger_n(x)a_m\phi_m(x)\right\rangle \\
&= -i\int d^4x\, e^{-\epsilon|t|}\sum_{nm} e^{i(E_n-E_m)t}\phi^\dagger_n(\vec{x})V(\vec{x})\phi_m(\vec{x})\langle a^\dagger_n a_m\rangle \\
&= -i\sum_{nm}\frac{2\epsilon}{(E_n-E_m)^2+\epsilon^2}\int d\vec{x}\,\phi^\dagger_n(\vec{x})V(\vec{x})\phi_m(\vec{x})\langle a^\dagger_n a_m\rangle \\
&= -\frac{2i}{\epsilon}\sum_{nm}\delta(E_n, E_m)V_{nm}\langle a^\dagger_n a_m\rangle + \mathcal{O}(\epsilon)
\end{aligned} \tag{44}
$$

where

$$\delta(E_n, E_m) = \begin{cases} 1 & \text{if}\quad E_n = E_m \\ 0 & \text{if}\quad E_n \neq E_m \end{cases} \tag{45}$$

and where V_{nm} is the matrix element of $V(\vec{x})$. The level shift is then given by

$$E^{(1)} = \lim_{\epsilon \to 0} \frac{1}{2} i\epsilon \langle S^{(1)}_\epsilon \rangle_c = \sum_{nm} \delta(E_n, E_m) V_{nm} \langle a^\dagger_n a_m \rangle \tag{46}$$

For one electron in the 1S state, the vector to which the expectation value refers is just

$$|1S\rangle = a^\dagger_{1S} |0\rangle \tag{47}$$

so the expectation value of the product of annihilation and creation operators simply yields

$$\langle a^\dagger_n a_m \rangle = \delta_{n,1S} \delta_{m,1S} \tag{48}$$

which gives rise to the standard first-order perturbation theory result for the energy-level shift

$$E^{(1)} = V_{1S,1S} \tag{49}$$

as expected.

Energy levels of one- and two-electron atoms

The principal application of QED in this review is the evaluation of energy levels of one- and two-electron atoms. Only the interaction of bound electrons with the virtual radiation field will be considered. In this case, the energy levels are in the form of a series of even powers of the coupling with electromagnetic radiation since only virtual photons, each of which enters with two powers of the coupling constant, are involved

$$E = E^{(0)} + E^{(2)} + E^{(4)} + \dots \tag{50}$$

The zeroth-order energy in this formulation is just the Dirac eigenvalue E_{nj} summed over the electrons in the unperturbed state. In particular

$$E^{(0)} = \langle H_0 \rangle \tag{51}$$

where the unperturbed states and the action of H_0 are explicitly given by

$$\begin{aligned} |nljm\rangle &= a^\dagger_{nljm} |0\rangle \\ H_0 |nljm\rangle &= E_{nj} |nljm\rangle \end{aligned} \tag{52}$$

for one-electron states, and by

$$|nljn'l'j'JM\rangle = \sum_{mm'} \langle jmj'm'|jj'JM\rangle a^\dagger_{nljm} a^\dagger_{n'l'j'm'} |0\rangle \tag{53}$$

$$H_0 |nljn'l'j'JM\rangle = (E_{nj} + E_{n'j'}) |nljn'l'j'JM\rangle$$

for two-electron states. Two-electron states are summed with vector addition coefficient weights to produce states of definite angular momentum. States that are sums over nondegenerate products of creation operators are not considered here.

Perturbation expansion

The first correction $E^{(2)}$, of order α, is given by

$$E^{(2)} = \lim_{\epsilon \to 0} \frac{i\epsilon}{2} \left[\langle S^{(1)}_\epsilon \rangle_c + 2 \langle S^{(2)}_\epsilon \rangle_c \right] \Big|_\alpha \tag{54}$$

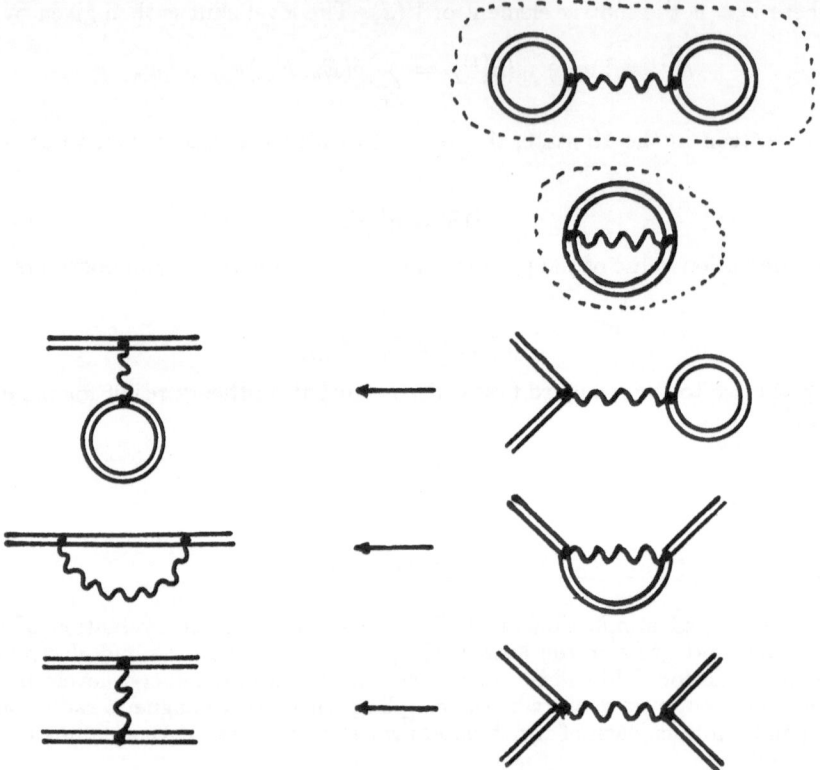

Figure 12. Feynman diagrams corresponding to
various contractions of the second-order correction

Note that the superscripts on E and on S have different meaning. The superscript on E refers to the order in e, where $\alpha = e^2/(4\pi)$, whereas the superscript on S denotes the order in perturbation theory. The orders are mixed, because the current is of order e and the mass renormalization term is of order e^2 in the interaction Hamiltonian (9). The α at the end of the right-hand-side in (54) indicates that terms of order higher than α are discarded.

The term $S_\epsilon^{(1)}$ in Eq. (54) arises from the mass renormalization term in (9), and its evaluation is similar to the evaluation of the external field perturbation discussed earlier

$$
\begin{aligned}
\langle S_\epsilon^{(1)} \rangle_c &= i \int d^4x \; e^{-\epsilon|t|} \langle \delta M(x) \rangle_c \\
&= \frac{2i}{\epsilon} \delta m \sum_{nm} \int d\vec{x} \; \phi_n^\dagger(\vec{x}) \gamma^0 \phi_m(\vec{x}) \delta(E_n, E_m) \langle a_n^\dagger a_m \rangle + \mathcal{O}(\epsilon)
\end{aligned}
\tag{55}
$$

The terms in $S_\epsilon^{(2)}$ that correspond to two vertices are expanded by Wick's theorem schematically in Fig. 12. In that figure the two vertices are shown with zero, one, and two contractions of the electron field operators on the left. The diagrams on the right are simply more conventional ways of showing the corresponding terms. The bottom two diagrams, circled by dotted lines, are disconnected and do not appear in the final expression. The S-matrix element corresponding to the connected diagrams in Fig. 12 is

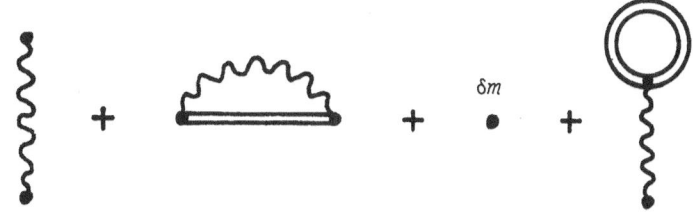

Figure 13. Operator Feynman diagrams for the terms in Eq. (57)

$$\langle S_\epsilon^{(2)} \rangle_c = -e^2 \int d^4x_2 \int d^4x_1 \; e^{-\epsilon|t_2|} e^{-\epsilon|t_1|} D_F(x_2 - x_1)$$

$$\times \left\{ \tfrac{1}{2} \sum_{nm} \bar{\phi}_n(x_2)\gamma_\mu \phi_m(x_2) \sum_{kl} \bar{\phi}_k(x_1)\gamma^\mu \phi_l(x_1) \langle a_n^\dagger a_k^\dagger a_l a_m \rangle \right.$$

$$+ \sum_{nm} \bar{\phi}_n(x_2)\gamma_\mu S_F(x_2, x_1)\gamma^\mu \phi_m(x_1) \langle a_n^\dagger a_m \rangle$$

(56)

$$\left. -Tr\left[\gamma_\mu S_F(x_2, x_2)\right] \sum_{nm} \bar{\phi}_n(x_1)\gamma^\mu \phi_m(x_1) \langle a_n^\dagger a_m \rangle \right\} + \mathcal{O}(\alpha^2)$$

Evaluation of the integrals over t_2 and t_1 and substitution into Eq. (54) yields an expression of the general form[6]

$$E^{(2)} = \sum_{nklm} B_{nklm} \delta(E_n + E_k, E_l + E_m) \langle a_n^\dagger a_k^\dagger a_l a_m \rangle$$

$$+ \sum_{nm} \Sigma_{nm} \; \delta(E_n, E_m)\langle a_n^\dagger a_m \rangle$$

(57)

$$+ \sum_{nm} U_{nm} \; \delta(E_n, E_m)\langle a_n^\dagger a_m \rangle$$

The operator diagrams corresponding the these corrections are shown in Fig. 13. The individual terms in this expression are examined in more detail in the following sections.

ONE-ELECTRON ATOMS

In a one-electron atom, the main QED effects are the self-energy and vacuum polarization. The two-particle exchanged photon operator in Eq. (57) vanishes.

Self-energy

In a one-electron state ϕ_0, the self energy operator reduces to

$$E_{SE}^{(2)} = \sum_{nm} \Sigma_{nm} \delta(E_n, E_m)\langle a_n^\dagger a_m \rangle = \Sigma_{00}$$

(58)

or

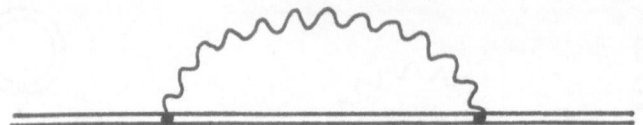

Figure 14. Feynman diagram for the second-order
self energy correction in a one-electron atom

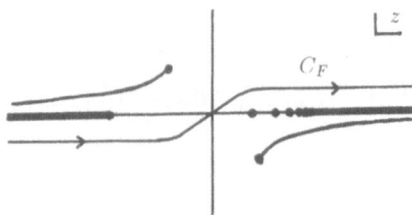

Figure 15. The contour of integration C_F and the singularities
of the integrand in the complex z plane in Eq. (59)

$$E_{SE}^{(2)} = -\frac{i\alpha}{2\pi} \int d\vec{x}_2 \int d\vec{x}_1 \phi_0^\dagger(\vec{x}_2)\alpha_\mu \int_{C_F} dz\, G(\vec{x}_2,\vec{x}_1,z)\alpha^\mu \phi_0(\vec{x}_1)\frac{e^{-bx_{21}}}{x_{21}}$$
$$- \delta m \int d\vec{x}\, \phi_0^\dagger(\vec{x})\beta\phi_0(\vec{x})$$

(59)

where

$$b = -i\left[(E_0 - z)^2 + i\delta\right]^{\frac{1}{2}} ; \qquad \text{Re}(b) > 0 \tag{60}$$

with the corresponding Feynman diagram shown in Fig. 14. The expression in Eq.
(59) follows from the appropriate terms in (56) integrated over t_2, t_1, and k_0, with
the aid of the definitions in (35), (38), and (39), in the limit $\epsilon \to 0$. The contour C_F
and the singularities of the integrand as a function of z in Eq. (59) are shown in Fig.
15.

Regularization

The integration over z in (59) is exponentially damped at large $|z|$, if $x_{21} \neq 0$,
however the integral is not convergent due to the contribution near $x_{21} = 0$. To make
the integral finite, a regularization scheme, such as dimensional regularization, Pauli-
Villars, etc., is needed. In this context, the Pauli-Villars method is implemented by
making the replacement[7]

$$\frac{1}{q^2 + i\delta} \to \frac{1}{q^2 + i\delta} - \frac{1}{q^2 - \Lambda^2 + i\delta} \tag{61}$$

in the integrand in the expression for the photon propagator in Eq. (38). In coordinate
space, this corresponds to the replacement

$$\frac{e^{-bx_{21}}}{x_{21}} \rightarrow \frac{e^{-bx_{21}}}{x_{21}} - \frac{e^{-b'x_{21}}}{x_{21}} \tag{62}$$

where

$$b' = -i\left[(E_0 - z)^2 - \Lambda^2 + i\delta\right]^{\frac{1}{2}}; \qquad \mathrm{Re}(b') > 0 \tag{63}$$

At the same time, the mass renormalization term δm is replace by the corresponding term calculated with the regularized photon propagator

$$\delta m \rightarrow \delta m(\Lambda) = \frac{\alpha}{\pi}\left(\frac{3}{4}\ln\Lambda^2 + \frac{3}{8}\right) \tag{64}$$

The full regularized expression is then

$$
\begin{aligned}
E_{SE}^{(2)} = \lim_{\Lambda\to\infty} \Bigg\{ &-\frac{i\alpha}{2\pi}\int d\vec{x}_2 \int d\vec{x}_1\, \phi_0^\dagger(\vec{x}_2)\alpha_\mu \int_{C_F} dz\, G(\vec{x}_2, \vec{x}_1, z)\alpha^\mu \phi_0(\vec{x}_1) \\
&\times \left[\frac{e^{-bx_{21}}}{x_{21}} - \frac{e^{-b'x_{21}}}{x_{21}}\right] - \delta m(\Lambda)\int d\vec{x}\, \phi_0^\dagger(\vec{x})\beta\phi_0(\vec{x}) \Bigg\}
\end{aligned}
\tag{65}
$$

which yields the finite physical result.

To evaluate this expression numerically, it is useful to deal with the singular terms separately. This can be done by isolating a piece of the expression that is simple to evaluate and contains the singular terms. Singular terms arise from the large $|z|$ region of the integration over z, which can be isolated in the first few terms in the expansion of the Green's function in powers of the external potential. In particular, in terms of the resolvent operator $G(z)$, Dirac Hamiltonian H, free Dirac Hamiltonian H_0, and the external potential V, that are related to the coordinate space expressions by

$$G(\vec{x}_2, \vec{x}_1, z) = \langle \vec{x}_2 | G(z) | \vec{x}_1\rangle \tag{66}$$

etc., the resolvent can be expanded as

$$
\begin{aligned}
G(z) = \frac{1}{H - z} &= \frac{1}{H_0 + V - z} = \frac{1}{H_0 - z} - \frac{1}{H_0 - z}V\frac{1}{H_0 - z} + \cdots \\
&= \frac{1}{H_0 - z} - V\frac{1}{(H_0 - z)^2} + \cdots
\end{aligned}
\tag{67}
$$

The terms on the second line of (67) give the leading behavior of $G(z)$ for large $|z|$, i.e., the definition

$$G_A(z) = \frac{1}{H_0 - z} - V\frac{1}{(H_0 - z)^2} \tag{68}$$

yields

$$G(z) = G_A(z) + \mathcal{O}(z^{-3}) \tag{69}$$

The asymptotic form G_A is written in terms of the free Green's function, so it is relatively simple to evaluate. The numerical evaluation in the region of large $|z|$ is thus implemented by making the separation[8]

$$G(z) = G_A(z) + [G(z) - G_A(z)] \tag{70}$$

for the Green's function in Eq. (65), evaluating the term corresponding to G_A analytically, and evaluating the finite remainder numerically. Some details of the numerical evaluation are discussed in the next section.

Numerical evaluation

Numerical calculations with the Dirac Green's function are facilitated by introducing radial Green's functions in analogy with the familiar two-component radial wave functions defined by

$$\phi_n(\vec{x}) = \begin{bmatrix} f_1(x)\chi_\kappa^\mu(\hat{x}) \\ if_2(x)\chi_{-\kappa}^\mu(\hat{x}) \end{bmatrix} \tag{71}$$

for solutions ϕ_n that are eigenfunctions of the Dirac spin angular momentum operator with eigenvalue $-\kappa$

$$\beta(\vec{\sigma}\cdot\vec{L}+1)\phi_n(\vec{x}) = -\kappa\phi_n(\vec{x}) \tag{72}$$

and of the z-component of the total angular momentum operator J_z with eigenvalue μ

$$(L_z + \tfrac{1}{2}\sigma_z)\phi_n(\vec{x}) = \mu\phi_n(\vec{x}) \tag{73}$$

Substitution of these functions in the eigenfunction expansion of the Dirac Green's function

$$G(\vec{x}_2, \vec{x}_1, z) = \sum_n \frac{\phi_n(\vec{x}_2)\phi_n^\dagger(\vec{x}_1)}{E_n - z} \tag{74}$$

yields

$$G(\vec{x}_2, \vec{x}_1, z)$$
$$= \sum_{n\kappa\mu} \frac{1}{E_{n,\kappa} - z} \begin{bmatrix} f_1(x_2)\chi_\kappa^\mu(\hat{x}_2)f_1(x_1)\chi_\kappa^{\mu\dagger}(\hat{x}_1) & -if_1(x_2)\chi_\kappa^\mu(\hat{x}_2)f_2(x_1)\chi_{-\kappa}^{\mu\dagger}(\hat{x}_1) \\ if_2(x_2)\chi_{-\kappa}^\mu(\hat{x}_2)f_1(x_1)\chi_\kappa^{\mu\dagger}(\hat{x}_1) & f_2(x_2)\chi_{-\kappa}^\mu(\hat{x}_2)f_2(x_1)\chi_{-\kappa}^{\mu\dagger}(\hat{x}_1) \end{bmatrix} \tag{75}$$

The radial Green's functions are simply defined as the radial portion of this expression, i.e.,

$$G_\kappa^{ij}(x_2, x_1, z) = \sum_n \frac{1}{E_{n,\kappa} - z} f_i(x_2)f_j(x_1) \tag{76}$$

and, in analogy with the full Green's function equation

$$\left[-i\vec{\alpha}\cdot\vec{\nabla}_2 + V(\vec{x}_2) + \beta - z \right] G(\vec{x}_2, \vec{x}_1, z) = \delta(\vec{x}_2 - \vec{x}_1) \tag{36}$$

the radial Green's functions satisfy the 2×2 matrix radial Green's function equation

$$\begin{bmatrix} V(x_2)+1-z & -\frac{1}{x_2}\frac{d}{dx_2}x_2 + \frac{\kappa}{x_2} \\ \frac{1}{x_2}\frac{d}{dx_2}x_2 + \frac{\kappa}{x_2} & V(x_2)-1-z \end{bmatrix} \begin{bmatrix} G_\kappa^{11} & G_\kappa^{12} \\ G_\kappa^{21} & G_\kappa^{22} \end{bmatrix} = I\frac{1}{x_2 x_1}\delta(x_2 - x_1) \tag{77}$$

Substitution of the right-hand-side of Eq. (75) for G in (65) yields an expression in which the integration over angles can be carried out completely with the result expressed in terms of radial Green's functions and Bessel functions. The finite remainder term, corresponding to the second term in Eq. (70) and denoted with a B, appears in this form as

$$E_{SE,B}^{(2)} = -\frac{i\alpha}{2\pi} \int_{C_F} dz \int_0^\infty dx_2 x_2^2 \int_0^\infty dx_1 x_1^2$$

$$\times \sum_\kappa \sum_{ij} f_i(x_2) \left[G_\kappa^{ij}(x_2, x_1, z) - G_{A,\kappa}^{ij}(x_2, x_1, z) \right] f_j(x_1) A_\kappa(x_2, x_1) \qquad (78)$$

+ similar terms.

where the summation over κ runs over all nonzero integers, and the summation over i and j runs over 1,2. For S-states, and for $\kappa > 0$, for example, the function A is

$$A_\kappa(x_2, x_1) = -\kappa b j_\kappa(ibx_<)h_\kappa^{(1)}(ibx_>) \qquad (79)$$

To numerically evaluate this expression, it is useful to take advantage of the fact that solutions of Eq. (77) may be constructed from appropriately normalized two-component solutions $F_<$ and $F_>$ of the homogeneous radial equation, regular at $x = 0$ and $x = \infty$, respectively, by writing[9]

$$G_\kappa(x_2, x_1, z) = \theta(x_2 - x_1) F_>(x_2) F_<^T(x_1) + \theta(x_1 - x_2) F_<(x_2) F_>^T(x_1) \qquad (80)$$

In general, the solutions F can be calculated by various methods, including direct numerical solution of the differential equation.

In the case of an external Coulomb potential, the solutions are known functions given by

$$F_<(x) = \begin{bmatrix} \frac{\sqrt{1+z}}{2ax^{3/2}} \left[(\lambda - \nu) M_{\nu-\frac{1}{2},\lambda}(2ax) - \left(\kappa - \frac{\gamma}{a}\right) M_{\nu+\frac{1}{2},\lambda}(2ax) \right] \\ \frac{\sqrt{1-z}}{2ax^{3/2}} \left[(\lambda - \nu) M_{\nu-\frac{1}{2},\lambda}(2ax) + \left(\kappa - \frac{\gamma}{a}\right) M_{\nu+\frac{1}{2},\lambda}(2ax) \right] \end{bmatrix}$$

$$F_>(x) = \frac{\Gamma(\lambda - \nu)}{\Gamma(1 + 2\lambda)} \begin{bmatrix} \frac{\sqrt{1+z}}{2ax^{3/2}} \left[\left(\kappa + \frac{\gamma}{a}\right) W_{\nu-\frac{1}{2},\lambda}(2ax) + W_{\nu+\frac{1}{2},\lambda}(2ax) \right] \\ \frac{\sqrt{1-z}}{2ax^{3/2}} \left[\left(\kappa + \frac{\gamma}{a}\right) W_{\nu-\frac{1}{2},\lambda}(2ax) - W_{\nu+\frac{1}{2},\lambda}(2ax) \right] \end{bmatrix}$$

$$(81)$$

where M and W are the Whittaker functions, and where $a = (1 - z^2)^{\frac{1}{2}}$, with $\text{Re}(a) > 0$, $\gamma = Z\alpha$, $\nu = \gamma z/a$, and $\lambda = (\kappa^2 - \gamma^2)^{\frac{1}{2}}$. In this case, numerical evaluation of the functions F can be based on power series and asymptotic expansions of the Whittaker functions. For example, for a wide range of the parameters α, β, and x, the function M can be evaluated efficiently by carrying out a numerical summation of the power series

$$M_{\alpha,\beta}(x) = x^{\beta+\frac{1}{2}} e^{-x/2} \sum_{n=0}^\infty T(n) \qquad (82)$$

with the terms $T(n)$ calculated by the simple algorithm

$$T(0) = 1$$

$$T(n+1) = \frac{\left(n - \beta + \frac{1}{2} - \alpha\right) x}{(n + 2\beta + 1)(n + 1)} T(n) \qquad (83)$$

The sum can be evaluated this way to a preassigned accuracy with the aid of the rigorous error bound[10]

$$R_N = \sum_{n=N+1}^{\infty} T(n) < \frac{N+2}{N+2-x} T(N+1) \qquad (84)$$

valid for $N+2 > x$. The remainder of the numerical evaluation consists of summation over κ and numerical integration over x_2, x_1, and z in Eq. (78).

Evaluation of the self energy based on this approach has been carried out for the states with principal quantum number $n = 1, 2$. Work is currently under way on such an evaluation for states with $n > 2$. Results for the 1S, 2S, and preliminary results for the 3S states are shown in Fig. 16. In that figure the numerical results are expressed in terms of a function $F(Z\alpha)$ defined by writing

$$E_{SE}^{(2)} = \frac{\alpha}{\pi} \frac{(Z\alpha)^4}{n^3} F(Z\alpha) m_e c^2 \qquad (85)$$

Calculations for an extensive set of excited states is currently under way.[11]

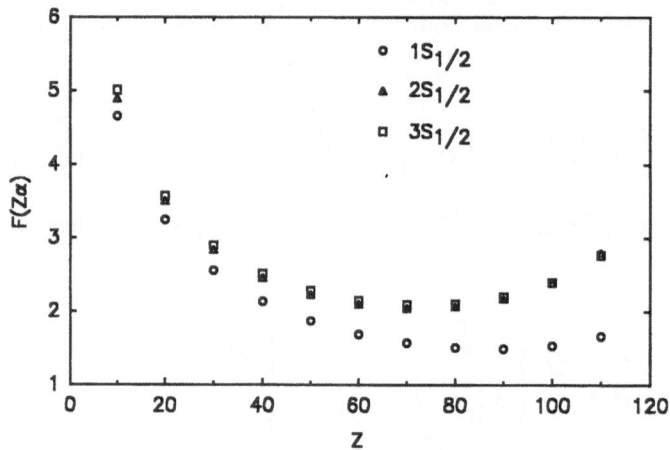

Figure 16. Self energy for S states with $n = 1, 2, 3$

Vacuum polarization

The vacuum polarization in state ϕ_0 is given by

$$E_{VP}^{(2)} = \sum_{nm} U_{nm} \delta(E_n, E_m) \langle a_n^\dagger a_m \rangle = U_{00} \qquad (86)$$

with the corresponding Feynman diagram shown in Fig. 17. From the appropriate term in Eq. (56), it follows that

$$E_{VP}^{(2)} = \int d\vec{x}_1 \phi_n^\dagger(\vec{x}_1) U(\vec{x}_1) \phi_n(\vec{x}_1) \qquad (87)$$

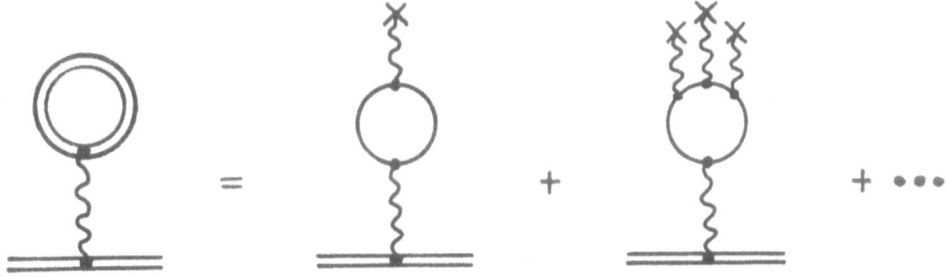

Figure 17. Feynman diagram for the vacuum
polarization correction in a one-electron atom

where the potential energy U is

$$U(\vec{x}_1) = \frac{i\alpha}{2\pi} \int d\vec{x}_2 \frac{1}{|\vec{x}_2 - \vec{x}_1|} \int_{C_F} dz \ Tr \ G(\vec{x}_2, \vec{x}_2, z) \tag{88}$$

An effective charge density ρ can be identified by writing

$$U(\vec{x}_1) = -e \int d\vec{x}_2 \frac{\rho(\vec{x}_2)}{|\vec{x}_2 - \vec{x}_1|} \tag{89}$$

which leads to

$$\rho(\vec{x}) = \frac{e}{2\pi i} \int_{C_F} dz \ Tr \ G(\vec{x}, \vec{x}, z) \tag{90}$$

for the vacuum polarization charge density.

Regularization

In order to isolate the infinite part of the charge distribution and carry out charge renormalization, the vacuum polarization charge density is expanded in powers of the external potential as illustrated in Fig. 17

$$\rho(\vec{x}) = \rho^{(1)}(\vec{x}) + \rho^{(3)}(\vec{x}) + \rho^{(5)}(\vec{x}) + \dots \tag{91}$$

Only odd powers appear as a consequence of Furry's theorem. The charge renormalization is thus isolated in the well-known Uehling term $\rho^{(1)}(\vec{x})$ which is first order in the external potential.[12,13] It is given in unrenormalized form explicitly by

$$\rho^{(1)}(\vec{x}) = \frac{e}{2\pi i} \int_{C_F} dz \ Tr \ G^{(1)}(\vec{x}, \vec{x}, z) \tag{92}$$

where

$$G^{(1)}(\vec{x}, \vec{x}, z) = - \int d\vec{y} \ F(\vec{x}, \vec{y}, z) V(\vec{y}) F(\vec{y}, \vec{x}, z) \tag{93}$$

and where F is the free Green's function

$$F(\vec{x}_2, \vec{x}_1, z) = \left[\left(\frac{a}{x} + \frac{1}{x^2} \right) i\vec{\alpha} \cdot \vec{x} + \beta + z \right] \frac{e^{-ax}}{4\pi x} \tag{94}$$

with $\vec{x} = \vec{x}_2 - \vec{x}_1$ and $x = |\vec{x}|$. Since the infinite charge renormalization is contained in $\rho^{(1)}(\vec{x})$, the difference

Figure 18. Light-by-light scattering graph
in the third-order vacuum polarization

$$\rho(\vec{x}) - \rho^{(1)}(\vec{x}) \tag{95}$$

is finite. However, regularization is still needed in the third-order term due to the spurious gauge dependent piece of the light-by-light scattering graph shown in Fig. 18. The physical third-order. term is given by

$$\rho^{(3)}(\vec{x}) - \tilde{\rho}^{(3)}(\vec{x}) \tag{96}$$

where

$$\tilde{\rho}^{(3)}(\vec{x}) = \lim_{M \to \infty} \rho_M^{(3)}(\vec{x}) = -\frac{e}{3\pi^2}[V(\vec{x})]^3 \tag{97}$$

The density $\rho_M^{(3)}(\vec{x})$ is the third-order charge density calculated with the electron mass in the electron propagator replaced by a hypothetical heavy mass M. This subtraction removes the spurious piece of the third-order charge density. The physical higher-order (third and higher-order) charge density $\rho^{(3+)}(\vec{x})$ is thus given by

$$\rho^{(3+)}(\vec{x}) = \rho(\vec{x}) - \rho^{(1)}(\vec{x}) - \tilde{\rho}^{(3)}(\vec{x}) \tag{98}$$

Numerical evaluation

Evaluation of the vacuum polarization charge density is facilitated by writing it in terms of the radial Green's functions described by Eqs. (75)-(77). Substitution of the expression for the Green's function in (75) into Eq. (90) yields

$$\rho(x) = \frac{e}{2\pi i} \int_{C_F} dz \sum_{\kappa} \frac{|\kappa|}{2\pi} \sum_{i} G_\kappa^{ii}(x, x, z) \tag{99}$$

with the analogous results

$$\rho^{(1)}(x) = \frac{e}{2\pi i} \int_{C_F} dz \sum_{\kappa} \frac{|\kappa|}{2\pi} \sum_{i} G_\kappa^{(1)ii}(x, x, z) \tag{100}$$

and

$$\tilde{\rho}^{(3)}(x) = \frac{e}{2\pi i} \int_{C_F} dz \sum_{\kappa} \frac{|\kappa|}{2\pi} \sum_{i} \tilde{G}_\kappa^{(3)ii}(x, x, z) \tag{101}$$

132

for the other terms in Eq. (98). These terms together yield an expression for the finite physical higher-order charge density in terms of a truncated sum over radial Green's functions

$$\rho^{(3+)}(x) = \frac{e}{2\pi i} \int_{C_F} dz \sum_{\kappa=\pm 1}^{\pm K} \frac{|\kappa|}{2\pi} \sum_i \left[G_\kappa^{ii}(x, x, z) - G_\kappa^{(1)ii}(x, x, z) \right]$$
$$+ R_K \tag{102}$$

where R_K is the remainder after truncating the sum over κ.

At first glance, it appears that the terms in (102) corresponding to $\tilde{\rho}$ are missing, but in fact, they are all zero. This peculiar phenomenon is linked with the spurious nature of the subtraction term. Despite the fact that the terms are all zero, the complete sum over κ is nonzero as indicated by (97).

It is interesting to illustrate this phenomenon with a simple example. Consider the expression S

$$S = \int_0^1 dx \frac{d}{dx} \sum_{n=1}^\infty x^n(1 - x) = \int_0^1 dx \frac{d}{dx} x = 1 \tag{103}$$

If the summation is truncated after any finite number terms, the partial sum vanishes

$$S_N = \int_0^1 dx \frac{d}{dx} \sum_{n=1}^N x^n(1 - x) = \left. \sum_{n=1}^N x^n(1 - x) \right|_0^1 = 0 \tag{104}$$

On the other hand, the remainder after any finite number terms contains the complete value of the sum

$$R_N = \int_0^1 dx \frac{d}{dx} \sum_{n=N+1}^\infty x^n(1 - x) = \int_0^1 dx \frac{d}{dx} x^{N+1} = 1 \tag{105}$$

These results are readily verified with the aid of the identity

$$\sum_{n=N+1}^\infty x^n = \frac{x^{N+1}}{1 - x} \tag{106}$$

The higher-order vacuum polarization was first studied by Wichmann and Kroll.[9] More recently, numerical evaluation of $\rho^{(3+)}(x)$ has been done for an electron in the external field of a finite nucleus. For the model of a shell of charge for the nucleus, radial Green's functions can be constructed with the aid of Eq. (80) from solutions F obtained by matching Coulomb field solutions outside the nucleus to free-electron solutions inside the nucleus. The solutions can be evaluated numerically to give the charge density. Gyulassy has carried out such a calculation for $|\kappa| = 1$,[14] and Soff and Mohr[15] have carried out a similar calculation that includes higher values of $|\kappa|$. Results of the latter calculation are shown in Fig. 19 and Fig. 20. In those figures, the higher-order vacuum polarization density $r^2 \rho^{(3+)}(r)$ and the individual radial components for $|\kappa| = 1, 2, 3, 4, 5$ are shown as a function of the radial coordinate r. The total density is the solid line, and the individual components are the dashed lines where the curve with the largest magnitude corresponds to $|\kappa| = 1$ and the successively smaller magnitude curves correspond to increasing values of $|\kappa|$.

The corresponding level shifts are obtained with the aid of Eqs. (89) and (87). The level shift is conveniently parameterized by writing

$$E_{VP(3+)}^{(2)} = \frac{\alpha}{\pi} \frac{(Z\alpha)^4}{n^3} H_{(3+)}(Z\alpha) m_e c^2 \tag{107}$$

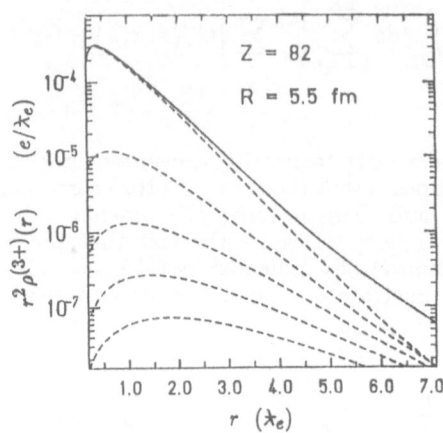

Figure 19.. Vacuum polarization charge density

Table I. $H_{(3+)}(Z\alpha)$

System	$R[fm]$	$1S_{1/2}$	$2S_{1/2}$	$2P_{1/2}$	$2P_{3/2}$
$_{30}Zn$	3.955	0.0020	0.0020	0.0000	0.0000
$_{54}Xe$	4.826	0.0059	0.0064	0.0004	0.0001
$_{82}Pb$	5.500	0.0150	0.0185	0.0035	0.0005
$_{92}U$	5.751	0.0207	0.0272	0.0068	0.0007
$_{100}Fm$	5.886	0.0269	0.0377	0.0118	0.0010
$Z = 170$	7.100	0.518	0.764	3.75	0.017

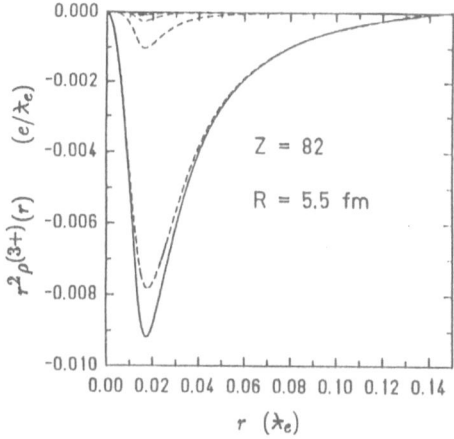

Figure 20. Vacuum polarization charge density at short distances

Results for the level shifts for various nuclei are shown in Table I.[15] In that table, the values assumed for the nuclear radius are denoted by R.

TWO-ELECTRON ATOMS

The QED formalism described above can be applied to atoms with more than one electron with a substantial increase in complexity due to electron interaction effects. The most simple extension is clearly to two-electron atoms, so that case is considered here in some detail. As discussed below, the high-Z two-electron atom provides a framework to study the combined effects of correlation, relativity, and QED effects, and presumably provides a useful prototype for the study of more complex atoms. To touch base with a more traditional framework, the expressions for the energy levels generated by QED are compared to the corresponding expressions generated by the nonrelativistic $1/Z$-expansion of the energy levels predicted by the Schrödinger equation in the next two sections.

Two-electron: nonrelativistic

The two-electron Schrödinger equation

$$\left(-\tfrac{1}{2}\nabla_2^2 - \tfrac{1}{2}\nabla_1^2 - \frac{Z\alpha}{r_2} - \frac{Z\alpha}{r_1} + \frac{\alpha}{r_{21}} - E\right)\phi(\vec{r}_2,\vec{r}_1) = 0 \qquad (108)$$

can be transformed with the substitutions

$$E = (Z\alpha)^2 W$$

$$r_i = \frac{x_i}{Z\alpha} \tag{109}$$

to the form

$$\left(-\tfrac{1}{2}\nabla_2^2 - \tfrac{1}{2}\nabla_1^2 - \frac{1}{x_2} - \frac{1}{x_1} + \frac{1}{Z}\frac{1}{x_{21}} - W \right)\phi'(\vec{x}_2, \vec{x}_1) = 0 \tag{110}$$

in which the eigenvalue W depends on Z but not on α, and the energy level is given by

$$E = W(Z)(Z\alpha)^2 m_e c^2 \tag{111}$$

If the term $\frac{1}{Z}\frac{1}{x_{21}}$ in Eq. (110) is treated as a perturbation, then the perturbation expansion for W yields

$$W(Z) = W_0 + W_1 Z^{-1} + W_2 Z^{-2} + \ldots \tag{112}$$

where W_0, W_1, ... are numerical constants. This expansion has been extended to the study of relativistic corrections.[16,17] The generalization of this formula to include all relativistic and QED effects is provided by the perturbation expansion in QED.

Two-electron: QED

The terms in the perturbation expansion provided by QED, as listed in Eq. (50), are of the form

$$E^{(2n)} = \alpha^n f_{2n}(Z\alpha) m_e c^2 \tag{113}$$

This form for the perturbation expansion follows from the fact that the diagrams corresponding to $E^{(2n)}$ contain n photons which give an explicit factor of α^n. That factor multiplies a function f_{2n} that depends only on $Z\alpha$ through the functional dependence of the propagation functions and wave functions in the Feynman diagram on the strength of the external potential. If we let

$$f_{2n}(Z\alpha) = (Z\alpha)^{2-n} W_n'(Z\alpha) \tag{114}$$

then

$$E = E^{(0)} + E^{(2)} + E^{(4)} + \ldots$$

$$= \left[f_0(Z\alpha) + \alpha f_2(Z\alpha) + \alpha^2 f_4(Z\alpha) + \ldots \right] m_e c^2 \tag{115}$$

$$= \left[W_0'(Z\alpha) + W_1'(Z\alpha)Z^{-1} + W_2'(Z\alpha)Z^{-2} + \ldots \right] (Z\alpha)^2 m_e c^2$$

For the orders studied so far in this expansion (up to $E^{(4)}$), and presumably to all orders, the QED calculation, in the limit of small $Z\alpha$, reproduces the perturbation expansion of the Schrödinger equation as expected.
More precisely,

$$\lim_{Z\alpha \to 0} W_i'(Z\alpha) = W_i \tag{116}$$

Zeroth order

In zeroth order, the energy level is just the sum of the Dirac one-electron energies. In particular

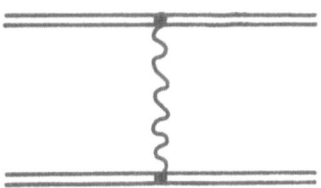

Figure 21. Feynman diagram for one
exchanged photon in a two-electron atom

$$E_D^{(0)} = E_a + E_b \tag{117}$$

for a two-electron state consisting of electrons in one-electron levels a and b. It is convenient numerically to not include the electron rest energy in the energy level so we define

$$E^{(0)} = E_D^{(0)} - 2m_e c^2 = \underbrace{[c_{ab} + d_{ab}(Z\alpha)^2 + \ldots](Z\alpha)^2 m_e c^2}_{f_0(Z\alpha)} \tag{118}$$

Second order

In order α, the exchanged photon graph in Fig. 21 gives the dominant contribution to two-electron energy levels at high Z. This diagram corresponds to the first line in Eq. (57), and is explicitly given by

$$E_{PE}^{(2)} = \alpha \int d\vec{x}_2 \int d\vec{x}_1 \Phi^\dagger(\vec{x}_2, \vec{x}_1) \frac{1}{x_{21}} \alpha_\mu^{(2)} \alpha^{\mu(1)} \Phi(\vec{x}_2, \vec{x}_1)$$

$$- \alpha \int d\vec{x}_2 \int d\vec{x}_1 \Phi^\dagger(\vec{x}_2, \vec{x}_1) \frac{\cos(\eta x_{21})}{x_{21}} \alpha_\mu^{(2)} \alpha^{\mu(1)} \Phi(\vec{x}_1, \vec{x}_2) \tag{119}$$

where $\eta = E_b - E_a$, is the difference between the eigenvalues of the zero-order hydrogenic states, and Φ is the unsymmetrized two-particle Dirac hydrogenic product wave function. Only the real part of the level shift is included in Eq. (119). In the nonrelativistic limit, $i.e.$, as $Z\alpha \to 0$

$$E_{PE}^{(2)} \to \alpha \int d\vec{x}_2 \int d\vec{x}_1 \Phi_{NR}^\dagger(\vec{x}_2, \vec{x}_1) \frac{1}{x_{21}} \Phi_{NR}(\vec{x}_2, \vec{x}_1)$$

$$- \alpha \int d\vec{x}_2 \int d\vec{x}_1 \Phi_{NR}^\dagger(\vec{x}_2, \vec{x}_1) \frac{1}{x_{21}} \Phi_{NR}(\vec{x}_1, \vec{x}_2) \tag{120}$$

where Φ_{NR} is the nonrelativistic hydrogenic wavefunction. The limiting form on the right-hand side of (120) is just the nonrelativistic first-order Coulomb interaction that yields the term W_1 in (112).

Exact values of the photon-exchange level shift $E_{PE}^{(2)}$ can be obtained by elementary numerical integration of the expression in (119).[6] The numerical values are conveniently expressed in terms of a function $P(Z\alpha)$ defined by

$$E_{PE}^{(2)} = \alpha(Z\alpha)P(Z\alpha)m_e c^2 \tag{121}$$

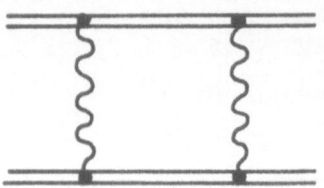 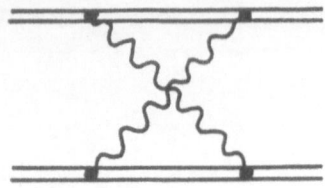

Figure 22. Feynman diagrams for two
exchanged photons in two-electron atoms

It is convenient to obtain a simple formula for the level shift for small $Z\alpha$ by evaluating the expansion of the function P in powers of $Z\alpha$.[18,6] For example, for the level separation $2^3S_1 - 2^3P_0$, the expansion is

$$E_{PE}^{(2)} = \alpha(Z\alpha)\left[\frac{248}{6561} + 0.142833(Z\alpha)^2\right.$$

$$\left. + 0.099668(Z\alpha)^4 + 0.0789(Z\alpha)^6 + \ldots\right]m_ec^2 \tag{122}$$

The remaining second-order graphs, shown in Fig. 13, are one-particle graphs in which only one of the electron lines interacts with the virtual radiation field. The corrections, from the second and third lines of Eq. (57), are the sums of the corresponding corrections for the Dirac hydrogenic orbitals. In particular

$$E_{SE}^{(2)} = \Sigma_{aa} + \Sigma_{bb} \tag{123}$$

and

$$E_{VP}^{(2)} = U_{aa} + U_{bb} \tag{124}$$

or

$$E_{SE}^{(2)} = \frac{\alpha}{\pi}\left[\frac{(Z\alpha)^4}{n_a^3}F_a(Z\alpha) + \frac{(Z\alpha)^4}{n_b^3}F_b(Z\alpha)\right]m_ec^2 \tag{125}$$

and

$$E_{VP}^{(2)} = \frac{\alpha}{\pi}\left[\frac{(Z\alpha)^4}{n_a^3}H_a(Z\alpha) + \frac{(Z\alpha)^4}{n_b^3}H_b(Z\alpha)\right]m_ec^2 \tag{126}$$

All graphs of order α can be combined into a single function of the form

$$E^{(2)} = \alpha f_2(Z\alpha)m_ec^2 = W_1'(Z\alpha)Z^{-1}(Z\alpha)^2 m_ec^2 \tag{127}$$

as discussed above, that has the appropriate nonrelativistic limit.

Fourth order

The largest contribution to the corrections of order α^2 are contained in

$$E^{(4)} = \lim_{\epsilon \to 0}\frac{i\epsilon}{2}\left\{4\langle S_\epsilon^{(4)}\rangle_c - 2\langle S_\epsilon^{(2)}\rangle_c^2 + \ldots\right\} \tag{128}$$

and arises from the Feynman diagrams in Fig. 22 in which two photons are exchanged between the two electron lines. To obtain operator versions of the two-exchanged

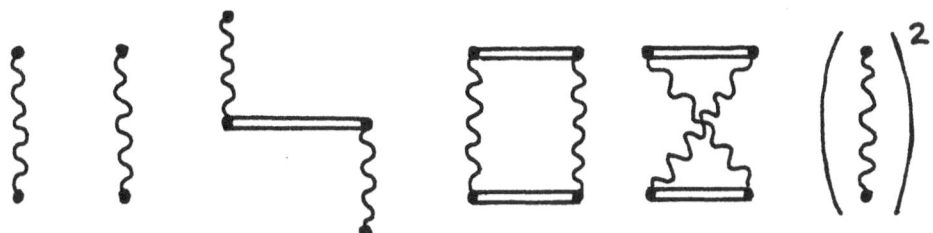

Figure 23. Feynman operator diagrams for the no-, one-, and
two- contraction terms, and the second-order-squared term

photon diagrams, it is necessary to consider a broader class of diagrams in order to cancel terms of order ϵ^{-1} that occur in individual diagrams but not in the total. Singularities arise from intermediate states that are degenerate with the unperturbed state. The necessary diagrams appear in Fig. 23. They consist of no-contraction, one-contraction, two-contraction diagrams, and the squared second-order diagram.

The two-exchanged photon contributions that correspond to the non-degenerate intermediate states are given by

$$\Delta E_{TP}^{(4)} = 4\pi i \alpha^2 \int dz \int d\vec{x}_4 \int d\vec{x}_3 \int d\vec{x}_2 \int d\vec{x}_1 \sum_{\substack{E_{n_4}+E_{n_2}\\ \neq E_{i_2}+E_{i_1}}}$$

$$\times \left\{ H\left(\vec{x}_4 - \vec{x}_2, E_{n_4} - z\right) H(\vec{x}_3 - \vec{x}_1, z - E_{n_3}) \right.$$

$$\times \phi_{n_4}^\dagger(\vec{x}_4)\alpha_\mu \frac{\phi_{i_2}(\vec{x}_4)\phi_{i_2}^\dagger(\vec{x}_3)}{z - E_{i_2}(1 - i\delta)}\alpha_\nu \phi_{n_3}(\vec{x}_3)$$

$$\times \phi_{n_2}^\dagger(\vec{x}_2)\alpha^\mu \frac{\phi_{i_1}(\vec{x}_2)\phi_{i_1}^\dagger(\vec{x}_1)}{E_{n_4} + E_{n_2} - z - E_{i_1}(1 - i\delta)}\alpha^\nu \phi_{n_1}(\vec{x}_1) \qquad (129)$$

$$+ H(\vec{x}_4 - \vec{x}_1, E_{n_4} - z)H(\vec{x}_3 - \vec{x}_2, z - E_{n_3})$$

$$\times \phi_{n_4}^\dagger(\vec{x}_4)\alpha_\mu \frac{\phi_{i_2}(\vec{x}_4)\phi_{i_2}^\dagger(\vec{x}_3)}{z - E_{i_2}(1 - i\delta)}\alpha_\nu \phi_{n_3}(\vec{x}_3)$$

$$\left. \times\phi_{n_2}^\dagger(\vec{x}_2)\alpha^\nu \frac{\phi_{i_1}(\vec{x}_2)\phi_{i_1}^\dagger(\vec{x}_1)}{z + E_{n_1} - E_{n_4} - E_{i_1}(1 - i\delta)}\alpha^\mu \phi_{n_1}(\vec{x}_1) \right\}$$

$$\times \delta(E_{n_4} + E_{n_2}, E_{n_3} + E_{n_1})\langle a_{n_4}^\dagger a_{n_2}^\dagger a_{n_1} a_{n_3}\rangle$$

Evaluation of these terms is facilitated by carrying out a procedure broadly analogous to the self-energy calculation. The contour of integration in the complex z plane is rotated, the Green's functions are expanded as a sum over angular momentum eigenfunctions, and the Green's functions are evaluated with the aid of a finite basis set. The preliminary results of such a calculation for the ground state of two-electron atoms are shown in Table II.[19] The column labeled ΔE gives the total level shift, ΔE_{CC} denotes the shift due to two relativistic Coulomb photons, ΔE_{BrC} gives the

Table II. Contribution of the box and crossed box diagrams
to energy levels of high-Z two-electron ions: units of a.u.

Z	ΔE	ΔE_{CC}	ΔE_{BrC}	ΔE_{QED}
25	−0.17785	−0.16070	−0.01640	−0.00075
50	−0.23957	−0.17301	−0.06098	−0.00558
75	−0.35402	−0.20297	−0.13493	−0.01612
100	−0.57786	−0.27968	−0.24422	−0.04962

Table III. Breakdown of the box and crossed box contributions
by energy of intermediate electron states: units of a.u.

Z	$\Delta E^{(+,+)}$	$\Delta E^{(+,-)} + \Delta E^{(-,+)}$	$\Delta E^{(-,-)}$
25	−0.17657	0.00035	−0.00164
50	−0.23457	0.00436	−0.00936
75	−0.34804	0.01849	−0.02447
100	−0.57834	0.05464	−0.05000

shift due to one transverse and one Coulomb photon, and ΔE_{QED} is the remainder.
The contributions of the various two-electron intermediate states to the two-photon
shift are shown in Table III. In that table, the superscripts on the column headings
indicate the level shift contribution from intermediate state energies that are positive
for both electrons $(+,+)$, positive for one and negative for the other $(+,-)$ and
$(-,+)$, or negative for both $(-,-)$. In both tables, the numbers are preliminary, and
are currently being improved in accuracy.[19] The calculated numbers determine the
dominant part of the function $W_2'(Z\alpha)$ that has the correct nonrelativistic limit W_2.

In addition to these terms, there are non-vanishing contributions from the degenerate intermediate states. The additional terms are the subject of current study.

EXTENSION TO MANY-ELECTRON ATOMS

The bound-interaction picture approach to relativistic atomic structure can be
extended to many-electron atoms in a way that is directly analogous to nonrelativistic
many-body perturbation theory. The formulation described above is correct for many
electrons, but due to the cumulative interactions of the electrons, the Coulomb field
solutions provide a poor zeroth-order approximation. To account for the cumulative
effects of the electrons, the zeroth-order states may be taken as solutions of the Dirac

equation with an effective potential $V(\vec{x}) + \delta V(\vec{x})$

$$\left[-i\vec{\alpha} \cdot \vec{\nabla} + V(\vec{x}) + \delta V(\vec{x}) + \beta - E_n \right] \phi_n(\vec{x}) = 0 \tag{130}$$

The theory is exact, although slowly convergent, without the δV correction in the Dirac equation, so the original result is restored by subtracting a corresponding term from the interaction Hamiltonian

$$H_I(x) = j^\mu(x)A_\mu(x) - j^0(x)\delta A_0(x) - \delta M(x) \tag{131}$$

where

$$\delta V(\vec{x}) = -e\delta A_0(x) \tag{132}$$

This formulation provides a framework for including QED effects in a many-electron calculation in a consistent way. A well-chosen δV can improve the convergence of perturbation theory. Further improvement is possible by summing subclasses of Feynman diagrams to all orders in perturbation theory.

REFERENCES

1. W. H. Furry, Phys. Rev. **81**, 115 (1951).
2. M. Gell-Mann and F. Low, Phys. Rev. **84**, 350 (1951).
3. J. Sucher, Phys. Rev. **107**, 1448 (1957).
4. F. Low, Phys. Rev. **88**, 53 (1952).
5. G. C. Wick, Phys. Rev. **80**, 268 (1950).
6. P. J. Mohr, Phys. Rev. A **32**, 1949 (1985).
7. W. Pauli and F. Villars, Rev. Mod. Phys. **21**, 434 (1949).
8. P. J. Mohr, Ann. Phys. (N.Y.) **88**, 26 (1974).
9. E. H. Wichmann and N. M. Kroll, Phys. Rev. **101**, 843 (1956).
10. P. J. Mohr, Ann. Phys. (N.Y.) **88**, 52 (1974).
11. Y.-K. Kim and P. J. Mohr, private communication (1988).
12. E. A. Uehling, Phys. Rev. **48**, 55 (1935).
13. R. Serber, Phys. Rev. **48**, 49 (1935).
14. M. Gyulassy, Nucl. Phys. **A244**, 497 (1975).
15. G. Soff and P. J. Mohr, Phys. Rev. A **38**, 5066 (1988).
16. D. Layzer and J. Bahcall, Ann. Phys. (N.Y.) **17**, 177 (1962).
17. A. Dalgarno and A. L. Stewart, Proc. Phys. Soc. London **75**, 441 (1960).
18. H. T. Doyle, *Advances in Atomic and Molecular Physics* (Academic, New York, 1969), Vol. 5, p. 337.
19. S. A. Blundell, W. R. Johnson, P. J. Mohr, and J. Sapirstein, private communication (1988).

RADIATIVE TRANSITIONS IN

ONE– AND TWO–ELECTRON IONS

Gordon W. F. Drake and A. van Wijngaarden

Department of Physics
University of Windsor
Windsor, Ontario N9B 3P4, Canada

INTRODUCTION

This series of lectures has a rather general title because it deals with a variety of topics, both theoretical and experimental, which are related to one another. A great deal of progress has been made over the past few years in the precision of measurements for one– and two–electron atoms. A parallel development of new techniques for high precision caculations is opening the way to a wide variety of comparisons between theory and experiment which are sensitive to higher order relativistic and quantum electrodynamic (QED) effects. There are close connections between these lectures, which focus primarily on the low to intermediate range of nuclear charge, and those of Peter Mohr and Berndt Müller, which describe effects in the high nuclear charge and super–critical field regimes.

The first lecture will begin with a review of the basic theory of radiative transitions in order to define the notation and lay the ground work for a discussion of the interference effects which occur in the electric field quenching of hydrogenic ions. I will then describe in some detail the quenching anisotropy method for measuring the Lamb shift, which has recently yielded the most accurate available determination of the Lamb shift in He⁺, thereby providing one of the most sensitive tests of higher order QED corrections.

The second lecture will begin with a brief description of closely related quenching asymmetry measurements which yield the level width of the 2p state, and the relativistic magnetic dipole matrix element for the $2\,^2S_{1/2} - 1\,^2S_{1/2}$ transition. I will then change topics to a discussion of new variational techniques for two–electron atoms. These techniques now make it possible to improve the precision of existing calculations by several orders of magnitude for nonrelativistic energies and relativistic corrections of $O(\alpha^2)$. This improvement is necessary in order to match the accuracy of existing measurements. An important advantage of these techniques is that they can be extended to higher members of the Rydberg series (at least up to n ~ 10) with no serious loss of accuracy. In contrast, standard variational calculations suffer a disasterous loss of accuracy for the more highly excited states.

The third lecture will begin with a summary of the numerous small corrections to the two–electron energy which must be included before a comparison with experiment becomes meaningful. These are finite nuclear mass effects of $O(\mu/M)$ and $O(\mu^2/M^2)$, relativistic corrections of $O(\alpha^2)$, relativistic reduced mass corrections of $O(\alpha^2\mu/M)$ and QED corrections of $O(\alpha^3Z^4)$ and higher. For the QED terms, the two–electron corrections of $O(\alpha^3Z^3)$ referred to as the "screening of the Lamb shift" are included in a nonrelativistic approximation in which the leading term is evaluated exactly, and the relativistic corrections of $O(\alpha^4 Z^4)$ and higher are estimated from the corresponding one–electron terms. An exact calculation of the two–electron relativistic corrections has not yet been done, although progress on this topic is reviewed in Peter Mohr's talk. A comparison with a wide variety of high precision transition frequency measurements yields well–defined discrepancies which can reasonably be accounted for by uncalculated terms of $O(\alpha^4 Z^4)$ and $O(\alpha^3 Z^2)$. Finally, a brief survey will be given of the comparison between theory and experiment for high–Z two–electron ions.

RADIATIVE TRANSITIONS

This section begins with an overview of the decay mechanisms for the low–lying states of one– and two–electron ions. Then the theory of spontaneous transitions is briefly reviewed. This establishes the basic concepts and notation for a more detailed discussion of relativistic magnetic dipole transitions and the quenching radiation asymmetries which allow one to measure the Lamb shift in one–electron ions.

Fig. 1 shows the low–lying states of one–electron ions together with their modes of radiative decay. As is well known, the $2s_{1/2}$ state is metastable because ordinary electric dipole (E1) transitions to the ground state are forbidden by the parity selection rule. For low Z ions, the dominant decay mechanisn is the simultaneous emission of two E1 photons, giving a decay rate of

$$w(2\mathrm{E}1) = 8.2293810 Z^6 \text{ s}^{-1} + \text{relativistic corrections} \qquad (1)$$

(Drake, 1986). However, the relativistic M1 mechanism discussed below increases in proportion to Z^{10} and eventually becomes dominant for $Z \geq 43$.

In the presence of an external electric field, the $2s_{1/2}$ state becomes mixed with the close–lying $2p_{1/2}$ and $2p_{3/2}$ states, making possible E1 and M2 transitions to the

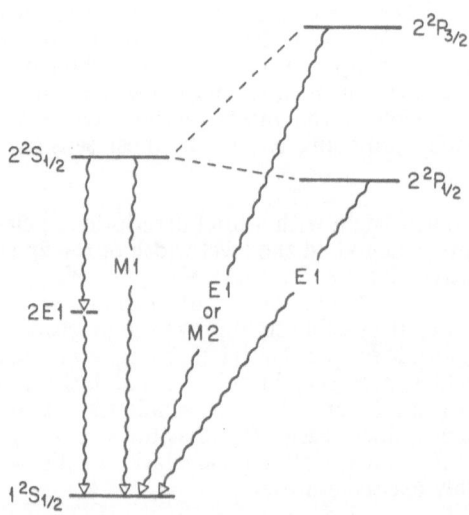

Fig. 1.   Energy level diagram for one–electron ions, showing the radiative decay modes. The dashed lines indicate electric field mixing of the $2s_{1/2}$ state with the $2p_{1/2}$ and $2p_{3/2}$ states, leading to field–induced E1 (electric dipole) and M2 (magnetic quadupole) transitions to the ground state.

ground state. The external field mixing is represented by dashed lines in Fig. 1. The rotational asymmetries discussed below arise from interference effects among the single photon decay channels.

Fig. 2 is a similar diagram of the radiative decay modes for two–electron ions. Here, both the states 1s2s $^1S_0$ and 1s2s $^3S_1$ are metastable. Calculations for the 2E1 decay mode from the 2 $^1S_0$ state have recently been done to improved precision by Drake (1986), and the decay rate has been measured in ions up to Kr^{34+} (Marrus, 1986). The relativistic M1 decay rate from the 2 $^3S_1$ state was first calculated by Drake (1971, 1972a), Beigman and Safranova (1971), and by Feinberg and Sucher (1971). These processes are of considerable astrophysical importance (see, for example, Drake and Robbins, 1972; Blumenthal et al. 1972).

As in the one–electron case, the 2 $^1S_0$ state of helium can be quenched by the application of an electric field due to field–induced mixing with the 2 $^1P_1$ state. However, fields on the order of 100 kV/cm are required because of the large electrostatic splitting between the states. The quench rate has been measured by Petrasso and Ramsey (1972). Their result of $0.926(20)F^2(\text{cm/kV})^2\text{s}^{-1}$ (F is the field strength) agrees with the theoretical value $0.932(1)F^2(\text{cm/kV})^2\text{s}^{-1}$ obtained by Drake (1972b).

The wavy dashed lines in Fig. 2 indicate more exotic decay modes which become increasingly important for the two–electron ions of higher Z. The 2 $^3P_1 - 1\ ^1S_0$ E1 transition is a spin–forbidden process which accurs through mixing between the 2 $^1P_1$ and n $^3P_1$ states due to the Breit interaction (Drake and Dalgarno, 1969; Drake, 1976). There is also a small contribution from doubly excited $npn'p\ ^3P_0^e$ intermediate states (Drake, 1976). The M2 decay rate from the 2 $^3P_2$ state is discussed by Drake (1969). Since it increases in proportion to Z^8 along the isoelectronic sequence, while the competing 2 $^3P_2 - 2\ ^3S_1$ E1 rate only increases as Z, the M2 process becomes dominant for ions beyond Cl^{15+}.

The very unusual E1M1 two–photon decay mode from the 2 $^3P_0$ state is discussed by Drake (1985). Even though it is strongly suppressed, the rate increases in pro-

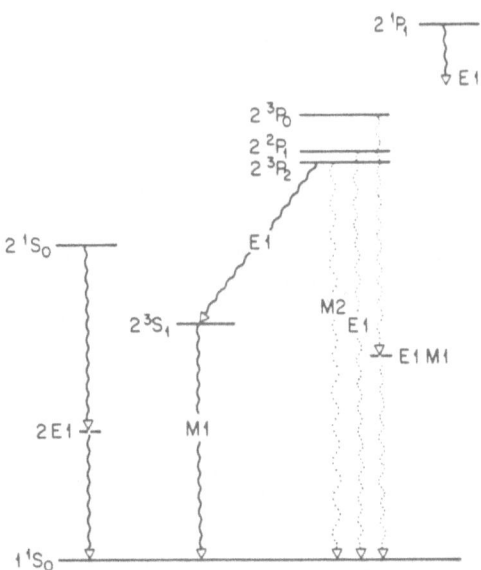

Fig. 2.   Energy level diagram for two–electron ions, showing the radiative decay modes for the low–lying states. The wavy dashed lines indicate inhibited transitions which become important for high–Z ions.

portion to Z^{12}, reaching about 46% of the allowed E1 rate in U^{90+}. The Munger and Gould (1986) measurement of the $2\,^3S_1 - 2\,^3P_2$ energy splitting in U^{90+} relies on the theoretical value for the E1M1 decay rate.

Theory of Spontaneous Transitions

Fig. 3 shows the basic Feynman diagram for the spontaneous emission of a single photon. The corresponding transition rate is given by Fermi's Golden Rule

$$w = \frac{2\pi}{\hbar} |\langle f | V_{int} | i \rangle|^2 \rho_f \tag{2}$$

where
$$V_{int} = \text{interaction energy operator}$$
$$\rho_f = \text{no. of final states per unit energy interval.}$$
The basic parameters which characterize the emitted photon are
$$\omega = \text{photon frequency}$$
$$\hat{e} = \text{photon polarization vector}$$
$$\vec{k} = \text{photon propagation vector } (|\vec{k}| = \omega/c).$$
Then

$$\rho_f = \frac{\mathcal{V} k^2 d\Omega}{(2\pi)^3 \hbar c} \tag{3}$$

is the number of photon states of polarization $\hat{e}$ per unit energy and solid angle in normalization volume $\mathcal{V}$. The interaction energy operator is

$$V_{int} = e\vec{\alpha} \cdot \vec{A}^* \ . \tag{4}$$

If the photon vector potential $\vec{A}$ is normalized to a field energy of $\hbar\omega$ per unit volume, then

$$\vec{A} = \frac{1}{k} \left[\frac{2\pi\hbar\omega}{\mathcal{V}} \right]^{1/2} \hat{e} e^{i\vec{k}\cdot\vec{r}} \ . \tag{5}$$

Collecting terms, eq. (2) becomes

$$w d\Omega = \left[\frac{e^2 k}{2\pi\hbar} \right] |\langle f | \vec{\alpha} \cdot \hat{e} e^{-i\vec{k}\cdot\vec{r}} | i \rangle|^2 d\Omega \quad \text{per unit time.} \tag{6}$$

In the nonrelativistic limit, $\vec{\alpha} \to \vec{p}/mc$, $e^{-i\vec{k}\cdot\vec{r}} \simeq 1$ and the above becomes the familiar dipole velocity form of the transition rate.

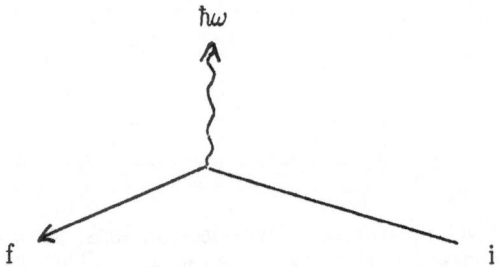

Fig. 3. Feynman diagram showing the basic lowest order process of spontaneous photon emission.

For the $2s_{1/2}$ state, E1 and M2 processes become allowed due to electric field mixing with the $2p_{1/2}$ and $2p_{3/2}$ states. A systematic way of incorporating all higher multipole moments is to use the standard multipole expansion

$$\hat{e}e^{-i\vec{k}\cdot\vec{r}} = \left[\frac{3}{8\pi}\right]^{1/2}\sum_{M}\{e_M\,\mathring{a}^{(1)*}_{1,M} + i[\hat{k}\times\hat{e}]_M\,\mathring{a}^{(0)*}_{1,M}$$

$$+ i\sqrt{10/3}\,[\hat{k},\hat{k}\times\hat{e}]_{2,M}\,\mathring{a}^{(0)*}_{2,M} + \cdots\} \tag{7}$$

where

$$e_{\pm 1} = \mp\frac{1}{\sqrt{2}}(e_x \pm ie_y), \qquad e_0 = e_z$$

and $[a,b]_{2,M}$ denotes the vector coupled product

$$[a,b]_{2,M} = \sum_{m_1,m_2}\langle 11m_1m_2|2M\rangle\,a_{m_1}b_{m_2}\ .$$

The $\mathring{a}^{(\lambda)}_{LM}$ are multipole operators with

$\lambda = 1$ for electric multipoles
$\lambda = 0$ for magnetic multipoles.

In the transverse or Coulomb gauge, they are given by

$$\mathring{a}^{(1)}_{LM} = \left[\frac{L}{2L+1}\right]^{1/2}g_{L+1}(kr)\,\vec{Y}^M_{LL+1}(\hat{r})$$

$$+ \left[\frac{L+1}{2L+1}\right]^{1/2}g_{L-1}(kr)\,\vec{Y}^M_{LL-1}(\hat{r}) \tag{8}$$

$$\mathring{a}^{(0)}_{LM} = g_L(kr)\,\vec{Y}^M_{LL}(\hat{r}) \tag{9}$$

where the $\vec{Y}^M_{Jl}(\hat{r})$ are vector spherical harmonics defined by

$$\vec{Y}^M_{Jl}(\hat{r}) = \sum_{m,q}Y^m_l(\hat{r})\hat{e}_q(l\,1\,m\,q|J\,M) \tag{10}$$

$$g_L(kr) = 4\pi i^L j_L(kr) \tag{11}$$

and

$$j_L(z) = \frac{z^L}{(2L+1)!!}\left[1 - \frac{z^2/2}{1!(2L+3)} + \frac{(z^2/2)^2}{2!(2L+3)(2L+5)} - \cdots\right] \tag{12}$$

is a spherical Bessel function. For low Z atoms, $kr = \omega r/c$ is small and one can make the long wavelength approximation in which only the leading term of eq. (12) is retained. In the nonrelativistic limit, the four component Dirac operators reduce to the equivalent nonrelativistic operators acting on two–component Pauli spinors

$$e\vec{\alpha}\cdot\mathring{a}^{(1)}_{1M} \longrightarrow e\sqrt{2}\,\Phi_{1M}, \tag{13}$$

$$e\vec{\alpha}\cdot\mathring{a}^{(0)}_{1M} \longrightarrow i(\nabla\Phi_{1M})\cdot\left[\frac{e\vec{L}}{mc\sqrt{2}} + \sqrt{2}\,\vec{\mu}\right] \tag{14}$$

and $\quad e\vec{\alpha}\cdot\mathring{a}^{(0)}_{2M} \longrightarrow i(\nabla\Phi_{2M})\cdot\left[\frac{e\vec{L}}{mc\sqrt{6}} + \sqrt{3/2}\,\vec{\mu}\right] \tag{15}$

where $\Phi_{LM} = g_L(kr)Y^M_L(\hat{r})$, $\vec{L} = \vec{r}\times\vec{p}$, and $\vec{\mu} = (e\lambdabar/2)\vec{\sigma}$. Here, $\vec{\mu}$ is the magnetic moment operator and $\lambdabar = \alpha a_0$ is the Compton wavelength.

Relativistic Magnetic Dipole Transitions

For the $2s_{1/2} \to 1s_{1/2}$ transition, only the M1 term $\mathbf{a}_{1M}^{(0)}$ contributes, but even this term vanishes in the nonrelativistic long wavelength approximation since $i\nabla\Phi_{1M} \longrightarrow k(4\pi/3)^{1/2}\hat{e}_M$, and matrix elements are proportional to the overlap integral. However, relativistic corrections of $O(\alpha^2 Z^2)$ and finite wavelength corrections of $O[(\omega r/c)^2]$ give

$$e\vec{\alpha}\cdot\mathbf{a}_{1M}^{(0)} \simeq -k\sqrt{8\pi/3}\, M_{1M} \tag{16}$$

where

$$M_{1M} = \mu_M\left[1 - \frac{2\,p^2}{3m^2c^2} - \frac{1}{6}\left[\frac{\omega r}{c}\right]^2 + \frac{Ze^2}{3mc^2 r}\right] \tag{17}$$

is the effective magnetic moment transition operator acting on nonrelativistic wave functions. For the $2s_{1/2} \to 1s_{1/2}$ transition, the matrix elements are

$$\langle 1s_{1/2,1/2}|M_{1,0}^{*}|2s_{1/2,1/2}\rangle = -\left[\frac{8\alpha^2 Z^2}{81\sqrt{2}}\right]e\lambda \tag{18}$$

$$\langle 1s_{1/2,-1/2}|M_{1,1}^{*}|2s_{1/2,1/2}\rangle = \left[\frac{8\alpha^2 Z^2}{81}\right]e\lambda \;. \tag{19}$$

The transition rate into solid angle $d\Omega$ is then

$$wd\Omega = \left[\frac{k^3}{2\pi\hbar}\right]|M|^2\{|[\hat{k}\times\hat{e}]_0|^2 + 2|[\hat{k}\times\hat{e}]_1|^2\}d\Omega \tag{20}$$

with $M = -\left[\dfrac{8\alpha^2 Z^2}{81\sqrt{2}}\right]e\lambda.$ \hfill (21)

Summing over any two linearly independent polarization vectors $\hat{e}$ perpendicular to $\hat{k}$ results in

$$\sum_{\hat{e}}|[\hat{k}\times\hat{e}]_0|^2 = \sin^2\theta \tag{22}$$

and

$$2\sum_{\hat{e}}|[\hat{k}\times\hat{e}]_1|^2 = 1 + \cos^2\theta \;. \tag{23}$$

Thus $\int d\Omega = \int\int \sin\theta d\theta d\varphi$ just gives a factor of 4π. Using the nonrelativistic value $k = \dfrac{3Z^2\alpha}{8\,a_0}$ yields the final decay rate

$$w(2s_{1/2} \to 1s_{1/2}) = 4k^3|M|^2/\hbar$$

$$= \left[\frac{\alpha^9 Z^{10}}{972}\right]\tau^{-1}$$

$$= 2.496\times10^{-6}\ \mathrm{s}^{-1} \tag{24}$$

for H. ($\tau = 2.41888\times10^{-17}$s is the atomic unit of time.) This is much less than the 2E1 decay rate for H. However, the M1 rate becomes dominant for $Z > 43$. Even for H and He$^+$, the M1 process produces observable interference effects in Stark quenching, as discussed in the following section.

Quenching Radiation Asymmetries

Basic formalism. The starting point for a discussion of quenching radiation asymmetries is the time dependent Schrödinger equation

$$i\hbar \frac{d\vec{a}}{dt} = H(t)\vec{a} \tag{25}$$

where

$$H(t) = E + F(t)V \tag{26}$$

is the time–dependent Hamiltonian matrix in a finite basis set of strongly interacting states, and

$\vec{a}$ = column vector of state amplitudes
E = diagonal matrix of field–free eigenvalues
V = interaction matrix with external field
$F(t)$ = time dependence of external field.

To a very good approximation, the decaying nature of the states can be incorporated by the use of the Bethe–Lamb phenomenological quenching theory in which one replaces each field free eigenvalue E_j by $E_j + i\Gamma_j/2$, where Γ_j is the field–free level width. In the absence of external field perturbations, the field–free occupation probabilities then decay exponentially with time as expected according to

$$|a_j(t)|^2 = |a_j(0)|^2 \exp(-\Gamma_j t/\hbar) \tag{27}$$

However, complex quantum beat patterns can be obtained if an external field is present to mix the decaying states (see, for example, van Wijngaarden et al. 1976; Drake 1976, 1988; Andrä 1974).

For the applications discussed here, assume that the external field $\vec{E}$ is switched on adiabatically; i.e. that $F(t)$ is slowly varying compared with $\hbar/\Delta E$, where ΔE is the energy splitting between the field–free states. For atoms or ions such as H and He$^+$ in fields up to several kV/cm, the only significant mixing is among the manifold of states $2s_{1/2}$, $2p_{1/2}$ and $2p_{3/2}$. For simplicity, we assume here that hyperfine structure is absent. Then the wave function can be expressed as the time–independent linear superposition

$$\Psi(2s_{1/2},m) = a(|\vec{E}|)\Psi_0(2s_{1/2},m)$$

$$+ \sum_{m'} [b^{(1/2)}_{m,m'}\Psi_0(2p_{1/2},m') + b^{(3/2)}_{m,m'}\Psi_0(2p_{3/2},m')] \ . \tag{28}$$

where the matrices $b^{(j)}$ ($j = 1/2, 3/2$) are given by

$$b^{(1/2)} = b^{(1/2)}(|\vec{E}|)\vec{\sigma}\cdot\hat{E} \tag{29}$$

$$b^{(3/2)} = b^{(3/2)}(|\vec{E}|)\begin{bmatrix} -\sqrt{3}\hat{E}_{-1} & \sqrt{2}\hat{E}_0 & -\hat{E}_1 & 0 \\ 0 & -\hat{E}_{-1} & \sqrt{2}\hat{E}_0 & -\sqrt{3}\hat{E}_1 \end{bmatrix} \tag{30}$$

and the $\hat{E}_q$ ($q = 0, \pm1$) are the irreducible tensor components of the unit vector $\hat{E}$ in the electric field direction.

Since the energies of the $2s_{1/2}$ ($m_j = \pm1/2$) states remain degenerate and are independent of the external field orientation, the form of equations (29) and (30) remains valid to all orders of perturbation theory. The only explicit dependence on field strength is through the overall multiplying factors $a(|\vec{E}|)$, $b^{(1/2)}(|\vec{E}|)$ and $b^{(3/2)}(|\vec{E}|)$. To lowest order in the external field, they are given by

$$a = 1 + O(|\vec{E}|^2) \tag{31}$$

$$b^{(1/2)} = \frac{e|\vec{E}|<2p_{1/2}||\vec{r}||2s_{1/2}>}{\sqrt{6}(\mathcal{L} + i\Gamma/2)} \tag{32}$$

$$b^{(3/2)} = \frac{e|\vec{E}|<2p_{3/2}||\vec{r}||2s_{1/2}>}{\sqrt{12}(\mathfrak{F} + i\Gamma/2)} \tag{33}$$

where $\mathcal{L} = E(2s_{1/2}) - E(2p_{1/2})$ is the Lamb shift, $\mathfrak{F} = E(2s_{1/2}) - E(2p_{3/2})$ is the Lamb shift minus the fine structure splitting, and Γ is the level width of the 2p state. The reduced matrix elements in the numerators of equations (32) and (33) are as defined by Edmonds (1960). The numerical values, including the leading relativistic corrections, are

$$\langle 2p_{1/2}||\vec{r}||2s_{1/2}\rangle = \frac{3\sqrt{2}a_0}{Z}(1 - \tfrac{5}{12}\alpha^2 Z^2) \tag{34}$$

$$\langle 2p_{3/2}||\vec{r}||2s_{1/2}\rangle = \frac{-6a_0}{Z}(1 - \tfrac{1}{6}\alpha^2 Z^2) \ . \tag{35}$$

Higher order perturbation corrections due to the external field can readily be calculated analytically, as discussed by Drake (1988a). Alternatively, one could perform an exact diagonalization of the Hamiltonian matrix in the $2s_{1/2}$, $2p_{1/2}$, $2p_{3/2}$ basis set to determine numerical values for the a and $b^{(j)}$ coefficients.

The properties of the quenching radiation are determined by the matrix elements

$$A_{m,m'} = \langle 1s_{1/2},m|\vec{\alpha}\cdot\hat{e}e^{-i\vec{k}\cdot\vec{r}}|2s_{1/2},m'\rangle \tag{36}$$

between the unperturbed $1s_{1/2}$ final state and the perturbed $2s_{1/2}$ initial state as given by eq. (28). The complete 2×2 transition matrix $\mathbf{A}$ with elements $A_{m,m'}$ is

$$\mathbf{A} = V_+ \hat{e}\cdot\hat{E}\mathbf{1} + \vec{\sigma}\cdot[iV_-(\hat{e}\times\hat{E}) + M(\hat{k}\times\hat{e})] \tag{37}$$

where

$$V_+ = V_{1/2} + 2V_{3/2} \tag{38}$$

$$V_- = V_{1/2} - V_{3/2} + M_{3/2} \tag{39}$$

$$M = M_{1/2} + 2i(\hat{k}\cdot\hat{E})M_{3/2}. \tag{40}$$

The V and M coefficients above can be expressed in terms of reduced transition matrix elements according to

$$V_{1/2} = -\frac{b^{(1/2)}}{4\sqrt{\pi}}\langle 1s_{1/2}||\vec{\alpha}\cdot\vec{a}_1^{(1)*}||2p_{1/2}\rangle \tag{41}$$

$$V_{3/2} = -\frac{b^{(3/2)}}{4\sqrt{2\pi}}\langle 1s_{1/2}||\vec{\alpha}\cdot\vec{a}_1^{(1)*}||2p_{3/2}\rangle \tag{42}$$

$$M_{1/2} = \frac{ia}{4\sqrt{\pi}} \langle 1s_{1/2} \| \vec{\alpha} \cdot \vec{a}_1^{(0)*} \| 2s_{1/2} \rangle \tag{43}$$

$$M_{3/2} = -\frac{b^{(3/2)}}{4\sqrt{2\pi/3}} \langle 1s_{1/2} \| \vec{\alpha} \cdot \vec{a}_2^{(1)*} \| 2p_{3/2} \rangle \ . \tag{44}$$

The reduced matrix elements, including the leading relativistic correction of order $\alpha^2 Z^2$, are

$$\langle 1s_{1/2} \| \vec{\alpha} \cdot \vec{a}_1^{(1)*} \| 2p_{1/2} \rangle = \frac{ika_0}{Z} \sqrt{2\pi/3}(2^9/3^5)[1-(\tfrac{11}{96}+\tfrac{3}{2}\ln 2 - \ln 3)\alpha^2 Z^2]$$

$$\langle 1s_{1/2} \| \vec{\alpha} \cdot \vec{a}_1^{(1)*} \| 2p_{3/2} \rangle = \frac{ika_0}{Z} \sqrt{4\pi/3}(2^9/3^5)[1-(\tfrac{11}{48}+\tfrac{5}{4}\ln 2 - \tfrac{3}{4}\ln 3)\alpha^2 Z^2]$$

$$\langle 1s_{1/2} \| \vec{\alpha} \cdot \vec{a}_1^{(0)*} \| 2s_{1/2} \rangle = ka_0 Z^2 \alpha^3 \sqrt{2\pi}(2^4/3^4)[1+0.4193\alpha^2 Z^2]$$

$$\langle 1s_{1/2} \| \vec{\alpha} \cdot \vec{a}_2^{(1)*} \| 2p_{3/2} \rangle = \frac{ik^2 a_0^2 \alpha}{Z} \sqrt{\pi}(2^8/3^5)[1-0.1821\alpha^2 Z^2] \ .$$

The physical sigificance of the terms in eq. (37) is as follows. $V_{1/2}$ and $V_{3/2}$ represent the amplitudes for electric field quenching of the $2s_{1/2}$ state via the admixture of $2p_{1/2}$ and $2p_{3/2}$ intermediate states respectively, with the emission of an E1 photon. $M_{3/2}$ is a small M2 correction. All three of these terms are proportional to the electric field strength through the $b^{(j)}$ coefficients. The combination V_+ comes from transitions with $\Delta m = 0$ in eq. (36), and the E1 part of V_- comes from transitions with $\Delta m = \pm 1$. $M_{1/2}$ is the amplitude for spontaneous M1 transitions as discussed in the previous section. This term is to a first approximation independent of field strength.

In addition to the vectors $\hat{e}$, $\hat{k}$ and $\hat{E}$, the quenching radiation also depends on the electron spin polarization of the $2s_{1/2}$ state. This is specified, in general, by the density matrix

$$\rho = \tfrac{1}{2}(1 + \vec{\sigma} \cdot \vec{P}) \tag{45}$$

where $\vec{P}$ is the polarization vector for the $2s_{1/2}$ state. The decay rate summed over final atomic states and averaged over initial states is then

$$wd\Omega = \frac{e^2 k}{2\pi\hbar} \operatorname{Tr}[\rho A^\dagger A] d\Omega \tag{46}$$

where Tr denotes the trace.

In order to describe the emitted radiation, it is necessary to introduce two orthogonal polarization vectors $\hat{e}_1$ and $\hat{e}_2$ both perpendicular to $\hat{k}$ such that

$$\hat{k} \times \hat{e}_1 = \hat{e}_2, \quad \hat{k} \times \hat{e}_2 = -\hat{e}_1 \ . \tag{47}$$

An arbitrary polarization vector for the general case of elliptical polarization is then given by

$$\hat{e} = \hat{e}_1 \cos\beta + i\hat{e}_2 \sin\beta \ . \tag{48}$$

In particular, $\beta = 0$, $\pi/2$, $\cdots$ corresponds to linearly polarized light, and $\beta = \pi/4$, $3\pi/4$, $\cdots$ corresponds to right or left circularly polarized light. With these conventions, eq. (46) becomes

$$w(\hat{e},\hat{k},\vec{P})d\Omega = \frac{e^2 k}{2\pi\hbar}[I_0 + \vec{P} \cdot \vec{J}_0 + \vec{P} \cdot \vec{J}_1 \sin 2\beta + \hat{E}(\hat{e}_1:\hat{e}_1 - \hat{e}_2:\hat{e}_2)\hat{J}_2 \cos 2\beta]d\Omega \tag{49}$$

where

$$I_0 = \tfrac{1}{2}|V_+|^2[1-(\hat{k}\cdot\hat{E})^2] + \tfrac{1}{2}|V_-|^2[1+(\hat{k}\cdot\hat{E})^2] + 2\mathrm{Im}(M^*V_-)(\hat{k}\cdot\hat{E}) + |M|^2 \tag{50}$$

$$\vec{J}_0 = (\hat{k}\times\hat{E})\{\mathrm{Re}[M^*(V_+ + V_-)] - \mathrm{Im}(V_-^*V_+)(\hat{k}\cdot\hat{E})\} \tag{51}$$

$$\vec{J}_1 = |V_-|^2(\hat{k}\cdot\hat{E})\hat{E} + \mathrm{Re}(V_-^*V_+)\hat{E}\times(\hat{k}\times\hat{E}) - \mathrm{Im}[M^*(V_+ + V_-)]\hat{E}$$
$$\quad + \mathrm{Im}[M^*(V_+ - V_-)](\hat{k}\cdot\hat{E})\hat{k} - |M|^2\hat{k} \tag{52}$$

$$\vec{J}_2 = \tfrac{1}{2}(|V_+|^2 - |V_-|^2)\hat{E} + \mathrm{Re}[M^*(V_+ - V_-)](\vec{P}\times\hat{k}) - \mathrm{Im}(V_-^*V_+)(\vec{P}\times\hat{E}) \; . \tag{53}$$

The dyadic notation $\vec{b}{:}\vec{c}$ in eq. (49) is defined by

$$\vec{a}(\vec{b}{:}\vec{c})\vec{d} = (\vec{a}\cdot\vec{b})(\vec{c}\cdot\vec{d}) \; . \tag{54}$$

Eq. (49) contains a complete description of all the angular and polarization dependent phenomena which can be observed in the Stark quenching of the $2s_{1/2}$ metastable state. Since M is smaller than V_+ by a factor of $O(\alpha^2 Z^2)$, the dominant terms are those containing only V_+ and V_-. The angular distribution of radiation from the main polarization–independent term I_0 is illustrated by the outer elliptical curve in Fig. 4. The anisotropy which this represents is approximately proportional to the Lamb shift

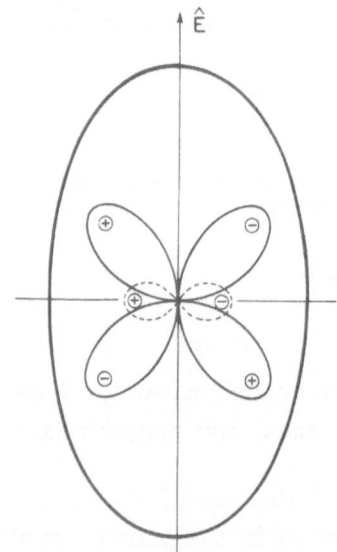

Fig. 4. Polar diagram (not to scale) showing the contributions to angular asymmetries in the field–induced quenching radiation emitted by the $2s_{1/2}$ state. The outer elliptical curve is the main Lamb shift anisotropy for unpolarized atoms. The clover leaf pattern is the E1–E1 damping asymmetry for atoms with a spin polarization vector $\vec{P}$ pointing into the page, and the inner dashed curve is the corresponding E1–M1 interference asymmetry.

as discussed in the following section. The term $-\vec{P}\cdot(\hat{k}\times\hat{E})\text{Im}(V_-^*V_+)(\hat{k}\cdot\hat{E})$ in $\vec{P}\cdot\vec{J}_0$ is sensitive to the imaginary level widths in the denominators of equations (32) and (33). This term produces the clover leaf pattern in Fig. 4. The remaining part $\vec{P}\cdot(\hat{k}\times\hat{E})\text{Re}[M^*(V_++V_-)]$ produces the innermost pattern in Fig. 4. Its measurement determines the 2s → 1s magnetic dipole matrix element M.

The $\sin2\beta$ and $\cos2\beta$ terms in eq. (49) vanish on summing over photon polarizations. but can be observed with a polarization–sensitive detector. Although the $\sin2\beta$ term has not been observed, the $\cos2\beta$ linear polarization term has been measured by Ott, Kauppila and Fite (1970).

As a final comment concerning eq. (49), the term involving $\text{Im}(M^*V_-)(\hat{k}\cdot\hat{E})$ in eq. (50) at first sight appears to violate time reversal invariance because $\hat{k}$ reverses direction for the time–reversed process of absorption. However, as discussed by Mohr (1978), $i\Gamma$ must also be replaced by $-i\Gamma$ in the denominator of V_-, making the product of both terms time–reversal invariant. The same sign reversals apply to similar terms in $\vec{J}_0$, $\vec{J}_1$ and $\vec{J}_2$, together with $\vec{P}\rightarrow-\vec{P}$.

It is instructive to consider the geometrical relationship between the polarization of the emitted radiation and the angular anisotropy. Suppose that the z–axis is an axis of rotational symmetry defined, for example, by an electric field. In the electric dipole approximation where only the V_+ and V_- terms are retained, the radiation intensities depend only on the orientation of $\hat{e}$ relative to the z–axis, and not on the direction of observation. Then the polarization in the x–direction, as illustrated in Fig. 5, is

$$P = \frac{I_y - I_z}{I_y + I_z} \tag{55}$$

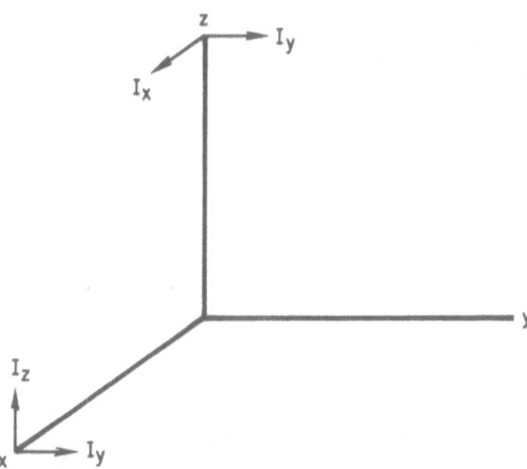

Fig. 5   Illustration of the relationship between the polarization of quenching radiation in the x–direction, and the anisotropy in the total intensity (summed over polarizations) in the x– and y–directions. The arrows indicate photon polarization vectors.

and the rotational anisotropy R, defined in terms of the total intensities emitted in the directions parallel $(I_{\parallel} = I_x + I_y)$ and perpendicular $(I_{\perp} = I_y + I_z)$ to the symmetry axis is

$$R = \frac{I_{\parallel}-I_{\perp}}{I_{\parallel}+I_{\perp}} = \frac{I_x-I_z}{I_x+2I_y+I_z} \ . \tag{56}$$

If the z–axis is a symmetry axis, then $I_x = I_y$, and it follows immediately that

$$P = \frac{2R}{1-R} \ . \tag{57}$$

The above applies to any system for which the electric dipole approximation is valid and an axis of rotational symmetry exists.

   The Lamb shift anisotropy. We now consider in detail the case of an unpolarized $(\vec{P}=0)$ beam of $2s_{1/2}$ ions quenched by an electric field. Only the I_0 term of eq. (49) survives. Among its contributions, the $|M|^2$ term is negligibly small, and the small $\text{Im}(M^* V_-)(\hat{k}\cdot\hat{E})$ term averages to zero on reversal of $\vec{k}$. Using (38) and (39) for V_+ and V_- and neglecting for the moment the $M_{3/2}$ contribution, the remaining terms in I_0 become

$$I_0(\theta) = |V_{1/2}|^2 + \text{Re}(V_{1/2}^* V_{3/2})(1 - 3\cos^2\theta) + \tfrac{1}{2}|V_{3/2}|^2(5 - 3\cos^2\theta) \tag{58}$$

where $\cos\theta = \hat{k}\cdot\hat{E}$ is the angle between the direction of observation and the electric field direction. For convenience in relating the measured anisotropy to the Lamb shift, define a parameter ρ by

$$\rho = V_{3/2}/V_{1/2} \tag{59}$$

so that

$$I_0(\theta) \propto [1 + \text{Re}(\rho)(1 - 3\cos^2\theta) + \tfrac{|\rho|^2}{2}(5 - 3\cos^2\theta)] \tag{60}$$

For sufficiently weak fields, ρ can be expanded in a perturbation series with the leading terms $\rho = \rho^{(0)} + O(|\vec{E}|^2)$, where (see equations 32 and 33)

$$\rho^{(0)} = \frac{\mathcal{L} + i\Gamma_p/2}{\mathfrak{F} + i\Gamma_p/2} \ . \tag{61}$$

The value of the real part of $\rho^{(0)}$ is approximately −0.1 for all low–Z ions. The corresponding zero–order anisotropy is

$$R^{(0)} = \frac{-3\,\text{Re}(\rho^{(0)}) - \tfrac{3}{2}|\rho^{(0)}|^2}{2 - \text{Re}(\rho^{(0)}) + \tfrac{7}{2}|\rho^{(0)}|^2} \ . \tag{62}$$

Since $\rho^{(0)}$ is small, the value of $R^{(0)}$ is approximately $-3\mathcal{L}/(2\mathfrak{F})$, which has a numerical value of ~ 0.1 for all low–Z ions. The higher order finite field strength corrections are of the form

$$R = R^{(0)} + R^{(2)}|\vec{E}|^2 + R^{(4)}|\vec{E}|^4 + \cdots. \tag{63}$$

If there is hyperfine structure, this expansion is only valid for very weak fields because the Stark shifts soon exceed the field–free energy splittings. For stronger fields, one must perform an exact diagonalization of the Hamiltonian in the basis set of strongly interacting hyperfine states (van Wijngaarden and Drake 1978).

Table 1. Input data for the analysis of the He$^+$ anisotropy experiment.

Quantity	Value
$E(2\,^2P_{3/2}) - E(2\,^2P_{1/2})$	175593.55(3) MHz
$\Gamma(2p)$	1.00307×10^{10} s^{-1}
$(\delta R/R^{(0)})_n$	-2.37×10^{-5}
$(\delta R/R^{(0)})_{\rm rel}$	0.64×10^{-5}
$(\delta R/R^{(0)})_{\rm M2}$	-6.54×10^{-5}
$R^{(2)}$	5.8467×10^{-4} (kV/cm)$^{-2}$
$R^{(4)}$	-3.80×10^{-6} (kV/cm)$^{-4}$

There is a number of other small corrections which must be taken into account in the interpretation of a high precision experiment. These are (i) contributions from intermediate p–states with n > 2, and from final state perturbations, (ii) relativistic and retardation corrections of order $(\alpha Z)^2$ to the E1 matrix elements and (iii) an M2 decay mode proceeding through the $2p_{3/2}$ intermediate state. All of the corrections can be summarized by the formula

$$R = R^{(0)}\left[1 + \left[\frac{\delta R}{R^{(0)}}\right]_n + \left[\frac{\delta R}{R^{(0)}}\right]_{\rm rel} + \left[\frac{\delta R}{R^{(0)}}\right]_{\rm M2}\right]$$
$$+ R^{(2)}|\vec{E}|^2 + R^{(4)}|\vec{E}|^4 \ . \tag{64}$$

Using the numerical values listed in Table 1 for the coefficients appearing in eq. (64), an effective experimental value for $R^{(0)}$ can be calculated from a measurement of R, and hence the Lamb shift from equations (61) and (62).

The experimental details of our anisotropy measurements for He$^+$ have recently been described by Drake et al. (1988) and van Wijngaarden et al. (1988). As shown in Fig. 6, a He$^+$ ion beam containing ~ 0.5% metastable ions is formed by passing 125 keV

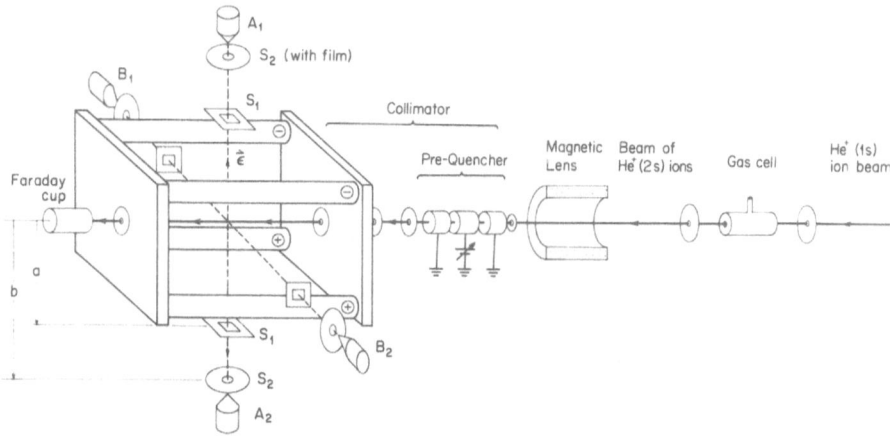

Fig. 6.   Schematic diagram of the apparatus for the He$^+$ anisotropy measurement. The four metal rods in the quenching cell are 1.2700(2) cm in diameter and are supported 4.064(1) cm apart on insulators. The length of the cell is 15.24 cm. S_1 and S_2 are collimating slits with c = 7.117 cm and s = 21.996 cm.

ground state He⁺ ions through a gas cell. The beam is then subjected to a static electric field in the quenching cell further down stream by supplying opposite polarities to two pairs of cylindrical rods mounted on insulators in a quadrupole arrangement. The resulting Ly–α intensities $I_\perp$ and $I_\parallel$ are detected simultaneously by measuring the photoelectric current from the photosensitive cones labelled A_1, A_2, B_1 and B_2. All data were taken at a single field strength of 631.04 V/cm at the center of the cell.

Beam contamination by ions other than He⁺(2s) and He⁺(1s) is kept small by the following three strategies. First, the long 900 ns flight time from the exit of the gas cell to the observation region allows the majority of highly excited ions to decay to the ground state. Second, small transverse electric fields are applied between the gas cell and the collimator to separate neutral atoms from the ion beam. Third, a small axial electric field of 100 V/cm is maintained between the electrodes of the prequencher to ionize any highly excited states which might survive the long flight path from the gas cell. (However, the main purpose of the prequencher is to apply a sufficiently strong electric field to depopulate the $2\,^2S_{1/2}$ state for purposes of noise determination.)

Each of the four photon detection systems shown schematically in Fig. 6 is actually a pair of identical detectors placed a small distance ℓ apart along the ion beam, as shown in Fig. 7. The reason for the doubling of the counters is to increase the total signal strength. The quench radiation passes through a photon collimator consisting of a rectangular entrance slit S_1 and a circular exit slit S_2, and then strikes a photosensitive cone P. The cylindrical holder C held at a positive potential $V_c = 67.5$ V collects the photoelectrons, and the high precision electrometer E (Keithley Model 642 LNFA) measures the photoelectron current. The shields Sh prevent photons from crossing between the two collimation systems.

The photoelectric yield of a cone is known to improve as its angle is decreased. However, in order to ensure that the detection system responds equally to photons of different polarization (Patel *et al.* 1987), the cone angles were left large ($\sim 100^0$). We increased the photoelectron efficiency back to the 20% level by coating them with a layer of MgF$_2$ of a few hundred Å thickness.

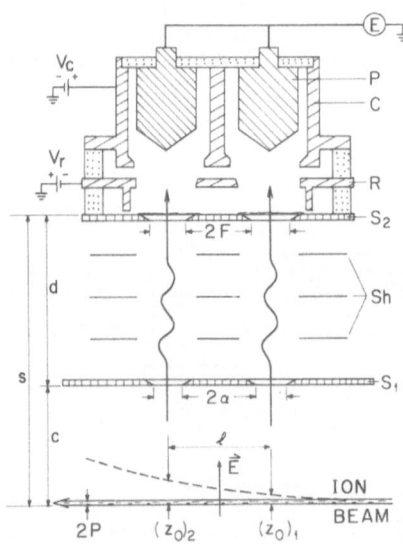

Fig. 7.   Details of the photon detection systems A_1, A_2, B_1, B_2 shown in Fig. 1. The beam diameter (2p) is 0.165 cm, the width (2α) of rectangular slit S_1 is 2,489 cm, the diameter (2F) of circular slit S_2 is 1.270 cm, and the cone angles are 100^0. The beam deflections $(z_0)_1$ and $(z_0)_2$ [see eq. (67)] due to the transverse quenching field have been exaggerated for clarity.

We have introduced the above photoelectric detection system in place of standard photon counting techniques because of non—linearities which are inherent in photon counting. The problem is that all photomultipliers produce a distribution of pulse heights which, for high photon fluxes, is weakly count—rate dependent due to variations in the electron multiplication process. In addition, dead—time corrections become increasingly troublesome for high count rates. Both problems are avoided by directly collecting and measuring the photoelectron current emitted from a large surface area without further amplification. The disadvantage is that the photocurrents produced are small. One must thus take care to ensure that stray electrons and low energy ions created by collisions of the fast ion beam with the residual gas ($\sim 7 \times 10^{-8}$ Torr) are not detected by the cones. We suppressed such stray particles by imposing an axial magnetic field of 11.8 G in the observation region, and by covering the exit slits S_2 of the collimators with thin aluminum films. The films are sufficiently thick to stop low energy particles, but are thin enough to transmit Ly—α photons. In addition, a repeller plate kept at −90 V was used to suppress electrons that might be ejected from the back surface of the aluminum films.

The four detector systems shown in Fig. 6 were electrically connected together into two opposite pairs with the total current from each pair being fed to one of the electrometers. The analogue voltage output of the electrometers was then connected to the input of a digital voltmeter (Hewlett Packard Model 3457A). The final output signal was normalized to the ion beam current and stored in a computer. The signal to noise ratio for each of the two detection systems was about 1000.

The accuracy of an anisotropy measurement is ultimately limited by the *linearity* (as opposed to absolute accuracy) of the photon detection system. As discussed above, a direct measurement of the photoelectron current was used in place of standard photon counting techniques because of non—linearities inherent in photon counting. However, it is now necessary to verify that the current measurements do not suffer from problems of their own.

One potential source of non—linearity is the finite voltage coefficient of the resistance in the input stage where the photocurrent is converted into a voltage signal. To minimize this voltage effect, the necessary input resistance of 10^{12} Ω was obtained by connecting four resistors in series so that for typical currents of 10^{-13} A, the potential across any component resistor would be small enough (50 mV) so that the deviation from linearity between the current and its voltage analogue at the input stage would be less than 1 ppm.

The most important test is to show that the output voltage of the electrometers varies linearly with the input voltage. This was checked by applying a dc voltage directly to the input of each electrometer. The small difference $V_i - V_o$ between the input potential V_i and the output potential V_o was then measured as V_i was varied from 0 to 400 mV. The difference, which should ideally be zero, was found to increase linearly with V_i at the rate of 48±2 μV/V for one electrometer and 25±2 μV/V for the other. A possible deviation from linearity as large as ±2 μV/V is still sufficiently small to determine the intensity ratio $r = I_\parallel / I_\perp$ to an accuracy of ±1 ppm. It is this high degree of linearity at high photon fluxes which has allowed a dramatic improvement in accuracy over our earlier measurements obtained by photon counting techniques (Drake *et al*. 1979).

The directly measured quantity is the intensity ratio $r = I_\parallel / I_\perp$, which is related to the anisotropy defined by eq. (56) according to

$$R = (r-1)/(r+1) \ . \tag{65}$$

A direct measurement of r would be greatly limited in accuracy because it requires a knowledge of the relative sentitivities of the detectors. However, this limitation can be avoided by measuring r for all four possible 90° rotations of the electric field in Fig. 6. The rotations are easily accomplished simply by switching potentials on the quadrupole rods. If θ is the angle between the electric field vector and, say, the A_1A_2 axis, then the intensity ratios $r(\theta)$ which can be measured are

$$r(0) = A(0)/B(0) \qquad r(\pi/2) = B(\pi/2)/A(\pi/2)$$

$$r(\pi) = A(\pi)/B(\pi) \qquad r(3\pi/2) = B(3\pi/2)/A(3\pi/2)$$

where $A = A_1 + A_2$ and $B = B_1 + B_2$ are the sums of currents from the opposite pairs of detectors. The currents are measured simultaneously, and represent time averaged values. One then forms

$$r = [r(0) \cdot r(\pi/2) \cdot r(\pi) \cdot r(3\pi/2)]^{1/4} \tag{66}$$

which is independent of the efficiencies of the detectors.

The above photocurrents must be corrected for the background noise ($\sim 0.5\%$), defined to be the quenching signal which still persists when the 2 $^2S_{1/2}$ ions are removed from the beam by prequenching. The noise consists of roughly equal parts of electronic noise and a photon noise which originates from a "glowing column" of excited atoms created in the residual gas by the passage of the fast primary He$^+$ ion beam. A further correction must be introduced for the small 2E1 two photon decay component present in the original quenching signal.

A grand average intensity ratio was obtained from 66 runs of about one day, each containing an average of 150 measurements of r for a total of 9930 individual measurements. The final result is is $r = 1.267651193(7176)$. The estimated error in the final figures obtained from a Gaussian fit to the data is given by the numbers in brackets. The corresponding R is listed in Table 2.

The measured R must be corrected for a number of small systematic effects in order to obtain the equivalent lowest-order field free $R^{(0)}$ which can be directly related to the Lamb shift. These are briefly discussed in the following paragraphs.

Table 2. Systematic and higher order corrections used to obtain the zero order anisotropy $R^{(0)}$ and the Lamb shift $\mathcal{L}$ from R_{exp}.

Quantity	Value
measured anisotropy R_{exp}	0.118030142(2791)
detector non–linearity	0.000000000(354)
2E1 two–photon decay	0.000002012(236)
finite solid angle of detectors and deflections of ion beam	0.000150115(259)
relativistic angular shift	0.000006980(56)
11.8 G Zeeman splitting	0.000000230(1)
$\vec{v} \times \vec{B}$ electric field	0.000000125(1)
finite quenching field effects	−0.000232222(60)
magnetic quadrupole transitions	0.000007715
mixing with higher np states and final state perturbations	0.000002796
$(\alpha Z)^2$ terms in matrix elements	−0.000000755
$R^{(0)}$ (sum of above)	0.117967137(2836)
$\mathcal{L}$ [from eq.'s (61) and (62)]	14042.220(349) MHz

First, the quenching signal contains a small isotropic background from the spontaneous decay of the $2\,^2S_{1/2}$ ions by 2E1 radiation, producing an apparent anisotropy which is too low. Because the 2E1 decay rate of $8.23Z^4$ s^{-1} is a factor of 4.74×10^4 smaller than the field–induced quenching rate at our field strength of 631 V/cm, the correction is small. It was estimated to the necessary $\pm10\%$ accuracy from the known 2E1 frequency distribution (Goldman and Drake 1980) and the known MgF$_2$ photoelectron yield (Lapson and Timothy 1973, 1976) for the photon detectors. The resulting correction is listed in Table 2.

Second, The intensities $I_{\parallel}$ and $I_{\perp}$ are signals which are observed over a small angular range defined by the finite slit sizes of the photon collimators. The solid angle correction must take this into account, along with the effects of beam deflection by the quenching field and the progressive depletion of the metastable state along the beam. Once the corrections have been obtained for a single detector, they must be averaged over the detector pairs shown in Fig. 7 with weighting factors equal to the relative radiation intensities. In addition, if the beam is assumed to be travelling in the y–direction, then the radiation intensity decays exponentially along the beam according to $I(y) = I(0)e^{-\gamma y}$, where $1/\gamma$ is the decay length due to quenching. The beam also bends due to the transverse electric field in the z–direction, giving it a parabolic trajectory of the form

$$z = z_0 + \lambda y + \mu y^2 \tag{67}$$

where z_0 is the beam deflection and $\lambda = v_z/v_y$ is the velocity ratio in the $z-$ and y–directions, all evaluated at the center of the detector viewing region. Finally, $\mu = |\vec{E}|/4V_a$, where V_a is the accelerating potential for the ion beam. In terms of these constants and the ones shown in Figs. 6 and 7, the observed anisotropy R is related to the solid angle corrected anisotropy R_c by

$$\frac{R}{R_c} = 1 - \frac{p^2}{2s^2} - \left[\frac{1 - R_c}{s^2}\right]t^2\left[\frac{\alpha^2}{3} + \frac{F^2}{4}\right] - \frac{F^2}{2s^2}$$

$$+ \frac{\tilde{z}_0^2}{R_c s^2}\left[\tfrac{9}{4}(1 - R_c^2) - R_c\right] - \frac{\tilde{\tilde{z}}_0^2}{R_c s^2}(1 - R_c^2) \tag{68}$$

where

$$\tilde{z}_0{}^2 = z_0{}^2 + [\lambda^2 + 2(\mu - \gamma)z_0]\left[\frac{\alpha^2 t^2}{3} + \frac{F^2(1 - t^2)}{4}\right],$$

$$\tilde{\tilde{z}}_0{}^2 = z_0{}^2 + [\lambda^2 + 2(\mu - \gamma)z_0]\left[\alpha^2 t^2 + \frac{F^2(3t^2 - 6t + 2)}{4}\right]$$

and $t = s/d$. Equation (68) assumes that the signals from opposite detectors are averaged so that first–order corrections from beam bending cancel out. For the present experiment, $V_a = 125$ kV, $|\vec{E}| = 631.04$ V/cm and $\gamma = 0.1017$ cm^{-1}. The above then yields a solid angle correction of $\delta R = 0.000150115(259)$.

Third, the observed intensity $I_{\parallel}$ emitted parellel to $\vec{E}$ in the laboratory frame by the fast moving ions corresponds to emission at a small angle $\theta = v/c$ to $\vec{E}$ in the co–moving atomic frame. There is a similar angular shift for $I_{\perp}$, but because of rotational symmetry about the field direction, this intensity is not affected. The net correction to the anisotropy is

$$\delta R = R_c(1 - R_c)(v/c)^2 \ . \tag{69}$$

Fourth, the Zeeman splittings of the $n = 2$ manifold of states in an axial magnetic field $\vec{B}$ produce corrections to the mixing coefficients $b(1/2)$ and $b(3/2)$ in eq. (28) which cancel out to first order in $|\vec{B}|$. However in second order, the net effect is to enhance the Stark coupling between the 2 $^2S_{1/2}$ and 2 $^2P_{1/2}$ sublevels, thereby decreasing the anisotropy. For our field strength of 11.8 G, the correction to R is $\delta R = 0.000000230(2)$. There is a further correction due to $\vec{B}$ which results from the fact that the transverse quenching field $\vec{E}$ deflects the ion beam, and hence it progressively acquires a velocity component $v_z = \lambda v_y$ (see eq. 67) in the transverse direction. The resulting $\vec{v} \times \vec{B}$ electric field is perpendicular to $\vec{E}$, and the vector sum produces a net effective quenching field which is slightly rotated. For our experiment, the correction is $\delta R = 0.000000125(1)$.

Comparison with other Lamb shift measurements. Numerical values for the above corrections, as summarized in Table 3, give a field–free anisotropy $R^{(0)}$ of 0.117967137(2836). Using equations (61) and (62), this corresponds to a Lamb shift of 14042.22(35) MHz. Table 3 presents a comparison of our value with theory and other measurements. Our value is consistent with the older microwave resonance measurement of Lipworth and Novick (1957) and the very recent one by Dewey and Dunford (1988), but not with that of Narasimham and Strombotne (1971), which lies about 4 MHz higher. It is also in excellent agreement with a calculation based on Mohr's value for the electron self-energy as discussed below, but not with Erickson's. The latter gives values which lie higher than Mohr's for all hydrogenic ions.

The significance of the results in terms of Lamb shift theory can be illustrated by writing the Lamb shift as an expansion in powers of αZ and α in the form (Drake 1982; Mohr 1976, 1982, 1983; Johnson and Soff 1985)

$$\mathcal{L} = \frac{8\alpha(Z\alpha)^4 mc^2}{6\pi n^3} \{A_{40} + A_{41}\ln(Z\alpha)^{-2} + A_{50}Z\alpha + (Z\alpha)^2[A_{62}\ln^2(Z\alpha)^{-2}$$
$$+ A_{61}\ln(Z\alpha)^{-2} + G(Z\alpha)] + (\alpha/\pi)[B_{40} + O(Z\alpha)] + O(\alpha^2/\pi^2)\}$$
$$+ \text{finite nuclear mass and size corrections} . \tag{70}$$

Each of the constants A_{ij} can be written as the sum of electron self–energy, vacuum polarization, and anomalous magnetic moment contributions. In particular, the term $G(Z\alpha)$, which represents the sum of all higher order terms in $Z\alpha$, consists of the parts

$$G(Z\alpha) = G_{SE}(Z\alpha) + G_{VP}(Z\alpha) . \tag{71}$$

The finite nuclear size correction is predominantly a non–QED effect resulting from the deviation of the nuclear potential from a pure Coulomb potential inside the nucleus. Its calculation, including relativistic corrections, is discussed by Mohr (1983) and Johnson and Soff (1985).

Table 3. Comparison of theory and experiment for the He$^+$ Lamb shift (in MHz).

Experiment	Theory
14042.220 ± 0.349[a]	14042.22 ± 0.5[e]
14042.0 ± 1.2[b]	14045.12 ± 0.5[f]
14046.2 ± 1.2[c]	
14040.2 ± 1.8[d]	

[a]Drake *et al.* (1988).
[b]Dewey and Dunford (1988).
[c]Narasimham and Strombotne (1971).
[d]Lipworth and Novick (1957).
[e]calculated with Mohr's (1976, 1982) electron self–energy.
[f]calculated with Erickson's (1977) electron self–energy.

The coefficients A_{ij} multiplying the various powers of $Z\alpha$ in eq. (70) are all well established. The term $G(Z\alpha)$ represents the remainder of all terms of order $\alpha(Z\alpha)^6 mc^2$ and higher. Calculations of $G(Z\alpha)$ by Erickson (1977), and by Mohr (1976, 1982) and Sapirstein (1981) differ by about 30%, which is much larger than the estimated ±4% uncertainty of their calculations. Aside from possible additional uncertainties arising from the value of the nuclear radius, the uncertainty in $G(Z\alpha)$ determines the theoretical uncertainty in the final calculated value of $\mathcal{L}$. Since the $G(Z\alpha)$ contribution scales as Z^6, as compared with the lower Z^4 scaling of $\mathcal{L}$ itself, the experimental precision required for an equally significant test of $G(Z\alpha)$ for ions of different Z scales as $Z^6/Z^4 = Z^2$. Hence less precision is required for a high–Z ion than for a low–Z ion.

The various QED corrections to the n = 2 states of He$^+$ have been recalculated by Drake *et al.* (1988) for a nuclear radius of 1.673(1) fm (Borie and Rinker 1978), using the other input data tabulated by Mohr (1983) and Johnson and Soff (1985). The results are summarized in Table 4. The relativistic recoil term contains the recently calculated corrections by Bhatt and Grotch (1987), and by Erickson and Grotch (1988). The latter decreases the Lamb shift for He$^+$ by 0.039 MHz. The uncertainty in the calculated $\mathcal{L}$ comes almost entirely from $G_{SE}(Z\alpha) = -22.9\pm1.0$ for He$^+$. Since the nuclear radius is exceptionally well known for He$^+$ from muonic fine structure measurements (Borie and Rinker 1978), the uncertainty due to the finite nuclear size correction is unusually small for this case.

Fig. 8 presents a plot of the observed deviation from the Mohr theory for a variety of hydrogenic ions where high precision measurements are available. The scaling of the vertical axis by Z^6 makes the size of the error bars roughly proportional to the sensitivity of the experiment to the value of $G(Z\alpha)$. The figure shows no serious disagreement with the Mohr theory over the entire range of Z, while Erickson's values lie substantially higher. The hydrogen measurement apparently lies two error bars below theory, but this result is clouded by uncertainties concerning the proton radius and the corresponding nuclear size correction (Lundeen and Pipkin 1986). There is some indication that the experimental values are falling consistently below theory for $Z \geq 15$, but there is as yet no clear disagreement.

Since the primary theoretical uncertainty comes from the G_{SE} part of $G(Z\alpha)$ in eq. (71), it is instructive to extract an experimental value for G_{SE} from the measurements by taking the other well–established terms in eq. (70) as correct and subtracting their contributions. The results are shown in Table 5, along with the values of the nuclear radii used and their uncertainties. [Not included in the table is the work of Sokolov and coworkers (Sokolov *et al.* 1982; Palchikov *et al.* 1983) who effectively measure the ratio $\mathcal{L}/\Gamma_p$ to very high precision.] Our value for He$^+$ of $G_{SE} = -22.99\pm0.76$, together with the value for S^{15+} (Georgiadis *et al.* 1986) of -19.45 ± 0.52, provide the most stringent tests of theory. While the former is in agreement with theory, the latter lies approximately two standard deviations below theory. This reflects the possible discrepancy between theory and experiment in the total Lamb shift for larger values of Z already noted above.

Table 4. Contributions to the Lamb shift of He$^+$ (in MHz).

Contribution	2 $^2S_{1/2}$	2 $^2P_{1/2}$
electron self–energy	14259.12±0.5	−204.05
vacuum polarization	−426.95	−0.02
(α/π) corrections	0.02	0.00
finite nuclear size	8.79	0.00
relativistic recoil	−2.92	−0.09
Total	13838.06	−204.16
Lamb shift ($\mathcal{L}$)	14042.22±0.5	

Table 5. Comparison of theory and experiment for the total Lamb shift, and the derived electron self–energy part G_{SE} of the term $G(Z\alpha)$ in eq. (1). R_N is the nuclear radius used.

Ion	R_N (fm)	$\mathcal{L}_{exp.}$	$\mathcal{L}_{theo.}$	$(G_{SE})^a_{exp.}$	$(G_{SE})^c_{theo.}$	$(G_{VP})^b_{theo.}$
^{1}H	0.862(20)	1057.845(9)[d]	1057.866(11) MHz	$-26.19 \pm 1.25 \pm 0.93$	$-23.4(1.2)$	$-0.517(20)$
^{4}He	1.673(1)	14042.22(35)[e]	14042.22(50)	$-22.91 \pm 0.76 \pm 0.05$	$-22.9(1.0)$	$-0.508(17)$
^{6}Li	2.56(5)	62765.(21)[f]	62737.(6)	$-17.12 \pm 4.0 \pm 0.77$	$-22.49(88)$	$-0.500(16)$
^{16}O	2.711(14)	2192.(15)[g]	2196.13(21) GHz	$-22.90 \pm 7.9 \pm 0.03$	$-20.72(45)$	$-0.473(12)$
		2215.6(7.5)[h]		$-10.44 \pm 4.0 \pm 0.03$		
		2203.(11)[i]		$-17.09 \pm 5.8 \pm 0.03$		
^{19}F	2.900(15)	3339.(35)[j]	3343.0(1.8)	$-21.46 \pm 9.0 \pm 0.03$	$-20.42(39)$	$-0.469(7)$
^{31}P	3.197(5)	20.17(7)[k]	20.25(1) THz	$-19.82 \pm 0.85 \pm 0.004$	$-18.81(14)$	$-0.449(2)$
^{32}S	3.247(4)	25.266(63)[l]	25.373(13)	$-19.45 \pm 0.52 \pm 0.004$	$-18.57(11)$	$-0.446(2)$
^{35}Cl	3.335(18)	31.19(22)[m]	31.34(2)	$-19.23 \pm 1.3 \pm 0.01$	$-18.34(8)$	$-0.444(1)$
^{40}A	3.428(8)	37.89(38)[n]	38.24(2)	$-19.55 \pm 1.6 \pm 0.005$	$-18.11(6)$	$-0.442(1)$
^{238}U	5.751(50)	70.4(8.3)[o]	75.3(4) eV	$-8.32 \pm 0.46 \pm 0.02$	$-8.050(4)$	$-0.600(8)$

[a]The first uncertainty listed is due to the experimental uncertainty in $\mathcal{L}$, and the second to the nuclear radius uncertainty. Nuclear size corrections to the self–energy and vacuum polarization terms have been subtracted for the high–Z ions.
[b]Includes Uehling and Wichmann–Kroll vacuum polarization contributions. The total $G(Z\alpha)$ for a point nucleus is $G_{SE} + G_{VP}$.
[c]Mohr (1976, 1982).
[d]Lundeen and Pipkin (1982).
[e]Drake et al. (1988).
[f]Leventhal (1975).
[g]Curnutte et al. (1981).
[h]Lawrence et al. (1972).
[i]Leventhal et al. (1975).
[j]Kugel et al. (1975).
[k]Müller et al. (1988).
[l]Georgiadis et al. (1986).
[m]Wood et al. (1982).
[n]Gould and Marrus (1983).
[o]Munger and Gould (1986).

162

The anisotropy method for He+ has not yet been persued to its ultimate limits of precision. Since the largest source of error in Table 2 is statistical, the precision can be improved by a factor of four simply by improving the signal strength and accumulating more data. We are currently modifying the apparatus to obtain such an improvement. This would bring the precision down to the 5 ppm level where other systematic effects would begin to play an important role. However at this level, the measurement would be an order of magnitude more accurate than existing theory. An improved He+ measurement, together with a similar improvement in precision for S15+, would severely test theoretical calculations in the low to intermediate range of Z.

E1–E1 damping interference. Consider next the case where the initial $2s_{1/2}$ state is spin–polarized, as described by the polarization vector $\vec{P}$. Then the term $\vec{P} \cdot \vec{J}_0$ of eq. (49) comes into play. Expanding the terms of eq. (51) in terms of the V_j's and M_j's yields

$$\vec{P} \cdot \vec{J}_0 = \vec{P} \cdot (\hat{k} \times \hat{E}) \mathrm{Re}[M^*(2V_{1/2} + V_{3/2})] - 3\vec{P} \cdot (\hat{k} \times \hat{E})(\hat{k} \cdot \hat{E}) \mathrm{Im}[V^*_{1/2}(V_{3/2} + M_{3/2})] \quad .$$
(72)

The second term above is called the E1–E1 damping interference term because it comes primarily from the imaginary part of the cross product $V^*_{1/2} V_{3/2}$. It therefore depends on the imaginary level widths contained in the denominators of $V_{1/2}$ and $V_{3/2}$. The term $M_{3/2}$ in (72) can be neglected for low Z because it is smaller ba a factor of $(\alpha Z)^2$.

The angular dependence of the E1–E1 damping term is

$$\vec{P} \cdot (\hat{k} \times \hat{E})(\hat{k} \cdot \hat{E}) = \tfrac{1}{2}|\vec{P}| \cos\theta \sin\theta$$
(73)

assuming that $\vec{P}$ is perpendicular to the $\hat{k}$, $\hat{E}$ plane, and θ is the angle between $\hat{k}$ and $\hat{E}$. The intensity distribution is illustrated by the clover leaf pattern in Fig. 4. Since the maximum intensity difference occurs in the directions $\theta = \pi/4$ and $\theta = 3\pi/4$, the E1–E1 damping asymmetry is defined to be

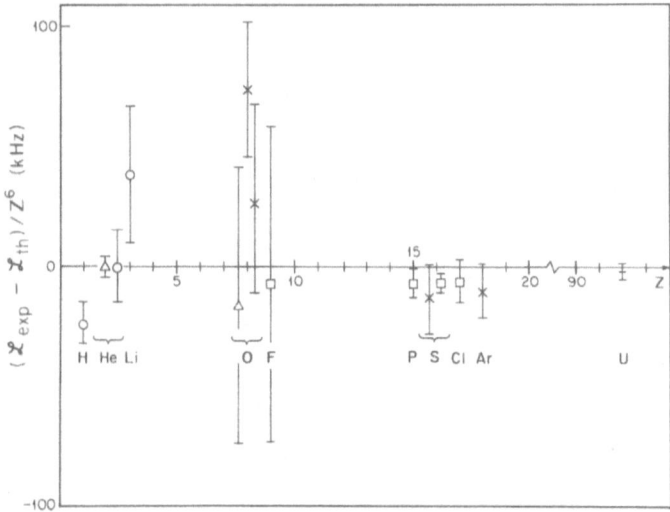

Fig. 8.   Experimental values for the Lamb shift, shown as scaled deviations from theory obtained with Mohr's electron self–energy. The methods of measurement are denoted by (o) for microwave resonance, (Δ) for anisotropy, (×) for quench rate and (□) for laser resonance. The U91+ result is deduced from a decay rate measurement in two–electron U90+ (Munger and Gould, 1986).

$$A = \frac{I(\pi/4)-I(3\pi/4)}{I(\pi/4)+I(3\pi/4)}$$

$$= \frac{3\mathrm{Im}[V^*_{1/2}(V_{3/2}+M_{3/2})]}{2I_0(\pi/4)} .\tag{74}$$

Neglecting the relativistic and $M_{3/2}$ corrections, this reduces in the limit of weak fields to

$$A = \frac{3\Gamma\,|\vec{P}|\,(\mathcal{L}-\mathfrak{F})}{4\mathcal{L}^2-2\mathcal{L}\mathfrak{F}+7\mathfrak{F}^2+11\Gamma^2/4}\tag{75}$$

independent of $|\vec{E}|$.

A high precision measurement of A for He$^+$ has been carried out by Drake, Patel and van Wijngaarden (1983), using essentially the same apparatus as for the Lamb shift anisotropy measurement shown in Fig. 6. Nearly complete spin polarization can easily be achieved by prequenching one of the $2s_{1/2}$ $(m_j=\pm1/2)$ magnetic substates in the presence of an axial magnetic field. As shown in Fig. 9, a magnetic field in the z–direction causes a well known crossing of the states $2s_{1/2,-1/2}$ and $2p_{1/2,1/2}$ near 7000 G. The application of a small transverse electric field causes the $2s_{1/2,-1/2}$ state to decay much more rapidly than the remaining $2s_{1/2,1/2}$ state.

The measured asymmetry at $|\vec{E}| = 246.71$ V/cm of 0.007603(20) for He$^+$ is in good agreement with the theoretical value of 0.0076209 (including finite field corrections). This provides direct experimental confirmation of the Bethe–Lamb phenomenological quenching formalism since the effect depends directly upon the damping terms. Alternatively, if the experiment is interpreted as a measurement of the 2p lifetime $\tau_{2p} = 1/(2\pi\Gamma)$, then the experimental value

$$\tau_{2p,\mathrm{exp.}} = (0.9992\pm0.0026)\times10^{-10}\ \mathrm{s}$$

falls within one standard deviation of the theoretical value

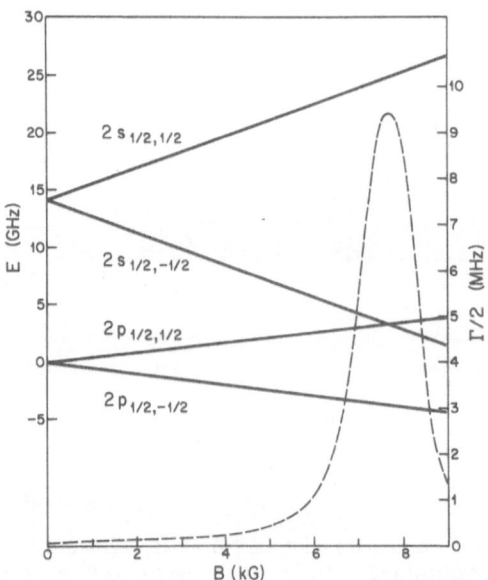

Fig. 9.   Graph of the He$^+$ $2s_{1/2}$ and $2p_{1/2}$ magnetic sublevel energies (left–hand scale) as a function of magnetic field strength. The dashed curve shows the $2s_{1/2}$ $(m_j=-1/2)$ level width (right–hand scale) for an electric quenching field strength of 100 V/cm.

$$\tau_{2p,theo.} = 0.9972 \times 10^{-10} \text{ s}.$$

The accuracy of this measurement improves substantially on the best previous beam–foil result of $(0.98\pm0.05)\times10^{-10}$ s (Lundin *et al.*, 1970). For non–hydrogenic systems, the most accurate measurements are for the resonance transitions of Li and Na by Gaup *et al.* (1982), who achieved a precision of $\pm0.16\%$. However, the Li result differs from the best theoretical calculations by a suprisingly large 0.8%. The origin of this discrepancy is not understood.

El–M1 interference. For a spin–polarized ion beam, the first term of eq. (72) introduces an additional rotational asymmetry. It is called the El–M1 interference term because it arises from the cross–product between the field–induced El and spontaneous M1 decay chanels. Its angular dependence is

$$\vec{P}\cdot(\hat{k}\times\hat{E}) = |\vec{P}|\sin\theta \qquad (76)$$

assuming that $\vec{P}$ is perpendicular to the $\hat{k}$, $\hat{E}$ plane, and θ is the angle between $\hat{k}$ and $\hat{E}$. The intensity distribution gives rise to the left–right asymmetry shown by the inner-most dashed curve in Fig. 4. Defining as before

$$A = \frac{I(\pi/2)-I(-\pi/2)}{I(\pi/2)+I(-\pi/2)} \qquad (77)$$

then

$$A = \frac{2}{I_0}\text{Re}[M^*_{1/2}(2V_{1/2}+V_{3/2}+M_{3/2})] \ . \qquad (78)$$

In the limit of weak quenching fields and small Z, A reduces to

$$A = 2Q\left[\frac{1+|\rho|^2\mathfrak{F}/(2\mathfrak{L})}{1+\text{Re}(\rho)+5|\rho|^2/2+Q^2}\right] \qquad (79)$$

where

$$Q = -\frac{243\sqrt{2}Z^2\mathfrak{L}M}{128e^2a_0^2|\vec{E}|}$$

$$= \left[\frac{\text{spontaneous M1 rate}}{\text{induced El rate}}\right]^{1/2} \qquad (80)$$

and M is the relativistic magnetic dipole matrix element defined by eq. (21). Since the M1 decay rate is only $2.496\times10^{-6}Z^{10}$ s^{-1}, Q^2 is much less than unity for low Z ions at practical quenching fields and may be neglected in the denominator of eq. (79). Then A increases in proportion to $1/|\vec{E}|$ as $|\vec{E}| \to 0$. It may also be necessary to include a contribution from two–photon transitions in the denominator of eq. (79) if the photon detectors are sensitive to a broad range of frequencies other than Ly–α radiation.

Van Wijngaarden *et al.* (1985) have recently reported a measurement of improved precision for the El–M1 asymmetry in He$^+$, using an apparatus similar to the one shown in Fig. 6. For a quenching field of $|\vec{E}| = 43.63$ V/cm, eq. (79) gives a theoretical asymmetry of 3.009×10^{-4}, in good agreement with the measured value of $(2.935\pm0.337)\times10^{-4}$. If the experiment is interpreted as a measurement of the M1 matrix element, then $M = -(0.273\pm0.031)\alpha^2e\lambda$ in agreement with the theoretical value $-0.2794\alpha^2e\lambda$ obtained from eq. (21).

The above is the only measurement of the relativistic M1 matrix element in one–electron ions. However, the M1 decay rate from the 1s2s $^1S_1$ state to the ground state has been measured in several two–electron ions, beginning with the first laboratory observation of the process by Marrus and Schmeider (1972). The two–electron ion measurements have recently been extended to Xe[54+] (Marrus *et al.*, 1988), where higher order relativistic corrections play an important role. For the two–electron case, the matrix element of the magnetic dipole moment operator in a nonrelativistic approximation is (Drake, 1971)

$$\langle 1\ {}^1S_0 | M_{10} | 2\ {}^3S_1 \rangle = \mu_B \langle 1\ {}^1S_0 | -\frac{2}{3m^2c^2}(p_1^2-p_2^2) - \frac{1}{6}\left[\frac{\omega}{c}\right]^2(r_1^2-r_2^2)$$
$$+ \frac{Ze^2}{3mc^2}(r_1^{-1}-r_2^{-1}) | 2\ {}^3S_1 \rangle \tag{81}$$

where the spin dependence of the wave functions and operators has already been factored out, and $\mu_B = e\hbar/2mc$ is the Bohr magneton. Using the Z^{-1} expansions for the matrix elements tabulated by Drake (1971), the nonrelativistic decay rate is

$$A(2\ {}^3S_1{\to}1\ {}^1S_0) = \frac{4}{3\hbar}\left[\frac{\omega}{c}\right]^3 | \langle 1\ {}^1S_0 | M_{10} | 2\ {}^3S_1 \rangle |^2 \tag{82}$$

$$= 3.2498{\times}10^{11}\ \text{s}^{-1}.$$

The relativistic corrections can be estimated from $\omega_{rel}/\omega_{nr} = 1.0348$ together with the one–electron relativistic correction to the M1 transition matrix element listed after eq. (44); i.e.

$$|\omega M|^2_{rel}/|\omega M|^2_{nr} \simeq (1+0.4193\alpha^2Z^2)^2 = 1.1345\quad.$$

The total relativistic correction factor is $1.0348{\times}1.1345 = 1.1739$, giving a decay rate of $A_{rel} = (3.82{\pm}0.03){\times}10^{11}\ \text{s}^{-1}$, and a lifetime of $\tau_{rel} = (2.62{\pm}0.02){\times}10^{-12}$ s. The uncertainty comes primarily from uncalculated terms of $O(\alpha^2Z)$. The relativistically corrected lifetime is in agreement with the value $(2.554{\pm}0.076){\times}10^{-12}$ s measured by Marrus *et al.* (1988).

NEW VARIATIONAL TECHNIQUES AND HIGH PRECISION EIGENVALUES FOR HELIUM

We turn now from one–electron atoms, which can be solved exactly in the nonrelativistic limit, to two–electron atoms such as helium. The theoretical study of helium goes back to the early days of quantum mechanics, where approximate solutions to the Schrödinger equation provided important confirmations that quantum mechanics gives a correct (nonrelativistic) description of atomic structure. Helium is the simplest system which incorporates the many–body nature of more complex atoms. The Hamiltonian

$$H = -\tfrac{1}{2}\nabla_1^2 - \tfrac{1}{2}\nabla_2^2 - \frac{Z}{r_1} - \frac{Z}{r_2} + \frac{1}{r_{12}} \tag{83}$$

would be separable if it weren't for the last term arising from the electron–electron Coulomb interaction, and one could then write the wave function in the product form

$$\Psi(\vec{r}_1,\vec{r}_2) = \varphi_1(\vec{r}_1)\varphi_2(\vec{r}_2) \pm \text{exchange}\quad. \tag{84}$$

The Hartree–Fock method in fact corresponds to finding the best possible solution that can be written in the form of (84) (with possibly some additional central field assumptions). For the $1s^2\ {}^1S$ ground state of helium, the Hartree–Fock energy is about -2.863 a.u., whereas the exact energy is $-2.903724\cdots$ a.u. The difference of about 0.04 a.u. or 1 eV is called the correlation energy.

Hylleraas (1928) first suggested writing the ground state trial wave function in the explicitly correlated form

$$\Psi(\vec{r}_1, \vec{r}_2) = \sum_{i,j,k} a_{ijk} \, r_1^i r_2^j r_{12}^k \, e^{-\alpha r_1 - \beta r_2} \pm \text{exchange} \tag{85}$$

where the a_{ijk} are linear variational parameters for each combination of powers in (85). The same trial function is often expressed in terms of the equivalent variables

$$s = r_1 + r_2, \quad t = r_1 - r_2, \quad u = r_{12}$$

with $r_{12} = |\vec{r}_1 - \vec{r}_2|$. The usual procedure is to include all combinations of i, j, k such that $i + j + k \le N$, where N is an integer, and then study the convergence of the calculation as N is increased. One can show that this basis set eventually becomes complete as $N \to \infty$ so that, unlike (84), the exact solution can be expressed in the form of (85). For any finite N, the a_{ijk} are determined by Schrödinger's variational principle

$$E(\Psi) = \frac{\langle \Psi | H | \Psi \rangle}{\langle \Psi | \Psi \rangle} = \text{Min.} \tag{86}$$

which gives the system of homogeneous linear equations

$$\frac{\partial E}{\partial a_{ijk}} = 0, \quad \text{all } a_{ijk} \quad . \tag{87}$$

The solutions to (87) can be regarded from a more general point of view. If we think of the functions in (85) as the members of a basis set

$$\chi_l = r_1^i r_2^j r_{12}^k \, e^{-\alpha r_1 - \beta r_2} \tag{88}$$
$$l = 1, 2, \cdots, P$$

where l denotes the l'th distinct combination of values for i, j, k, then the solutions to eq. (85) correspond to finding the linear combinations

$$\Phi_m = \sum_{l=1}^{P} a_l^{(m)} \chi_l \tag{89}$$

which satisfy

$$\langle \Phi_m | \Phi_n \rangle = \delta_{m,n} \tag{90}$$

$$\langle \Phi_m | H | \Phi_n \rangle = \varepsilon_m \delta_{m,n} \quad . \tag{91}$$

Thus the solutions to eq. (87) are the same as what one would obtain by diagonalizing the Hamiltonian in the orthonormal basis set constructed from the same set of functions. If there are P linearly independent functions, then one obtains P variational eigenvalues ε_m ($m=1,2,\cdots,P$).

An important property of the eigenvalues obtained above follows from the MATRIX INTERLEAVING THEOREM, which says that when an extra row and column is added to a matrix, the new eigenvalues fall between the old, with the new highest higher than the old highest and the new lowest lower that the old lowest. Since by eq. (86) the lowest eigenvalue is bounded from below by the true ground state, the higher eigenvalues must similarly lie above the corresponding excited states, and move progressively downward as P is increased. The result is summarized by the HYLLERAAS–UNDHEIM THEOREM (1930):

> When a Hamiltonian operator whose spectrum is bounded from below is diagonalized in a P–dimensional finite basis set, then the P eigenvalues are upper bounds to the first P energies of the actual spectrum.

Table 6.   Results of the calculation of Accad *et al.* (1971) for the nonrelativistic ionization energy (J_{nr}), relativistic correction (ΔJ_{rel}) and $<\delta(\vec{r}_{12})>$ for the n ^{1}P states of helium.

State	J_{nr} (cm⁻¹)	ΔJ_{rel} (cm⁻¹)	$<\delta(\vec{r}_{12})>$ (a.u.)
2 ^{1}P	27176.688	0.46772	0.000736
3 ^{1}P	12101.57	0.1734	0.000253
4 ^{1}P	6818.1	0.080	0.00011
5 ^{1}P	4368.2	0.04	0.0001

The optimization of the nonlinear parameters in eq. (88) provides a more difficult problem because the equations

$$\partial E/\partial \alpha = 0, \quad \partial E/\partial \beta = 0$$

are transcendental. One must resort to a process of recalculating the variational eigenvalues for different values of α and β in order to locate the variational minimum for a given state. This problem is further discussed below.

The Hylleraas method has been applied with great success by many authors to the low–lying states of helium and helium–like ions, culminating in the 1960's and early 1970's with the extensive calculations of C. L. Pekeris and co–workers (Accad *et al.* 1971). Their work covers the singlet and triplet n S and n P states up to $n = 5$ for nuclear charge Z in the range $2 \leq Z \leq 10$.

Despite this large body of work, there remain important problems to be solved as follows.
1. The calculations of Pekeris *et al.* give energies for the lowest–lying states accurate to about 10^{-9} a.u., using basis sets containing up to 560 terms. This leads to uncertainties in energy differences of ± 0.0002 cm⁻¹, which is larger than the current levels of experimental precision (Martin 1987). The comparison between theory and experiment is limited as much by a lack of knowledge of the nonrelativistic energies as it is by higher order relativistic and QED effects. One might try to obtain higher accuracy simply by increasing N. The problem is that the number of terms is given by

$$P = (N+1)(N+2)(N+3)/6 \tag{92}$$

and so grows rapidly with N. In addition, numerical problems of near linear dependence in the basis set become progressively more severe as N increases.
2. An even more serious problem is that the accuracy of Pekeris' calculations (as measured by the rate of convergence with N) seriously deteriorates as one goes up the Rydberg series to more highly excited states. The point is illustrated by the data in Table 6. Approximately one significant figure after the decimal is lost each time n (the principal quantum number) is increased by one. There is clearly no point going beyond the 5 ^{1}P and 5 ^{3}P states, and even here, the accuracy is far short of what is required for spectroscopic precision. The large body of high precision experimental data for transitions among high n states (Farley *et al.* 1979; Hessels *et al.* 1987) remains unanalysed in terms of *a priori* calculations. The asymptotic calculations of Drachman (1985) become useful when the angular momentum L is also large, but there is as yet no overlap region where the asymptotic expansions can be compared with the bounds provided by direct variational calculations.

We have recently developed some new variational techniques which now make it possible to improve the precision of calculated eigenvalues by several orders of magnitude. An important advantage of these techniques is that they do not suffer from a loss of accuracy as one goes up the Rydberg series to moderately high values of n (~10). The following Sections will describe these techniques, together with the calculation of relativistic, QED and finite nuclear mass corrections to the nonrelativistic eigenvalues. Of particular interest here is a newly derived (Au 1988) connection between two–electron QED corrections for low–lying states, and long range Casimir–Polder type retardation corrections for high–lying states.

New Variational Techniques

Various strategies have been tried in the past to improve the convergence of variational calculations beyond the usual limit of $\pm 10^{-10}$ a.u. For example, Fock (1954) showed that an analytic expansion of the wave function about the points $r_1 = 0$ and $r_{12} = 0$ contains logarithmic terms and half–integral powers. Recently, Freund et al. (1984) obtained ground state energies accurate to a few parts in 10^{13} by including terms of this type in relatively small 230–term wave functions. For the 2 ^3P state, Schwartz (1964) found a significant improvement in convergence when he included terms of the form $(r_1 + r_2)^{1/2}$, but his final accuracy was limited to about one part in 10^{10} with 439 terms. In addition, he found it necessary to carry 52 decimal digits in the calculations, indicating possible problems of near linear dependence in the basis set. These methods have not yet been extended to more highly excited states or states of higher angular momentum.

A key ingredient of our new techniques is to double the basis set so that eq. (85) becomes

$$\Psi(\vec{r}_1, \vec{r}_2) = \sum_{i,j,k} [a_{ijk}\, \chi_{ijk}(\alpha_1, \beta_1) + b_{ijk}\, \chi_{ijk}(\alpha_2, \beta_2)] \times (\text{angular function})$$

$$\pm \text{ exchange} \qquad (93)$$

Here, (angular function) denotes a vector coupled product of solid spherical harmonics for the two electrons to form a state of total angular momentum L, and, as before,

$$\chi_{ijk}(\alpha, \beta) = r_1^i r_2^j r_{12}^k\, e^{-\alpha r_1 - \beta r_2} \; .$$

Each combination of powers i, j, k is now included twice in eq. (93) with different nonlinear parameters α_1, β_1 and α_2, β_2. At first sight, one might think that this would lead to problems of linear dependence, but in fact a complete optimization of the energy with respect to all four nonlinear parameters leads to well–defined and numerically stable values for the parameters, with the two sets being well separated from each other. For the first set of terms in eq. (93), the optimum values of α_1 and β_1 are close to their screened hydrogenic values $\alpha_1 \simeq Z$ and $\beta_1 \simeq (Z{-}1)/n$. These terms describe the asymptotic behavior of the wave function. For the second set of terms the optimum values of α_2 and β_2 are much larger. These terms describe the complex inner correlation effects. The complete optimization therefore has the effect of producing a natural division of the basis set into two sectors with quite different distance scales – an asymptotic sector (Sector A) and an inner correlation sector (Sector B). Recent work by Kono and Hattori (1986) also makes use of doubled basis sets, but the lack of complete optimization and other constraints they place on the basis set limits their accuracy to a few parts in 10^{10}.

Nonrelativistic Eigenvalues

I will now describe in somewhat greater detail how the basis sets are constructed and the nonlinear parameters optimized. One can take advantage of the near screened hydrogenic nature of the excited states of helium by writing the nonrelativistic Hamiltonian for infinite nuclear mass in the form $H = H_0 + V$ where (in atomic units)

$$H_0 = -\tfrac{1}{2}\nabla_1^2 - \tfrac{1}{2}\nabla_2^2 - \frac{Z}{r_1} - \frac{(Z-1)}{r_2} \tag{94}$$

$$V = \frac{1}{r_{12}} - \frac{1}{r_2} \tag{95}$$

and $Z-1$ is the screened nuclear charge. Although the above decomposition is unsymmetric in r_1 and r_2, the total Hamiltonian H remains symmetric. One simply interchanges r_1 and r_2 in H_0 and V when operating on the exchange part of the wave function. The advantage gained is that the solutions to the zero order problem

$$H_0\psi_0(1snl) = E_0\psi_0(1snl) \tag{96}$$

are known exactly, and the eigenvalues

$$E_0 = -\frac{Z^2}{2} - \frac{(Z-1)^2}{2n^2} \tag{97}$$

give correctly the first few figures of the true energy E. It is numerically advantageous to include $\psi_0(1snl)$ in the basis set and to cancel algebraically the E_0 contribution to the matrix elements so that the variational principle applied to $H - E_0$ yields directly the correction to E_0. For example, matrix elements involving ψ_0 are simply

$$\langle \chi_{ijk}|H - E_0|\psi_0\rangle = \langle \chi_{ijk}|V|\psi_0\rangle . \tag{98}$$

The final results would be the same without the subtraction, but the above procedure saves several significant figures in the evaluation of matrix elements, particularly for the more highly excited states.

With the ψ_0 term included, the complete variational trial function becomes

$$\Psi_{tr} = \psi(\vec{r}_1, \vec{r}_2) \pm \psi(\vec{r}_2, \vec{r}_1) \tag{99}$$

with

$$\psi(\vec{r}_1, \vec{r}_2) = a_0\psi_0(1s, nl)$$

$$+ \sum_{i,j,k} [a_{ijk}\,\chi_{ijk}(\alpha_1, \beta_1) + b_{ijk}\,\chi_{ijk}(\alpha_2, \beta_2)] \times (\text{angular function}) . \tag{100}$$

The a_{ijk} terms of (100) represent the A Sector and the b_{ijk} terms the B Sector as described earlier. Not all terms need be included in both Sectors such that $i+j+k \leq N$. In fact, one can show with some experimentation that truncations of the form $i \leq N_1, k \leq N_2$ in Sector A have little effect on the energies. It is physically reasonable that high powers of r_1 and r_{12} are not important in the asymptotic region, and one can progressively reduce the maximum values of N_1 and N_2 as one goes to more highly excited states. Also the single term χ_{000} should be omitted from the first summation since it is nearly the same as $\psi_0(1s, nl)$ when α_1 and β_1 are close to the screened hydrogenic values. No significant truncation appears to be possible for Sector B without loss of precision, and so all terms are included.

We next describe an efficient scheme for the complete optimization of all four α's and β's. (There may be more than four for states of higher angular momentum where additional types of angular functions are required). Since differentiation of eq. (100) with respect to an α or β just brings down a factor of $-r_1$ or $-r_2$, the derivatives of the variational energy with respect to the α's and β's are

$$\frac{\partial E}{\partial \alpha_t} = -2\langle \Psi_{tr}|H - E|r_1\psi(\vec{r}_1,\vec{r}_2; \alpha_t) \pm r_2\psi(\vec{r}_2,\vec{r}_1; \alpha_t)\rangle \qquad (101)$$

$$\frac{\partial E}{\partial \beta_t} = -2\langle \Psi_{tr}|H - E|r_2\psi(\vec{r}_1,\vec{r}_2; \beta_t) \pm r_1\psi(\vec{r}_2,\vec{r}_1; \beta_t)\rangle \qquad (102)$$

$$t = 1,2$$

with the assumed normalization $\langle \Psi_{tr}|\Psi_{tr}\rangle = 1$. Here $\psi(\vec{r}_1,\vec{r}_2; \alpha_t)$ denotes the terms in (100) which depend explicitly on α_t. An important simplification occurs because there is no contribution to the derivatives from the implicit dependence of E on α_t or β_t through the linear coefficients a_{ijk} and b_{ijk}. This follows because the energy is already stationary with respect to first order variations of the linear coefficients. Once the first derivatives are known, the second derivatives can be estimated by changing the α_t's and β_t's in the direction of lower energy and taking differences. Newton's method can then be applied to locate the zero's of the first derivatives. Provided that the initial α's and β's are chosen close to a minimum, the procedure converges in a few iterations. As an indication of the numerical stability, ordinary double precision arithmetic (about 16 decimal digits) is just adequate for the evaluation of derivatives and optimization of the basis sets up to the largest ones used. However as a check, the final calculations of wave functions and energies are done in quadruple precision (about 32 decimal digits). This does not typically change the variational eigenvalues by more than a few parts in 10^{14}.

Mass Polarization Corrections.

   To account for mass polarization effects, all of the calculations (including the nonlinear optimization) are repeated with the Hamiltonian (94) and (95) extended to include the mass polarization operator $(\mu/M)\vec{p}_1\cdot\vec{p}_2$, where μ is the electron reduced mass and M is the nuclear mass. First order perturbation theory gives the linear dependence of the energy on μ/M, but higher order terms are too large to be neglected at this level of accuracy. The higher order terms could be calculated directly, but it is simpler just to include the mass polarization term in the Hamiltonian from the outset, and extract the higher order dependence on μ/M by differencing.

Numerical Results.

   As a typical example, the variational eigenvalues for the 1s2p ^{1}P and ^{3}P states are listed in Table 7. It can be seen that the accuracy of the Schiff et al. (1965) 560–term calculation is achieved with only 240 terms. The results are also substantially more accurate than those of Kono and Hattori (1986) for basis sets of comparable size. The extrapolated values are obtained by assuming that ratios of successive differences continue decreasing at a constant rate. The uncertainty of 1 part in 10^{14} for the 2 ^{1}P state and 5 parts in 10^{14} for the 2 ^{3}P state is conservatively estimated to be the entire amount of the extrapolation.

   Concerning the nonlinear parameters, the optimum values are sharply defined and independent of the values used to start the iterative optimization process. It is inportant to carry the optimization up to the largest basis sets for two reasons. First, the change is large enough in going from one basis set to the next that it does affect the energy, and may produce a false impression of convergence if the optimization is not done. Second, while α_1 and β_1 remain nearly constant, α_2 and β_2 increase nearly linearly with N. This behavior ensures that the two sectors of the doubled basis set remain linearly independent, and is in fact necessary to preserve the numerical stability of the wave function.

   The optimum nonlinear parameters with mass polarization included differ only slightly from the infinite nuclear mass case and the eigenvalues show a parallel pattern of convergence. For the P–state example discussed above, the energy shifts due to mass polarization are conveniently expressed in the form

Table 7. Variational Eigenvalues for the P States of Helium (a.u.)

N	Number of terms	2 ¹P	2 ³P
4	70	−2.1238430609942	−2.1331641578114
5	112	−2.1238430837930	−2.1331641842173
6	168	−2.1238430861444	−2.1331641903258
7	240	−2.1238430864417	−2.1331641906805
8	328	−2.12384308648797	−2.13316419076479
9	432	−2.12384308649648	−2.13316419077680
10	552	−2.12384308649750	−2.13316419077784
11	688	−2.12384308649800	−2.13316419077883
12	840	−2.12384308649807	−2.13316419077905
extrapolation		−2.12384308649808(1)	−2.13316419077910(5)
Lewis and Serafino (455 terms)			−2.1331641814
Schiff *et al.* (560 terms)		−2.1238430858	−2.1331641905
Kono and Hattori (284 terms)		−2.1238430862	−2.1331641906
Schwartz (439 terms)			−2.133164190626
Morgan (213 terms)		−2.1238430684025	−2.1331641906735

$$\Delta E_M(\text{He } 2\ ^1P) = 0.0460445247(3)(\mu/M) - 0.168271(1)(\mu/M)^2 \tag{103}$$

$$\Delta E_M(\text{He } 2\ ^3P) = -0.0645724250(1)(\mu/M) - 0.204958(1)(\mu/M)^2 \tag{104}$$

in units of $2R_M$, where $R_M = [1 - (\mu/M)]R_\infty$ and μ is the reduced electron mass $mM/(m+M)$. For ⁴He, $(\mu/M) = 1.370745663 \times 10^{-4}$. The leading coefficient above is the first order perturbation coefficient $\langle \vec{p}_1 \cdot \vec{p}_2 \rangle$ calculated for infinite nuclear mass, and the next coefficient is obtained by subtracting the leading term from the directly calculated total energy shift due to mass polarization. This second order term may therefore contain a small contamination from the next term of order $(\mu/M)^3$. However, the above formulas are certainly adequate to calculate accurate mass polarization shifts for values of μ/M in the neighborhood of the one used for ⁴He.

As shown by Drachman (1987), the order of magnitude of the average of the second order terms in equations (103) and (104) can be understood as a kinematical effect as follows. The outer electron is pictured as moving in an approximately hydrogenic orbit about an effective nucleus with charge $+e$ and mass $M+m$ consisting of the actual nucleus of mass M together with the inner electron. This results in a slightly larger effective reduced mass for the outer electron and hence stronger binding.

Relativistic Corrections.

Corrections of $O(\alpha^2)$. The relativistic corrections of $O(\alpha^2)$ are calculated by evaluating the matrix elements of the nonrelativistic Pauli form of the Breit interaction $H_1, H_2, \cdots, H_5$ as given by Bethe and Salpeter (1957). Since these terms are well known and have been extensively discussed previously, the detailed formulas will not be repeated here. However, it is important to realize that they contain one–electron contributions from the nonrelativistic reduction of the Dirac equation, as well as two–electron terms from $e^2/r_{12} + B$, where B is the full 16–component Dirac form of the Breit interaction. For S–states, the only terms which contribute are

$$\Delta E_{rel} = -\frac{\alpha^2}{4}\langle p_1^4 \rangle + \langle H_2 \rangle + \pi\alpha^2\langle Z\delta(\vec{r}_1) + \delta(\vec{r}_{12})\rangle \tag{105}$$

where H_2 is the orbit–orbit interaction. It arises in part from retardation effects in B.

For states of higher angular momentum, one requires the spin–orbit, spin–other–orbit and spin–spin matrix elements from the Breit interaction terms H_3 and H_5. These are conveniently expressed in terms of reduced matrix elements according to (Edmonds 1960)

$$\langle \gamma' L'S'JM \,|\, H_{\mathrm{so}} \,|\, \gamma LSJM \rangle = (-1)^{L+S'+J} \begin{Bmatrix} J & S' & L' \\ 1 & L & S \end{Bmatrix} \langle \gamma' L'S' \| H_{\mathrm{so}} \| \gamma LS \rangle$$

(106)

and similarly for H_{soo}. For P–states, the values of the multiplying factor are $-1/3$, $-1/6$ and $1/6$ for the diagonal elements with $S = S' = 1$ and $J = 0,1,2$ respectively. For the off–diagonal element with $S' = 1$, $S = 0$ and $J = 1$, the factor is $-1/3$. For the spin–spin interaction, which can be written in terms of irreducible tensors of rank 2, the corresponding formula is

$$\langle \gamma' L'S'JM \,|\, H_{\mathrm{ss}} \,|\, \gamma LSJM \rangle = (-1)^{L+S'+J} \begin{Bmatrix} J & S' & L' \\ 2 & L & S \end{Bmatrix} \langle \gamma' L'S' \| H_{\mathrm{ss}} \| \gamma LS \rangle$$

(107)

with the P–state multiplying factor being $1/3, -1/6$ and $1/30$ for $J = 0,1,2$. The reduced matrix elements are therefore related to the quantities C_Z, C_e and D defined by Schwartz (1964) according to

$$\langle 2\ ^3\mathrm{P} \| H_{\mathrm{so}} \| 2\ ^3\mathrm{P} \rangle = 6C_Z$$

$$\langle 2\ ^3\mathrm{P} \| H_{\mathrm{soo}} \| 2\ ^3\mathrm{P} \rangle = 6C_e$$

$$\langle 2\ ^3\mathrm{P} \| H_{\mathrm{ss}} \| 2\ ^3\mathrm{P} \rangle = -10D \ .$$

The diagonal relativistic correction for infinite nuclear mass is then

$$\Delta E_{\mathrm{rel}} = -\frac{\alpha^2}{4} \langle p_1^4 \rangle + \langle H_2 \rangle + \pi \alpha^2 \langle Z \delta(\vec{r}_1) + \delta(\vec{r}_{12}) \rangle$$
$$+ \langle H_{\mathrm{so}} \rangle + \langle H_{\mathrm{soo}} \rangle + \langle H_{\mathrm{ss}} \rangle \ .$$

(108)

Relativistic Reduced Mass Corrections. The relativistic reduced mass corrections are corrections of $O(\alpha^2 m/M)$ arising from two sources. The first is from a transformation to center–of mass and relative co–ordinates in the Breit interaction itself, together with the mass scaling of these terms as discussed by Stone (1963), and Douglas and Kroll (1974). The second can be thought of as a second–order cross term between the Breit interaction and mass polarization corrections to the wave functions. This contribution is straight–forward in principle to evaluate since one need only repeat the calculation of matrix elements with mass polarization included and subtract the results to obtain the higher order correction. However, one important point concerns the calculation of

$$H_1 = - (p_1^4 + p_2^4)/(8m^3c^2)$$

(109)

when mass polarization effects are included. The usual procedure is to use the fact that the nonrelativistic wave function satisfies

$$\frac{1}{2m}(p_1^2 + p_2^2)\Psi = f\Psi$$

(110)

with $f = E + \dfrac{Ze^2}{r_1} + \dfrac{Ze^2}{r_2} - \dfrac{e^2}{r_{12}}$

(111)

to transform $\langle H_1 \rangle$ into $\langle H_1'(\infty) \rangle$, where

$$H_1^!(\infty) = \frac{1}{4m^3c^2}\left[p_1^2 p_2^2 - 2m^2 f^2\right] \tag{112}$$

is the form appropriate for infinite nuclear mass. However, if finite mass effects are included, there is an additional contribution. It is convenient first to transform to reduced mass atomic units where distance is measured in units of $a_\mu = (m/\mu)a_0$ and energy in units of $e^2/a_\mu = 2(\mu/m)R_\infty$. Then eq. (110) becomes

$$\left(-\tfrac{1}{2}\nabla_1^2 - \tfrac{1}{2}\nabla_2^2\right)\Psi = \left(f + \tfrac{\mu}{M}\nabla_1\cdot\nabla_2\right)\Psi \tag{113}$$

and

$$H_1^!(M) = \frac{\alpha^2}{4}\left[\frac{\mu}{m}\right]^3\left[\nabla_1^2\nabla_2^2 - 2\left[f + \tfrac{\mu}{M}\nabla_1\cdot\nabla_2\right]^2\right]\frac{e^2}{a_\mu} . \tag{114}$$

If terms quadratic in (μ/M) are neglected, then the additional contribution is

$$(\Delta H_1^!)_M = -\frac{\alpha^2}{2}\left[\frac{\mu}{m}\right]^3\frac{\mu}{M}(f\nabla_1\cdot\nabla_2 + \nabla_1\cdot\nabla_2 f)\frac{e^2}{a_\mu} . \tag{115}$$

In calculating $\langle(\Delta H_1')_M\rangle$, delta function singularities involving $\delta(\vec{r}_{12})$ can be avoided by letting $\nabla_1\cdot\nabla_2$ operate to the right in the first term of eq. (115) and to the left in the second term so that f is just a multiplicative factor. One can show by direct calculation that this yields the same expectation values for the complete matrix element $\langle H_1^!(M)\rangle$ as the equivalent form (Bethe and Salpeter 1957)

$$\langle H_1^{!!}\rangle = -\frac{\alpha^2}{8}\left[\frac{\mu}{m}\right]^3\left[(\nabla_1^2\Psi,\nabla_1^2\Psi) + (\nabla_2^2\Psi,\nabla_2^2\Psi)\right]\frac{e^2}{a_\mu} . \tag{116}$$

However, eq. (116) appears to be much more slowly convergent with basis set size than (114), and so (114) is the more useful form. As an example, Table 8 gives the expectation values of the operators

$$F = (2f^2 - \nabla_1^2\nabla_2^2)/4 \tag{117}$$

and

$$G = (f\nabla_1\cdot\nabla_2 + \nabla_1\cdot\nabla_2 f)/2 \tag{118}$$

with $\langle F\rangle$ expressed in the form

$$\langle F\rangle = \langle F_0\rangle + \langle F_1\rangle(\mu/M) . \tag{119}$$

The first term of eq. (119) is the matrix element calculated for infinite nuclear mass, and the second is the change in $\langle F\rangle$ when the mass polarization term is included in the Hamiltonian. Then, in atomic units

$$\langle p_1^4\rangle/4 = \langle F_0\rangle + (\mu/M)[\langle F_1\rangle + \langle G\rangle] \tag{120}$$

up to terms linear in (μ/M). The $\langle G\rangle$ term raises the ground state energies of H^- and He by 0.000152 cm^{-1} and 0.001455 cm^{-1} respectively. For the 1s2s ^{1}S and ^{3}S states, the shifts are 0.000128 cm^{-1} and 0.000025 cm^{-1}.

To summarize, the finite nuclear mass corrections can be written in the form

$$\Delta E_{RR} = (\Delta E_{RR})_M + (\Delta E_{RR})_X \tag{121}$$

where

Table 8. Contributions to the matrix elements of $\langle p_1^4\rangle/4 = \langle F_0\rangle + (\mu/M)[\langle F_1\rangle + \langle G\rangle]$ as defined by equations (117) and (118) the text (in atomic units).

State	$\langle F_0\rangle$	$\langle F_1\rangle$	$\langle G\rangle$
H⁻ 1s² ¹S	0.61563964(3)	−0.00063343(3)	−0.0238770(2)
He 1s² ¹S	13.5220168(1)	−0.1758280(1)	−0.9082723(1)
1s2s ¹S	10.2796689(1)	0.0200726(2)	−0.0797993(1)
1s2s ³S	10.45888519(1)	0.01062391(4)	−0.01558432(1)
1s2p ¹P	10.0292513215(1)	0.3393(1)	−0.0767530999(1)
1s2p ³P	9.912093697(4)	−0.5923(1)	0.11975737(1)

$$(\Delta E_{RR})_M = \Delta_1 + \Delta_2 - (m/M)\{-\tfrac{3\alpha^2}{4}\langle p_1^4\rangle + 2[\langle H_2\rangle + \pi\alpha^2\langle Z\delta(\vec{r}_1) + \delta(\vec{r}_{12})\rangle$$
$$+ \langle H_{soo}\rangle + \langle H_{ss}\rangle]\} \qquad (122)$$

with $\Delta_1 = \displaystyle\sum_{k\neq l} \left[\frac{Ze^2}{mMc^2 r_k^3}\right] \vec{r}_k\times\vec{p}_l\cdot\vec{s}_k$ \qquad (123)

$$\Delta_2 = -\sum_{k,l} \left[\frac{Ze^2}{2mMc^2 r_k^3}\right] [r_k^2\vec{p}_k\cdot\vec{p}_l + \vec{r}_k\cdot(\vec{r}_k\cdot\vec{p}_k)\vec{p}_l] \qquad (124)$$

and $(\Delta E_{RR})_X$ is the second–order cross term induced in ΔE_{rel} by mass polarization corrections to the wave function. For the 2 ³P state, the various contributions to $(\Delta E_{RR})_X$ listed in Table 9 agree with the $E^{(i,7)}$ $(i=1,...,6)$ calculated by Lewis and Serafino (1978), but the results here converge to several more significant figures . The terms Δ_1 and Δ_2 are finite mass corrections to the Breit interaction discussed above.

It is instructive to investigate what happens to the above terms in the one–electron case. Here, Δ_1 does not contribute and Δ_2 reduces to

$$\Delta_2 = -(Z/2mMc^2)r^{-1}(p^2 + p_r^2) \qquad (125)$$

where $p_r^2 = r^{-2}\vec{r}\cdot(\vec{r}\cdot\vec{p})\vec{p}$. This term, when combined with the other terms in (3.18), gives the one–electron operator H_b in eq. (42.2) of Bethe and Salpeter (1957). Its expectation value is the well–known relativistic reduced mass shift

$$E_b = -\left[\frac{Z\alpha}{2n}\right]^2 \frac{m}{M}|E_0| \quad . \qquad (126)$$

The values of the various quantities appearing in equations (108) and (122) are summarized in Table 9. The uncertainties are estimated from the degree of convergence with basis set size. All of these quantities are substantially more accurate than any previous calculation. for $\delta(\vec{r}_1)$, the expectation values are calculated by means of the global operator derived by Hiller et al. (1978), modified to take account of the $(\mu/M)\vec{p}_1\cdot\vec{p}_2$ term in the Hamiltonian (Drake 1988b). The modified operator is

Table 9. Values for various matrix elements required to calculate relativistic and QED corrections to the energy. Each quantity is expressed in the form $<T> = <T_0> + <T_1>(\mu/M)$ a.u., where $<T_1>(\mu/M)$ is the change in the matrix element when the mass polarization term $\vec{p}_1 \cdot \vec{p}_2 (\mu/M)$ is included explicitly in the Hamiltonian $H = -(\nabla_1^2 + \nabla_2^2)/2 - Z/r_1 - Z/r_2 + 1/r_{12}$.

matrix element	$<T_0>$	$<T_1>$
	He 1s2p ^{1}P	
$<p_1^4>/4$	10.0292513215(1)	0.2625(1)
$<H_2>/\alpha^2$	−0.02033047408(2)	0.1045056(3)
$\pi<\delta(\vec{r}_1)>$	4.00362331908(2)	0.12376(3)
$\pi<\delta(\vec{r}_{12})>$	0.002309601(1)	−0.010853(2)
$\Delta_2/(\alpha^2 m/M)$	−16.286503967(1)	
Q	0.003374498(1)	
	He 1s2p ^{3}P	
$<p_1^4>/4$	9.912093697(4)	0.4728(1)
$<H_2>/\alpha^2$	0.03508088684(1)	0.1523726(2)
$\pi<\delta(\vec{r}_1)>$	3.954827224(1)	−0.22525(3)
$<^3P\|H_{so}\|^3P>/\alpha^2$	0.207955244(1)	0.69869(1)
$<^3P\|H_{soo}\|^3P>/\alpha^2$	−0.3088684536(5)	−0.959469(3)
$<^3P\|H_{ss}\|^3P>/\alpha^2$	−0.1351209953(1)	−0.3244130(1)
$<^3P\|\Delta_1\|^3P>/(\alpha^2 m/M)$	−0.59732701(1)	
$\Delta_2/(\alpha^2 m/M)$	−15.75868217(1)	
Q	0.00381391791(1)	
	He 1s2p ^{3}P − 1s2p ^{1}P	
$<^3P\|H_{so}\|^1P>/\alpha^2$	0.1073194807(3)	−0.054970(1)
$<^3P\|H_{soo}\|^1P>/\alpha^2$	−0.03876204224(2)	−0.0320412(3)
$<^3P\|\Delta_1\|^1P>/(\alpha^2 m/M)$	−0.07646528(2)	

$$\langle \delta(\vec{r}_1) \rangle = \frac{1}{2\pi} \left\langle \frac{Z}{r_1^2} - \frac{1}{r_{12}^2} \frac{\partial r_{12}}{\partial r_1} - \frac{1}{r_1^3} l_1^2 \right.$$

$$\left. + \frac{\mu}{Mr_1} \left[\nabla_1 \cdot \nabla_2 - \frac{1}{r_1^2} \vec{r}_1 \cdot (\vec{r}_1 \cdot \nabla_1)\nabla_2 - \frac{1}{r_1^2} \vec{r}_1 \cdot \nabla_2 \right] \right\rangle \tag{127}$$

It can be seen from Table 10 that for the 1s3p ^{3}P and ^{1}P states the global operator (127) gives about one more significant figure than a direct evaluation of $\langle \delta(\vec{r}_1) \rangle$.

For consistency with one–electron calculations, we will include the anomalous magnetic moment corrections to H_{so}, H_{soo} and H_{ss} along with the other QED terms described in the following section.

Table 10. Comparison of matrix elements of $\delta(\vec{r}_1)$ for infinite nuclear mass obtained by use of the Hiller *et al.* (1978) global operator (127), and by direct calculation. The quantity tabulated is $(\pi<\delta(\vec{r}_1)> - 4) \times 10^3$.

N	No. of terms	1s3p ^{3}P global	1s3p ^{3}P direct	1s3p ^{1}P global	1s3p ^{1}P direct
4	70	−1.293291	−1.290964	0.122234	0.120471
5	110	−1.293440	−1.292690	0.121486	0.120348
6	160	−1.293611	−1.293491	0.121852	0.121968
7	220	−1.293755	−1.294345	0.121982	0.122600
8	290	−1.293663	−1.293936	0.121898	0.122212
9	345	−1.293676	−1.293892	0.121790	0.121645
10	411	−1.293639	−1.293731	0.121820	0.121832
11	489	−1.293634	−1.293672	0.121805	0.121743
12	580	−1.293632	−1.293668	0.121812	0.121796
13	685	−1.293629	−1.293637	0.121811	0.121788
14	805	−1.293630	−1.293646	0.121813	0.121808
Accad *et al.* (1971).		−1.2942		0.1220	

Quantum Electrodynamic Corrections

Results for low–lying states. The analysis of Araki (1957), Kabir and Salpeter (1957) and Sucher (1958) shows that in the nonrelativistic limit, the two–electron QED energy shift can be expressed as the one–electron energy shift corrected for the electron density at the nucleus, together with explicit two–electron terms dependent on $\langle \delta(\vec{r}_{12}) \rangle$ and Q. In this approximation, the energy shift is

$$\Delta E_L = \Delta E_{L,1} + \Delta E_{L,2} \tag{128}$$

where $\Delta E_{L,1} = \frac{4}{3} Z\alpha^3 \{ \ln(Z\alpha)^{-2} + \ln[Z^2 \mathrm{Ryd}/\epsilon(nLS)] + \frac{19}{30}$

$$+ 3\pi Z\alpha(\tfrac{427}{384} - \tfrac{1}{2}\ln 2) + (Z\alpha)^2[-\tfrac{3}{4}\ln^2(Z\alpha)^2 + C_{61}\ln(Z\alpha)^2 + C_{60}]$$

$$+ \frac{\alpha}{\pi} 0.4042 \} \langle \delta(\vec{r}_1) + \delta(\vec{r}_2) \rangle \tag{129}$$

$$\Delta E_{L,2} = \alpha^3(\tfrac{14}{3}\ln\alpha + \tfrac{164}{15})\langle \delta(\vec{r}_{12}) \rangle - \tfrac{14}{3} \alpha^3 Q \tag{130}$$

and $Q = \frac{1}{4\pi} \lim_{a \to 0} \langle r_{12}^{-3}(a) + 4\pi(\gamma + \ln a)\delta(\vec{r}_{12}) \rangle$. $\tag{131}$

Here, γ is Euler's constant and a is the radius of a sphere centered at $r_{12} = 0$ which is excluded from the integration over r_{12}. We have calculated the Bethe logarithms correct to terms of relative order Z^{-1} (Goldman and Drake, 1983, 1984) and expressed the results in the screened hydrogenic form

$$\ln[\epsilon(nLS)/Z^2\mathrm{Ryd}] = \ln[\epsilon_0(nLS)(Z - \sigma)^2/Z^2\mathrm{Ryd}] \tag{132}$$

where $\ln\epsilon_0(nLS)$ is determined from the hydrogenic Bethe logarithms according to

$$\ln\epsilon_0(nLS) = [\ln\epsilon(1s) + n^{-3}\ln\epsilon(nl)]/(1 + n^{-3}\delta_{0,l}) \ . \tag{133}$$

The hydrogenic Bethe logarithms are tabulated by Klarsfeld and Maquet (1973). All of the screening constants turn out to be quite small, indicating that eq. (132) may be quite accurate even for neutral helium. As an example, the final results for the P–states are

177

Table 11. Contributions to the P–state energies of helium (cm⁻¹), using R_∞ = 109737.31569 cm⁻¹, α^{-1} = 137.03596, μ/M = 1.370745633×10⁻⁴ and R_M = 109722.273495 cm⁻¹. $\Delta E_M^{(1)}$ and $\Delta E_M^{(2)}$ are the first and second order mass polarization corrections given by equations (103) and (104), and ΔE_{st} is the singlet–triplet mixing correction.

	$2\,^1P_1$	$2\,^3P_0$	$2\,^3P_1$	$2\,^3P_2$
E_{nr}	−27176.690015	−29222.155521	−29222.155521	−29222.155521
$\Delta E_M^{(1)}$	1.385032	−1.942356	−1.942356	−1.942356
$\Delta E_M^{(2)}$	−0.000694	−0.000845	−0.000845	−0.000845
ΔE_{rel}	−0.467728	1.300852	0.314817	0.237533
ΔE_{anom}	0.0	0.001203	−0.000620	0.000131
ΔE_{st}	0.000158	0.0	−0.000158	0.0
$(\Delta E_{RR})_M$	−0.000284	−0.000014	0.000208	0.000132
$(\Delta E_{RR})_X$	0.000126	0.000593	0.000263	0.000228
ΔE_{nuc}	0.000002	−0.000026	−0.000026	−0.000026
$\Delta E_{L,1}$	0.002009	−0.041087	−0.041095	−0.041111
$\Delta E_{L,2}$	−0.002096	−0.001518	−0.001518	−0.001518
Total	−27175.773489	−29222.838720	−29223.826851	−29223.903353

$$\ln[\epsilon(2\ ^1P)/\text{Ryd}] = \ln[19.6952298(Z + 0.00600)^2] \tag{134}$$

$$\ln[\epsilon(2\ ^3P)/\text{Ryd}] = \ln[19.6952298(Z + 0.00475)^2] . \tag{135}$$

The coefficints C_{61} and C_{60} in eq. (129) are state dependent. The values of C_{61} for the 1s, 2p$_{1/2}$ and 2p$_{3/2}$ states are 3.964530, 0.429167 and 0.241667 respectively. In parallel with eq. (133) for the Bethe logarithm, the two–electron value for C_{61} is estimated to be

$$C_{61}(nLS) = [C_{61}(1s) + n^{-3}\,C_{61}(nl)]/(1 + n^{-3}\delta_{0,l}) . \tag{136}$$

After transforming from jj to LS coupling, this gives the values

$$C_{61}(2\ ^3P_0) = 3.964530, \quad C_{61}(2\ ^3P_1) = 4.010363, \quad C_{61}(2\ ^3P_2) = 3.994738$$

and $C_{61}(2\ ^1P_1) = 4.002551.$

For C_{60}, the values obtained from an equation analagous to (136) do not differ significantly from the value −24 for the 1s state.(Mohr 1974, Sapirstein 1981).

In addition to the above, there are anomalous magnetic moment corrections to the spin–dependent operators H_{so}, H_{soo} and H_{ss} . One advantage of calculating these corrections separately is that the statistically weighted sum over states vanishes for each manifold of fine structure states. The diagonal contribution to the energy is (Stone 1963)

$$\Delta E_{anom} = (\alpha/\pi)(\langle H_{so}\rangle + \tfrac{2}{3}\langle H_{soo}\rangle + \langle H_{ss}\rangle) . \tag{137}$$

This replaces the usual anomalous magnetic moment correction which normally appears in the one–electron Lamb shift for $l \neq 0$.

Finally, the correction due to finite nuclear size can be estimated from the lowest order expression

Table 12. Comparison of calculated energy levels (in cm⁻¹) with Martin's (1987) tabulation of experimental values. The calculated energies are adjusted by 198310.773489 cm⁻¹ to bring the 2 ¹P reference state into correspondence.

Term	calculated value	Martin	difference
1 $^1S_0$	0.110688	0.00±0.15	0.110688
2 $^1S_0$	166277.541501	166277.542(3)	−0.0005(30)
2 $^3S_1$	159856.078387	159856.07760(50)	0.00077(50)
2 $^1P_1$	171135.000000	171135.00000(11)	0.00000(11)
2 $^3P_2$	169086.870151	169086.869782	0.00037
2 $^3P_1$	169086.946636	169086.946208	0.00043
2 $^3P_0$	169087.934769	169087.934120	0.00064
3 $^1D_2$	186105.071296	186105.06984(9)	0.00146(9)
3 $^3D_3$	186101.650790	186101.64950(3)	0.00129(3)
3 $^3D_2$	186101.653305	186101.65204(3)	0.00126(3)
3 $^3D_1$	186101.697503	186101.69622(3)	0.00128(3)
4 $^3D_1$	191444.605211	191444.60399(6)	0.00122(6)
5 $^3D_1$	193917.265921	193917.26469(8)	0.00123(8)

$$\Delta E_{nuc} = \frac{2\pi Z (R/a_0)^2}{3} \langle \delta(\vec{r}_1) + \delta(\vec{r}_2) \rangle \tag{138}$$

where R is the root–mean–square radius of the nuclear charge distribution and a_0 is the Bohr radius. The accurate value $R = 1.673$ fm (Borie and Rinker 1978) is known from fine structure measurements in muonic helium.. Numerical values for the above corrections are summarized in Table 11 for the P–states.. All quantities are given relative to the corresponding term for the He⁺(1s) state so that its negative is a contribution to the ionization potential.

The final results of the new high precision calculations for the low–lying S– and P–states, and the D–states up to $n = 5$ are summarized in Table 12. For ease of comparison with the tabulations of Martin (1987), the calculated energies are re–expressed with the 1s2p ¹P state fixed at 171135 cm⁻¹ in conformity with his convention. The required constant shift for all the states is 198310.773489 cm⁻¹. Since all quantities have converged to more than the number of figures quoted, the differences shown in the last column of the table must be attributed either to terms not included in the calculation, or to errors in the measurements. For example, it is known from the calculations of Lewis and Serafino (1978) that there are spin–dependent corrections of about 0.0004 cm⁻¹ to the 1s2p ³PJ fine structure from the second–order mixing effects of the Breit interaction, and other higher order terms. These terms account for the differences between theory and experiment for the 2 ³PJ fine structure. A full derivation of all the $O(\alpha^4)$ spin–independent operators in the two–component nonrelativistic approximation has not yet been carried out, but these terms could reasonably account for the remaining discrepancies among the $n = 2$ states. What is harder to understand is why the discrepancies seem to increase for the higher–lying states, tending to a constant value of about 0.0013 cm⁻¹ for the n ³D sequence. The fine structure splittings for these states are well reproduced but there is a further displacement of the 3 ¹D₂ state relative to 3 ³D₂ of 0.0002 cm⁻¹. Since uncertainties arising from uncalculated relativistic and QED effects are much reduced for these states, their positions can be taken as reliable relative to the lower–lying states. It remains to be seen if a downward shift as large as

0.0013 cm⁻¹ will emerge from future calculations of higher order relativistic and QED corrections for the 2 ¹P state.

Relationship to Asymptotic Expansions for Rydberg States. It is often useful to regard a Rydberg atom as a distinguishable outer electron moving in the field of a polarizable core consisting of the inner electron and the nucleus. In this picture, the instantaneous Coulomb interactions lead to an effective potential of the form (Drachman 1982)

$$V_{eff} = -\frac{1}{2}\left[\frac{a_0}{R}\right]^4\left[\frac{\alpha_d}{a_0^3}\right] + \frac{1}{2}\left[\frac{a_0}{R}\right]^6\left[\frac{-\alpha_q + 6\beta}{a_0^5}\right] + O(R^{-7})$$

+ finite nuclear mass corrections   (139)

in units of e^2/a_0. Here, α_d is the dipole polarizability, α_q is the quadrupole polarizability, β is a nonadiabatic correction etc. All of these instantaneous effects are automatically included in a complete Hylleraas–type variational calculation for a Rydberg state if done to sufficient accuracy.

In studies of retardation effects in the long range interactions of Rydberg electrons with the inner core (Kelsey and Spruch 1978), Au, Feinberg and Sucher (1984), and more recently Au (1988), have shown that there are important corrections to V_{eff}. The form of the corrections depend on the dominant range of $R \simeq r_2$ values for the outer electron. For $R < a_0/\alpha$, the correction is

$$\Delta V = \frac{\alpha^2}{4}\left[\frac{a_0}{R}\right]^4 - \frac{7\alpha^3}{6\pi}\left[\frac{a_0}{R}\right]^3 + O[\alpha^4(a_0/R)^2]$$ (140)

The first term is automatically included in a Hylleraas–type calculation as a correlation correction to matrix elements of the Breit interaction. The second term is identical to the "Q" term in equations (130) and (131) if one makes the replacement $r_{12} \rightarrow R$ and neglects the $\delta(\vec{r}_{12})$ contact terms. The latter give rise to exponentially decreasing potentials in the asymptotic region, and so do not contribute to a $1/R^n$ type expansion. However, the contact terms may still make a significant contribution in this inner asymptotic range. Also the replacement $r_{12} \rightarrow R$ introduces errors of several percent for only moderately excited states. For example, the percentage difference between $\langle 1/r_{12}^3 \rangle$ and $\langle 1/R^3 \rangle$ increases slowly from 4.36% for the 1s3d state to 5.70% for the 1s8d state.

For $R >> a_0/\alpha$, the terms in eq. (140) display what Au calls "Casimir behavior" due to the long range retardation effects.. Each term changes its R–dependence and the two combine to give the single Kelsey–Spruch correction

$$\Delta V = \frac{11\alpha}{4\pi}\left[\frac{a_0}{R}\right]^5\left[\frac{\alpha_d}{a_0^3}\right] .$$ (141)

The difference between this and eq. (140) is what one might define as a genuine long range retardation correction. The difference represents effects which would not be included in a straight–forward Hylleraas–type calculation for very high n states, with the Breit interaction included in the usual way as a perturbation.

ENERGY LEVELS FOR HIGH–Z TWO–ELECTRON IONS

A detailed and in principle exact theory of relativistic and QED effects in high–Z two–electron ions is described in Peter Mohr's lectures. However the calculations required are very lengthy, and final results may not be available for some time. In view of the importance of the spectra emitted by two–electron ions in plasmas and astrophysical objects, and the large body of laboratory measurements which now exists, it is desirable to have approximate methods for extending the low–Z results described in the previous sections into the high–Z region. The present section describes calculations which take into account nonrelativistic and lowest order relativistic corrections to all orders in Z^{-1}, together with the higher order relativistic and QED corrections in a one–electron approximation. The QED terms include two–electron corrections to the Bethe logarithm and to the electron density at the nucleus. The leading terms not included are corrections of $O(\alpha^4 Z^4)$ a.u. and $(\alpha^2 Z^4 \mu/M)$ a.u., arising from relativistic correlation and relativistic reduced mass effects.

As described previously (Drake 1979, 1982, 1985, 1988c) the calculation of energy levels is based upon a diagonalization of the martix

$$H = (H_{nr} + B_P + H_{mp} + E_L + H_{ns})_{LS} + R(H_D + V_{12} + B)_{jj} R^{-1} - \Delta \qquad (142)$$

in the basis set of zero–order degenerate states of the same parity. For example, the states $1\ ^1S_0$, $2\ ^1S_0$, $2\ ^3S_1$, $2\ ^3P_0$ and $2\ ^3P_2$ are all nondegenerate in zero order, and eq. (142) is just a simple scalar equation. Only the states $2\ ^3P_1$ and $2\ ^1P_1$ are degenerate, making (142) 2×2 matrix equation for this pair of states. The subscript LS on the first group of terms means that these terms are evaluated in LS–coupling using high precision correlated variational wave functions obtained from the nonrelativistic Schrödinger equation. The subscript jj on the second group of terms means that these terms are calculated with antisymmetrized products of one–electron Dirac spinors for nuclear charge Z, which are most naturally expressed in jj–coupling. R is the $jj \rightarrow LS$ recoupling transformation, and Δ subtracts the contributions of order Z^2, Z, $\alpha^2 Z^4$ and $\alpha^2 Z^3$ which are counted twice in the first two groups of terms. The net effect is that eq. (142) sums to infinity the one– and two–electron relativistic corrections of the form $\alpha^2 Z^4$, $\alpha^4 Z^6$, $\cdots$ and $\alpha^2 Z^3$, $\alpha^4 Z^5$, $\cdots$. The meaning of the individual terms in eq. (142) is as follows:

H_{nr} = diagonal matrix of nonrelativistic eigenvalues.

B_P = Pauli form of the Breit interaction matrix, including off–diagonal terms.

E_L = diagonal QED corrections (see eq. 128–137).

H_{ns} = nuclear size corrections.

H_D = matrix of one–electron Dirac eigenvalues.

$V_{12} + B$ = matrix of relativistic electron–electron interactions.

The nuclear size corrections are the one–electron terms tabulated by Johnson and Soff (1985) with two–electron effects incorporated in analogy with eq. (133) for the Bethe logarithm.

A detailed tabulation of energies for the $n = 1$ and 2 states of helium–like ions up to $Z = 100$ has recently been given by Drake (1988c). A summary of the comparison with experiment for the $2\ ^1S_0 - 2\ ^3P_1$ and $2\ ^3S_1 - 2\ ^3P_J$ transitions is shown in Table 13. In a few cases, results from recent multiconfiguration Dirac–Fock (MCDF) calculations (Indelicato, 1987) is also available for comparison.

The recent high precision measurement of the $2\ ^1P_1 - 2\ ^1S_0$ transition in Be^{++} Scholl et al. 1988) is in excellent agreement with theory. However, because of uncertainties due to $\alpha^4 Z^4$ contributions, the measurement confirms the calculated Lamb shift for the $2\ ^1S_0$ state of 2.926 cm^{-1} only to an accuracy of 10%. A similar situation exists for the high precision $2\ ^3P_J - 2\ ^3S_1$ measurements in Li$^+$ (Holt et al. 1980, 1984; Riis et al. 1986). Here, the calculated Lamb shift for the $2\ ^3S_1$ state of 1.013 cm^{-1} is confirmed to an accuracy of about 10%, although the potential accuracy is much greater.

For the higher members of the isoelectronic sequence, no marked systematic discrepancies are apparent for the $2\ ^3P_0 - 2\ ^3S_1$ and $2\ ^3P_1 - 2\ ^3S_1$ transitions. Agreement is satisfactory except for the single measurement at $Z = 26$, which is of low accuracy. The MCDF calculations give substantially larger transition energies than the present results.

TABLE 13. Comparison of theoretical and experimental transition energies. Numbers in brackets indicate the uncertainties in the final figures quoted. Units are cm^{-1} or eV as noted with 1 eV = 8065.5410 cm^{-1}.

Z	present work	experiment	MCDF[a]
		$2\,^1P_1 - 1\,^1S_0$	
18	3139.557(2) eV	3139.553(36)[b]	3139.65
26	6700.404(7)	6700.90(25)[c]	6700.75
36	13114.337(25)	13115.45(30)[c]	13114.80
		$2\,^1P_1 - 2\,^1S_0$	
4	16276.766(30) cm^{-1}	16276.772(7)[d]	
		$2\,^3P_0 - 2\,^3S_1$	
3	18231.312(10) cm^{-1}	18231.303(1)[e]	
		18231.30188(19)[f]	
5	35393.70(8)	35393.2(0.6)[h]	
6	43898.96(16)	43899.0(0.1)[i]	
7	52420.97(29)	52413.9(1.4)[j]	
		52420.0(1.1)[k]	
8	60979.65(49)	60978.2(1.5)[j]	
		60978.4(0.6)[k]	
9	69592.5(0.8)	69586.0(3.0)[l]	
		69590.9(3.5)[m]	
10	78265.9(1.2)	78266.9(2.4)[l]	
		78265.0(1.2)[k]	
12	95853(2)	95851(8)[m]	
13	104787(3)	104778(11)[m]	
14	113820(5)	113815(4)[n]	
15	122970(6)	122940(30)[o]	
16	132238(8)	132198(10)[n]	
		132219(5)[p]	
17	141640(10)	141643(40)[n]	
18	151186(13)	151204(9)[q]	151277
26	233604(57)	232558(550)[r]	233643
92	256.6(1.0) eV	260.0(7.9)[s]	
		$2\,^3P_1 - 2\,^3S_1$	
3	18226.107(10) cm^{-1}	18226.108(1)[e]	
		18226.11206(21)[f]	
4	26853.039(30)	26853.1(0.2)[g]	
5	35377.40(8)	35377.2(0.6)[h]	
6	43866.22(16)	43866.1(0.1)[i]	
7	52429.20(29)	52429.0(0.6)[j]	
		52428.2(1.1)[k]	
8	61037.67(49)	61036.6(3.0)[j]	
		61037.6(0.9)[k]	
9	69742.4(0.8)	69743.8(3.0)[l]	
		69739.9(3.5)[m]	
10	78564.7(1.2)	78566.3(2.4)[l]	
		78565.7(1.8)[k]	
12	96683(2)	96683(3)[m]	
13	106028(3)	106023(7)[m]	

Table 13 (continued)

Z	present work	experiment	MCDF[a]
		$2\ ^3P_2 - 2\ ^3S_1$	
3	18228.194(10) cm^{-1}	18228.198(1)[e]	
		18228.19935(25)[f]	
4	26867.917(30)	26867.9(0.2)[g]	
5	35430.02(8)	35429.5(0.6)[h]	
6	44021.94(16)	44021.6(0.1)[i]	
7	52720.06(29)	52719.5(0.6)[j]	
		52720.2(0.7)[k]	
8	61588.98(49)	61588.3(1.5)[j]	
		61589.7(0.6)[k]	
9	70699.8(0.8)	70700.4(3.0)[l]	
		70697.9(3.5)[m]	
		70705.8(3.0)[s]	
10	80121.6(1.2)	80120.5(1.3)[l]	
		80123.3(0.8)[k]	
12	100253(2)	100263(6)[m]	
13	111152(3)	111157(6)[m]	
14	122743(5)	122746(3)[n]	
15	135152(6)	135153(18)[o]	
16	148497(8)	148493(5)[n]	
		148494(4)[p]	
17	162923(10)	162923(6)[n]	
18	178577(13)	178591(31)[q]	178684
20	214172(19)	214225(45)[u]	
22	256685(28)	256746(46)[v]	256807
26	368745(47)	368960(125)[r]	368878
29	483664(85)	483910(200)[w]	
36	900012(202)	900034(160)[x]	900179

[a]Indelicato et al. (1987).
[b]Deslattes et al. (1984).
[c]Indelicato et al. (1986).
[d]Scholl et al. (1988).
[e]Holt et al. (1980).
[f]Riis et al. (1986).
[g]Löfstrand (1973).
[h]Edlén (1934).
[i]Edlén and Löfstrand (1970).
[j]Baker (1973).
[k]Stamp (1983).
[l]Engelhardt and Sommer (1971).

[m]Klein et al. (1985).
[n]DeSerio et al. (1981).
[o]Livingston and Hinterlong (1982).
[p]Galvez (1986).
[q]Beyer et al. (1986).
[r]Buchet et al. (1981).
[s]Munger and Gould (1986).
[t]Stamp et al. (1981).
[u]Hinterlong and Livingston (1986).
[v]Galvez et al. (1986).
[w]Buchet et al. (1985).
[x]Martin (1988).

The measurement for U^{90+} by Munger and Gould (1986) requires some additional discussion. The experimental $2\ ^3P_0 - 2\ ^3S_1$ energy difference was obtained indirectly from a measurement of the $2\ ^3P_0$ decay rate, with allowance for the theoretical E1M1 two—photon decay rate to the ground state (Drake, 1985). The calculated energy difference of 256.6 eV consists of a 330.97 ± 1.0 eV contribution from electrostatic terms (the first three columns in Table 2) and a −74.34 ± 0.40 eV contribution from QED and finite nuclear size terms. The uncertainties of ±1.0 eV and ±0.40 eV come from the uncalculated $\alpha^4 Z^4$ contribution and the nuclear radius uncertainty respectively. The above numerical values are slightly different from the 330.4 eV and −75.3 eV used by Munger and Gould because the present results include higher order two—electron corrections. The one—electron input data of Johnson and Soff (1985) is the same in both calculations. Subtracting the 330.97 ± 1.0 eV electrostatic contribution from the measured energy difference yields a total Lamb shift (including finite nuclear size effects) of

−71.0 ± 8.3 eV. The two–electron corrections included here somewhat improve the agreement between theory and experiment over that reported by Munger and Gould (1986). Their measurement provides one of the most stringent tests of the higher order contributions to the one–electron Lamb shift.

The $2\,^3P_2 - 2\,^3S_1$ transitions present an interesting comparison with experiment because of the rich availability of measurements up to $Z = 36$. The calculated values appear to be falling progressively further below the measurements between $Z = 18$ and 29, but then come back into remarkably good agreement at $Z = 36$. This last measurement confirms the calculated Lamb shift for the $2\,^3S_1$ state of 12703 cm^{-1} to an accuracy of 2%. However, one cannot exclude the possibility that the good agreement with theory is due to a fortuitous cancellation of uncalculated terms.

OUTLOOK FOR THE FUTURE

High precision calculations for neutral helium of the kind described here have now been completed for the low–lying S– and P–states (Drake 1988b, Drake and Makowski 1988), and the D–states up to $n = 8$ (Drake 1987). Preliminary results have been obtained for the higher–lying P–states up to $n = 10$ with no significant loss of accuracy. This indicates that there should be no serious problems in extending the D–state calculations up to $n = 10$, and then proceeding on to study the G– and H–states where a direct comparison with the asymptotic calculations of Drachman will become meaningful, in addition to very interesting comparisons with Lundeen's high precision measurements. However, there is clearly a need for a more thorough study of the $O(\alpha^4 Z^4)$ relativistic corrections. The spin–dependent terms of this order are known in a nonrelativistic Pauli form from the work of Douglas and Kroll (1974), but no similar analysis exists for the spin–independent parts. For the higher Z ions, it will be interesting to compare the results obtained in the previous section with formally exact two–electron QED calculations as they become available.

ACKNOWLEDGEMENTS

One of us (GWFD) is grateful to the Santa Barbara Institute of Theoretical Physics for its hospitality during the preparation of these lectures. Research support by the Natural Sciences and Engineering Research Council of Canada and by the National Science Foundation under grant No. PHY82–17853, supplemented by funds from the National Aeronautics and Space Administration, is gratefully acknowledged.

REFERENCES

Accad, Y., Pekeris, C. L., and Schiff, B., 1971, Phys. Rev. A, 4:516.
Andrä, J., 1974, Phys. Scr., 9:257.
Araki, H., 1957, Prog. Theo. Phys., 17:619.
Au, C. K., 1988, Phys. Rev. A (submitted) and private communication.
Au, C. K., Feinberg, G., and Sucher, J., 1984, Phys. Rev. Lett., 53:1145.
Baker, S. C., 1973, J. Phys. B, 6:709.
Beigman, I. L., and Safranova, U. I., 1971, Zh. Eksp. Teor. Fiz., 60:2025 [Sov. Phys. – JETP, 33:1102].
Bethe, H. A., and Salpeter E. E., 1957, "Quantum Mechanics of One– and Two–Electron Atoms", Springer–Verlag, Berlin.
Beyer, H. F., Folkmann, F., and Schartner, K. H., 1986, Z. Phys. D, 1:65.
Bhatt, G. C., and Grotch, H., 1987, Ann. Phys. (N.Y.), 178:1.
Blumenthal, G. R., Drake, G. W. F., and Tucker, W. H., 1972, Astrophys. J., 172:205.
Borie E., and Rinker, G. A., 1978, Phys. Rev. A, 18:324.
Buchet, J. P., Buchet–Poulizac, M. C., Denis, A., Desesquelles, J., Druetta, M., Grandin, J. P., and Husson, X., 1981, Phys. Rev. A, 23:3354.
Buchet, J. P., Buchet–Poulizac, M. C., Denis A., Desesquelles, J., Druetta, M.,

Grandin, J. P., Husson, X., Lecler, D. and Beyer, H. F, 1985, Nucl. Instrum. and Meth. B, 9:645.

Curnutte, B., Cocke, C. L., and DuBois, R. D., 1981, Nucl. Instrum. and Meth. 202;119.

DeSerio, R., Berry, H. G., Brooks, R. L., Hardis, H., Livingston, A. E., and Hinterlong, S., 1981, Phys. Rev. A, 24:1872.

Deslattes, R. D., Beyer, H. F., and Folkmann, F., 1984, J. Phys. B 17:L689.

Dewey, M. S., and Dunford, R. W., 1988, Phys. Rev. Lett., 60:2014.

Douglas, M., and Kroll, N. M., 1974, Ann. Phys. (N.Y.), 82:89.

Drachman, R. J., 1982, Phys. Rev. A., 26:1228.

Drachman, R. J., 1985, Phys. Rev. A., 31:1253 and earlier references therein.

Drachman, R. J., 1987, Phys. Rev. A, 37:979.

Drake, G. W. F., 1969, Astrophys. J., 158:1199.

Drake, G. W. F., 1971, Phys. Rev. A, 3:908.

Drake, G. W. F., 1972a, Phys. Rev. A, 5:1979.

Drake, G. W. F., 1972b, Can J. Phys., 50:1896.

Drake, G. W. F., 1976, J. Phys. B, 9:L169.

Drake, G. W. F., 1977, J. Phys. B 10:775.

Drake, G. W. F., 1979, Phys. Rev. A, 19:1387.

Drake, G. W. F., 1982, Adv. At. Mol. Phys., 18:399.

Drake, G. W. F., 1985, Nucl. Instrum. and Meth. B, 9:465.

Drake, G. W. F., 1986, Phys. Rev. A, 34:4.

Drake, G. W. F., 1987, Phys. Rev. Lett., 59:1549.

Drake, G. W. F., 1988a, in "Spectrum of Atomic Hydrogen: Advances", Edited by G. W. Series, World Scientific, Singapore.

Drake, G. W. F., 1988b, Nucl. Instrum. and Meth. B, 31:7.

Drake, G. W. F., 1988c, Can. J. Phys., 66:in press.

Drake, G. F. W., and Dalgarno, A., 1969, Astrophys. J., 157:456.

Drake, G. W. F., Goldman, S. P. and van Wijngaarden, A., 1979, Phys. Rev. A, 20:1299.

Drake, G. W. F., and Makowski, A. J., 1988, J. Opt. Soc. Am., (in press).

Drake, G. W. F., Patel, J., and van Wijngaarden, A., 1988, Phys. Rev. Lett., 60:1002.

Drake, G. W. F., and Robbins, R. R., 1972, Astrophys. J., 171:55.

Drake, G. W. F., Patel, J., and van Wijngaarden, A., 1988, Phys. Rev. Lett., 60:1002.

Edlén, B., 1934, Nova Acta Regiae Soc. Sci. Ups., 9.

Edlén, B., and Löfstrand, B., 1970, J. Phys. B, 3:1380.

Edmonds, A. R., 1960, "Angular Momentum in Quantum Mechanics", Princeton University Press, Princeton.

Engelhardt, W., and Sommer, J., 1971, Astrophys. J., 167:201.

Erickson, G. W., 1977, J. Chem Phys. Ref. Data, 6:831.

Erickson, G. W., and Grotch, H., 1988, Phys. Rev. Lett., 60:2611.

Farley, J. W., MacAdam, K. B., and Wing, W. H., 1979, Phys. Rev. A, 20:1754.

Feinberg, G., and Sucher J., 1971, Phys. Rev. Lett., 26:681; see also Sucher (1977).

Fock, V. A., 1954, Izv. Akad. Nauk SSSR Ser. Fiz., 18:161.

Freund, D. E., Huxtable, B. D., and Morgan, J. D. III, 1984, Phys. Rev. A, 29:980.

Galvez, E. J., 1986, Ph.D. Thesis, University of Notre Dame, unpublished.

Galvez, E. J., Livingston, A. E., Mazure, A. J., Berry, H. G., Engström, L., Hardis, J. E., Somerville, L. P., and Zei, D., 1986, Phys. Rev. A, 32:3667.

Gaup, A., Kuske, P., and Andrä, J., 1982, Phys. Rev. A, 26:3351.

Georgiadis, A. P., Müller, D., Strätter, H. D., Gassen, J., van Brentano, P., Sens, J. C., and Pape, A., 1986, Phys. Lett. A, 115:108.

Goldman, S. P., and Drake, G. W. F., 1980, Phys. Rev. A, 24:183.

Goldman, S. P., and Drake, G. W. F., 1983, J. Phys. B, 16:L183.

Goldman, S. P., and Drake, G. W. F., 1984, J. Phys. B, 17:L197.

Hessels, E. A., Sturrus, W. G., Lundeen, S. R., and Cok D. R., 1987, Phys. Rev. A, 35:4489.

Hiller, J., Sucher J., and Feinberg, G., 1978, Phys. Rev. A, 18:2399.

Hinterlong, S. J., and Livingston, A. E., 1986, Phys. Rev. A, 33:4378.

Holt, R. A., Rosner, S. D., Gaily, T. D., and Adam, A. G., 1980, Phys. Rev. A, 22:1563.

Holt, R. A., Rosner, S. D., Gaily, T. D., and Adam A. G., 1984, Phys. Rev. A, 29:1544.

Hylleraas, E. A., 1928, Z. Phys., **48**:469.

Hylleraas, E. A., 1929, Z. Phys., **54**:347.

Hylleraas, E. A., and Undheim, B., 1930, Z. Phys., **65**:759.

Indelicato, P, Gorceix, O., and Desclaux, J. P., 1987, J. Phys. B, **20**:651.

Indelicato, P., Briand, J. P., Tavernier, M., and Liesen, D., 1986, Z. Phys. D 2:249.

Johnson, W. R., and Soff, G., 1985, At. Data Nucl. Data Tables, **33**:405.

Kabir, P. K., and Salpeter, E. E., 1957, Phys. Rev A, **108**:1256.

Kelsey, E. J., and Spruch, L., 1978, Phys. Rev. A, **18**:15, 845 and 1055.

Klarsfeld, S., and Maquet, A., 1973, Phys. Lett., **43B**:201.

Klein, H. A., Moscatelli, F., Myers, E. G., Pinnington, E. H., and Silver, J. D., 1985, J. Phys. B, **18**:1483.

Kono, A., and Hattori, S., 1986, Phys. Rev. A, **34**:1727.

Kugel, H. W., Leventhal, M., Murnick, D. E., Patel, C. K. N., and Wood, O. R. II, 1975, Phys. Rev. Lett., **35**;647.

Lapson, L. B., and Timothy, J. G., 1973, Appl. Opt. **12**:388.

Lapson, L. B., and Timothy, J. G., 1976, Appl. Opt. **15**:1218.

Lawrence, G. P., Fan, C. Y., and Bashkin, S., 1972, Phys. Rev. Lett., **28**;1612.

Leventhal, M., 1975, Phys. Rev. A, **11**:427.

Leventhal, M., Murnick, D. E., and Kugel, H. W., 1972, Phys. Rev. Lett., **28**;1609.

Lewis, M. L., and Serafino, P. H., 1978, Phys. Rev. A, **18**:867.

Lipworth, E., and Novick, R., 1957, Phys. Rev., **108**:1434.

Livingston, A. E., and Hinterlong, S. J., 1982, Nucl. Instrum. and Meth., **202**:103.

Löfstrand, B., 1973, Phys. Scr., **8**:57.

Lundeen, S. R., and Pipkin, F. M., 1986, Metrologia, **22**:9

Lundin, L., Oona, H., Bickel, W. S., and Martinson, I., 1970, Phys. Scr., **2**:213.

Marrus, R., San Vicente, V., Charles, P., Briand, J. P., Bosch, F., Liesen, D., and Varga I., 1986, Phys. Rev. Lett., **56**:1683.

Marrus, R., and Schmieder, R. W., 1972, Phys. Rev. A, **5**:1160.

Marrus, R., Charles, P., Indelicato, P., DeBilly, Lucile, Tazi, C., Briand, J. P., Simionovici, A., Detreich, D., Bosch, F., and Liesen D., 1988, Phys. Rev. A, to be published.

Martin, W. C., 1987, Phys. Rev. A, **36**:3575.

Martin, S., 1987, Thése d'Etat, Université Claude–Lyon I, unpublished.   Result quoted by Gorceix O. and Indelicato, P.,1988, Phys. Rev. A, **37**:1087.

Mohr, P. J., 1974, Ann. Phys. (N.Y.), **88**:26 and 521.

Mohr, P. J., 1976, in "Beam Foil Spectroscopy", Edited by I. A. Sellin and D. J. Pegg, Plenum. New York.

Mohr, P. J., 1978, Phys. Rev. Lett., **40**:854.

Mohr, P. J., 1982, Phys. Rev. A, **26**:2388.

Mohr, P. J., 1983, At. Data Nucl. Data Tables, **29**:453.

Morgan, J. D. III, 1988, private communication.

Müller, D., Gassen, J., Kremer, L., Pross, H. J., Scheuer, F., Sträter, H. D., von Brentano, P., Pape. A., and Sens, J. C., 1988, Europhys. Lett., 4;503.

Munger, C. T., and Gould, H., 1986, Phys. Rev. Lett., **57**:2927.

Narasimham, M., and Strombotne, R., 1971, Phys. Rev. A, **4**:14.

Ott, W. R., Kauppila, W. E., and Fite, W. L., 1970, Phys. Rev. A, **1**:1089.

Palchikov, V. G., Sokolov, Yu. L., and Yakovlev, V. P., 1983, Lett. J. Tech. Phys., **38**:347.

Patel, J., van Wijngaarden, A., and Drake, G. W. F., 1987, Phys. Rev. A, **36**:5130

Petrasso, R., and Ramsey, A. T., 1972, Phys. Rev. A, **5**:79.

Riis, E., Berry, H. G., Poulsen, O., Lee, S. A., and Tang, S. Y., 1986, Phys. Rev. A, **33**:3023.

Sapirstein, J., 1981, Phys. Rev. Lett., **47**:1723.

Schiff, B., Lifson, H., Pekeris, C. L., and Rabinowitz P., 1965, Phys. Rev., **160**:A1104.

Scholl, T. J., Holt, R. A., and Rosner, S. D., 1988, in "Eleventh International Conference on Atomic Physics Abstracts", Paris.

Schwartz, C., 1964, Phys. Rev., **134**:A1181.

Sokolov, Yu. L. and Yakovlev, V. P., 1982, Zh. Eksp. Teor. Fiz., **83**:15 [Sov. Phys. JETP, **56**:7]

Stamp, M. F., 1983, D. Phil. Thesis, University of Oxford, unpublished.

Stamp, M. F., Armour, I. A., Peacock, N. J., and Silver, J. D., 1981, <u>J. Phys. B</u>, 14:3551.

Stone, A. P., 1963, <u>Proc. Phys. Soc. (Lond.)</u>, 81:868.

Sucher, J., 1958, <u>Phys. Rev.</u>, 109:1010.

Sucher, J., 1977, in "Atomic Physics 5" Ed. by R. Marrus, M. Prior and H. Shugart, Plenum, New York.

van Wijngaarden, A., and Drake, G. W. F., 1978, <u>Phys. Rev. A</u>, 17:1366.

van Wijngaarden, A., Goh, E., Drake, G. W. F., and Farago, P. S., 1976, <u>J. Phys. B</u>, 9:2017

van Wijngaarden, A., Patel, J., and Drake, G. W. F., 1986, <u>Phys. Rev. A</u>, 33:312.

van Wijngaarden, A., Patel, J., and Drake, G. W. F., 1988, in "Proceedings of the Eleventh International Conference on Atomic Physics", Paris.

Wood, O. R. II, Patel, C. K. N., Murnick, D. E., Nelson, E. T., Leventhal, M., Kugel, H. W., and Niv, Y., 1982, <u>Phys. Rev. Lett.</u>, 48;398.

HIGHLY CHARGED IONS IN ASTROPHYSICS

John C. Raymond

Harvard-Smithsonian Center for Astrophysics
60 Garden St.
Cambridge, MA 02138 USA

ABSTRACT

High temperatures or intense photon fluxes create highly charged ions in astrophysical plasmas. The atomic properties of these ions determine heating and cooling rates and opacities. The intensities of X-ray and ultraviolet spectral lines provide diagnostics for plasma temperature, density, ionization state and elemental composition. Here we discuss the solar corona, X-ray emitting binary stars and supernova remnants, emphasizing the kinds of observations made, the important ions and atomic processes, and the current and desirable levels of accuracy needed for the atomic rates.

INTRODUCTION

Atomic processes are important in astrophysics both because they determine the heating and cooling rates of many astrophysical plasmas and because they create the spectral features which we observe. They provide diagnostics from which the temperature, density, elemental composition and ionization state may be inferred.

Two very general considerations determine which ions are most important in astrophysics. The first is elemental abundance. Table 1 shows the relative abundances by number of the most common elements. Excluding hydrogen and helium, oxygen is most abundant. Nickel is three hundred times less common than oxygen. While the neutral or singly ionized species of elements much less abundant than these can be observed by means of their absorption lines in the spectra of cool stars, it is difficult enough to detect the highly charged ions of the abundant elements. X-ray lines of multiply charged ions of elements less abundant than chromium have not been observed, and their contributions to total heating and cooling rates of hot astrophysical plasmas are negligible.

The second consideration is interstellar opacity. Photoabsorption by neutral hydrogen and helium makes the interstellar gas opaque between about 200 Å and 912 Å toward all but the nearest stars. Few stars beyond 100 parsec can be observed at these wavelengths, while our galaxy is nearly 10,000 parsec in radius. Lines in this spectral range are still quite important for solar studies, since interstellar opacity is not a problem for observing the sun. Lines in this range are also extremely important to the cooling rate of plasmas between about 10^5 and 10^6 K. The cooling curve, assuming ionization equilibrium and the abundances listed in Table 1 is shown in Figure 1. The power emitted in ergs cm^{-3} s^{-1} is given by $\Lambda(T)n_e n_H$. The large peak in cooling rate between 10^5 and 3×10^5 K is due to 2s - 2p transitions in C III, C IV, N IV, N V, O IV,

Table 1

Elemental Abundances from Allen (1973) modified
to fit Solar Flare Spectra

Element	Abundance	Element	Abundance
H	1.00	Si	.000040
He	.085	P	.00000033
C	.00033	S	.000016
N	.000091	Cl	.000040
O	.00066	Ar	.0000079
F	.000000040	Ca	.0000040
Ne	.000083	Cr	.00000070
Na	.0000018	Fe	.000040
Mg	.000052	Ni	.0000040
Al	.0000025		

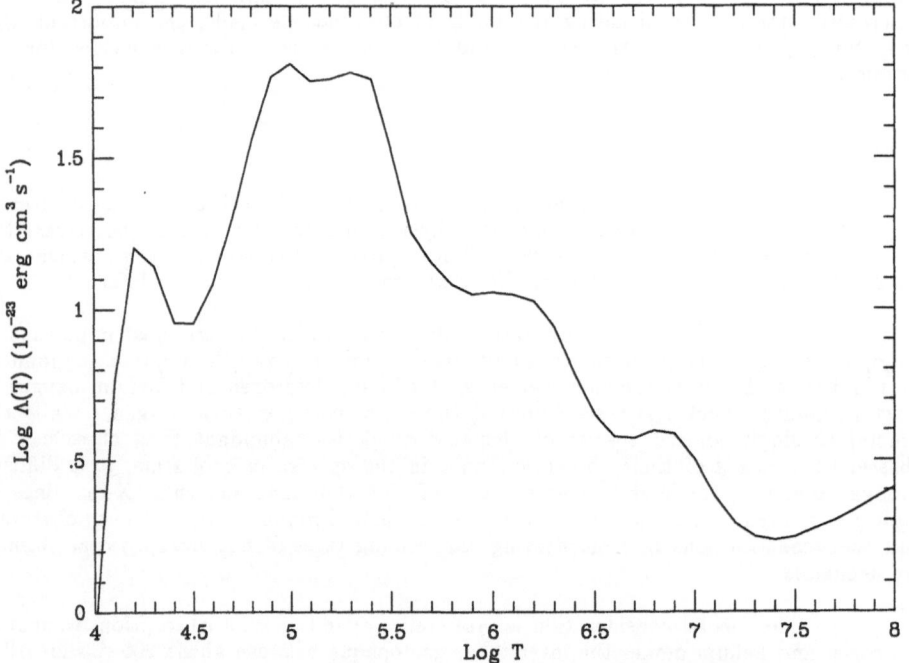

Figure 1. The cooling coefficient Λ as a function of temperature. It is somewhat density dependent at low temperatures, and an electron density of 10^{10} cm^{-3}, typical of the solar transition region, was assumed.

O V and O VI. The cooling rate determines the temperature and gasdynamics of both interstellar plasmas and the coronae of stars. For example, the peak in the cooling curve is reflected in a deep minimum in the amount of gas as a function of temperature in stellar coronae. In general, however, more astrophysicists care about X-ray, optical or ultraviolet lines than about XUV lines.

At shorter wavelengths the opacities of other elements dominate. In particular, the K absorption edges of C, O and Fe are prominent. For most purposes the absorption by neutral atoms in the interstellar gas is most important, but some bright X-ray sources contain enough highly ionized gas that the photoabsorption cross sections of highly charged ions must be considered (Krolik and Kallman 1984).

The following sections discuss three types of astrophysical plasmas and the atomic processes important in each. The types of instruments used to observe these plasmas are described to provide a basis for discussion of the accuracy needed for interpretation of astronomical observations.

BINARY X-RAY SOURCES

a) Description

Highly charged ions can be found where temperatures are high or photon fluxes are large, and X-ray binaries provide both conditions. A normal star and a compact star orbiting one another produce hard X-rays if the normal star transfers mass to its compact companion. When matter falls onto the compact star its gravitational potential energy becomes kinetic energy corresponding to the free fall velocity of thousands of kilometers per second. That kinetic energy must be thermalized, then radiated away if the accreting matter is to settle onto the surface of the compact star.

If the compact star is a white dwarf -- a star of roughly one solar mass and a size comparable to the Earth -- and the normal star is a rather small, ordinary star, the system is called a cataclysmic variable. Its luminosity is on the order of the solar luminosity, but about half the energy is emitted as X-rays. It may brighten every few months by a factor of 10 - 100 (dwarf nova outbursts), or a nuclear explosion on the white dwarf surface may cause a brightening by 5 orders of magnitude as a nova.

On the other hand, if the compact object is a neutron star -- again a solar mass of material, but a radius on the order of 10 km -- the system is called an X-ray binary. Nearly all the accretion luminosity is emitted as X-rays, with a typical energy of 10 keV. Most of these systems have orbital periods of a few hours to a few days. Mass transfer can occur as matter is drawn off the normal star through the Lagrangian point between the stars. In other cases the normal star emits a wind which the neutron star can capture. Thermonuclear flashes can occur on the neutron star surface, but instead of nova explosions they produce extremely intense X-ray bursts lasting just a few seconds.

Finally, the compact object could be a black hole. Only three X-ray binaries are considered strong black hole candidates, Cygnus X-1, LMC X-3 and A0620-00. In all three cases, the mass of the compact object derived from the orbital period and velocity seems to be larger than the theoretical upper limit for a stable neutron star.

There are several very hot regions in binary X-ray systems, though they are not all present in all systems. In most systems, the angular momentum of the accreting material forces it to flatten into a disk. The gas must lose angular momentum to migrate inwards, and in doing so it dissipates energy, heating the disk to $\sim 10^5$ K in cataclysmic variables and 10^7 K in neutron star systems. The disk material is still orbiting very rapidly just above the surface of the compact star. The associated kinetic energy, which is half the total energy released, must be dissipated in a thin boundary layer. Depending on the accretion rate, the boundary layer material may be anywhere between 10^5 and 10^8 K. Both cataclysmic variables and neutron star systems may also contain winds or static coronae above the accretion disk. In both cases the gas is heated and ionized by X-rays from the central source, but the ions in the winds of CV's are typically six times ionized, while in the neutron star systems most of the elements up to iron are likely to be stripped.

The above description probably applies to the majority of X-ray binary systems, but the compact component in some neutron star and white dwarf systems has a very strong magnetic field. Fields up to about 3×10^7 Gauss are seen in some cataclysmic variables, and 10^{12} Gauss in some neutron stars. This field channels the accreting material directly onto the magnetic pole. The accreting material then passes through a standing shock front, which heats the gas to 10^8 K.

b) Types of observations

The X-ray and ultraviolet radiation from these systems cannot penetrate the Earth's atmosphere, so they must be observed from satellites. The IUE satellite has been operating for a decade now, observing the 1200 - 3200 Å band at 6 Å resolution or, for stars which are bright in the ultraviolet, 0.1 Å resolution. It consists of a small (45 cm diameter) telescope with two echelle spectrographs, UV to optical converters, and vidicon detectors. In most X-ray binaries, IUE spectra show the resonance doublets of C IV, Si IV and N V. In a few cases, such as recurrent novae, forbidden lines such as [Fe XII] $\lambda\lambda$ 1240, 1354 and [Si VIII] λ 1440 are seen.

Most of the observations at X-ray energies have been made with proportional counters. The X-rays pass through a thin plastic window into a gas filled chamber, where they ionize the gas. Thin wires in the gas are held at high positive potential, so electrons produced by an X-ray are accelerated toward the nearest wire, producing a cascade of secondary electrons. Since the final signal (pulse height) is more or less proportional to the energy of the incident X-ray, one can infer its energy. The efficiency of these devices is very high, but the energy resolution is quite poor, typically $\Delta E/E = 1$ at 1 keV. In fact, one generally can't uniquely determine the incident photon spectrum from the pulse height spectrum. Rather, a model spectrum must be folded through the instrument response and compared with the pulse heights. For X-ray binaries, a blackbody spectrum or an optically thin thermal bremsstrahlung continuum is generally used, with the interstellar absorbing column as an additional parameter. For most observations, the limitations in signal-to-noise due to particle background and the limited number of photons detected, together with the limited energy range of the data, make it impossible to determine any spectral information beyond a continuum temperature. As one gets better data, it usually turns out that more components, either continuua of different temperatures or spectral lines, must be added to the model spectra to obtain an acceptable fit to the data.

Many of the best observations at modest X-ray energies have been made with grazing incidence telescopes aboard the *Einstein* and EXOSAT satellites. They are nested sets of polished mirrors which efficiently reflect X-rays at angles of a few degrees. The advantage of such a telescope is not an increase in collecting area, as it is for large ground-based telescopes. Many proportional counters with areas larger than those of the telescopes have been flown. Rather, the advantage comes from the reduction of the background noise level by orders of magnitude, since the background is spread over the entire detector, while the image of the X-ray source is much smaller. So far these telescopes have observed up to 2 keV (EXOSAT) and 4 keV (*Einstein*) photon energies. Most of these observations were made with proportional counters or microchannel plates. In order to achieve better spectral resolution, both X-ray telescopes carried transmission diffraction gratings which could be placed across the mirrors. Spectral resolution of about 0.5 -1 Å was obtained between 44 and 95 Å in the EXOSAT observations of the magnetic cataclysmic variable AM Her.

In the not too distant future, greatly improved X-ray spectra are expected from NASA's AXAF satellite and ESA's High-Throughput X-ray Spectroscopy Mission to be launched in the mid 1990's. These will offer greater collecting areas and extend the energy range to include the important Fe K feature at 6.7 keV. This will be coupled with lower background, higher sensitivity detectors. At ultraviolet wavelengths the

Hubble Space Telescope will offer a much larger telescope area and better detectors, improving the quality of the spectra and making it possible to study time variablity of the UV spectra on short time-scales.

c) Atomic processes

Where does atomic physics fit into all of this? Iron is the abundant element of highest atomic number, and even iron can be completely stripped in the hottest parts of the accretion flow. Bremsstrahlung emission from electron collisions with hydrogen and helium ions dominates the emission, and the spectrum is a continuum, perhaps modified by electron scattering.

In many X-ray binaries, however, a corona above the accretion disk envelopes the compact star and extends to about half the disk radius. In most models it is supported by thermal pressure. The temperature is determined by a balance of radiative cooling against compton heating (photons having $h\nu > kT$ colliding with electrons) and photoelectric absorption. Temperatures of a few million Kelvin are expected. The ionization state of this gas is given by a balance between photoionization and radiative recombination. Thus the most important atomic parameters are the photoionization cross sections from the ground and low-lying metastable levels. Photoionization cross sections of excited states are also crucial in determining the radiative recombination rates and the intensities of emission lines resulting from the decay of the excited states. For the most part, only hydrogenic cross sections are available for the excited levels. Computed photoionization cross sections are available for the ground levels of most astrophysically important ions (Reilman and Manson 1979; Clark, Cowan and Bobrowicz 1986), but it is difficult to assess their accuracy. Only a few calculations and no measurements are available for the highly ionized species. An accuracy level of around 30% seems plausible for the photoionization continuum, but resonances can be important for all but the simplest ions (e.g. Drew and Storey 1982).

The accretion disk corona may also contain enough hot gas that optical depths to photoabsorption may be significant at some wavelengths. Cygnus X-3 is a likely example of such absorption (L. Molnar, private communication). The depth of eclipse as a function of energy does not match that expected for absorption by cold gas, but it does show a feature of about the right strength to be the iron K edge.

The strongest X-ray emission feature is iron Kα. It is a blend of lines between 6.4 and 6.7 keV ranging from the resonance transitions of hydrogenic and helium-like Fe XXVI and Fe XXV through the inner-shell transitions of Li- through Ne-like ions, all the way down to Fe II. The lines of the higher ions are formed by direct collisional excitation, while lines of the lower ions are produced by radiative decay following K shell photoionization. The competing Auger ionization process is, of course, important in determining the intensities of the inner-shell transitions. Thus one needs to know collisional excitation cross sections, radiative and Auger rates, and recombination rates.

So far, these iron features have been measured only with proportional counters ($\Delta E/E \sim 0.2$) or gas scintillation proportional counters ($\Delta E/E \sim 0.1$) in X-ray sources other than the Sun. The equivalent width, defined as the ratio of line to continuum intensity, can be measured quite accurately even at low spectral resolution because of the lack of other significant spectral features in that energy range. It is also possible to measure the centroid of the emission line energy to an accuracy far beyond the spectral resolution if the statistical quality of the data is high enough. This makes it possible to tell whether the feature arises from Fe XXVI (6.97 keV), Fe XXV (6.70 keV), Fe XVIII-XXIV innershell transitions (6.43 - 6.69 keV) or fluorescence of low ionization stages (6.4 keV). As an example, Pravdo et al. (1977) found that the iron feature in the X-ray binary HZ Her arises mostly from fluorescence, while Sco X-1, for example, seems to show the Fe XXV line expected from hot plasmas.

More detailed information can, of course, be obtained with higher spectral resolution observations. Kahn et al. (1984) and Vrtilek et al. (1986) have analyzed spectra of Sco X-1 and a few other bright X-ray sources obtained with the Objective Grating Spectrometer on the *Einstein* satellite. The resolution of 0.4 Å makes it possible to pick out individual spectral lines over the 7 to 46 Å wavelength range. The N VII line at 24.8 Å is strong enough to suggest an abnormally high abundance of nitrogen, which indicates that the 'normal' star has evolved significantly by burning most of its available nuclear fuel. The data are quite noisy, but there are several statistically significant features at shorter wavelengths. The strongest, at 16 Å, is a blend of Fe XVII, Fe XVIII, and O VIII Lyβ lines. Tim Kallman has been attempting sophisticated models of the emission from hot, photoionized gas, but so far the fits are not adequate. His models include recombinations to excited levels, which is a minor part of the emission in collisionally ionized plasmas, but quite important in photoionized plasmas. It may be that radiative transfer in the emission lines must be considered in detail, since the optical depths to resonance line scattering may be substantial.

Cataclysmic variables are much less luminous in X-rays, and therefore few high quality X-ray spectra exist. Swank, Ross and Fabian (1984) analyzed the 6.7 keV iron feature in the spectrum of AM Her. They used models of the effect of electron scattering and resonance scattering in the line to find the geometry of the X-ray emitting region.

Finally, we should say something about quasars. These may be a lot like enormous X-ray binaries, with a 10^8 solar mass black hole replacing the compact object and accreting mass supplied by stars which are torn appart when they come too near. The physical conditions a short distance from the central energy source may be quite similar to those encountered in X-ray binaries, though the intrisic X-ray emission spectrum has a non-thermal (F$_\nu$ goes as $\nu^{-\alpha}$), instead of a thermal shape. Iron K emission features are not generally seen, possibly because at low spectral resolution they are overwhelmed by the bright power law continuum. An extremely broad absorption feature is seen at higher resolution in the spectrum of a BL Lac object (sort of a quasar without optical or UV emission lines). It has been attributed to O VIII resonance line absorption in a relativistic jet (Krolik *et al.* 1985).

At UV wavelengths, Lyα and lines of C IV, N V and O VI are observed. These can be modeled with clouds of gas heated and ionized by the quasar continuum. K shell photoionization follwed by Auger ionization dominates the heating, and collisional excitation of the observed UV lines dominates the cooling. At present the uncertainties in the atomic rates are dwarfed by uncertainties in the geometrical structure and radiative transfer in these models.

SUPERNOVA REMNANTS AND HOT INTERSTELLAR GAS

a) Description

A star must support its own weight against gravitational collapse, so gravity must be balanced by a pressure gradient, either normal thermal pressure or degenerate electron pressure. If the stellar mass is less than 1.4 solar masses, or if a high central temperature can be maintained by nuclear burning, there is no problem. However, once a star has burned the material in its core up to iron, any further nuclear fusion would be endothermic. This explains the relatively low abundances of elements heavier than iron. It also implies that after a massive star exhausts its nuclear fuel, the core will collapse to form a neutron star or black hole, depending on the core mass. This collapse liberates about 10^{53} ergs of gravitational energy, mostly in the form of neutrinos. The small fraction of the energy which is transmitted to the outer layers of the star drives off a few solar masses of material at velocities around 10,000 km/s for a total kinetic energy of $\sim 10^{51}$ ergs. The optical display, which makes the supernova comparable in brightness to an entire galaxy for several months, accounts for only a percent or so of the kinetic energy of the explosion.

The expanding gas cools by adiabatic expansion to very low temperatures and remains cold until it encounters the ambient interstellar gas. When that happens, a blastwave moves out into the interstellar gas and a 'reverse shock' propagates back into the ejected material. Last year's supernova in the Large Magellanic Cloud provides an excellent example. Before it exploded, the precursor star was a relatively hot star with a strong stellar wind. Very early on, the interaction of the ejected material with this wind produced some radio emission, but basically the wind just cleared out a large cavity around the star. Thus the expanding gas is now heated only to a few thousand K by residual radioactivity of ^{56}Co. A few months ago, though, IUE observations detected emission lines of N III] λ1750, N IV] λ1486, and N V λ1240, He II λ1640, O III] λ1664, and C III] λ1909 which reveal the presence of a circumstellar shell about a lightyear (10^{18}cm) from the star. The shell was photoionized by an intense burst of UV radiation from the exploding star, and it is now at a temperature of around 40,000 K (Fransson *et al.* 1988). In 10 - 30 years the fastest stellar ejecta will reach the circumstellar shell. Depending on the relative densities, both the shell and the ejecta will be shock heated to 10^{7-9} K. The gas will be rapidly ionized, producing strong UV and X-ray line emission along the way. Figure 2 shows a possible X-ray spectrum one year after the ejected gas encounters the dense shell. This particular model assumes that the X-ray emission is dominated by 10^8 K emission, which is expected for a density contrast of about 10 between the shell and the ejected gas.

Similar temperatures, though generally at lower densities, should be found in most young supernova remnants, roughly up to 1000 years of age. The shocked stellar ejecta usually dominate the emission at all wavelengths. By an age of 10,000 years, so much interstellar gas has been swept up that shocked interstellar material dominates X-ray and UV emission. Most of the volume in such middle-aged supernova remnants is filled with hot, low density gas, typically 1 cm^{-3} and 2×10^6 K. Slower shocks in denser regions of the interstellar gas heat material to $\sim 10^5$ K. This gas can radiatively cool to about 1000 K, producing optical and UV emission lines.

It is also possible that supernova remnants do not evolve in isolation. For the interstellar pressure and supernova rate of our own galaxy, the cooling time of million degree gas is comparable to the interval between successive encounters with supernova shocks. It is therefore likely that something like half of the volume of the galactic disk is filled with million degree gas at a density of 10^{-2} cm^{-3}. This hypothesis is supported by observations of interstellar O VI absorption lines toward distant stars and a diffuse background of very soft (1/4 keV) X-rays seen over the entire sky (Cox and Reynolds 1987). The hot interstellar gas may extend up into a halo above the disk of the galaxy. Still hotter gas is found in rich clusters of galaxies. A cluster of hundreds of galaxies is likely to contain a comparable mass of low density gas at temperatures near 10^8 K. The gas in some clusters is low enough in density and high enough in temperature that its radiative cooling time is longer than the age of the universe, but in many clusters the relatively dense gas near the cluster center is able to cool, producing emission over the entire temperature range from 10^4 K to the cluster temperature.

In all these examples, one wants to determine the abundances of the elements to test models of supernova explosions and nucleosynthesis, or to test hypotheses for the origin of the hot gas in galaxy clusters. It is also necessary to find the temperature, density and ionization state of the gas, as these determine the cooling time and the energetics of the gas. These are also crucial to understanding the origin of hot cluster gas and the structure of the interstellar medium.

b) Observations

Broad band X-ray images of many supernova remnants and galaxy clusters are available, and the soft X-ray background has been mapped over the entire sky in several X-ray bands. Interpretation of these images rests heavily on a huge collection of atomic rates used to predict the X-ray line and continuum emission from hot, optically thin gas. One of the most important spectral features is again the iron K feature at 6.7 keV, which occurs in a clean part of the spectrum. The continuum in this region is produced by bremsstrahlung, so a measurement of its shape determines the electron temperature. This can be combined with the iron line equivalent width and the model calculations discussed below to find the abundance of iron relative to hydrogen. High spectral resolution is not needed, and proportional counter or Gas Scintillation Proportional counters have been extensively used for iron abundance determinations (cf. Rothenflug and Arnaud 1985).

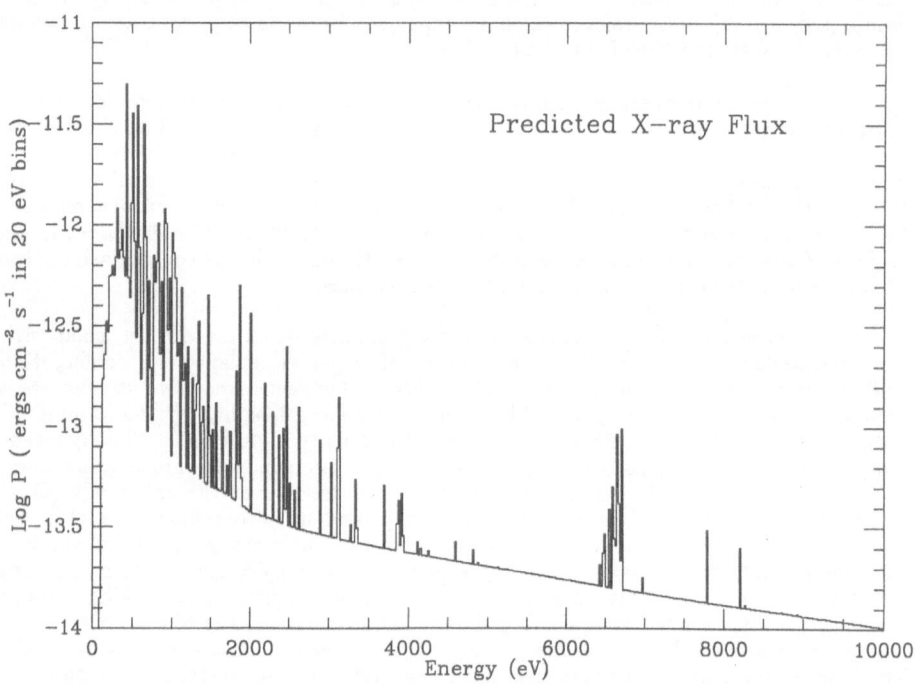

Figure 2. The predicted UV and X-ray emission one year after the
material ejected by supernova SN1987A strikes the circumstellar shell,
assuming a density constrast of about 10.

At lower photon energies, $h\nu < 4$ keV, there are so many strong spectral lines that they are completely blended together at proportional counter resolution. An exciting improvement came with the Solid State Spectrometer aboard the *Einstein* satellite. The *Einstein* telescope focussed X-rays onto a small, cooled semiconductor chip. The number of electron-hole pairs produced by an X-ray gave a measure of its energy. The device had an energy resolution of about 160 eV over the 0.5 - 4 keV range. It produced spectacular spectra of supernova remnants, galaxy clusters and other X-ray sources. Many of its most important results came from measurements of the Si XIII and XIV lines (He- and H-like ions) at 1.8 keV. Though they are blended at SSS resolution, the line centroid can be measured accurately enough to tell which dominates, and the statistical accuracy is good enough to constrain models very tightly (Becker *et al.* 1980).

Hot gas can also be studied from the ground by means of a few forbidden transitions in coronal ions. The best known is [Fe XIV] $3p^2P_{3/2} \rightarrow 3p^2P_{1/2}$ at 5303 Å. These lines are difficult to study because they are faint, even in bright suprenova remnants, but recent developments in Charge-Coupled Device detectors make them much more accessible. They now provide an excellent complement to X-ray surface brightness maps.

The next decade should see vast improvements in the X-ray spectroscopy of extended astronomical objects. A charge-coupled device being developed for the Advanced X-ray Facility will provide spectral resolution comparable to that of the SSS, but extending to higher energies and coupled with good spatial resolution. Another exciting development is the X-ray calorimeter being studied at the Goddard Space Flight Center and the University of Wisconsin. This is a tiny semiconductor chip at very low temperature whose conductivity changes in response to the temperature increase caused by the absorption of a single X-ray. A laboratory version has attained 17 eV resolution, and the goal is several times better (Holt 1987).

c) Atomic Processes

To compute the emission line spectrum of a hot, collisionally ionized gas, we must know its ionization state. This is determined by a set of equations for each element

$$\frac{1}{n_e}\frac{d}{dt}\,n_i = q_{i-1}n_{i-1} - (q_i + \alpha_i)\,n_i + \alpha_{i+1}n_{i+1} \tag{1}$$

Here n_e is the electron density, n_i the fraction of the element in ionization state i, and q_i and α_i are the collisional ionization and recombination rate coefficients, both of which are functions of temperature. For the hot gas in galaxy clusters, the time derivative in equation (1) is zero, since the gas is fairly quiescent over the age of the universe (10^{10} years), while the time constants $1/(n_e q_i)$ and $1/(n_e \alpha_i)$ are merely 10^{7-8} years. The solution of the equations with the left hand side set to zero, called the 'ionization equilibrium', is a function of temperature alone.

A recent tabulation of ionization and recombination rate coefficients and the ionization balance is given by Arnaud and Rothenflug (1985). The strongest temperature sensitivity enters through the Boltzmann factor $e^{-\chi/kT}$ in the ionization rate. It is easy to see that $n_i = n_{i+1}$ if $q_i = \alpha_{i+1}$, and this tends to happen around $\chi/kt \sim 5$. Therefore the ionization fractions of most ions peak at $\chi/kt = 5 - 10$. Elements are typically 3 times ionized at 10^5 K, 8 times ionized at around 10^6 K, and 15 - 20 times ionized at 10^7 K. It is also worth keeping in mind that the dramatic increase in χ from the Li-like ion (2s) to the He-like ion (1s), together with a relatively low recombination rate for the He-like ion, makes the He-like ion the dominant ionization stage over a wide range of temperature for all elements.

In supernova remnants, on the other hand, the time constants for ionization or recombination are likely to be longer than the age of the remnant until the supernova remnant reaches an age of 1000 - 10,000 years. Therefore, one must use an appropriate initial ionization state and solve the equations as an initial value problem for each element of shock heated gas. Thus the ionization state is specific to a particular set of supernova and interstellar gas parameters, and to the age of the supernova remnant. It cannot be conveniently tabulated, but models for individual remnants must be compared in detail with observed spectra (e.g. Hamilton, Sarazin and Chevalier 1983).

The collisional ionization rates in equation (1) have been computed with various methods. For nearly all astronomical purposes, the rate coefficient given by integrating electron velocity times ionization cross section over a Maxwellian distribution is used, if only because of our ignorance of the velocity distribution in greater detail. Cross section calculations based on the Distorted Wave approximation (e.g. Younger 1983a) seem to be the best available. Measurements exist for a number of ions, and the level of agreement is generally $\sim$ 20-30% (e.g. Gregory *et al.* 1987). The contribution due to inner shell excitation followed by autoionization can be quite important, especially for Na-like ions (Cowan and Mann 1979).

The recombination rate is made up of radiative recombination, discussed above, and dielectronic recombination, which tends to dominate in collisionally ionized plasmas. The theory of dielectronic recombination is reviewed by Hahn (1985). As an example, consider a Li-like ion in its 2s ground state which is struck by an electron on energy ϵ and angular momentum $\hbar l$. If ϵ is greater than the threshold energy E, the electron can excite the ion to its 2p state and continue on with energy E-ϵ. If ϵ is less than E, the excitation may still occur, but now the electron is left with negative energy, meaning that it is bound to the ion in a state nl'.

Once the doubly excited $2pnl'$ level has been formed, the ion has two options. It can autoionize, exactly reversing the capture process which created the doubly excited ion. In that case, the net result of the whole encounter is an elastic scattering. On the other hand, the doubly excited ion may emit a photon, most likely a $2p \rightarrow 2s$ transition. In that case, the system has reached the $2snl'$ bound level of the Be-like ion.

The cross section for capture to the $2pnl'$ state is related to the autoionziation rate A_a by detailed balance. The branching ratio for the stabilizing emission of a photon is $A_r/(A_a+A_r)$, where A_r is the radiative transition probability. Therefore, the dielectronic recombination rate by way of $2pnl'$ is proportional to

$$\frac{A_a A_r}{A_a+A_r} \tag{2}$$

The A_a can be found by extrapolating the 2s-2p excitation cross section below threshold in the Quantum Defect treatment. They vary as n^{-3}, and they are small for l larger than about 6. For modestly charged Li-like ions, n levels up to a few hundred must be included in the sum for the total dielectronic recombination rate.

Burgess (1965) computed dielectronic recombination rates for many ions and fit the results to a simple function of oscillator strength and excitation energy. This General Formula seems to be good to around 30% accuracy for many ions, though modifications are needed at low temperatures (Nussbaumer and Storey 1983) and for dielectronic recombination by way of transitions involving a change in principal quantum number, $\Delta n \neq 0$ (Merts *et al.* 1976). It is also necessary to include other decay channels in some cases. Autoionization to intermediate excited levels can drastically reduce α_{di} (Jacobs *et al.* 1977). As an example for the Li-like ion, consider $2s + \epsilon l \rightarrow 3pnl'$. For large enough n, autoionizations to $2p + \epsilon' l$ and $3s + \epsilon'' l$ are energetically allowed, and they are typically an order of magnitude faster than the inverse of the process which formed the doubly excited level. Thus, they can reduce α_{di} by an order of magnitude. The relative importance of this effect tends to decrease with increasing ionic charge. This is because A_r increases as Z^4, while A_a stays fairly constant. As a result, α_{di} is dominated by lower n doubly excited levels which are energetically incapable of autoionizing

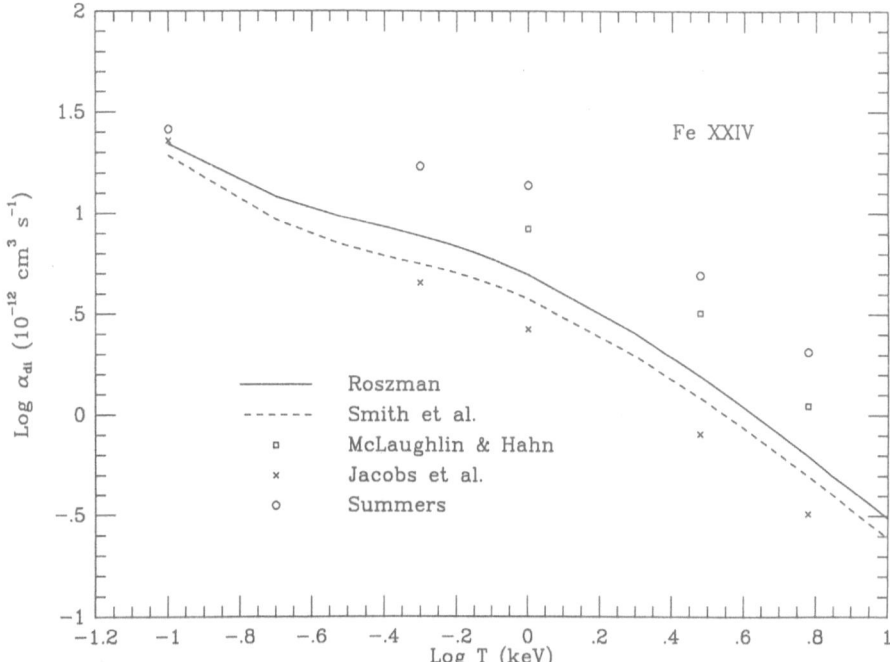

Figure 3. Dielectronic recombination rates of Fe XXIV predicted by Roszman (1987), Nasser and Hahn (1983), Jacobs *et al.* (1977), and Smith *et al.* (1985).

to intermediate levels (e.g. Smith *et al.* 1985 for Fe XVII). No complete set of α_{di} values is available which properly treats autoionization to excited levels, but calculations do exist for several iso-electronic sequences (e.g. Younger 1983b; Roszman 1987).

Unfortunately, there is less experimental guidance than there is for ionization rates, mostly because the small cross sections make measurements difficult. Dielectronic recombination cross sections have been accurately measured for Mg^+ and C^{3+} (Müller *et al.* 1987), but under the influence of strong electric fields which field ionize ions above n ~ 40, and which mix the l values for lower n. A comparison of various computed dielectronic recombination rates for Fe XXIV is shown in Figure 3. The agreement is very good at the lowest temperature, where the $\Delta n = 0$ transition dominates, but factor of five discrepancies arise for higher temperatures, where $\Delta n \neq 0$ is important. One would expect the Summers calculation, which did not include autoionization to excited levels, to be an overestimate, while Jacobs *et al.* (1977) overestimated this effect, giving too small a rate. This still leaves a factor of two difference between the Roszman and the McLaughlin and Hahn calculations, however. Considering the many ions for which the better understood $\Delta n = 0$ transitions dominate, 40% may be a reasonable estimate of the uncertainty in dielectronic recombination rates.

What do the uncertainties in ionization and recombination rates mean for the accuracy of spectral predictions? Given 20 - 40% uncertainties in both ionization and recombination, we expect the predicted ratio of adjacent ionization states n_i/n_{i+1} to be off by as much as a factor of 1.5. Some cases are better determined. For instance the rates for H- and He-like ions are better known than most, and the lines of these ions are often the strongest lines in the X-ray spectrum. In some other cases it doesn't matter very much; nearly all the silicon is in the He-like ion between about 3×10^6 and 6×10^6 K, and for many observations it doesn't really matter whether the fraction in the Li-like ion is 4% or 6%, since few astronomical observations have the dynamic range to detect the weak lines of minority ions.

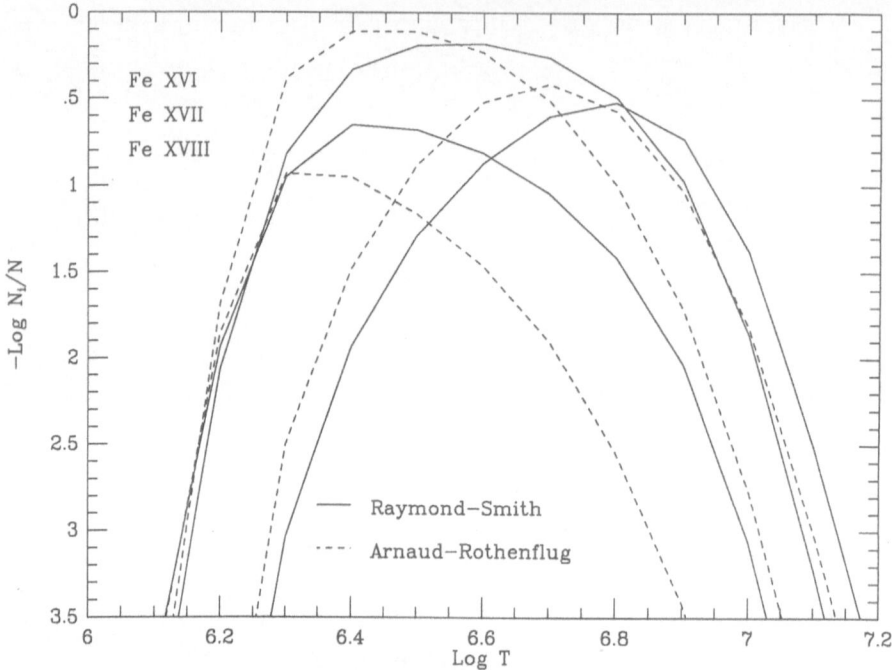

Figure 4. The ionic concentrations of Fe XVI, Fe XVII, and Fe XVIII as predicted by Arnaud and Rothenflug (1985) and the current version of the Raymond and Smith (1977) X-ray emission code.

In general, though, the uncertainties in ionization state do matter. The ionization rates increase rapidly with temperature due to the Boltzmann factor, $e^{-x/kT}$, and comparison of various ionization balance calculations shows that the uncertainties in the rates lead to ionization balance curves of similar shapes, but shifted by about 0.1 in log T relative to each other. Figure 4 shows an example, the ionization fractions of a few iron ions as computed by the current version of the Raymond and Smith (1977) code and by Arnaud and Rothenflug (1985). The neon-like ion Fe XVII dominates over a wide temperature range in both cases because its closed shell structure leads to a relatively small ionization rate compared with the loosely bound Na-like ion Fe XVI. Also, the lack of a $\Delta n = 0$ transition in the neon-like ion implies a relatively small dielectronic recombination rate. Computed intensities of Fe XVII lines at temperatures near the peak of the ionization curves agree reasonably well for the two ionization balance calculations. However, the Fe XVI abundance is very sensitive to the Fe XVII dielectronic recombination rate, and emissivity models with these two ionization calculations will differ by a factor of three in the Fe XVI line intensities predicted. Many of the astronomical objects observed possess gas over a wide temperature range, and a shift of the ioniozation curves by 0.1 in log T doesn't greatly affect the interpretation. However, an overall change in the height of a curve, such as that of Fe XVI is bound to be important for any observation of Fe XVI lines.

Models for the emission of hot gas in galaxy clusters have now been extensively used to derive iron abundances for many clusters. About half the solar abundance is found in most cases (Rothenflug and Arnaud 1985). Most of the clusters are hot enough that direct collisional excitation of Fe XXV and XXVI dominates the 6.7 keV iron feature, and the uncertainties in the atomic data are small. In some of the cooler clusters, inner shell excitation of Fe XX-XXIV and dielectronic recombination satellites to the Fe XXV lines are important, so the atomic data uncertainties are correspondingly greater. The observations have also revealed bright X-ray emission from the centers of some clusters. Both the spatial and spectral distributions indicate cooler gas than the mean cluster temperature. This is generally taken as a signature of a flow of radiatively cooling gas, as up to a few hundred solar masses per year emit their thermal energy, settle into the gravitational potential of the cluster, and perhaps form low mass stars.

As mentioned above, supernova remnant models are quite individual, since the ionization state depends not only on the temperature of the shock heated gas, but also on its history. It is necessary to start with a model structure and composition for the supernova precursor and an interstellar density. A hydrodynamics code is then used to compute the positions and strengths of the shocks and to follow the temperature and ionization structure till the present age of the remnant. It is possible to acheive quite good agreement with the available data, especially for the remnants which are simple in appearance, since the 1 D treatments of the models are most reasonable in those cases, but more detailed observations with future spacecraft are sure to require more detailed models.

IV THE SOLAR CORONA

a) Description

The Sun produces energy at its center and radiates it away into space, so from simple thermodynamics its temperature ought to monotonically decline outward. It is observed, however, that above the photosphere (the surface of the sun as seen at visible wavelengths), the temperature declines about 1000 K to a minimum of about 4600 K, then rises again. It first rises slowly to about 10,000 K (the chromosphere), then abruptly rises to 10^6 K through a narrow transition zone only 1000 km thick. Higher up, the solar wind emerges from the corona at a few solar radii. Typical densities are 10^{12} cm^{-3} in the chromosphere, 10^{10} cm^{-3} in the transition region, and 10^9 cm^{-3} in the corona.

These are only average conditions, and a good deal of spatial and temporal variation is seen down to the smallest observable scales, around 1000 km in lateral extent and a few seconds in time. Large scale structures of interest include coronal holes, which are regions of open magnetic field configuration associated with high speed solar wind streams. Densities and maximum temperatures are a little lower than in the quiet Sun. Solar active regions are areas of strong, closed magnetic field loops which confine coronal plasma. They are often associated with sunspots. Densities in the transition zone and corona are a few times larger than in the quiet Sun, and maximum temperatures in the coronal loops are around 3×10^6 K.

Solar flares are energetic events which last a few minutes to a few hours. Densities are 1-2 orders of magnitude larger than in the quiet Sun, and temperatures reach $2-3 \times 10^7$ K. Far more energetic particles are also present, as inferred from hard nonthermal X-ray emission and, in the most energetic flares, gamma ray emission, but these are less important from the point of view of interesting atomic physics.

The basic question about all of this hot gas is "Why is it hot?" As mentioned above, thermodynamics implies that it cannot be heated by radiation. In general terms, the answer is clear. Convection, rather than radiation, transports most of the energy flux through the outer third of the Sun. Some of the mechanical energy associated with that convection must reach the solar surface, and as it dissipates in very low density gas, it heats the gas to very high temperatures. Most mechanisms proposed for carrying this mechanical energy involve acoustic waves steepening into shocks as they travel down a density gradient, various sorts of MHD waves, or direct liberation of magnetic field energy by reconnection of field lines. Acoustic shocks may dominate the low chromospheric heating, but they do not carry enough energy to heat the layers above. Reconnection is almost certainly responsible for solar flares and other transient events, but the plasma physics of this process is not understood. Alfven waves almost certainly supply momentum and energy to the solar wind, and they may be important in heating the quiet sun and active region coronae, but neither the flux nor the dissipation rate of these waves is reliably known. Redistribution of energy by thermal conduction is quite efficient above 10^5 K, and this greatly complicates attempts to understand the location and nature of the heating.

In order to understand the heating, we must understand the cooling rate, $\Lambda(T)n_e n_H$, since that is what the heating must balance. The dependence on density squared simply reflects the fact that binary collisions between electrons and ions excite the ions to states which radiate away the energy. In addition to the energy balance, we must find out about the mass balance, because material seems to circulate rapidly through the whole temperature range. To assess the terms in these balances, it is important to have accurate diagnostics for electron density and temperature. We would dearly love to have reliable diagnostics for electric and magnetic fields, but so far they are available only for a few low ionization lines corresponding to the temperature minimum region and chromosphere. Zeeman splitting (Chang and Noyes 1983) and Stark broadening (Hinata 1987) are observed in Fe II, Mg II and H I.

b) Types of Observations

The immense photon flux from the sun makes high spectral and spatial resolution observations possible. At X-ray wavelengths, Bragg crystal spectrometers use the very sharp peak in reflectance at the Bragg angle and various crystals to achieve .0008 - .02 Å resolution over the 1.8 - 22 Å range. Proportional counters or microchannel plates can be used as detectors. Despite the very low efficiency of these devices, due to the roughly 1% peak reflectivity, they obtain thousands of counts in the brighter spectral lines during solar flares. The spectrometers aboard P78-1 (McKenzie *et al.* 1980) and the Solar Maximum Mission (Phillips *et al.* 1982) returned many excellent spectra. Calibration is quite good as well, because the Bragg reflection efficiency changes slowly with wavelength. However, many of the Bragg crystal spectrometers are scanning devices, sampling one wavelength at a time, and a solar flare may change considerably during the scan time. Besides the instruments which covered the 5 - 22 Å range, both P78-1 and SMM carried high resolution spectrometers which observed the K lines of the He-like ions of Fe and Ca and their satellite lines. These have been extensively used to infer temperatures and ionization states of flare plasmas.

At longer wavelengths, say 20 - 300 Å, grazing incidence devices with diffraction gratings are used (e.g. Acton *et al.* 1985). Most such instruments used so far use photographic film to record the spectra. They generally have excellent spectral resolution, but the intensity calibration is unlikely to be better than 30%. At wavelengths longer than 300 Å normal incidence spectrographs are used (e.g. Vernazza and Reeves 1978).

A major set of new solar observations should come from the Solar Heliospheric Observatory (SOHO). An ultraviolet coronagraph will measure intensities and line profiles of Lyα, the N V, O VI and Mg X resonance lines, and [Fe XII] $\lambda 1240$ as far as 9 solar radii above the Sun's surface to investigate the acceleration and heating of the solar wind. A normal incidence spectrograph will measure line intensities and line profiles in the 500 - 2000 Å range. A grazing incidence spectrograph will observe shorter wavelength lines.

In summary, the entire wavelength range from UV to X-rays can be observed at spectral resolutions high enough to distinguish individual lines and measure line profiles, though no single instrument covers the whole wavelength range. Absolute intensity calibrations tend to be good to around 30%, but the relative intensities of lines at neighboring wavelengths can be measured more accurately.

c) Atomic Processes

As mentioned earlier, the ionization state of the X-ray emitting plasma must be computed from ionization and recombination rate coefficients. At coronal densities, the ionization and recombination times are roughly 10-100 seconds, so ionization equilibrium is usually a good approximation. The major exceptions are the initial rise phase of solar flares (e.g. Mewe *et al.* 1985) and some steady flows up or down temperature gradients (Mariska 1984; Doyle *et al.* 1985).

The ionization and recombination rates are as described in the previous sections except that at solar densities the dielectronic recombination rates are somewhat reduced.

To see this, recall that the recombined ion is initially formed in a doubly excited state, such as 2pnl of a Be-like ion. If n is large, the rates for collisional excitation to other $n'l'$ levels can be larger than the radiative decay rate of the nl electron. The effects are complex, because two processes compete. First, the doubly excited 2pnl levels are initially formed at low l, typically $l < 7$. This is because A_a in expression 2 is small for larger l. However, collisions with protons can rapidly shuffle the excited electron from one l to another. When the electron is shifted from low l to high l, the ion can no longer autoionize, and the branching ratio for recombination goes to one, tending to increase the dielectronic recombination rate. The second process generally overcomes this increase, however. Even after the stabilizing 2p $\rightarrow$ 2s + hν transition of the inner electron, the outer nl electron is susceptible to direct collisional ionization and to collisional excitation to higher or lower n. These collisions effectively remove the dielectronic recombination contribution from high n. Summers (1974) has made extensive calculations of the density dependence of dielectronic recombination rates based on matrix equations for the level populations. As an example, the dielectronic recombination rate of O VI is reduced by 30% at $n_e = 10^8$ cm^{-3} and a factor of 2 at 10^{10} cm^{-3}.

Electromagnetic fields can have similar effects, both mixing l levels and imposing a high n cutoff. These effects are more difficult to compute, and calculations have so far been made only for a few ions. A field of a few Volts/cm can impose a cutoff around $n = 40$, but α_{di} by way of lower levels is increased by a factor of 3 - 6 (Müller et $al.$ 1987). In sunspots, magnetic fields of 1800 Gauss are inferred from Zeeman splitting of Fe II lines, and thermal motions of ions at 10^5 K imply effective electric fields of a few Volts/cm.

Both density and field effects are smaller for $\Delta n \neq 0$ inner transitions and for high ionic charge. This is because the nl levels which dominate in these cases have low n, and are less susceptible to collisions or fields. The basic reason is that the expression 2 is largest for $A_a \sim A_r$. A_r increases roughly as Z^4 (for $\Delta n \neq 0$), while A_a stays roughly constant with Z, but varies as n^{-3}. Thus, the peak n declines with Z. At the same time, a given nl level is more tightly bound for higher Z.

d) Collisional Excitation

To predict emission line intensities, we need to know collisional excitation rates. These are given by

$$q = 8.63 \times 10^{-6} \frac{\Omega}{\omega\sqrt{T}} e^{-E/kT} \text{ cm}^3 \text{s}^{-1}$$

The collision strength, Ω, is a slowly varying function of temperature, and ω is the statistical weight of the ground state. The wavelengths and excitation energies of the strong transitions are generally known from laboratory studies, so the major effort lies in determining the collision strength.

Seaton (1975) reviews theoretical methods for computing collision strengths. Well above threshold, the collision strength can be reliably computed in the Coulomb-Born approximation. Unfortunately, it is almost always the collision strengths near threshold we need. Ions are typically found at $\chi/kT \sim 7$. The excitation thresholds are typically $E/\chi \sim 1/4$ for $\Delta n = 0$ and 2/3 for $\Delta n \neq 0$. Therefore, for the important spectral lines $E/kT \sim 2$-5, and electrons just above threshold dominate the excitation.

Coulomb-Born calculations tend to overestimate collision strengths near threshold. Better results are obtained with the Distorted Wave approximation, in which the wavefunction of the free electron is computed with the potential of the target ion, rather than a simple point charge. An accuracy of around 20% is expected for uncomplicated ions, and this is confirmed by laboratory measurements with crossed beam techniques (e.g. Gregory et $al.$ 1987) though the agreement is not as good for complicated ions (Lafyatis and Kohl 1987). Calculations are available for most $\Delta n = 0,1$ transitions in isoelectronic sequences up through Ar-like (e.g. Mason et $al.$ 1979).

The most sophisticated calculations include the coupling among all the important reaction channels. These Close Coupling calculations naturally handle the resonances in excitation cross sections which correspond to the autoionization to intermediate levels discussed in connection with dielectronic recombination. This method requires relatively large amounts of computer time, so it is used mostly when very high accuracy is needed for line ratio diagnostics (e.g. Kingston and Tayal 1983) or when resonances make especially large contributions. In general, the resonance contribution to the excitation rate of a strong allowed line is no more than 20%, but resonances often double the total excitation of forbidden or intercombination transitions (e.g. Dufton *et al.* 1978; Bhadra and Henry 1980; Smith *et al.* 1985).

Another contribution which is sometimes important comes from dielectronic recombination satellite lines. The stabilizing photon emitted during the dielectronic recombination process is close in wavelength to the resonance transition, but slightly shifted by the presence of the outer nl electron. In collisional ionization equilibrium, the recombinations balance the ionizations of the next lower ion. Therefore, the rate of production of the satellite lines can be no larger than the ionization rate of the next lower ion, which is generally much smaller than the direct collisional excitation rates of the strong permitted lines. Exceptions occur for ions in the He-like sequence. The ionization potentials of the next lower ions, which are Li-like, are quite a lot smaller than the excitation potentials of the He-like resonance lines. The ionization rates of the Li-like ions, and therefore the excitation rates of dielectronic recombination satellites, are comparable to the direct excitation rates, especially at the low end of the temperature range where the He-like ion is found. Similar comments apply to the closed-shell Ne-like ions and the loosely bound Na-like ions just below them.

Some intensity ratios are especially useful because they can be used as direct diagnostics for important physical parameters of the emitting gas. Density diagnostics are especially crucial, because the density must be known to evaluate the energy balance of the emitting gas. The radiative losses are given by the cooling coefficient times $n_e n_H$, and the energy transport by thermal conduction is computed from the temperature gradient, which in turn is computed from the absolute line intensities and $n_e n_H$.

A line ratio will be density sensitive to some extent if it involves a metastable level whose radiative decay rate is comparable to the collisional deexcitation rate in the density range of interest. For example, the $1s2s^3S$ levels of He-like ions are metastable, and therefore the ratios of the forbidden $1s2s^3S \rightarrow 1s^2\,^1S$ line to the resonance $1s2p^1P \rightarrow 1s^2\,^1S$ line should be constant at low density, but decline for densities above a critical n_c. The radiative decay rates of these metastable levels go as a high power of Z and therefore n_c increases rapidly with the charge. Typical quiet solar corona and active region densities are comparable to n_c for O VII, making this the best diagnostic, while Ne IX is appropriate for flares (e.g. McKenzie and Landecker 1982).

Other popular density diagnostics include line ratios dependent on the populations of the $2s2p^3P$ and $2s2p^2\,^4P$ levels of Be-like and B-like ions such as C III and O IV (e.g. Dufton *et al.* 1978). These are appropriate for transition zone densities, and are widely used in analysis of EUV spectra. The corresponding levels in highly charged ions such as Fe XXIII have very short lifetimes, so the line ratios tend to be in the low density limit even in solar flares.

Other important density diagnostics can be obtained from the populations of states within of the ground terms of the B-like through F-like ions of iron. Ratios among either 2s-2p transtions near 100 Å or among 2p-3d transitions near 10 Å depend on the relative populations of, for instance, the $^3P_{0,1,2}$ states of the C-like or O-like ion (Mason *et al.* 1984).

In general, the density diagnostics yield results consistent with each other, but with uncertainties approaching a factor of two. This is because there are generally three or so collision cross sections and a radiative transition probability which go into computing the line ratio as a function of density, and if each is uncertain at the 20% level, they add up. A further problem is that one is often limited to bright lines for reliable intensity measurements, and the ratio often varies more slowly than linearly with density. In the case of the $2p^2$ $^3P \rightarrow 2s2p^3P$ to $2s2p^1P \rightarrow 2s^2$ ^1S ratio of Be-like ions, the ratio changes by only a factor of three over about an order of magnitude density range, so the uncertainty in the line ratio measurement is magnified.

Temperature sensitive line ratios have recieved less attention, but they are also important. They have been neglected mostly because few observations cover a wavelength range which provides lines with a broad range of excitation potential, and also because people generally expect that a temperature diagnostic should just yield the temperature at which the ion peaks in ionization equilibrium. However, temperature diagnostics have recently proven useful both in checking the ionization equilibrium predictions (Doschek and Feldman 1981). They also imply departures from equilibrium which must be considered for a correct interpretation of emission line intensities and which provide important insight into the energetics of the observed gas (Doyle et al. 1985).

One type of temperature diagnostic is the ratio of satellite to resonance lines for He-like ions. The excitation energy of the satellite line goes as

$$T^{-3/2} e^{-\bar{E}/kT}$$

while the excitation rate of the resonance line goes as

$$T^{-1/2} e^{-E/kT}$$

so that the ratio goes as

$$T^{-1} e^{(E-\bar{E})/kT},$$

where the excitation energy of the satellite line, $\bar{E}$, is lower than that of the resonance line by the binding energy of the nl electron which is captured. Since the observed satellites involve $n=2$, this energy difference is considerable, and the temperature sensitivity of the line ratio is quite strong. Doschek, Feldman and Cowan (1981) have investigated the line ratios of Fe XX-XXIII. They find that current ionization balance calculations can account for most of the solar flare data with the assumption of ionization equilibrium.

The most obvious type of temperature diagnostic relies on the Boltzmann factor in the excitation rate for lines of differing excitation potential. Doyle et al. (1985) compared EUV spectra of sunspot plumes with spectra of the quiet Sun. They found that ratios of high to low excitation lines were smaller in the sunspot spectra by amounts corresponding to lower temperature by about $\Delta \log T \sim 0.1$. This is about what is expected for the shift in ionization balance at transition region temperatures for radiatively cooling gas at constant density. It suggests that the sunspot plume is composed of material condensing out of the corona and cooling without energy input. Unfortunately, the uncertainties in atomic rates and in instrumental calibration made the result somewhat ambiguous. The data can be interpreted either as a shift from ionization equilibrium to lower temperatures in the sunspot, or as a shift from equilibrium to higher temperatures in the quiet sun. The former is expected for a cooling flow above sunspots, while the latter could arise from diffusion or from the net outflow of gas needed to supply the solar wind (Rousel-Dupré and Beerman 1981).

e) Comparisons

Two types of comparisons can be made to assess the accuracy of the model predictions. One can compare different theoretical calculations, or one can compare models with observed spectra. Figure 5 shows the emissivity computed with the current version of the Raymond and Smith (1977) X-ray emission code for a temperature of 10^6 K and

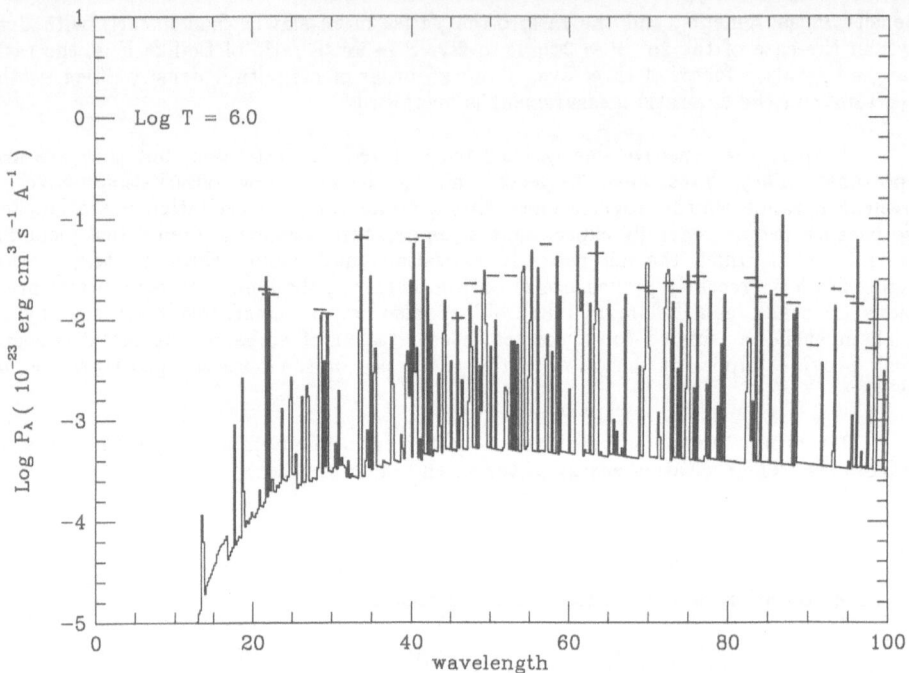

Figure 5. X-ray emissivity predicted by the current version of the Raymond and Smith (1977) X-ray code for 10^6 K. Emissivities of strong lines predicted by Mewe, Gronenschild and van den Oord are indicated by horizontal dashes.

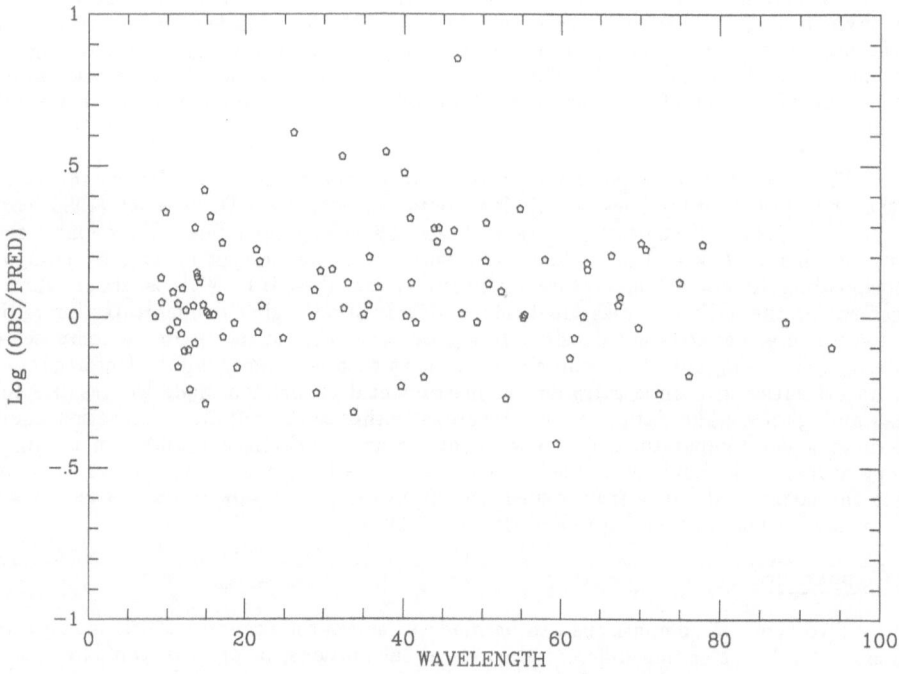

Figure 6. Ratios of observed line intensities to those predicted by flare model.

0.2 Å resolution. Horizontal lines denote the emissivities of the strong lines computed by Mewe, Gronenschild and van den Oord (1985). The strong lines of H-like and He-like ions (near 22, 28,33 and 40 Å) agree quite well. In fact, the agreement is probably better than the actual uncertainties in the calculations due to very similar choices of atomic rates. At longer wavelengths, the strong lines include many lines of moderately ionized iron, and both the ionization balance and the excitation cross sections are less well known.

In principle, a stronger test of the calculated emissivity is a comparison with observation. In practice, this is difficult because one has no independent way of knowing the temperature and density, and because there is no way to discriminate between observational errors and theoretical errors.

X-ray observations of high quality exist only for the Sun. McKenzie *et al.* (1980) present the 8 - 22 Å spectrum of a solar flare observed with the SOLEX experiment, a Bragg crystal spectrometer flown on the P78-1 satellite. A different flare was observed by Acton *et al.* (1985) over the 15 - 100 Å range with a grazing incidence spectrograph aboard a rocket. A composite spectrum can be obtained by scaling the spectra to the same O VII λ21.6 intensity.

Plasma throughout the 10^4 to $10^{7.5}$ K range is present in a solar flare. A model spectrum is computed by specifying the amount of emitting material at each temperature, then adjusting the amount of material until the strongest, most reliable lines are matched. Figure 6 shows the ratio of observed intensities to predicted intensities. The mean error is nearly 50%. This is probably composed of equal parts atomic rate errors, observational errors, and errors due to the joining of spectra of two entirely different flares by a single scaling parameter. Thus this comparison is consistent with a typical error of 30% for the theoretical line intensities, as might be expected from the excitation rates and ionization balances.

SUMMARY

A huge number of atomic rate coefficients is required for accurate interpretation of astrophysical X-ray and EUV spectra, in particular excitation, ionization, dielectronic recombination and photoionization cross sections and radiative transition probabilities for all the ions of the dozen or so most abundant elements. The current level of uncertainty in these rates seems to range from less than 20% for some of the simple, well studied ions to worse than 50% for more complicated ions. The level of uncertainty in the atomic physics has more or less kept pace with the level of observational uncertainty, but experiments planned for the next decade, including AXAF, XMM, and SOHO will require more accurate theoretical models.

This work has been supported by NASA Grant NAGW-528.

REFERENCES

Acton, L.W., Bruner, M.E., Brown, W.A., Fawcett, B.C., Schweizer, W., and Speer, R.J. 1985, *Ap. J.*, **291**, 865.
Allen, C.W. 1973, *Astrophysical Quantities*, (3rd ed.; London: Athlone).
Arnaud, M,, and Rothenflug, R. 1985, *Astr. Ap. Suppl.*, **60**, 425.
Becker, R.H., Holt, S.S., Smith, B.W., White, N.E., Boldt, E.A.,Mushotzky, R.F., and Serlemitsos, P.J. 1980, *Ap. J.*, **235**, L5.
Bhadra, K., and Henry, R.J.W. 1980, *Ap. J.*, **240**, 368.
Burgess, A. 1965, *Ap. J.*, **141**, 1588.
Chang, E.S., and Noyes, R.W. 1983, *Ap. J.*, **275**, L11.
Clark, R.E.H., Cowan, R.D., and Bobrowitz, F.W. 1986, *At. Data and Nucl. Data Tables*, **34**, 415.
Cowan, R.D., and Mann, J.B., Jr. 1979, *Ap. J.*, **232**, 940.
Cox, D.P., and Reynolds, R.J. 1987, *Ann. Revs. Astr. Ap.*, **25**, 303.

Doschek, G.A., and Feldman, U. 1981, *Ap. J.*, **251**, 792.

Doschek, G.A., Feldman, U., and Cowan, R.D. 1981, *Ap. J.*, **245**, 315.

Doyle, J.G., Raymond, J.C., Noyes, R.W., and Kingston, A.E. 1985, *Ap. J.*, **297**, 816.

Drew, J.E., and Storey, P.J. 1982, *J. Phys. B*, **15**, 2357.

Dufton, P.L., Berrington, K.A., Burke, P.G., and Kingston, A.E. 1978, *Astr. Ap.*, **62**, 111.

Fransson, C., Cassatella, A., Gilmozzi, R., Panagia, N., Wamsteker, W., Kirshner, R.P., and Sonneborn, G. 1988, *Ap. J.*, submitted.

Gregory, D.C., Meywe, F.W., Muller, A., and Defrance, P. 1987, *Phys. Rev. A*, **34**, 3657.

Gregory, D.C., Wang, L.J., Meyer, F.M., and Rinn, K. 1987, *Phys. Rev. A*, **35**, 3256.

Hahn, Y. 1985, *Adv. Atomic and Mol. Phys.*, **21**, 123.

Hamilton, A.S., Sarazin, C.L., and Chevalier, R.A. 1983, *Ap. J. Suppl.*, **51**, 115.

Hinata, S. 1987, *Solar Physics*, **109**, 321.

Holt, S.S. 1987, *Astrophys. Lett. & Comm.*, **26**, 61.

Jacobs, V.L., Davis, J., Kepple, P.C., and Blaha, M. 1977, *Ap. J.*, **211**, 605.

Kahn, S.M., Seward, F.D., and Chlebowski, T. 1984, *Ap. J.*, **283**, 286.

Kingston, A.E., and Tayal, S.S. 1983, *J. Phys. B*, **16**, 3465.

Krolik, J.H., and Kallman, T.R. 1984, *Ap. J.*, **286**, 366.

Krolik, J.H., Kallman, T.R., Fabian, A.D., and Rees, M.J. 1985, *Ap. J.*, **295**, 104.

Lafyatis, G., and Kohl, J. 1987, *Phys. Rev. A*, **36**, 59.

Mariska, J.T. 1984, *Ap. J.*, **281**, 435.

Mason, H.E. Bhatia, A.K., Kastner, S.O., Neupert, W.M., and Swartz, M. 1984, *Solar Physics*, **92**, 199.

Mason, H.E., Doschek, G.A., Feldman, U., and Bhatia, A.K. 1979, *Astr. Ap.*, **73**, 74.

McKenzie, D.L., Landecker, P.B., Broussard, R.M., Rugge, H.R., Young, R.M., Feldman, U., and Doschek, G.A. 1980, *Ap. J.*, **241**, 409.

Merts, A.L., Cowan, R.D., Magee, N.H. 1976, LA-6220-MS, Los Alamos.

Mewe, R., Gronenshcild, E.H.B.M., and van den Oord, G.H.J. 1985, *Astr. Ap. Suppl.*, **62**, 197.

Mewe, R., Lemen, J.R., Peres, S., Schrijver, J., and Serio, S. 1985, *Astr. Ap.*, **152**, 229.

Müller, A., Belic, D.S., DePaola, B.D., Djuric, N., Dunn, G.H., Mueller, D.W., and Timmer, C. 1987, *Phys. Rev. A*, **36**, 599.

Nasser, I., and Hahn, Y. 1983, *JQSRT*, **29**, 1.

Nussbaumer, H., and Storey, P.J. 1983, *Astr. Ap.*, **126**, 75.

Phillips, K.J.H., Leibacher, J.W., Wolfson, C.J., Parkinson, J.H., Fawcett, B.C., Kent, B.J., Mason, H.E., Culhane, J.L., and Gabriel, A.H. 1982, *Ap. J.*, **256**, 774.

Pradhan, A.K., Norcross,D.W., and Hummer, D.G. 1981, *Ap. J.*, **246**, 1031.

Pravdo, S.H., Becker, R.H., Boldt, A.E., Holt, S.S., Serlemitsos, P.J., and Swank, J.H. 1977, *Ap. J.*, **216**, L23.

Raymond, J.C., and Smith, B.W. 1977, *Ap. J., Suppl.*, **35**, 419.

Reilman, R.F., and Manson, S.T. 1979, *Ap. J. Suppl.*, **40**, 815.

Roszman, L.J. 1987, *Phys. Rev. A.*, **35**, 3368.

Rothenflug, R., and Arnaud, M. 1985, *Astr. Ap.*, **144**, 431.

Rousel-Dupré, R., and Beerman, C. 1981, *Ap. J.*, **250**, 408.

Seaton, M.J. 1975, *Adv. Atomic and Mol. Phys.*, **11**, 83.

Smith, B.W., Raymond, J.C., Mann, J.B., Jr., and Cowan, R.D. 1985, *Ap. J.*, **298**, 898.

Summers, H.P. 1974, *Culham Laboratory Internal Memo* IM-367, Culham Laboratory, Ditton Park, Slough, England.

Swank, J.H., Fabian, A.C., and Ross, R.R. 1984, *Ap. J.*, **280**, 734.

Vernazza, J.E., and Reeves, E.M. 1978, *Ap. J. Suppl.*, **37**, 485.

Vrtilek, S.D., Helfand, D.J., Halpern, J.P., Kahn, S.M., and Seward, F.D. 1986, *Ap. J.*, **308**, 644.

Younger, S.E. 1983a, *JQSRT*, **29**, 61.

Younger, S.E. 1983b, *JQSRT*, **29**, 67.

X-RAY LASERS

P. Jaeglé

LULI* and Laboratoire de Spectroscopie Atomiquet et Ionique,
associé au CNRS, Université Paris-Sud, 91405 Orsay, France

ABSTRACT: Multicharged ions in laser-produced plasmas are able to provide soft X-ray amplification. Conditions under which hot plasma columns act as X-ray amplifyers are investigated in several laboratories. Plasma optical properties, especially in relation with refractive index gradient and radiation trapping, are of importance to reach large gain-length product values. Multilayer mirrors are now available for designing X-ray laser cavities. Regarding X-ray laser pumping, the theory of which requires detailed atomic-level population calculation to be joined to hydrodynamical plasma modelling, experiments set off now two types of mechanisms amongst many possible theoretical schemes. On the one hand, recombination, during fast plasma cooling, populates high-lying ion levels more rapidly than the first resonance levels. This is observed for the 2-3 transition of H-like ions, especially for the line at 182-A wavelength in carbon which has shown gain with a coefficient attaining 4 cm^{-1}. This is also observed for 3-5 and 3-4 transitions of lithium-like ions, near 100-A wavelength. Gain coefficients are between 0.5 cm^{-1} and 2.5 cm^{-1}. On the other hand, the balance between collisional excitation by plasma free electrons and fast radiative decay produces population inversions between 3s- and 3p-levels in Ne-like ions and between 4p- and 4d-levels in Ni-like ions. Ne-like selenium has provided a gain coefficient of 5 cm^{-1} at 206.3 A. Ni-like ions allow to extend this scheme to wavelenghts below 100 A. In view of their possible applications, X-ray lasers are expected to be by far the brightest X-ray sources in laboratory.

I - INTRODUCTION

During the last years, conclusive experiments have demonstrated that stimulated emission from multicharged ions is able to produce X-ray amplification /1,2/. That such ions are emitters of extreme ultraviolet and X-ray radiation is of course a knowledge familiar to physicists. Taking the simplest case of hydrogenic ions as example, the energy of the optical levels reads:

$$E_n = -13.6 \times Z^2/n^2 \text{ ev} \tag{1}$$

Z being the charge of the core and n the main quantum number. For ten times

---

*National Facility for Use of Intense Lasers from CNRS and Ecole Polytechnique, 91128 Palaiseau Cedex.

charged ions, say, the transition between levels n=3 and n=4 will have an energy of 66 ev, the corresponding wavelength being of 188 A, a value which belongs to the soft X-rays. But what has posed problem during tens of years is the no less familiar relation between the spontaenous and stimulated emission coefficients which, for transition between levels m and n, can be written as:

$$A_{mn} = 8\pi h.(v/c)^3.B_{mn} \qquad (2)$$

wherefrom spontaneous emission was meant to overlay completely the stimulated processes in the short wavelength range. Then the X-ray laser was regarded as definitively impracticable. However it must be kept in mind that the radiation intensity is proportionnal to the emission coefficient A_{mn} if and only if the emitted photons do escape from the medium without interaction with other emitters, i.e. if the emitting medium is very dilute or of very small dimension. In any other cases the outgoing intensity is to be found on integrating an equation of radiative transfer which expresses both absorption and stimulated emission due to photon-ion interactions. For photon frequency v, this equation takes the form :

$$dI/dx =(hv/4\pi)A_{ji}N_j\phi(v) +$$

$$+ (hv/c).I.B_{lj}g_l(N_i/g_i - N_j/g_l)\phi(v) \qquad (3)$$

where the N's are level populations, the g's, statistical weights, $\phi(v)$, a profile function. The medium acts as an amplifier once the second term on the left hand side is negative. Then the ability of the medium to amplify radiation is not a matter of radiation frequency. Whatever is the spectral range under consideration, the only condition to be satisfied for achieving amplification is that there exists a population inversion between the levels j and i. As a matter of fact, to use population inversions between multicharged ion levels for producing X-ray amplification was suggested for the first time in 1967 /3/.

In the case of X-rays, however, the true difficulty was that, owing to the short life-time of excited levels, a large amount of pumping energy must be brought into the active medium in a very short time. Therefore a huge power was required if the usual pumping techniques were to be used. Beside the lack of laboratory X-ray source of power large enough for carrying out an effective optical pumping, such a power would certainly have destructive consequencies for the lasing device. Other methods of producing population inversion had to be found.

The fruitful approach has been opened by the study of the radiation produced by dense hot laser-plasmas. Plasmas of temperature between 100 ev (10^6 K) and 10 Kev (10^8 K) are produced since many tens of years in laboratory for the thermonuclear fusion research or merely for basic studies of atomic data or plasma processes. They are made up of multicharged ions as well as of the free electrons which have been pulled off from neutral atoms. There is a well known equilibrium relation between electronic density N_e, electronic temperature T_e , ionization energies E_{Z+1}, E_Z and ion abundances N_{Z+1} and N_Z, that is the Saha-Boltzman relation:

$$N_{Z+1}/N_Z = 2(Q_{Z+1}/Q_Z)(1/N_e)(2mkT_e/h^2)^{3/2} \text{ x}$$

$$x \exp-(E_{Z+1}-E_Z)/kT_e) \qquad (4)$$

where the Q's are the partition functions. If the plasma density is too small for the collisional processes to ensure this statistical distribution, the ratio of ion abundances will depend of atomic coefficients in a way which is simple only in special cases, like in the corona model where:

$$N_{Z+1}/N_Z = <\sigma_R v>/<\sigma_I v> \qquad (5)$$

the quantities between brackets being the averaged cross sections of collisional ionization and photorecombination.

One pecularity of laser-produced plasmas is that they have regions of very large density, because they are produced in focusing an intense laser beam onto a solid material. The early spectroscopic studies of laser-produced plasmas, in the last 60's, revealed that, for many lines, the photons do not escape freely from the medium. Moreover the observation of intensity anomalies led to thinck the populations of some levels to depart from the statistical distribution given by Boltzman's law. Then absorption, stimulated emission and population anomalies were to be taken into account to explain the observed spectral features /4/. These ideas did not meet immediatly a general agreement for, at the beginning, no detailed population calculations existed. The large density of laser-plasmas prompted to guess populations to be naturally statistically distributed rather than swept away from equllibrium.

Various ways of working out this difficulty were conceived /5-8/ until the development of calculation models showed that, in suitable conditions, population inversions may occur in plasmas, without outer pumping system, but just as the result either of plasma cooling during the recombination /9/, or of ion excitation by free electron collisions at peculiar density /10/. Such calculations have been extensively developped since this time so that, from the theoretical point of view, the X-ray laser research is now one of the most important application of the multicharged-ion physics in plasmas.

II – HOT-DENSE-PLASMA COLUMNS FOR SOFT-X-RAY LASERS

Beside the multicharged ion atomic properties, which have an essential part in the ability of the plasma to produce population inversions, there are other general plasma features, regarding temperature and density distribution, time evolution, integrated radiative properties. line shape perturbation and others, that are to be known for the clarity of the presentation of calculations and experiments.

1. Production of plasma columns

The more usual technique of laser-plasma production consists in using spherical lenses for focusing one or more beams of a powerful laser, like neodymium or CO_2 laser, at the surface of a solid target. If this is made for inertial confinement experiments, the target is spherical and its illumination must be as smooth as possible on the whole surface. As for the geometry requiered for X-ray laser experiments, it is cylindrical and the focusing devices have generally to make use of cylindrical lenses. Figure 1 shows the principle of the system in the case of a single laser beam and a massive target. If a Nd-laser is used, the 1.06 μ laser wavelength can be reduced by cristal frequency doubling.

The plasma being expanding from the target surface towards the free vacuum space, i.e roughly perpendiculary to the target, temperature and density vary along the expansion direction. On the left of fig. 1, two curves display the shapes of these variations. The electronic density N_e is maximum (10^{23} cm^{-3}) very near the surface and falls in quickly, so that it is about 10^{19}–10^{18}cm^{-3} at a few hundreds of microns. Let us notice the density region of interest for X-ray lasers to be between 10^{18}cm^{-3} and 10^{21}cm^{-3}. The maximum of the electronic temperature T_e is located at small distance from the target because the laser beam cannot propagate beyond the critical density fixed by the equality of plasma frequency and laser beam frequency (10^{21}cm^{-3} for 1.06 μ). The maximum temperature is the important parameter with regard to the highest ion charge produced in the plasma. It is around 800 ev in the collisionally pumped laser and 200ev-300ev in the recombination laser. It is to be mentionned that, in the second case, amplification does not take place at the temperature maximum but during the plasma cooling. Temperatures, densities and other parameters are obtained from hydrodynamical

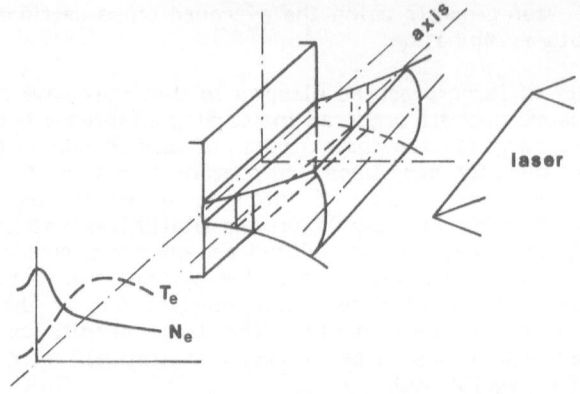

Figure 1. Cylindrical focusing of a powerful laser beam onto the target produces a hot plasma "column". The plasma properties are approximately constant along an axis parallel to the surface but vary strongly from the target surface to the empty region, as shown by the left curves. Density and temperature are suitable for soft X-ray amplification at short distance (i.e. $\sim$100 μ) from the target. The plasma life-time is of a few nanoseconds or less.

numerical codes and inserted in atomic calculations of level populations in order to obtain the conditions of population inversion occurence.

Variation on time of plasma parameters is another important feature which is closely related to the laser pulse duration. This one may be less than 100 ps and goes up to a few ns. As an illustration of atomic and plasma parameter variation, we present, in figure 2, some characteristics of an aluminum plasma produced by a 2.7 ns laser pulse. The figure displays the temperature variation together with the Li-like

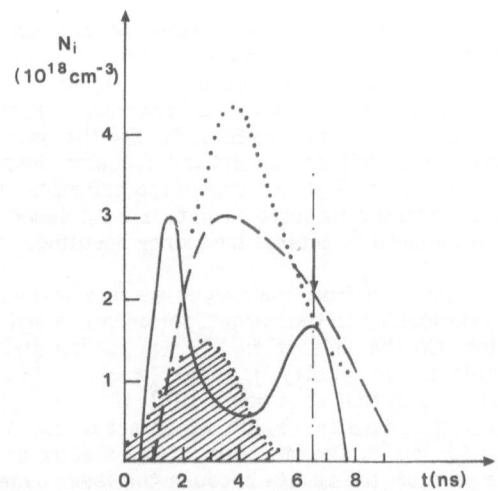

Figure 2.   Example of calculated time-dependent ion abundances in aluminum laser-produced plasma. Dashed aera: laser-pulse. Solid curve: Li-like ions. Dashed curve: He-like ions. Dotted curve: electron temperature (max. at 210 ev). The vertical arrow points at the predicted moment of the 3d - 5f population inversion outset.

and He-like ion densities as functions of time. An arrow points at the beginning of the calculated population inversion between levels of Li-like ions. More details on this calculation are given further.

A very different way of producing plasma columns, using Z-pinch implosions, has been proposed /11-13/. We just mention it here because, as far as we know, amplification is not yet proved in this type of plasma.

2. Gain coefficient

The equation (3), in the first section, recalled the base of line radiation transfer calculation. On account of the way in which plasma columns are produced, it makes sens to assume the plasma to be homogeneous in the direction of its axis (see fig. 1). Then the integration of equation (3) over the column length L gives an expression of the intensity which can be written:

$$I = S(1 - e^{-kL}) \qquad (6)$$

k being the absorption coefficient, S, the source function, or:

$$I = S'(e^{GL} - 1) \qquad (7)$$

for wavelengths at which the medium does amplify with a gain coefficient G. Both expressions reduces to each other with the substitution :

$$G = -k \quad ; \quad S = -S' \qquad (8)$$

Expression (7) is valid in the regime of weak amplification, often called the regime of amplification of spontaneous emission (ASE), for which no saturation at all does occur in beam propagation. Weak saturation may appear when $e^{GL} \gg 1$. Strong saturation requires $GL \gg 1$.

Now, let i and j be the quantum numbers of two excited levels, the level j lying above the level i; let v be the j-i transition frequency, $\phi(v)$ the profile function of the transition, g_i and g_j the statistical weights, $B_{i,j}$, Einstein's absorption coefficient. If one knows how to produce a population inversion in the plasma, the gain coefficient will be:

$$G = (hv/c) \; \phi(v) \; g_i \; B_{i,j} \times \Delta N \qquad (9)$$

where ΔN is the inversion density defined by:

$$\Delta N = (N_j/g_j - N_i/g_i) \qquad (10)$$

From (7), one sees that measuring the intensity for two different lengths, L and l, of the plasma column enables to determine the gain coefficient by the equation:

$$(e^{G L} - 1)/(e^{G l} - 1) = I_L/I_l \qquad (11)$$

It is also possible to compare the axial and transverse intensities I_a and I_t. If the plasma transverse inhomogeneity can be neglected and if the transverse measurement integrates the whole of the plasma length, one has:

$$(e^{G L} - 1)/GL = I_a/I_t \qquad (12)$$

In non-equilibrium plasmas used as radiation amplifiers, population rate equations have to be explicitly solved in order to calculate the upper and lower level populations appearing in (10). These equations express the balance between populating and depopulating processes. The values of their coefficients are fixed with the help of separated calculations supplying the large number of collisional and radiative transition probabilities which are involved in the calculation. The radiation intensities appear also in several terms of the rate equations. They are themselves functions of population distributions. Then, in a general treatment, the set of rate equations must be coupled with a system of radiative transfer equations of the form (6) or (7). If photoexcitation and photoionization contribute weakly to the populations, they can be neglected and the populations are no longer coupled to radiation. In practical cases, this assumption is valid for most levels but the radiation trapping, the possible effect of which is explained below, can change significantly the population of a few levels. Often the balance between population and de-population of levels is a result from radiative decay and excitation or de-excitation by electron impact only. Then the rate equation for the ith ion level has the form:

$$dN_i/dt = \sum_{k>i} A_{ki} N_k - N_i \sum_{k<i} A_{ik} + N_e \sum_k N_k <\sigma_{ki}\cdot v> - $$

$$- N_e \cdot N_i \sum_k <\sigma_{ik}\cdot v> \qquad (13)$$

where A_{ki} and A_{ik} are Einstein's coefficients for spontaneous emission, $<\sigma_{ki}\cdot v>$ the collision rate for the $k \to i$ transition, and N_e, the plasma electron density. In section III, the calculation results of rate equations for multicharged ions giving experimentally observed amplification will be presented.

The second important quantity in the gain coefficient definition (9) is the line profile function $\phi(v)$. In first approximation $\phi(v)$ may be replaced by the inverse of the line width $1/\delta v$. In plasmas, the spectral lines are broadened by the ion thermal motion as well as by the collisions of emitting particles with other ions and with free electrons. The Doppler effect due to the ion thermal motion at temperature T yields the familiar Gaussian profile:

$$\phi(v) = (1 /\sqrt{\pi} \xi v_0) \times \exp(- (v - v_0)/\xi v_0)^2 \qquad (14)$$

where:

$$\xi = (2kT/Mc^2)^{1/2} \qquad (15)$$

M being the atomic mass of the emitting ions. It is convenient to calculate the full-width at half-maximum of this profile in using the formula:

$$\delta v/v = 7.67 \times 10^{-5} (kT/M)^{1/2} \qquad (16)$$

where M is in grams and kT in ev's. In addition, the Doppler shift due to the plasma expansion velocity is sometimes important but it can be calculated only by numerical plasma modelling.

There is no general analytical expression of line profiles resulting from collisional processes. Extensive theoretical studies have been performed only for hydrogenic ions. Otherwise it is usual to approximate the collisionally broadened profile by a Lorentzian curve, the width of which is difficult to estimate in many cases. For hydrogenic ions, the ion-collision broadening prevails over the thermal broadening at plasma electron density such that:

$$N_e => 4.48 \times 10^{16} (z^4/n^3) (kT/M)^{3/4} \qquad (17)$$

This criterion holds mostly in the line wings. The center of lines is dominated by electron-collisions when:

$$N_e => 6.58 \times 10^{18} (z^4/n^4) (kT/M^{1/2})(1/Q) \qquad (18)$$

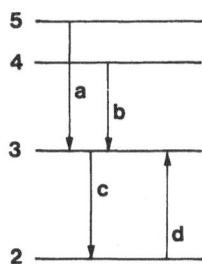

Figure 3. Mechanism of radiation trapping. a and b are
lasing transitions. c represents the radiative decay of the
lower-level 3 to the ion ground level 2. Radiation trapping
occurs via the transition d which reduces the depopulation
rate of level 3.

where Q is a Gaunt factor of the order of the unity; n is the main quantum number
of the line upper level; N_e is expressed in cm^{-3}, kT, in ev's, M, in grams /14/. It is
to be noticed that, when the width of the instrumental function is larger than the
profile function, the expression (7) of intensity is modified in a way which
depends on the shape of the profile. The modification is generally to be
determined by numerical integration.

3. Radiation trapping

If the physics of the the amplifying medium was perfectly plain, the plasma
column length increase should merely result in a proportional gain-length product
growth. However various impediments, like radiation trapping or variable
refractive index, are known to make the course toward X-ray lasers more difficult.
Refractive index will be discussed in the next paragraph. Here we explain why the
radiation trapping may set up a limitation of the gain when the plasma size
increases.

Radiation trapping is said to occur when photoexcitation either increases the
population of the lower level, or decreases the population of the upper level, of
the lasing transition. In both cases this leads to destroy the population
inversion. We will take the example of recombination lasers where the lower level
may be affected by such a trapping. An example involving the upper level can be
found in ref. /11/.

Let us look at the diagramm of figure 3. For the sake of simplicity it is reduced to
one level for each main quantum number. Moreover the radiative transitions alone
are taken into consideration. The 5-3 and 4-3 transitions are assumed to yield
laser lines. The 3-2 transition drives the emptying of the level 3. If the
population of the level 2 is large, what occurs for instance if it is the ground
level, the probability of photon reabsorption from level 2 to level 3 can be large,
then reducing the 4-3 and 5-3 population inversion. Now the reabsorption rate is
a function of the plasma size. Let us consider first an isotropic plasma, not a
column. The radiation intensity for the 3-2 transition is in all directions:

$$I = 2 \; (g_2/g_3)(h\nu^3/c^2) \; (N_3/(N_2 - (g_2/g_3)N_3)) \; (1 - e^{-kX}) \qquad (19)$$

where k is given by relations (8),(9 and (10) with j = 3 and i = 2. X is the plasma
radius. A criterion of the trapping efficiency is the intensity (19) to be large

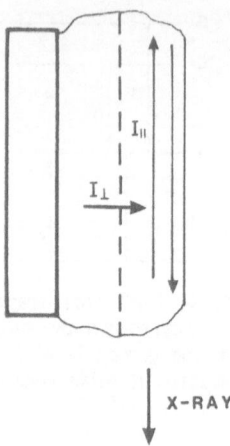

Figure 4. Rough representation of strongly anisotropic radiative transfer in a plasma column produced from a massive target (see equation 23).

enough to balance the radiative de-excitation of the level 3 when this last is populated by the decay of levels 4 and 5. Then we writte:

$$dN_3/dt = - N_3A_{32} + N_5A_{53} + N_4A_{43} + 4\pi I(B_{23}/c)(N_2 - (g2/g3)N_3) = 0 \qquad (20)$$

After a short calculation, we deduce from (20) trapping to re-populate the level 3 once:

$$e^{-kX} <= (N_5A_{53} + N_4A_{43})/N_3A_{32} \qquad (21)$$

If the populations are just close to the inversion, this reduces to:

$$e^{-kX} <= (A_{53} + A_{43})/A_{32} \qquad (22)$$

In the case of Li-like aluminum, this gives $e^{-kX} \lesssim 0.3$, i.e. $kX \gtrsim 1.2$. Owing to the small broadening of the 2-3 line, such an optical depth is reached for plasma radius of 100 µ or less.

Now we take back the column of plasma. If it is symmetric around its axis, like in the fiber technique used at the Rutherford Laboratory /15/, the above treatment remains valid, X being the transverse radius of the column. But if there is a strong asymmetry, like in some of our experiments using massive targets, the rate of the trapping depends also on the length of the column. When the plasma id produced from massive target, between the active part having an electronic density around 10^{19} cm$_{-3}$ and the target surface, takes place a dense plasma which does not contribute to amplification. This is schematically shown on figure 4. Then the trapped intensity cannot be represented by the isotropic term $4\pi I$ in equation (20). Let $I_{//}$ and $I_{\perp}$ be respectively the intensity propagating along the axis and perpendiculary to it. An estimation of the trapping can be made be made from the new equation:

$$- N_3A_{32} + N_5A_{53} + N_4A_{43} + (2I_{//} + I_{\perp})(B_{23}/c)(N_2 - (g2/g3)N_3) = 0 \qquad (23)$$

If $I_{\perp}$ is dominated by the radiation coming from the dense plasma, it will not depend on the size of the active plasma. On the other hand, since we have:

$$I_{//} = 2 (g2/g3)(h^3/c^2) N_3/(N_2 - (g2/g3)N_3)(1 - e^{-kL}) \qquad (24)$$

there is a value of L, the column length, which is solution of (23). For Li-like aluminum, in using population densities calculated with the help of atomic and hydrodynamical numerical codes /16/, we found that the lowering of the 3d-5f population inversion (λ= 105.7 A), due to the 2p-3d radiation trapping (λ= 52.4 A) , starts at L~4 mm.

Before to leave this topic, it is worth to notice the important part of the target design in the cure of radiation trapping. For instance, for long plasma comlumns, massive supports may be used if and only if they are free from the element used in the active plasma, as well as from any strong emission in coincidence with the radiation trapping wavelength.

4. Gradient of refractive index

The plasma refractive index n is related to the electronic density, N_e, and to the wavelength of the radiation by the relation:

$$n = 1 - N_e \lambda^2 (e^2/2\pi mc^2) \tag{25}$$

We have seen that, in general, N_e varies rapidly in the direction perpendicular to the target surface (see fig. 1). Therefore the plasma column has a transverse gradient of refractive index $\vec{\Delta n}$. It is known that this feature entails the curvature of the radiation path, the curvature being:

$$1/R = \vec{N} (\vec{\Delta n}/n) \tag{26}$$

where $\vec{N}$ is the unit vector normal to the trajectory. Let assume the electronic density to vary as:

$$N_e = N_c e^{(-x/\Lambda_c)} \tag{27}$$

where Λ_c is the characteristic length of density decreasing ("gradient length") and N_c, the critical density. Then the radius of curvature will be:

$$R = (2\pi mc^2/e^2 \lambda^2)(\Lambda_c/N_e) \tag{28}$$

that is to say, if the curvature radius and the gradient length are in cm, the wavelength λ in angström ms and the density in cm^{-3}:

$$R = 2.25 \times 10^{29} (1/\lambda^2)(\Lambda_c/N_e) \tag{29}$$

After a path of length L in the plasma, the deviation will reach a value approximately given by (in cm):

$$\delta x = 2.22 \times 10^{-30} \lambda^2 (N_e/\Lambda_c) L^2 \tag{30}$$

This effect is important in experiments using plasma columns of several centimeters and requiring a large electronic density for producing the population inversion /17/. The deviation can be of several hundreds of microns. This results in the beam leaving the amplifying region and the lowering of the net gain. In recombination lasers, the electronic density is generally too low for this deviation to be of consequence.

5. Multilayer mirrors

Ordinary mirrors are not available in the soft X-ray range. The refractive index of solid materials is very near the unity and the reflection coefficients in normal incidence are of the order of 10^{-4}. Realistic cavities for lasers can be designed only since multilayer mirrors have been developped /18,22/. This, in turn, needed a great deal of detailed data on optical constants that have been measured only recently. These mirrors consist in piles of thin layers (see figure 5), alternatively

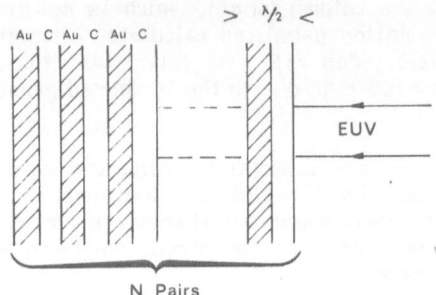

Figure 5. Example of multilayer mirror for extreme ultraviolet and soft X-ray radiation.

absorbing and transparent for soft X-rays, such that constructive interferences occur and provide reflection like any interferential mirror. The reflection coefficients are strongly dependent on the wavelength. The substratum surface polish must be of high quality for achieving large efficiencies. Reflection coefficients up to 50% have been reported. 5% to 20% is more usual.

These values, which are much lower than generally required for laser cavities, could be regarded as of little interest in practical applications. However a propitious feature of plasmas is to supply fairly large gain coefficients. Then, let I_1 the intensity of the amplified beam at the output of the plasma column when no mirror is used, I_2, the intensity with the multilayer mirror set perpendiculary to the propagation axis, R the reflection coefficient of the mirror, G an L as before. It is easy to calculate that, for a parallel beam:

$$I_2 = I_1(1 + R.e^{GL}) \qquad (31)$$

From this expression we see that, even with a 5% reflection coefficient, a single reflection on the mirror will double the intensity once the gain-length product GL reaches a value of 3. Observed gain coefficients are, at the present time, between 0.5 cm^{-1} and 5 cm^{-1}, while the length of plasma columns varies from a few millimeters up to 6 cm. Thus multilayer mirrors can be used with large advantages in many experiments.

III - AMPLIFICATION IN RECOMBINING PLASMAS

Regarding the production of population inversions, we have already mentionned two types of mechanisms which are extensively studied from both experimental and theoretical points of view. On the one hand, the rapid cooling of a plasma, after the end of the laser pulse, gives rise to a large departure from thermodynamical equilibrium, as a consequence of which population inversions appear between some levels of the multicharged ions (see figure 2). Many results have been obtained according to this scheme in H-like and Li-like ions. They will be summarized in the present section. On the other hand, steady state population inversions are also produced in some particular multicharged ions, as a result of the balance between the collisional ground state excitation by the plasma free electrons and the fast radiative decay of low lying levels. This is known as the collisional pumping scheme, used with Ne-like and Ni-like ions. It needs well defined values of plasma density and temperature. The results obtained in this way will be presented in section IV.

Though gain coefficients be smaller in the recombination scheme than for Ne-like ions pumped by plasma free-electron collisions /23,24/, the former is worthy of deep investigation on account of its comparatively moderate power requirements. From wavelength scaling studies one expects lasing conditions to be

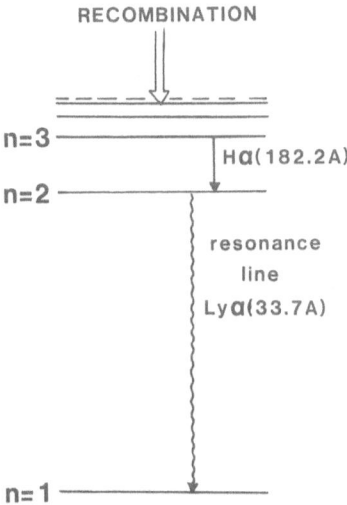

RECOMBINATION

n=3 ————————————————
            Hα(182.2A)
n=2 ————————————————

                resonance
                line
                Lyα(33.7A)

n=1 ————————————————

Figure 6. Diagram of hydrogenic C VI; amplification occurs
for the Hα-line, at 182.2 A.

reached near λ= 40 A with power density no larger than 10^{13} W/cm"^{2}, while at least
one more magnitude order is required in the collisional scheme. Amplification of
spontaneous emission is already demonstrated in recombining plasmas up to 81 A
in hydrogenic fluorine /25/ and 65 A in lithium-like sulfur /26/.

Early considered at long wavelengths in neutral hydrogen /27/ , later on this
scheme has been extensively studied in H-like ions /28-33/.

1. Hydrogenic ions '

Let us consider a plasma, initially completely ionized. Due to the fast cooling, the
plasma gets out of equilibrium. Three body recombination populates strongly the
most excited levels of H-like ions and is followed by collisional-radiative cascades
to intermediate levels. At the same time, owing to the free-electron low
temperature, electron-ion collisions from the ground level cannot balance the
radiative decay of the lowest lying levels. As a consequence, transient population
inversions occur between intermediate and lower levels. Inversions with respect to
the ground level have also been predicted, but the necessary cooling rate seems
too high to be obtained in experiments.

Adiabatic expansion. It has been shown the adiabatic expansion of a freely
expanding plasma to be able to produce population inversion between n=3 and n=2
levels of H-like carbon, provided that the expansion rate of the plasma is
optimized by choosing an appropriate target geometry . The use of thin carbon
fibers allows a symmetrical illumination by the Nd-laser beams and suits well
rapid cooling requirements /28,29/. The main transitions of interest in H-like
carbon are shown in the diagram figure 6. The 3-2 lasing-transition wavelength is
182.2 A.

Gain coefficient calculation needs two different kinds of calculations to be
brought together. First, the time history of the plasma is described by an
hydrodynamical model which includes only the gross feature of the atomic
composition. Second, the detailed atomic level populations are obtained by solving

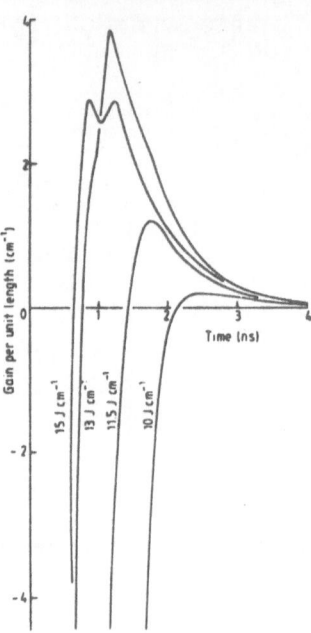

Figure 7. Gain coefficient calculated as a function of time
for the 182.2-A line of C VI. Calculations are made for 4
different Nd-laser energies (ref. /29/).

a system of coupled rate equations. In figure 7 are displayed examples of the final
result of such calculations, after substracting the population of n=2 to the
population of n=3 level and making use of expression (9) above. The gain coefficient
is given as a function of time for various laser energies. This calculation has been
performed for the analysis of experimental results obtained with a 180 ps laser
pulse duration. One can see that the time scale of the population inversion is in
the nanosecond range. In fact, during the laser pulse, the 2-3 transition is
absorbing and it comes to be amplifying in a late stage of the plasma life, after the
end of the laser pulse. This behaviour is characteristic of the recombination
scheme.

A laser-beam focusing system especially suitable for thin fibre irradiation has
been developped at Rutherford Appleton Laboratory (UK). It consists in a
combination of focusing lens and off-axis spherical mirror which produces a line
focus free of transverse aberration. Up to six laser beams can be brought round on
the same thin fibre of 7-mm length. Various options in beam arrangement enable to
choose among different lengths. 50 - 200 ps laser pulse duration are used. The
frequency of the Nd-laser is doubled. A sketch of the interaction chamber can be
seen in figure 8. This system has been successfully used for measuring the
variation of the carbon 182.2-A line intensity as a function of the length of the
fibre. The result is the curve shown in figure 9 which exhibits an exponential
increasing with a gain coefficient about 4 cm^{-1} /34/. In the same conditions, the
line at 135 A of the same ion has a linear variation, what corresponds to the
optically thin line behaviour.

A similar work, recently performed in fluorine /25/, has revealed three different
regimes according to the laser energy per unit length. A gain maximum of 5.5 cm^{-1}
has been observed at λ= 81 A for E/L = 9 ± 2 J/cm. At E/L = 16 ± 1 J/cm the gain is
reduced to 3 cm^{-1} while at E/L = 4 ± 2 J/cm there is no gain at all. These three
regimes are well accounted by the numerical simulations which suggests that, in
the low energy case, the plasma ionization is too weak and, in the high energy case,
the plasma is too hot.

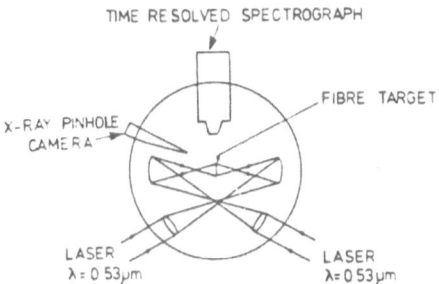

TIME RESOLVED SPECTROGRAPH

X-RAY PINHOLE
CAMERA

FIBRE TARGET

LASER
λ = 0.53 μm

LASER
λ = 0.53 μm

Figure 8. Sketch of the Rutherford Laboratory experiment, illustrating the focusing optics designed for thin fibre irradiation.

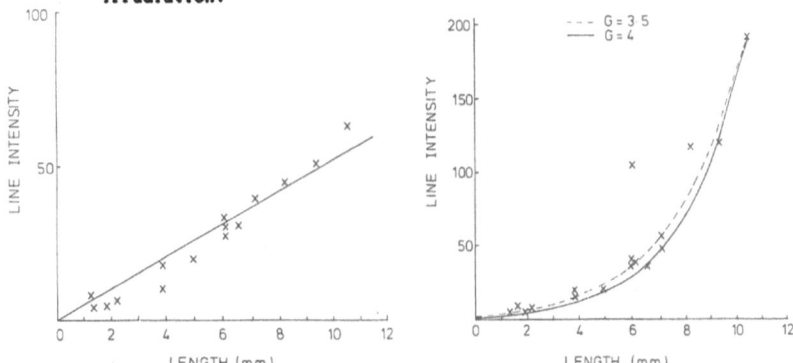

Figure 9. On the left, line intensity of the 2-3 transition, at 182.2 A, of carbon C VI, versus the length of plasma column. The exponential increasing is in agreement with a gain coefficient of 4 cm^{-1}. On the right, for comparison, intensity of the 4-2 line at 135 A, in the same conditions.

Magnetically confined plasma   An alternate scheme for obtaining population inversions in a recombining plasma has been proposed and developped by Suckewer et al. at Princeton University /30-32/. In order to increase the duration of the population inversion up to several tens of nanaoseconds, the plasma is confined by a strong solenoidal magnetic field, so that the plasma density is kept almost constant while the electronic temperature is falling down, due to plasma radiation losses.

A magnetic coil can take place around the plasma column because the target illumination is obtained at the focus of a long-focal spherical mirror, the room around the column remaining free for the confinement device. However, in this system the column length cannot vary without changing other plasma parameters. Therefore the gain-length product has to be deduced from line intensities simultaneously measured in axial and transverse directions, for one and the same laser shot (see equation (12)).

It has been shown that conditions for maximum gain are better in the off-axis region of the plasma column than in the center. In the center the temperature is maximum. It decreases rapidly in the outer region whereas the electron density reaches a maximum off axis. Thus, recombination of bare nucleus of carbon to H-like ions is maximum in the outer region. Figure 10 shows the gain coefficient calculated as a function of plasma radius. The maximum gain occurs in a few-tens-of-microns-wide annulus whose the radius is of 1.4 mm /32,35/.

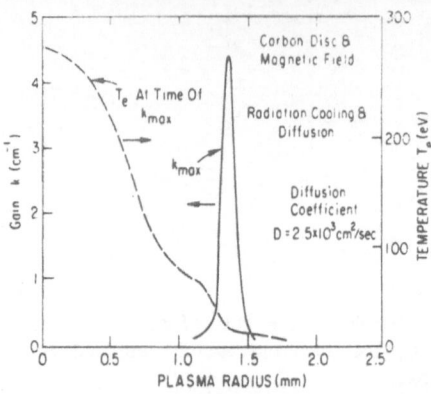

Figure 10. Predicted radial profiles of hydrogenic-carbon 182-A gain coefficient (k) and electron temperature (T_e) in the magnetically confined plasma column. The figure shows that there is a narrow high gain region /35/.

EXPERIMENTAL ARRANGEMENT WITH XUV MIRROR

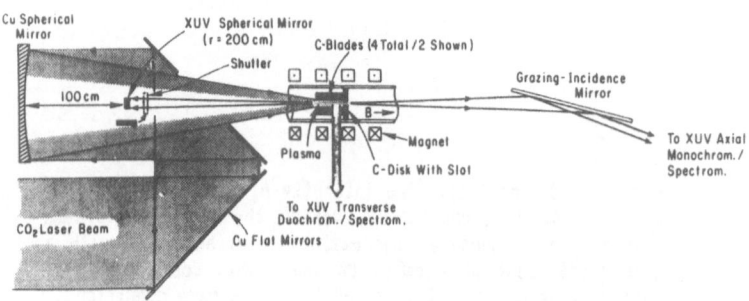

Figure 11. Device producing a magnetically confined laser-plasma at Princeton University. X-rays amplified during the plasma cooling are received on the entrance slit of the the axial monochromator. A multilayer mirror, represented in the left part of the figure, can be used for increasing the XUV path length through the plasma.

The sketch of the experimental set up is shown in figures 11 and 12. The plasma is produced, along the axis of a 4-m radius spherical mirror, by a high power CO_2 laser. The target consists in a solid carbon disc, possibly completed by blades making the plasma more uniform (figure 11). A small hole in the center of the disc enables to measure the soft X-ray axial intensity. Slots are set in the two observation directions in order to select the high gain region of the plasma (see figure 10). The length of the plasma column is around 2 cm.

With a magnetic field of 90 kG, a gain-length product, G.L, as large as 6.5 has been obtained at 182 A in the high gain region shown in figure 10. The CO_2-laser pulse parameters were: E = 0.3 KJ, FWHM = 75 ns, power density = 5×10^{12} W/cm^2. In the recombination phase, during which the peak of the line was observed, the electron temperature was 10-20 ev and the electron density, $(6-7) \times 10^{18}$ cm^{-3}. More recently a still larger gain (G.L = 8) has been reported /35/.

A further and very impressive proof of amplification has been given by using an X-UV multilayer spherical mirror. The reflectivity of the mirror at 182 A was measured to be 12 %, whereas the increasing of the line intensity along the axial direction, due to the mirror was about 120 % (see equation (31)). This result is shown in figure 13.

222

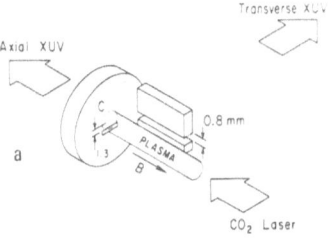

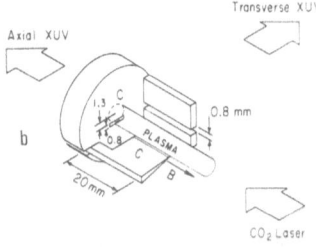

Figure 12. Sketch showing the disc, the magnetic field B, one blade of carbon (in b) and the diaphragms allowing the emission spatial resolution in the high gain region of plasma.

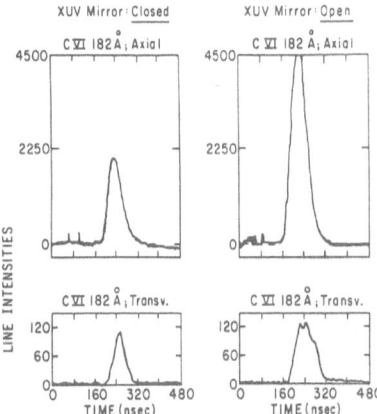

Figure 13. On the left, enhancement of the 3-2 line of C^{5+} from transverse to axial observation. The gain G.L deduced from this measurement is of 4.3. On the right, additional enhancement due to the XUV spherical mirror ($R = 12\% \pm 4\%$) represented in figure 11.

Measurements of the angular divergence of the 182.2-A lasing line emission have been also made with the same experimental set up /35/. To screen reflections from the walls of the vacuum chamber, a 1-mm-wide and 10-mm-high collimating slit was installed ahead of the spectrograph entrance slit,. A device for precise horizontal scanning of the axial spectrometer was added. Taking into account the plasma-to-spectrometer distance, the horizontal displacement of the spectrometer can be expressed in terms of angular deviation of the line of sight. Results are displayed for two values of the magnetic field, 35 kG and 50 kG, in figure 14. The widths of the peaks at half maximum are about 5 mr. This value gives confirmation that it is a narrow high gain region, like shown in figure 10, which supplies practically all the 182-A line emission. It is suggested the shift at higher magnetic field to be caused by a small tilt of magnet at high current, It is

223

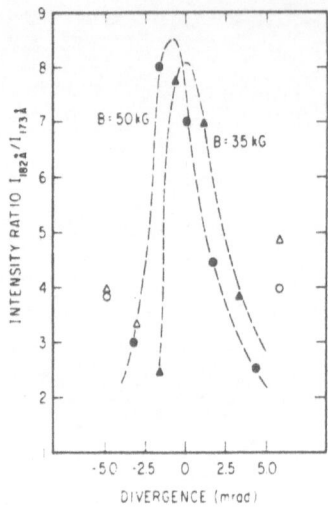

Figure 14. Divergence of the CVI 182-A line emission for two values of the magnetic field.

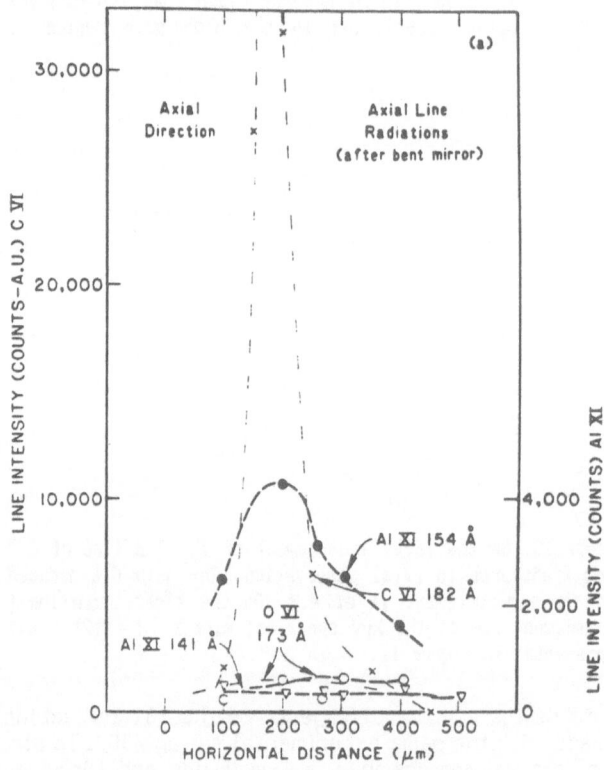

Figure 15. Comparison of the divergences exhibited by various lines belonging to different elements.

interesting to point out that at lower magnetic field, 20 kG say, the divergence is found to be significantly larger, namely 9 mr. The divergences of several lines, belonging to various elements, have been compared in using relative angular distribution measurements. In figure 15 one sees the curves obtained for the 182-A line of carbon (thin dashed line) as well as for some Li-like lines of oxygen and aluminum /36/. The 141-A line of aluminum and the 173-A line of oxygen do not show any angular concentration. However, as for the 154-A line of aluminum, though much less peaked that the carbon line, it exhibits also emission directivity. As a matter of fact, it is one of the Li-like lines which may be amplified in aluminum plasma. This is the subject of the next paragraphs.

2. Lithium-like ions

Besides plasma expansion, the emission of line radiation may also take part to plasma cooling. In this respect, lithium-like ions are interesting because the line emission of the parent ion (He-like) is very intense. Another interesting feature of Li-like electronic structure for producing population inversions is the exceptionally fast radiative decay of $3d_{5/2,3/2}$ to $2p_{3/2,1/2}$ levels (10^{12} sec^{-1} in Al^{10+}) owing to the large overlapping of the corresponding nodeless radial wavefunctions . Hence, when the high-lying level populations grows large due to radiativo-collisional cascades occuring in the course of plasma recombination, the 3d levels will easily turn out to be the lower levels of lasing transitions.

Calculations   Calculations have been developped for ions from aluminum (Z = 13) to calcium (Z = 20) /26,37-39/ and also, to some extend, to titanium (Z = 22) and copper (Z =27) /40/.   Aluminum has been extensively investigated on account of his importance in many experiments. That is why, results for this element are presented below as typical examples of the scheme.

Lithium-like Al^{10+} ions have to be considered in an   intermediate plasma-density range, comprised between the very high densities (10^{23} cm^{-3}), for which purely collisional LTE assumptions are satisfied, and the coronal range (10^{16} – 10^{17} cm^{-3}), where de-excitations are purely radiative. At intermediate densities, the dominant populating and de-populating processes are not the same for low lying and high lying levels. The highest excited levels of Al^{10+} tend to have collisionally thermalized populations whereas the populations of low lying levels are much more affected by radiative decays. In addition, at electron densities of interest, the lowering of the ionization limit makes it unneccessary to consider the levels above a maximum value of the main quantum number (n_{max}~ 6-8) because they are undistinguishable from the continuum levels. These are the proper conditions for using a collisional-radiative (C-R) model.

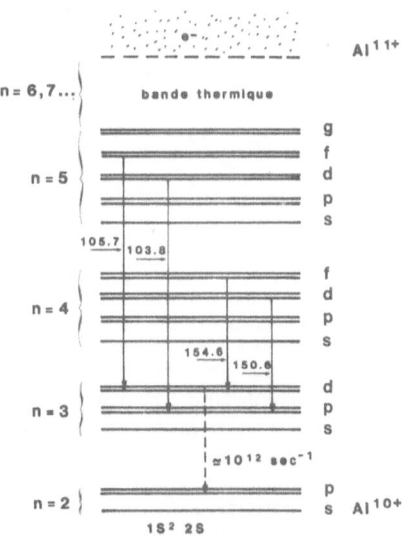

Figure 16. Level diagram of lithium-like aluminum ion, with the wavelengths of some lines of interest for soft X-ray lasers.

In practice, calculations of inversion densities in realistic conditions are carried out by using the Al^{10+} C-R model in as a post-processor of a 1D – lagrangian hydrodynamical code which includes a description of the ionization dynamics. The code is used in cylindrical geometry. An example of lithium-like and helium-like ion abundances calculated as functions of time for a 2.7 ns laser-pulse has been given in figure 2 above.

The population distribution of excited states of lithium-like ions is obtained at each moment by the solution of a system of coupled rate equations. The optically thin approximation is made. It has been pointed out in Section II (&.3) this approximation not to be valid for any target gemoetry because the 2p – 3d radiation trapping may reduce the population inversion. The level diagram of Al^{10+} is displayed in figure 16 and the main features of the model are summarized as follows:

a) The most excited states, which are assumed in local thermodynamic equilibrium (LTE) with each other and with the ground state of the helium-like ion, constitute the "thermal band" which, for the present calculations, includes all states of the $1s^2nl$ ^2L_J levels with n = 6 and n = 7. Higher excited levels are neglected because their coupling with low-excited levels becomes negligible and also because the ionisation limit, as already mentioned, is rapidly reached in hot dense plasmas. Nevertheless, the choice of the limits of the thermal band is, to a certain extent, arbitrary.

b) The low-lying first excited levels $1s^22p$ of lithium-like ions behave as quasi-degenerate with the ground level $1s^22s$ rather than as ordinary excited levels, especially when the electron temperature is large compared with the 2s-2p energy difference (~22 ev). Due to high collision rate, the population ratios between these three levels are likely to keep close to the LTE values, i.e. given by Boltzmann's relation.

c) Since the levels which are the most involved in the direct ionisation or recombination processes are mainly those of the thermal band (by collisional processes) on one part, and the ground and quasi-degenerate low-lying levels (by radiative recombination) on other part, it is assumed that, concerning the intermediate bound levels, the difference between the ionisation and the recombination rates is negligible compared to other processes (radiative decay, excitation or de-excitation by collision with free electrons). Hence the set of equations which determine the population densities of intermediate bound levels may be written in the following form:

$$dN_i/dt = \sum N_j \; A_{ji} + N_e \sum N_j < \sigma_{ji}.v> -$$
$$- N_i x \left(\sum A_{ij} + N_e . \sum < \sigma_{ij}.v> \right) \qquad (32)$$

where the summation indexes must be limited as indicated in (13). The index i refers successively to each level from $1s^23s$ $^2S_{1/2}$ to $1s^25g$ $^2G_{9/2}$ included. Within the mentionned limits, the index j refers to any Li-like level including those of the thermal band as well as the n = 2 levels

The major result of the full calculation is to show population inversions to occur for several optically allowed transitions, i.e. 5f – 3d, 4f – 3d, 5d – 3p, 5f – 4d and 5g – 4f in the plasma corona and after the end of the laser pulse, that is, in the cooling expanding plasmas. In figures, the results will be displayed in terms of the population inversion density, ΔN, according to equation (10). Figure 17 presents typical 3d – 5f density calculation for 0.6 ns laser pulse duration. In this figure, ΔN is plotted versus the distance to the target at various times.

Figure 18 gives the temporal variation of 3d – 4f and 3d – 5f population inversions as well as N_e and T_e at a fixed distance from the target, for 2-ns

226

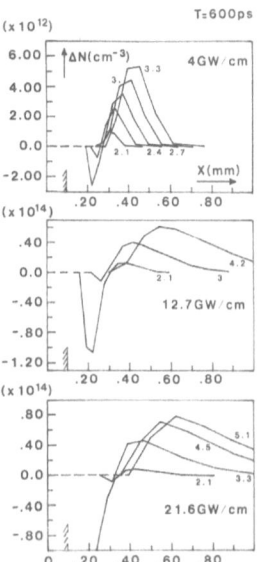

Figure 17. Calculated 3d - 5f inversion density versus the distance to the target at various times (indicated in ns near each curve) for three laser-flux linear densities. Laser-pulse duration: 0.6 ns. The initial target surface position is plotted on the left (hatching).

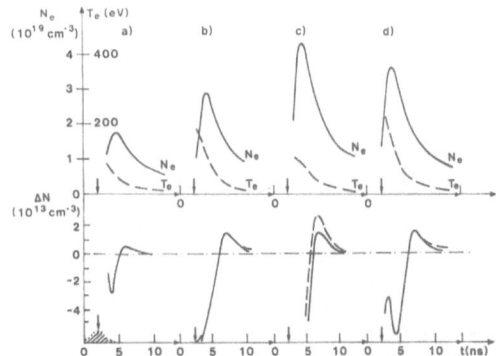

Figure 18. Bottom: time-variation of 3d - 5f (solid curves) and 4f - 3d (dashed curves) inversion densities in the space region of the maximum (0.45 mm from the target). Top: plasma electron density and temperature at the same place. Laser-pulse duration: 2 ns. The arrow shows the top of pulse. a) laser wavelength: 1.06 μ; flux density: 4 GW/cm; b) laser wavelength: 1.06 μ; flux density: 12.7 GW/cm; c) laser wavelength: 0.53 μ; flux density: 12.7 GW/cm; d) laser wavelength: 1.06 μ; flux density: 21.6 GW/cm

laser-pulse. If the profile is dominated by thermal Doppler broadening, the gain coefficients, G, can be deduced of such curves in using equations (9) and (14). In assuming the ion temperature to be about 15 ev at the moment of the gain peak, the highest gain coefficient predicted by calculations reported in figure 18 should be of 1 per centimeter for the unresolved $5f_{7/2}-3d_{5/2}$ and $5f_{5/2}-3d_{3/2}$ transitions. In similar conditions, the plasma being produced by green (instead of infrared) laser light, the predicted gain could be as large as 4.5 per centimeter for the 4f - 3d transitions.

Table 1. Gain coefficients, at optimal density and temperature, for 3-4, 3-5 and 4-5 transitions of elements of increasing atomic number, from ref. /41/.

Ion	Al^{10+}		S^{13+}		Ca^{17+}	
N_e cm^{-3} :	10^{19}	:	10^{19}	:	5×10^{19}	
T ev :	17.4	:	19.1	:	50	
Transition	A	G cm^{-1}	A	G cm^{-1}	A	G cm^1
3d - 5f :	105.7	1.4 :	65.2	1.3 :	39.5	0.8
3d - 4f :	154.7	8.6 :	95.5	3.5 :	57.9	2
4d - 5f :	334.4	1.6 :	206.5	4.4 :	125	7.2

An important question of each pumping scheme is its ability of being extended to shorter wavelengths, especially up to wavelengths about 40 A, that is, in the so-called "water window" which is of great interest for biological applications. Table 1 gives information in this respect for the Li-like recombination scheme. In the table, the electron densities and the temperatures are the values providing the largest gain for each ion. It must be noticed that the temperatures are not the maxima at the top of the laser pulse but the temperatures required at the moment of the optimal density, during the cooling phase. The table shows that the gain for the 3-5 and 3-4 transitions decreases slowly when the atomic number increases. On the contrary, the gain should increase rapidly for the 4-5 transitions.

Experiments. Experiments on lithium-like ions have been mainly developped by the Laboratoire de Spectroscopie atomique et Ionique (Université Paris-Sud), in using the laser facilities of Palaiseau (LULI) /23, 42-46/. The main characteristic of the experimental technique consists in the single-side illumination of the target (like in the sketch of figure 1). This enables to make experiments even if only one laser beam is available. Another advantage is that solid target plate supports can be used, so that target preparation and adjustment in the vacuum chamber, are easier and faster than in the case of thin fibers or foils. In particular, large size targets can be easily prepared for the production of very long plasmas. But on the other hand, the plasma expansion has not exactly the cylindrical symmetry of the calculation model. Hence the comparison between experiment and theory may be more difficult. An important example of that has been given in section II (&.4), in connection with radiation trapping in the case of massive target (figure 4). Most of the experiments in aluminum have been made in using 2 - 3 nanosecond Nd-laser pulse duration. The power density on the trarget surface is of $2 - 5\times10^{12}$ W/cm^2.

The experimental device involves a grazing-incidence X-UV spectrometer with either a scintillator plus an optical multichannel analyser (OMA) or a X-ray streak camera as a detector. Thus time-integrated as well as time-resolved recordings can be made. The relationship between both types of results is brought in by numerical integration which involves the simulation of plasma temporal evolution. An example is given in figure 19 which presents the result of an integration performed in relation with the analysis of preliminary time-integrated measurements on sulfur, at 65 A /26/. One is starded from a trial time-dependent function of the gain, preserving the general features of recombination scheme shown, for instance, in figure 19. Line profile functions, depending on time and space, have been introduced. Then gain (absorption) is calculated at wavelengths separated by small intervals, so that one obtains a time-integrated absorption spectrum of the plasma in the neighbourhood of the line of interest. This is represented by the middle curve of figure 20. One sees that the temporal gain peak (1 cm^{-1}), exhibited by the upper curve, turns to a deep breaking in of the time-integrated absorption spectrum. Finally the instrumental function changes

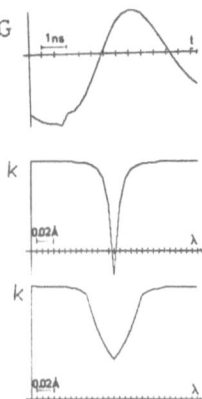

Figure 19. Relation between the time-dependent gain (upper curve) and the observable time-integrated absorption spectrum (middle curve). The time-averaged gain reduces in a line of absorption breaking in. The lower curve shows the intrumental broadening of the absorption line.

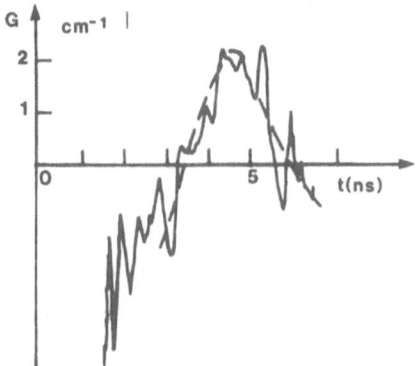

Figure 20. Gain as a function of time for the 3d – 5f transition of Al^{10+} ($\lambda = 105.7$ Å) measured for a 0.7 cm long plasma column. Laser-pulse half-maximum duration: 2.5 ns; flux density: 5×10^{12} W/cm^2

this curve in the lower one, which fits fairly well the experimental results reported in /26/.

Examples of time-integrated and time-resolved measurements performed on the 3d – 5f line of aluminum are shown in figures 20 and 21. In both cases relation (11) has been used for pairs of lengths either for time-dependent or for time-integrated I_L and I_1.

Experiments about the 3d – 5f and 3d – 4f lines have also been carried out at Rutherford Laboratory in using the thin fiber technique (same section, &. 1) /47/. A thin aluminum layer was deposited on the carbon fiber. The laser pulse duration was of 100 ps and the laser frequency was doubled (green light). Notwithstanding these differences, the gain coefficient at 105.7 A has been found to be around 1.6 cm^{-1} for a 12 mm long plasma column, that is similar to the gain obtained at Palaiseau. At 154.7 A, the gain coefficient is of 2.5 cm^{-1}, what seems to be a little larger than Palaiseau's value /48/. This could be due to the shorter laser wavelength used at Rutherford, as suggested by the calculation of figure 18 c).

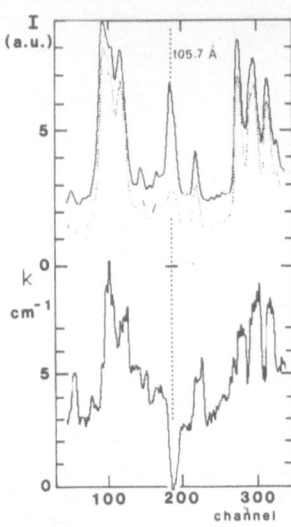

Figure 21. Time-integratred gain measurement for the same line as in figure 20. Emission spectra for the two plasma lengths are displayed at the top of the figure. One sees strong absorption peaks for most of the lines and the absorption breaking in at 105.7 A, the wavelength of the 3d - 5f line.

As previously mentionned, the magnetically confined plasma technique has also been used for aluminum Li-like ion study /36/. It has proved gain and small beam divergence at 154.7 A (see figure 15). The analysis of the results leads to a value of the G.L product about 3.7.

In the experiments described above, the length of the plasma column was of the order of 1 cm and always shorter than 2 cm. The goal of the experiment presented now was to investigate the soft X-ray lasing properties of a 6-cm long recombining plasma column /49/. If the physics of the amplifying medium was really plain, the plasma lengthening will result in a proportional gain-length product growth. However we have seen, in the first section, that various impediments make the course towards X-ray lasers more difficult. As a matter of fact, one can observe that, in the case of plasma columns produced from massive targets, the gain gnefficient decreases when the column length increases. This is explained by the anisotropic radiation trapping discussed in the first section (&.3). The new experiment has shown indeed that no gain can be observed with a 6-cm long plasma column produced from a massive polish slab of aluminium.

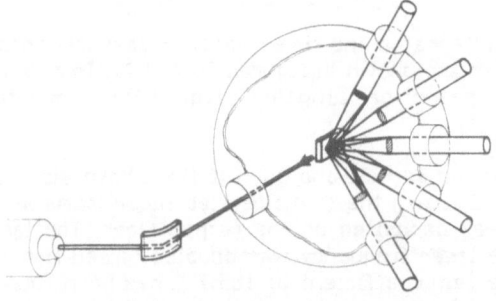

Figure 22. Sketch of the experimental set up carried out at LULI (Palaiseau) for producing a 6-cm long amplifying plasma column. Five beams are focused on the same line.

Figure 23.   Photography of the vacuum chamber schowing the
five laser-beam entrance tubes.

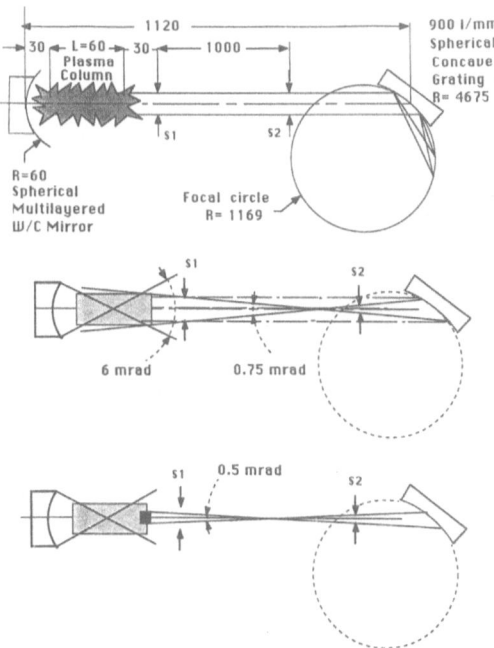

Figure 24. Top: simplified diagram showing the multilayer
mirror and a parallel X-ray beam focused on the focal
circle of the grazing incidence grating. S_1 and S_2 are two
slits limiting the beam; distances are given in mm. Middle:
diagram showing the maximum angular divergence accepted by
the slits. Bottom: diagram showing a possible reduction of
the divergence due to the reduction of the beam cross
section at the end of the plasma.

The cure for radiation trapping needed to remove the dense plasma standing on the side of the active section in the massive target case. This led to use as few aluminium as possible for the target, that is to say a thin Al-layer (1000 A). The support of the layer was still massive but made of material free of aluminium.

The framework of the experimental set-up is shown in figure 22 which represents the arrangement of the five laser beams converging to the target.The focal spot is a 6.4 cm x 200 u rectangular surface. The vacuum chamber and the beam entrance tubes can be seen on the photography of figure 23.

The result presented below has been obtained in using a multilayer W/C mirror (see section 1, &.5) which was set perpendiculary to the axis, at the plasma end opposite to the detector. The mirror is spherical with a 6 cm curvature radius, so that the mirror sphere centre is just in the middle of the plasma column. The reflexion coefficient of the mirror is put at the value of 5% deduced from calculation for the 105.7 A wavelength. The scheme of the device, including the focal-circle mounting of the spectrometer, is shown in figure 24. It is designed with a view to select plasma emission having a strong directivity in axial direction. Only parallel beams falling on the grating are perfectly focused on the detector surface. Small beam divergence entails line broadening, given by relation:

$$\Delta\lambda_A = 2.08 \times \theta_{mrad} \qquad (33)$$

Now, figure 25 shows the comparison between the spectra corresponding to single-pass beam (mirror closed) and double-pass beam (mirror open) in the region of the 3d – 5f (λ = 105.7 A) and 3p – 5d (λ = 103.8 A) transitions. One sees double-passing to enhance the 105.7 A-line intensity much more than the 5% reflexion would do alone. It can be calculated from the experimental data that the gain-length product is about 3, i.e. the gain coefficient about 0.5 cm^{-1}. In using the relation (31) one can also calculate an "effective" gain-length product for the double-pass beam. Its value is 4.

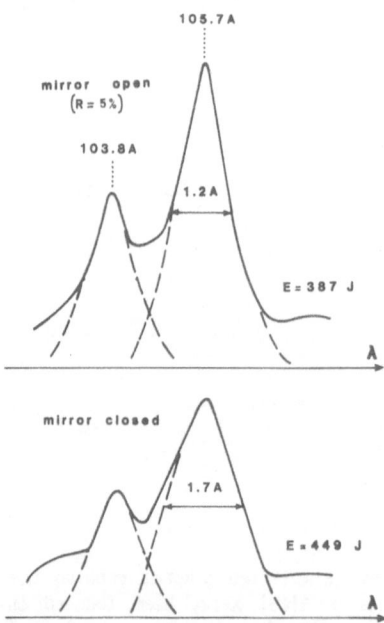

Figure 25. Intensity increasing produced by a multilayer mirror (r = 5%) for the 3d-5f transition at 105.7 A, in using a thin layered aluminium target (1000 A). Mirror closed: single-pass beam. Mirror open: double pass beam. The line narrowing is discussed in text.

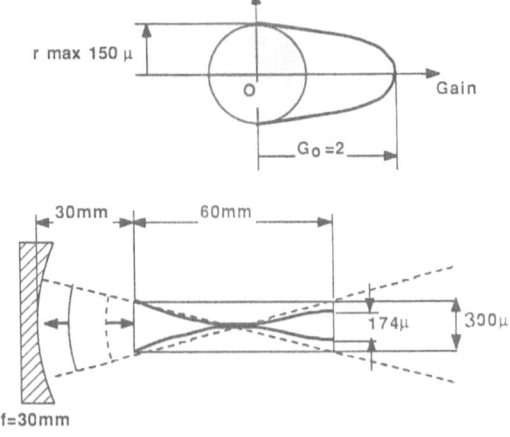

Figure 26. Top: gain profile assumed for the calculation of a lenslike effect in the plasma. Bottom: Reduced section of the reflected beam (solid line) inside the plasma and at its boundary.

In addition a 30%-reduction of line width is observed in the case of the double-pass beam. In accordance with equation (33), this means that the divergence of the beam at the entrance of the spectrometer has been reduced. This suggests in turn that a lenslike effect of the amplifying medium could reduce the cross section of the beam at the end of the column /50/, so leading to smaller angular beam spread, as shown in the lower diagram of figure 24. For illustrating this process, figure 26 shows the reduction of the beam diameter produced by double-passing in a column of 300 u diameter having the gain profile displayed at the top of the figure.

The gain measured in this experiment with mirror leads to think that a laser pumped by recombination should be feasible at 105.7-A wavelength in using a multilayer mirror cavity. In improving the reflection coefficient by a better choice of mirror components, the saturation regime could be achieved in a six-pass cavity. The amplification life-time in the Li-like recombination scheme should be long enough for a 12-cm long cavity, consistent with the plasma column length.

IV - COLLISIONALLY PUMPED LASERS

With the collisional pumping scheme, we reach the world of geant devices for producing plasmas, of huge powers for plasma heating and also of large soft X-ray amplifications. Most of the experiments have been performed with the laser NOVA from Lawrence Livermore Laboratory, the largest Nd-glass laser in the world at the present time. The power density available for producing plasma columns of several centimeter length is about 10^{14} W/cm^2 or more, that is, an order of magnitude almost two times higher than in other experiments. From the very first experiment on selenium /51/, large gain coefficients have been obtained in 1-cm long plasmas. They have been maintained later in 4-cm long plasmas /24/, and afterwards extended to heavier elements /17/. Working with lighter elements, which indeed require a little less heating, Naval Research Laboratory has produced gain for the same transitions with laser power density less than 10^{13} W/cm^2 /53,54/. This large difference may pose problem about the total similarity of the population inversion mechanisms in both works. Experiments are also in progress at the Centre d'Etudes de Limeil /55/.

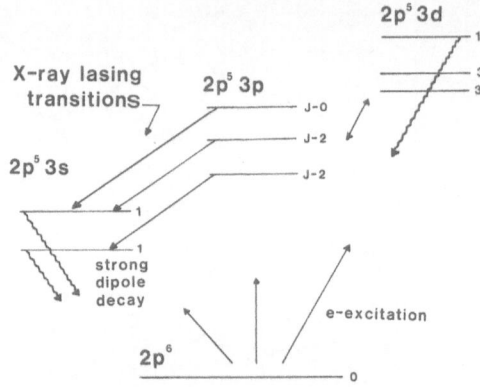

Figure 27. Simplified level diagram of the n = 2 and n = 3
levels of Ne-like ions, showing the main transitions
producing population inversions between 3p and 3s levels.

Collisional pumping has been mostly considered for the multicharged ions of the
neon sequency. It must be kept in mind that there is no external electron source.
Collisional processes under consideration are uniquely due to plasma free
electrons. A simplified diagram of Ne-like ion level is plotted in figure 27. The
general features of population transfer between the various levels can be roughly
summarized as follows.

The collisional excitation rate from the ground level is approximately the same
for $1s^2$ $2s^2$ $2p^5$ 3s and $1s^2$ $2s^2$ $2p^5$ 3d levels. At the same time, owing to
selection rules, the radiative decay from the $1s^2$ $2s^2$ $2p^5$ 3p to the ground level is
strictly forbidden, whereas it is quite large from the $1s^2$ $2s^2$ $2p^5$ 3s level.
Moreover there is a large rate of population transfer from 3d to 3p level by
radiative and collisional cascades. As a consequence, population inversion can
occur between the 3p and 3s levels provided that plasma free electron density is
chosen in a suitable range. Quantitative predictions must include dielectronic
recombinations. Rate equation calculations show that population inversions occur
only in a rather narrow range of electron density. At too low density, there is no
collisional pumping and, when the density becomes too large, collisional mixing
between the levels thrusts populations to statistical equilibrium.

Calculations need a very large number of atomic data of good accuracy /56/.
Having them, the rate equations of the collisional-radiative model are solved in
close connection with plasma hydrodynamics simulation. The first quantitative
predictions of gain as a function of plasma density at various temperatures, for Mg
III, Ca XI and Fe XVII, have been reported in 1980 , the wavelengths lying mostly
between 500 a and 1500 A /57/. A great deal of work has been developped during
the last years at Livermore for Se XXV, Y XXX, Mo XXXIII and so on, the lasing
wavelengths being between 100 A and 200 A /24,58-60/.

We show below a theoretical result obtained at Palaiseau for Ne-like strontium,
the lasing wavelengths of which are expected near 160 A /61/. The calculation
leading to gain curves plotted in figure 28 treats simultaneously 5 ionization
stages, from Sr XXVII to Sr XXXI. The Ne-like strontium (Sr XXIX) is described by
more than 50 levels. A less number of levels is used for neighbour ions.

It can be noticed, in figure 28, that the electron density as well as the
temperature needed for high gain production are significantly larger than in the
case of the recombination scheme (see for instance section III, table 1). The high
temperature requirement leads, as already pointed out, to that experiments can be
made only with very large laser power. On the other hand, the large density will
give more importance to beam refraction in plasma than in recombination lasers.

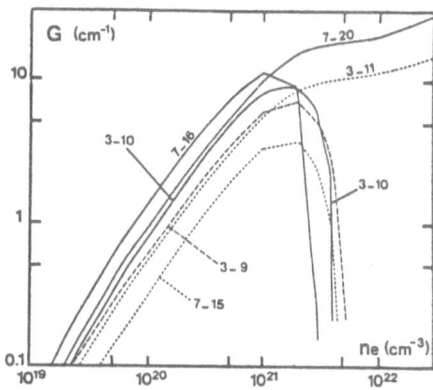

Figure 28. Ne-like strontium 3s - 3p predicted inversion densities versus electron density, for a 800-ev temperature. Level designations: n° 3 and 7, 3s $^1P_1$ and $^3P_1$; n° 9, 10, 11, 15, 16 and 20, 3p $^1P_1$, $^3P_2$, $^3P_0$, $^3P_1$, $1D2$, $^1S_0$.

1. Ne-like ions

The experimental set up used at Livermore for gain measurements in Ne-like ions is represented in figure 29. The length scale can be estimated from the diameter of NOVA beams, i.e. 74 cm. Targets are prepared for being used as exploding foils /58/.

A number of results have been obtained with this device, especially for selenium and molybdenium. The highest gain-length product that has been ever achieved at the present time in X-rays is schown in figure 30. The intensity increasing of the 3s $^3P_1$ - 3p $^3P_2$ line, at 206,3 A, is plotted against the plasma length. One sees that multiplying the plasma length by a factor 4 increases the line intensity by 5 orders of magnitude. The gain-length product is about 15. This value seems to be the limit laid down by the inhomogeneity of refractive losses.

Time-resolved recordings, associated to beam divergence measurements, have shown that the history of the beam refractive deflection is complex. As a matter of

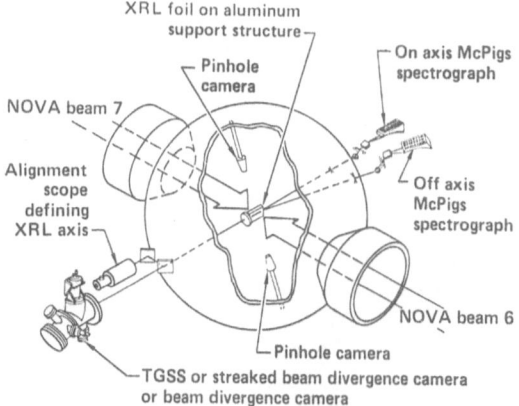

Figure 29. Nova two-beam facility chamber set up /24/. XRL: X-ray laser. TGSS: transmission grating streaked spectrometer. McPigs spectrographs are grazing incidence grating devices with a gated microchannel plate of large size as the detector /24/.

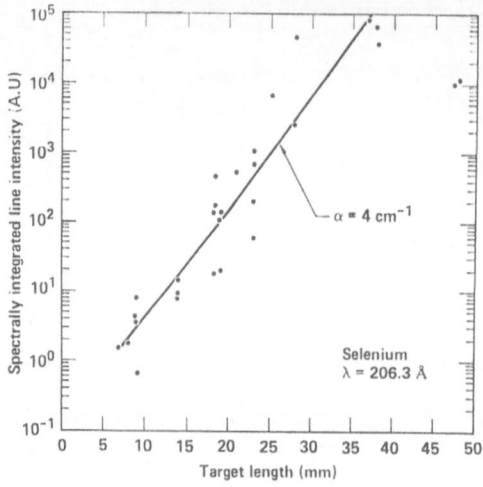

Figure 30. Line intensity of 3s - 3p transition (J= 2 to 1) of selenium Ne-like versus the length of plasma column. The gain coefficient is of 4 cm^{-1} /24/.

fact, the X-ray laser pulse is found at different times in different lines of sight. This can be seen in figure 31 where the temporal evolution of the 206.3-A beam of a 3-cm long plasma is displayed for two directions. The common time scale is provided by the NOVA pulse. In 31 a) there is only a small deflection of 1 mr with respect to the axis of the plasma column, whereas it is of 12.5 mr in 31 b). The difference in occurence time between the two directions is 250 ps, i.e. an interval whose the magnitude is similar to the x-ray pulse duration itself. This variation is assigned to the steeper plasma density gradient at early time. Photon trajectories propagating the plasma at that time are more refracted than later on, when inhomogeneity has been smoothed by plasma expansion.

Measured and predicted gain coefficients for Ne-like selenium lines are summarized in figure 32 /62/. Calculations and experiment are in good agreement with the well known exception of the J = 0 to 1 lines. Notwithstanding some improvements in the introduction of dielectronic recombination in atomic

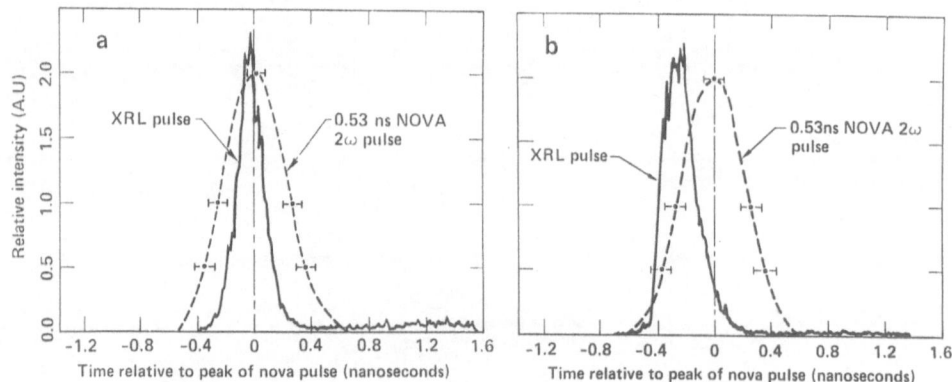

Figure 31. Measured time history of the X-ray laser pulse in two different lines of sight. a) within 1 mr of the geometrical axis of plasma column; b) at 12.5 mr of the same axis.

236

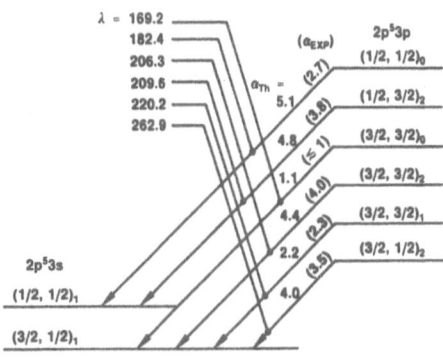

Figure 32. Diagram showing the lasing lines of Ne-like selenium with observed (in parentheses) and predicted gain coefficients.

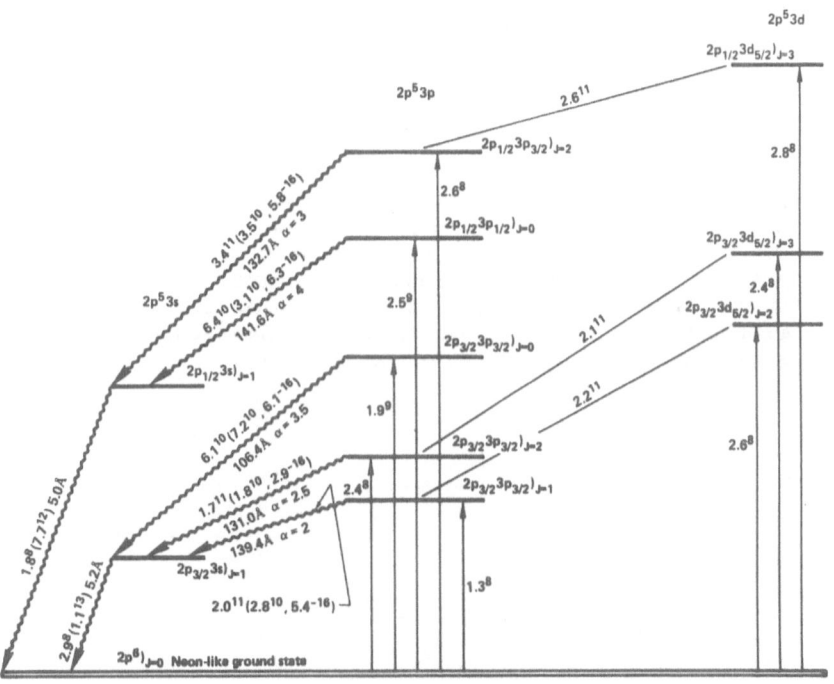

Figure 33. Simplified level diagram and transition probabilities of neon-like molybdenium /17/.

calculation, the observed discrepancy for these lines is not yet completely elucidated.

Ne-like molybdenium provides lasing lines of shorter wavelengths, with gain coefficients similar to the ones of selenium /17/. The diagram of figure 33 shows typical values of the transition probabilities used in the inversion population calculation. Plasma parameters are assumed to be $N_e = 5 \times 10^{20}$ cm^{-3}, $T_e = 2$ Kev, $T_i = 500$ ev. Experimental results are presented in figure 34. Ne-like ytterbium as also shown gain at 155.0 A and 157.1 A.

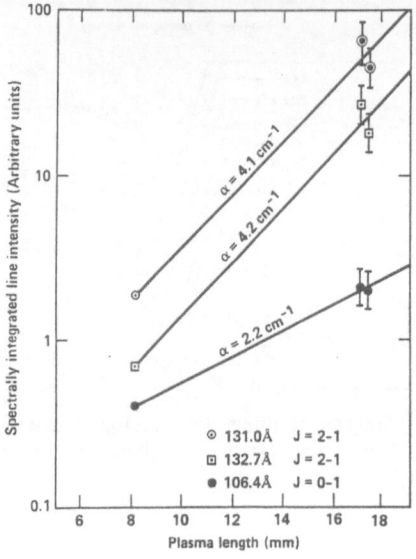

Figure 34. Measurement of gain for 3s - 3p lines of Ne-like molybdenium

2. Ni-like ions

The objective of producing an X-ray laser in the "water window" between the K-edges of carbon and oxygen, i.e. near 40 A, for applications to biological specimen imaging, has led to develop the theory of a Ni-like analog to Ne-like scheme /59,63/. This was necessary because, at so short wavelength, the Ne-like scheme would require Gd^{54+}, for instance, to be produced, and the necessary irradiance should be more than 10^{16} W/cm². In the Ni-like sequency, the population inversions are produced between the 4p and 4d levels /52/.

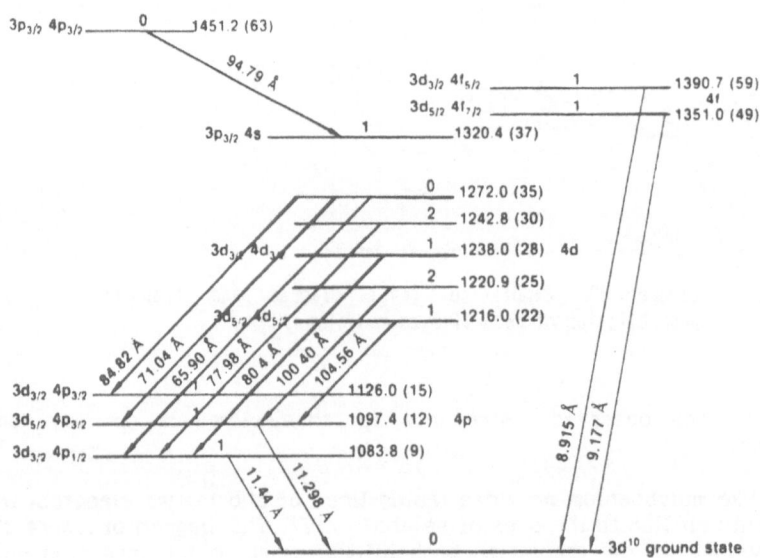

Figure 35. Simplified diagram of Ni-like Eu energy levels. Population inversions occur between 4d- and 4p- levels.

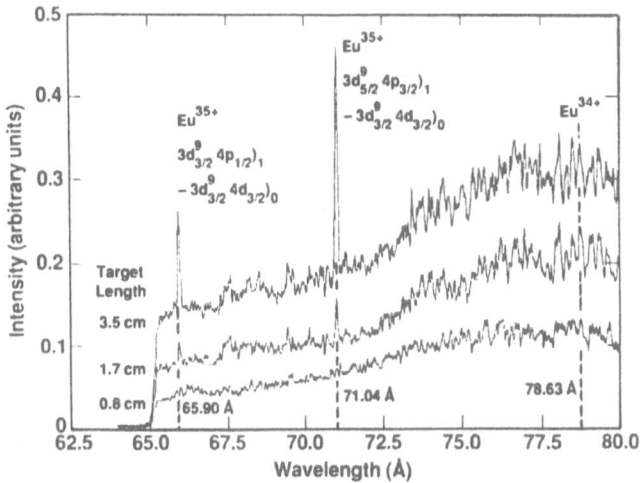

Figure 36. Observation of gain at 65.60 and 71.04 A in Ni-like europium /64/.

The ground level is the $3d^{10}$ closed shell. The dominant mechanism of 4d-level excitation is the collisional excitation from the ground state. A rapid radiative decay from the 4p to the 3d level maintains the population inversion. A level diagram of Ni-like europium is shown in figure 35. Lasing lines are expected between 65 A and 105 A.

In the experiments, the target was an EuF^2 thin foil irradiated with a power density of 7×10^{13} W/cm^2. The non-linear intensity increasing, for the 65.90-A and 71.04-A lines, is shown as a function of the plasma length in figure 36. The data are fitted by gains of 0.6 and 1.1 cm^{-1} respectively. The extrapolation of this scheme has been explored in looking at Yb, in the spectrum of which a 4p-4d line, at 50.26-A, has shown a non-linear increasing consistent with a gain coefficient of 1 cm^{-1} /64/.

V - OUTLOOK

We have described in detail the two chief pumping methods used in the experimental works intending to carry out X-ray amplifiers from hot plasmas. In the future, these methods will be very likely extended to new multicharged ions. It must also be kept in mind that many other schemes are under investigation. So, photoionization pumping of metastable levels could open the way to lasers having low threshold pumping energy /65/. Pumping with synchrotron radiation has been calculated for a Li-laser /66/. Large excited populations, leading to X-UV fluorescence enhancement, have been produced by multiphoton excitation /67/. Also charge-exchange excitation in ion-atom collision is a possible mechanism for soft X-ray laser pumping /68,69/. It will belong to future experiments to provide arguments for or against each possible scheme, in showing at first the effective production of gain.

We emphasized already that an important property of each pumping scheme is its ability of being extended to shorter wavelengths, up to about 40 A, where interesting biological applications are possible. The schemes described above can be extended to ions of higher Z provided that plasma temperature is large enough. The wavelength-to-temperature scaling differs considerably according to the population inversion scheme. This is shown in terms of density flux deposited on

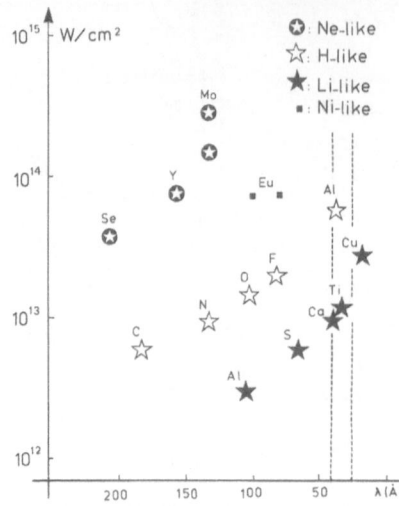

Figure 37. In order to decrease the lasing wavelength more
energy must be spent for producing the plasma. This figure
shows how the density power onto the target does scale with
the wavelength for the collisional schemes (Ne-like and
Ni-like ions) and the recombination schemes (H-like and
Li-like ions). The dashed vertical lines delimits the
"water-window" which is of a great interest for biological
applications.

target surface in figure 37 where the points below 50 A are calculated.
Lithium-like ion plasma allows an especially economical extension to short
wavelengths because, in this case, lasing wavelength and plasma temperature have
approximately the same Z^2-scaling. The figure displays the "water-window" near
40 A. Inside this window, radiation is absorbed by the K-edge of carbon atoms of
organic molecules but not by the k-edge of oxygen of water molecules. Thus using
an X-ray laser in the water-window would make possible holographic imaging of
living samples.

As regards the comparison of X-ray lasers with other sources, besides
coherence and directivity, a very large enhancement will occur in their brightness.
Assuming "realistic" experimental conditions in which the observed emitting area
should be 10^{-3} cm^2, the solid angle at the entrance of the measurement device
being 10^{-6} steradian, one finds approximately that the same number of $1 - 5 \times 10^4$
photons per pulse (for a 120-A line) are emitted by laser-plasmas and

Table 2. Comparison of soft X-ray sources

Source	: Pulse duration(sec) :	Photons per pulse	: Brightness (a.u.)
Tokamak plasma	10^{-1}	$1 - 5 \times 10^4$	1
Laser plasma	10^{-9}	$1 - 5 \times 10^4$	10^8
Storage ring	2×10^{-10}	2×10^3	10^8
Undulator	2×10^{-10}	2×10^8	10^{13}
Soft X-ray laser	5×10^{-10}	10^{12}	2×10^{16}

tokamak-plasmas. But the pulse duration is 10^8 times shorter for laser than for tokamak, leading to the same ratio, to laser advantage, between their brihtnesses. Typical photon number and brightness are compared in table 2 for various sources, in letting in a thermal line broadening of 0.02% for plasmas and using the same band width for other sources. The estimate displayed in table 2 is made out in assuming a population inversion density of $10^{15} cm^{-3}$ in a 10 cm long amplifying plasma column.

However, as regards X-ray lasers, though large amplifications are observed in plasma columns, it remains to reach the saturation theshold, at which there will be approximately as much photons as inverted atoms in the plasma. Before this point, the characteristic optical properties of laser-beam cannot be completely achieved. It follows from the analysis presented in the previous sections, especially section II, that this still requires improvements in target manufacturing and, on the other hand, the development of multilayer mirror cavities coupled to long plasma columns.

A number of applications are waiting for these achievements. In addition to holography of living biological samples, interferometry, microscopy, X-ray imaging, surface physics and physical-chemistry , are to be mentionned at once.

ACKNOWLEDGEMENT

This work has been partially supported by DRET under contract 85/1437.

REFERENCES

1. Proceedings of International Colloquium on X-ray Lasers, ed. by P. Jaeglé and A. Sureau, Journal de Physique, 47, Colloque C6, Supplement N° 10, 1986.
2. The Generation of Coherent XUV and Soft-X-Ray Radiation, feature editions of Journal of Optical Society of America, ed. by D.L. Matthews and R.R. Freeman, 4, N° 4, 1987
3. M. Dugay , P.M. Rentzepis, Appl. Phys. Lett., 10 (1967), 350.
4. P. Jaeglé, A. Carillon, P. Dhez, G. Jamelot, A. Sureau, M. Cukier, Phys. Lett. A36 (1971) 167
5. I.N. Knyasev, V.S Letokhov, Optics Communications, 3 (1971) 332
6. T.C. Bristow; M.J. Lubin, J.M. Forsyth, E.B. Goldman, J.M. Soures, Optics Communications, 5 (1972) 315
7. R.C. Elton, App. Optics, 14 (1975) 97.
8. J. Davis, K.G. Withney, App. Phys. Letters, 29 (1976) 419.
9. G.J. Pert, J. Phys. B, 9 (1976) 3301.
10. A.V. Vinogradov, I.I. Sobelman, E.A. Yukov, Soviet J. Quant. Electron., 7 (1977) 32.
11. E.J McGuire et al., ref. 1, p. C6-81.
12. M.K. Matzen et al., ref. 1, p. C6-135
13. S. Maxon, P. Hagelstein, K. Reed, J. Scofield, J. Appl. Phys. 57 (1985) 971.
14. P. Jaeglé, Line Formation in Laboratory Plasmas, in Progress in Stellar Spectral Line Formation Theory, ed. by J.E. Beckman and L. Crivellari, Reidel Publishing Company, (1985) 239
15 M.H. Key, et al., ref. 1, p. C6-71.
16. H. Guennou, A. Sureau, J. Phys. B. , 20 (1987) 919
17. B.J. MacGowan et al., J.Appl.Phys. 61 (1987) 5243
18. A. V. Vinogradov, B. Ya. Zeldovich, App. Optics,16 (1977) 89.
19. E. Spiller, AIP Conference Proceedings N° 175, ed by D.T. Atwood and B.L. Hencke, 1981, p. 124, and AIP Conference Proceedings N° 119, ed. by S.E. Harris and T.B. Lucatorto, (1984) 312.
20. T.W. Barbee, AIP Conference Proceedings N° 175, ed by D.T. Atwood and B.L. Hencke, (1981) 131.
21. P. Dhez, G. Jamelot, A. Carillon, P. Jaeglé, P. Pardo, D. Naccache, AIP Conference Proceedings N° 119, ed. by S.E. Harris and T.B. Lucatorto, (1984) 199.
22. Multilayer Structures and Laboratory X-Ray Laser Research, Proceedings of

SPIE, Vol. 688, ed. by N. M. Ceglio and P. Dhez, Bellingham, Washington 98277-0010 USA (1988).

23. P. Jaeglé, G. Jamelot, A. Carillon, A. Klisnick, A. Sureau, H. Guennou, in ref. 2, p. 563.

24. D. Matthews, M. Rosen, S. Brown, N. Ceglio, D. Eder, A. Hawryluk, C. Keane, R. London, B. MacGowan, S. Maxon, D. Nilson, J. Scofield, J. Trebes, in ref. 2, p. 575.

25. Rutherford Appleton Laboratory Annual Report to the Laser Facility Commitee (1988) 1.

26 - A. Carillon, F. Gadi, B. Gauthé, H. Guennou, P. Jaeglé, G. Jamelot, A. Klisnick, C. Moller, A. Sureau, Journal de Physique, 48 (1987) C9-375.

27. Gudzenko, L.I., and Shelepin, L.A., Sov. Phys. Doklady, 10 (1965) 147.

28. G.J. Pert, J. Phys. B, 9 (1976) 3301; 12 (1979) 2067.

29. Jacoby, D., Pert, G.J., Shorrock, L.D., and Tallents, G.J., J. Phys. B: Atom. Molec. Physics, 15 (1982) 3557.

30. S. Suckewer, H. Fishman, J. Appl. Phys., 51 (1980) 1922.

31. S. Suckewer, C.H. Skinner, D. Voorhees, H.M. Milchberg, C. Keane, A. Semet, IEEE J. Quantum Electron.,QE-19 (1983) 1855.

32. S. Suckewer, C.H. Skinner, H.M. Milchberg, D. Voorhees, Phys. Rev. Lett., 55 (1985) 1753.

33. C. Chenais-Popovics et al., Phys. Rev. Lett., 59 (1987) 2161.

34. M. Key et al., Vol. Ref. 1, p. 71.

35. S. Suckewer, C.H. Skinner, D. Kim, E. Valeo, D. Voorhees, A. Wouters, Phys. Rev. Lett., 57 (1986) 1004.

36. C.H. Skinner, D. Kim, A. Wouters, D. Voorhees, S. Suckewer, Proceedings of SPIE 31st Annual International Technical Symposium, San Diego, Ca , 831, (1987), in press.

37. H. Guennou, A. Sureau, C. Moller , Journal de Physique, C6-351, 47, 1986.

38. A. Klisnick, H. Guennou, J. Virmont, Journal de Physique, 47 (1986) C6-345.

39. H. Guennou, A. Sureau, J. Phys. B, 20 (1987) 919

40. A. Sureau, H. Guennou, C. Moller, Journal de Physique, 49 (1988) C1-195; Europhys. Lett., 5 (1988) 19.

41. A. Sureau, H. Guennou, private communication.

42. G. Jamelot, P. Jaeglé, A. Carillon, A. Bideau, C. Moller, H. Guennou, A. Sureau, Proceedings of the International Conference on Lasers' 81, ed by C.B. Collins (STS, Mc Lean, Va.), (1981) 178.

43. P. Jaeglé, G. Jamelot, A. Carillon, A. Klisnick, A. Sureau, H. Guennou, AIP Conference Proceedings N° 119, ed. by S.E. Harris and T.B. Lucatorto, (1984) 468.

44. A. Klisnick, P. Jaeglé, G. Jamelot, A. Carillon, in Spectral Line Shapes (de Gruyter, Berlin),3 (1985) 157

45. P. Jaeglé, A. Carillon, A. Klisnick, G. Jamelot, H. Guennou, A. Sureau, Europhys. Lett., 1 (1986) 555.

46. P. Jaeglé, G. Jamelot, A. Carillon, A. Klisnick, ref 22, p.

47. A. Carillon, P. Jaeglé, G. Jamelot, M. Key, G. Kiehn, A. Klisnick, G. Pert, S. Ramsden, C. Regan, S. Rose, R. Smith, T. Tomie, O. Willy, Invited Conf. at the XVIIIth ECLIM, PRAGUE (Tchécoslovaquie), 4-8 mai 1987

48 C. Lewis et al, to be published

49. P. Jaeglé, A. Carillon, P. Dhez,, B. Gauthé, F. Gadi, G. Jamelot, A. Klisnick, to be published in Europhysics Letters, .

50. H. Kogelnik, Applied Optics, 4 (1965) 1562.

51. D.L. Matthews et al. , Phys. Rev. Lett., 54 (1985) 110.

52. S. Maxon et al.. Phys. Rev. A, 37 (1988) 2227.

53.T.N. Lee, E.A. McLean, R.C. Elton, Phys. Rev. Lett., 59 (1987) 1185.

54. E.A. McLean, T.N. Lee, R.C. Elton, Journal de Physique, 49 (1988) C1-51.

55. E. Berthier, J.L. Bourgade, P. Combis, S. Jacquemot, J.P. Le Breton, M. Louis-Jacquet, D. Naccache, M. Nail, O. Peyrusse, Journal de Physique, 49 (1988) C1-123.

56. M. Cornille, J. Dubeau, M. Loulergue, S. Jacquemot, Journal de Physique, 49 (1988) C1-95.

57. A. V. Vinogradov, V.N. Shlyaptsev, Sov. J. Quant. Electron., 10 (1980) 754.

58. M.D. Rosen et al., Phys. Rev. Lett., 54 (1985) 106.

59. S. Maxon, P. Hagelstein, J.. Scofield, Y. Lee, J. Appl. Phys.,59 (1986) 293.

60. M.D. Rosen, P.L. Hagelstein, Proceedings ot the Topical Meeting, Monterey, CA, Mar. 24-26, 1986, New York, American Institute of Physics, p. 110.

61. P. Monier, Thèse d'Etat, 30 Avril 1987, Université Paris-Sud, Orsay

62 M. D. Rosen et al., invited paper to the Conference on Atomic Processe in Plasma, Santa Fe, N.M. October 1987, UCRL Preprint 97954.

63. P.L. Hagelstein, Phys. Rev. A, 34 (1986) 874.

64. B.J. MacGowan, S. Maxon, P.L. Hagelstein, C.J. Keane, R.A. London, D.L. Matthews, M.D. Rosen, J.H. Scofield, D.A. Whelan, Phys. Rev. Lett., 59 (1987) 2157.

65. S.E Harris, J.F. Young, in ref. 2, p.547.

66. B. Rozsnai, H. Watanabe, P.L. Csonka, Phys. Rev. A, 32 (1985) 357.

67. A. McPherson et al., in ref. 2, p. 595.

68. W.H. Louisell, M.O. Scully, W.B. McKnight, Phys. Rev. A, 11 (1975) 989.

69. S.P. Zhukov, V.V. Korukov, N.G. Nikulin, B.I. Troshin, A.A. Chernenko, in ref. 1, p. C6-171.

SELECTED TOPICS IN X-RAY SPECTROSCOPY

FROM HIGHLY IONIZED ATOMS IN HOT PLASMAS

Elisabeth Källne

Department of Physics I
Royal Institute of Technology
S 10044 Stockholm, Sweden

1. INTRODUCTION

The purpose of these lectures is to highlight results achieved so far
in the field of X-ray spectroscopy from hot plasmas. The lectures are
divided into four parts: 1) general description of the plasma source, i.e.,
the environment in which the highly ionized atoms are embedded,
2) discussion of dominant atomic processes in the center of the plasma and
spectroscopic results related to the hot, quasi-stationary region,
3) discussion of results related to the plasma regions showing gradi nts in
the key parameters, i.e., when changes in radial profiles of electron
temperature and density are important and the atomic processes must be
studied in detail, and 4) discussion of fast time-varying phenomena
other spectroscopic results possibly related to processes not usually
accounted for.

In the two decades which have passed since X-ray spectroscopy gave the
first reliable results on ion temperature of hot laboratory plasmas from
line profile analysis of individual X-ray lines [1] we have seen both a fast
development and establishment of the experimental techniques to extract with
better accuracy and sensitivity some of the key parameters for mapping the
plasma behaviour. Additionally, the observations have contributed with
important results to the physics of highly ionized atoms. In this paper, we
will briefly discuss the first aspect and emphasize the questions related to
the atomic physics issues. Some reviews have recently been published on the
role of atomic physics in fusion research [2,3] and the diagnostic aspects
of spectroscopy in general [4-7]. These lectures emphasize results related
to X-ray spectroscopy still achieving attention through new observations or
theoretical calculations. For more basic information on spectroscopy from
tokamaks the reader is referred to earlier references [8-10].

In fusion research of today the parameters of the plasmas have reached
new domains compared to those achieved in previous machines. For example,
electron temperatures in excess of 10 keV have been reached at JET (Joint
European Torus, Abingdon, UK), ion temperatures of 15 keV and above have
been reached both at JET and TFTR (Tokamak Fusion Test Reactor, Princeton,
US), while electron densities still are in the range of 10^{13}-10^{14} cm^{-3}. The

volume of the plasma is also much larger than that of previous machines; at JET the plasma volume is ~140 m^3. These parameters imply that light atoms are completely stripped of all electrons, while medium Z atoms (with higher Z than for earlier machines; e.g., Ni^{27+}, Kr^{33+} and Mo^{39+} for JET) are stripped to few electron systems. Doppler broadening of spectral lines is substantial for these hot ion temperatures. Furthermore, instrumental distances are long compared to any atomic collision lengths and the discharge times are long compared to any atomic collision times.

Thus, we expect the conditions in the central part of the plasma to be determined mainly by atomic processes. On the other hand, boundary conditions, i.e., effects of plasma-wall interactions [11], have proven to greatly influence and determine the plasma behaviour also in the center of the plasma. Therefore, there is an increased emphasis on experiments to map and also to control and influence the boundary conditions and to explore e.g., possible dependencies on confinement time for different boundary conditions. In these lectures these issues will not be discussed but must be referred to only [11,12].

2.   THE PLASMA SOURCE

a)   Ohmic Heating Mode

The plasma source for the topics discussed in these lectures is always a magnetically confined plasma produced by a tokamak. The plasma is thus of low density (n_e ~ 10^{13}-10^{14} cm^{-3}) and high temperature (T_e ~ 2-10 keV, T_i ~ 1-15 keV). The plasma is produced by an inductive current through a transformer action (see Fig.1) and the plasma discharge is pulsed. Therefore, all the plasma parameters will show a time dependence following the initial current ramp-up phase, a stable flat-top phase and a current ramp-down phase. Figure 2 shows a typical plasma discharge time trace with (from top to bottom) plasma current, electron density, additional heating power of radio-frequency antennae, electron temperature and ion temperature.

The working gas, usually hydrogen or deuterium, is let into the vacuum chamber in the beginning of the discharge. Typical filling pressures are in the range of mTorr while the vacuum chamber initially is at a vacuum of 10^{-9} Torr. The cleanliness of the vacuum chamber is important and the chamber is therefore baked regularly and also kept at an elevated temperature of ~150°C during discharges. The interaction of the plasma edge with the vacuum chamber is minimized through introduction of limiters. The vessel chamber at JET is made out of inconel alloy (mainly nickel and chromium) and the inner wall as well as the limiters are covered with graphite tiles. Thus, the main impurity elements entering the plasma will be carbon and oxygen (to a few percentage level of concentration), while metal impurity concentrations are lower by one to two orders of magnitude.

In a typical plasma discharge the elements deposited on the limiter will re-enter the plasma and become successively ionized during the radial diffusion of particles. Thus, radial distributions of the different ionization stages of the impurity elements will evolve and reach an equilibrium distribution determined both by atomic collisions and by the particle transport processes. Figure 3a shows a radial distribution of the main plasma parameters (electron and ion temperature, T_e and T_i, and electron and ion density, n_e and n_i) at one time in the discharge. Figure 3b shows a three-dimensional plot of the electron temperature with time and radius. Under such stable conditions we can also describe the evolution of the ionization stages and we obtain equilibrium radial shell distributions such as shown in Figs.4a and b. These figures are examples of the shell distributions for a light element, oxygen, and a heavier element, nickel.

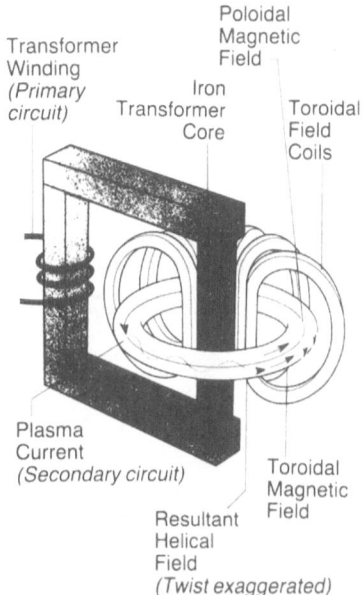

Fig.1   Schematic of the principle
         of a tokamak.

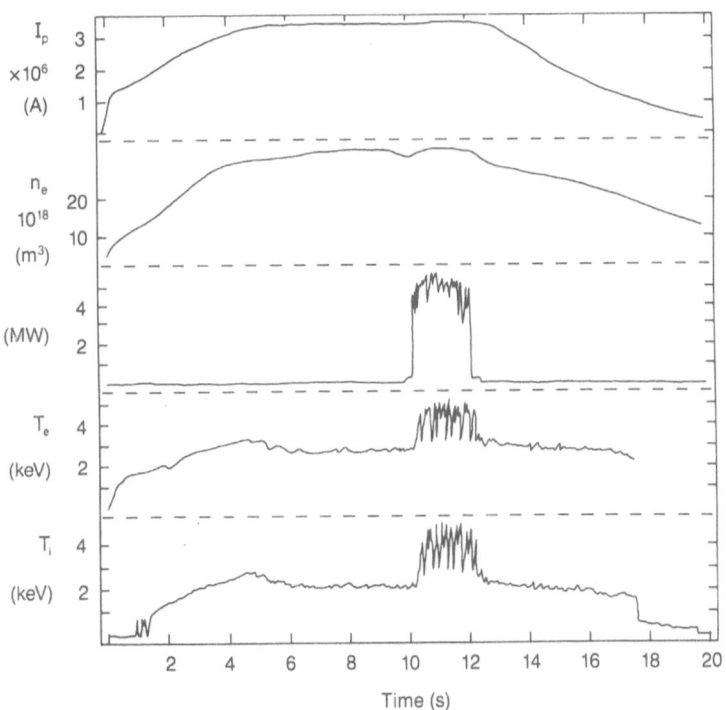

Fig.2   Time evolution of plasma
         current, electron density, RF
         power, electron temperature
         and ion temperature during a
         pulse discharge.

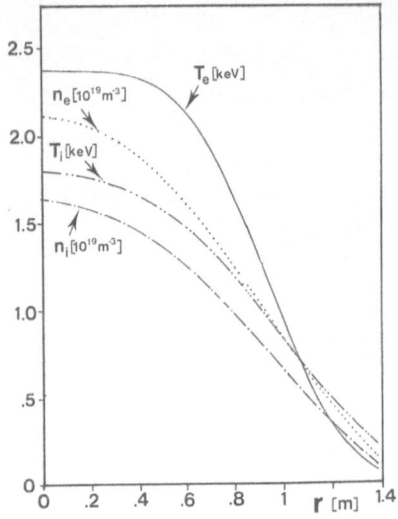

Fig.3a Radial distribution of
        electron and ion temperature,
        electron and ion density in
        JET.

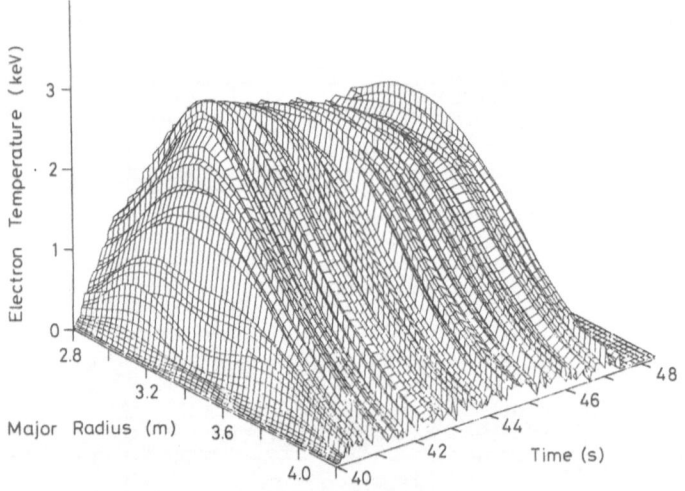

Fig.3b Electron
        temperature
        from JET as
        a function
        of radius
        measured
        with
        electron
        cyclotron
        emission.

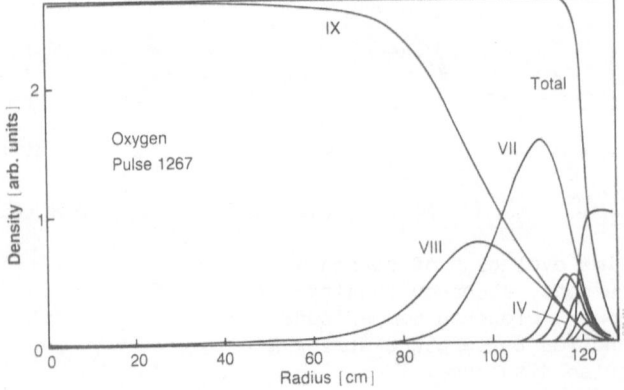

Fig.4a Predicted radial
        shell distribution
        for oxygen.

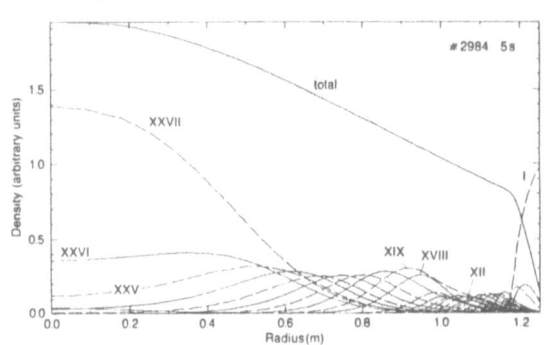

Fig.4b Predicted radial shell
distribution for nickel.

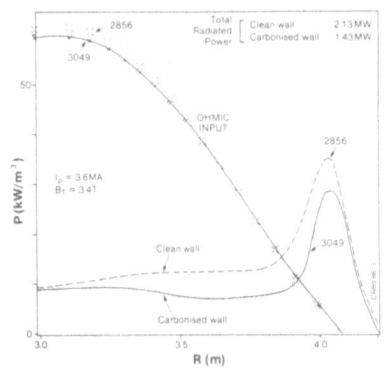

Fig.5a Radial distributions of
total Ohmic input power and
radiative power for two
different discharges.

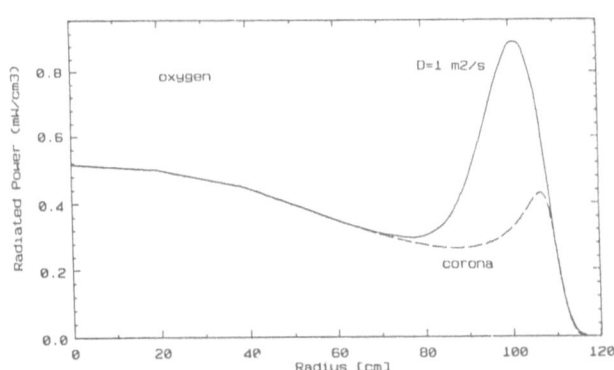

Fig.5b Radial distribution of
emission for oxygen predicted
from a corona model and from
a transport model [51].

At these high electron temperatures low-Z impurities are fully ionized
and give contribution to the radiative power only in the very outer edge of
the plasma. The metal impurities, however, will be in the H- and He-like
stages in the center of the plasma and give rise to X-radiation from the
center and to VUV from the outer regions of the plasma. Measurements of the
radial distribution of the total radiated power show these features. An
example of results from such measurements is shown in Fig.5a where the Ohmic
input distribution and the total radiated power are shown. A clear radial
dependence is found and it can be seen from Fig.5b that the radiation from
the outer edge of the plasma can be explained as caused by radiation from
lower-Z elements such as oxygen. From analysis of these data we can not
only evaluate the total radiated power, its radial distribution and its
energy dependence, but we can also try to control the plasma discharge to
prevent detrimental losses to occur.

b) Additional Heating Mode

As the plasma current is increased the heating efficiency with Ohmic
heating only is decreased since the plasma resistivity decreases.
Instabilities are likely to occur with increasing current and, therefore,
additional heating by other methods than Ohmic heating is necessary in order
to reach the conditions for fusion and an ignited plasma.

At JET, ion cyclotron resonance heating (ICRH) [13] and neutral beam
injection (NBI) [14] are both used in combination and with Ohmic heating.
Presently, plasma currents up to 7 MA, NBI power up to ~ 10 MW and ICRH
power up to ~ 10 MW have been applied. The maximum recorded power with a
combination of the three heating methods is ~ 20 MW.

In ICRH, the resonance heating is applied directly to the bulk ions.
However, it might be necessary to specially inject minority ions, e.g., ^{3}He,
to match the resonance frequency for ion cyclotron heating. Through collisions the various ion species will equilibrate in velocity distribution
although high energy tails of ions might persist in the plasma. The ICRH
usually causes large fluctuations in the electron temperature, so-called
sawtooth oscillations. These can be both of a regular periodic structure
(50-200 ms period) and evolve with a much longer duration sawteeth (monster)
oscillations [15].

During NBI high energy deuterium particles (presently 80 keV at JET)
are injected to the plasma. Through collisions with the bulk plasma ions
the NBI energy is deposited. In addition to heating the bulk plasma ions
the momentum of the beam particles will also induce a toroidal rotation of
the plasma. For NBI power of some 10 MW the ion temperature increases from
~ 3 keV to over 10 keV and the toroidal rotation velocity might be as high
as $(2-4)\cdot10^5$ m/s, i.e., of the same order of magnitude as the thermal
velocity.

c) Plasma Parameters

In a fusion project like JET, the goal is to reach plasma parameters to
achieve conditions for an ignited plasma with the reaction d + τ → ^{3}He + n.
Thus, optimization of parameters, combination of different operating scenarios for a plasma discharge, is the main effort. Presently at JET, the best
achieved confinement criteria, i.e., $n_e\cdot\tau_e\cdot T_i$ (where n_e is the electron
density, τ_e is the energy confinement time and T_i is the ion temperature)
has a value of $3\cdot10^{20}$ m^{-3} s   keV; to reach ignition a value of $5\cdot10^{21}$ is
required. The latest results obtained for the different fusion projects are
usually published at the plasma physics conferences [16-18] and will not be
discussed here.

At issue for these lectures, however, is that due to the optimization
and variation of parameters in the quest to produce the fusion plasma
follows, that the quasi-stationary period of the plasma might be short, also
in comparison with atomic times due to the occurrence of e.g., MHD instabilities. Thus, spectroscopic studies of atomic physics phenomena might be
possible only with the more modest operating scenarios of the plasma.
Therefore, more basic information on the physics of highly ionized atoms in
hot plasmas will be obtained from (smaller) tokamak machines which are not
necessarily in the forefront of national or international fusion research,
while results obtained from machines like JET are more of direct relevance
for diagnostics of fusion plasmas. For example, observations from JET will
reveal much of the interplay between atomic physics and the phenomena related more to physics of impurities in fusion plasma such as particle transport, recycling, impurity accumulation; some of these issues will also
become visible in these lectures.

3. X-RAY SPECTROSCOPY INSTRUMENTATION

The instrumental requirements at a pulsed plasma discharge are both survey recordings to identify spectral line features over a broad energy range and high resolution line recordings for line shape measurements. Furthermore, in the search for optimum plasma parameters for ignition, the plasma conditions will vary considerably and spectrometers must be able to cover a large variation in intensity (dynamic range of 3-4 orders of magnitude or more). Additionally, one of the major new aspects of instrumentation is that it should be able to give spatial resolution in elements of a few cm:s from an extended source of height 4 m and temporal resolution to ms level in a pulse which can be 30 s long. Much of the spectroscopic results obtained so far from tokamak plasmas have been obtained in a parasitic mode, i.e., spectroscopy being one of the many diagnostics around the machine. Usually, a large set of instruments are installed around the tokamak. Fig.6 shows the diagnostic equipment installed around the JET tokamak with the purpose to measure the many different plasma parameters and also to measure, if possible, the same plasma parameter with several different independent techniques. However, just because of the large variation in plasma parameters, it is usually very difficult to have several independent measuring techniques active simultaneously.

Spectroscopic instrumentation should, besides having the capability of spatially and temporally resolved measurements, also cover the wavelength range from visible to X-rays in order to observe the emission from the different ionization stages evolving during the plasma discharge. Table I shows an example of spectroscopic instrumentation installed to operate simultaneously during a plasma discharge at JET.

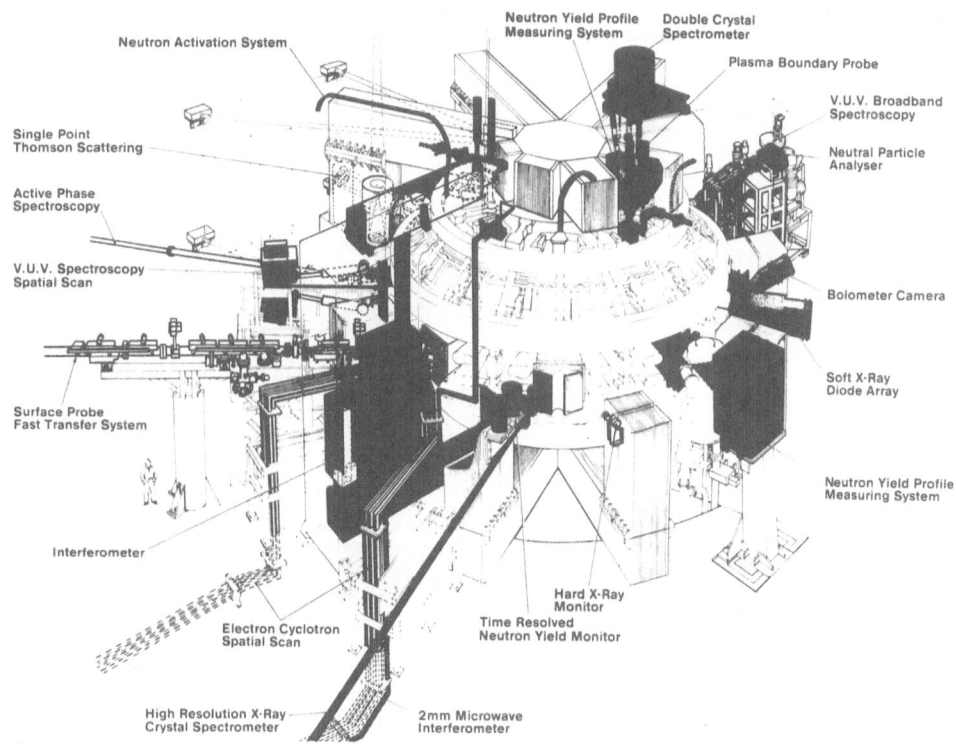

Fig.6 Schematic of the diagnostic systems around the JET tokamak.

TABLE I
Spectroscopy instrumentation at JET

	Wavelength Range (Å)	Instrument	Information
Visible	4000-8500	1m Czerny Turner + OMA interference filter+PM poloidal array of PM:s Czerny Turner + OMA	Survey individual lines H,D,He,C,O, Hα,bremsstrahlung charge exchange spectroscopy
VUV	10-40 20-1200	Grazing Incidence Flat-field, toroidal grating	Line shape, Impurity Survey Survey, Impurity concentrations
X-ray	1-24 1-24 ~ 2	Double Crystal Double Crystal Rowland Circle	Spatial Scan Survey Line Shape Analysis

   X-ray instruments can either be coupled via a thin window (if λ < 25 Å) or directly to the vacuum chamber. Most X-ray spectrometers in use have been of traditional Rowland circle type with a large Be window being the interface at the tokamak machine. Figure 7 shows two set-ups of X-ray spectrometers presently in use at the large tokamaks JET [19] and TFTR [20]. The Bragg crystal diffracts the X-rays onto position sensitive detectors. In present day applications stringent requirements are put on the detectors to give good position resolution, large area and high count rate capability [21-23]. Presently, resolution elements of less than 100 μm over a length of 10 cm have been achieved at count rates above several MHz.

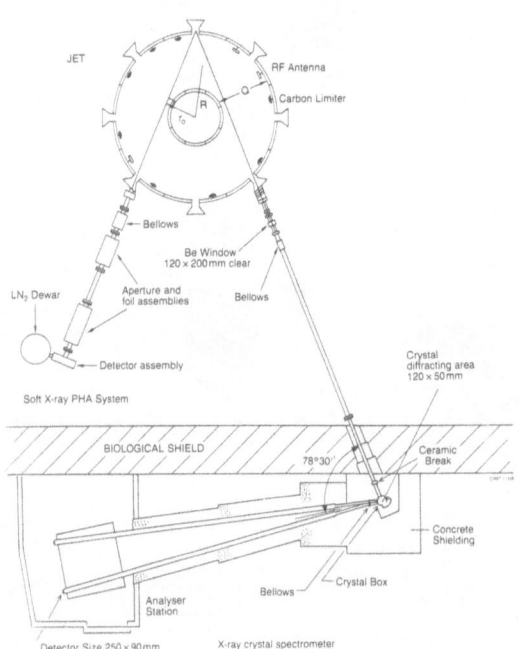

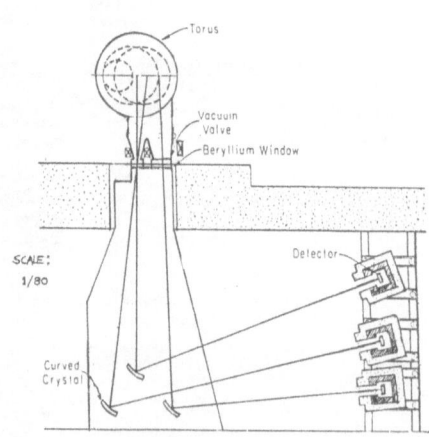

Fig.7a Schematic outline of the X-ray crystal spectrometer at JET [19].

Fig.7b Schematic outline of the X-ray crystal spectrometer at TFTR [20].

4. BASIC DIAGNOSTIC INFORMATION FROM X-RAY SPECTROSCOPY

Before discussing results from X-ray spectroscopy in relation to atomic physics of highly ionized atoms we will briefly discuss the basic diagnostic information obtained from X-ray spectroscopy. In addition to the parameters mentioned below, X-ray spectroscopy can give important information on other plasma parameters such as electron temperature and density, neutral particle density, magnetic and electric fields. These subjects require more discussion of the atomic processes and will be discussed in Sections 5 and 6.

a) Ion Temperature and Toroidal Velocity

The line profile (the width and position) of an individual spectral line gives data on the collective motion of the emitting particles provided other broadening mechanisms are negligible. The X-ray line of e.g., He-like nickel (Ni^{26+}, 1s-2p transition) at 1.58 A is well separated from the resonance lines of other ionization stages and its width is determined by the Doppler motion of the nickel ions in the hot plasma since other broadening mechanisms such as Stark and Zeeman effect are several orders of magnitude smaller at the conditions for a tokamak plasma produced in JET. Additionally, the natural width of the transition is small. Typically, for a plasma at an ion temperature of 5 keV, the Doppler broadening is 5 eV for Ni^{26+}, while instrumental and natural broadening is 0.5 eV and other broadening mechanisms are another order of magnitude smaller. However, there are other factors caused by for example the line-of-sight through the plasma with radial variations in e.g., electron and ion temperature, rotation velocity etc. [24-26] which can cause extra broadening. Therefore, even the rather straightforward measurement of an individual line profile requires plasma modelling for analysis and preferably other independent measurements should be performed to support the analysis.

A high resolution instrument is required ($E/\Delta E > 5000$, typically instruments in use have $E/\Delta E = 10$-20000) to measure the line profiles of individual X-ray lines. Figure 8 shows examples of results on ion temperature measurements at tokamaks. In Fig.8a an individual line profile from the resonance line (1s-2p transition) in Ni^{26+} is shown during Ohmic and NBI heating, Fig.8b shows the time evolution of the ion temperature during additional heating, clearly showing the requirement of high temporal resolution (in the figure a time resolution of 20 ms is obtained) and Fig.8c shows a comparison of results from different techniques of measuring the ion temperature [27]. Although excellent data has been obtained on ion temperatures from the hot tokamak plasmas, we still lack instrumentation to give sufficient temporal and spatial resolution, simultaneously.

b) Impurity Identification and Measurement of Concentrations

From a spectral survey measurement the impurity content of the plasma can be determined. Firstly, there are normally occurring impurities whose concentration should be monitored from shot-to-shot and time resolved during the discharge. Secondly, occurrence of a new impurity should be easily observed to give precursors for any possible disruptive behaviour due to unusual conditions in the vacuum chamber.

To follow the impurity behaviour in the plasma, instrumentation is needed which can give spectral surveys both spatially and temporally resolved. Presently, there are only two approaches used for this task in the X-ray region, spatially scanning rotating crystal [28] and spatially scanning double crystal instruments [29,30]. Both of these approaches are mechanically very requiring and there is no instrument of this type as yet working automatically at a tokamak. On the other hand, the first

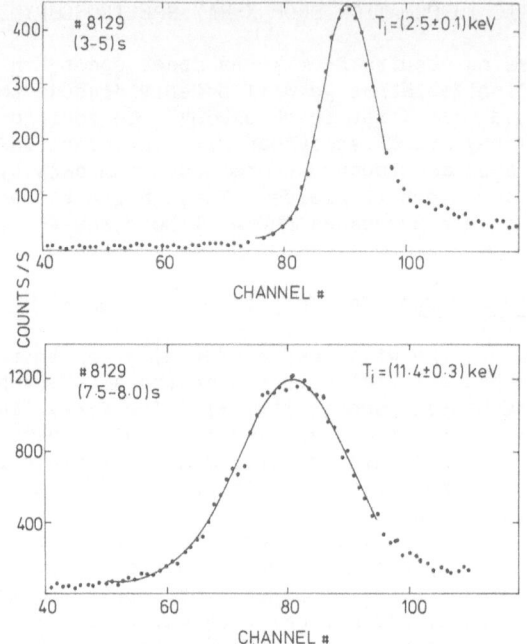

Fig.8a Line profiles of (1s-2p)
transition for Ni²⁶⁺ emitted
for the JET plasma.
The upper trace is before
the NBI, the lower trace
during NBI.

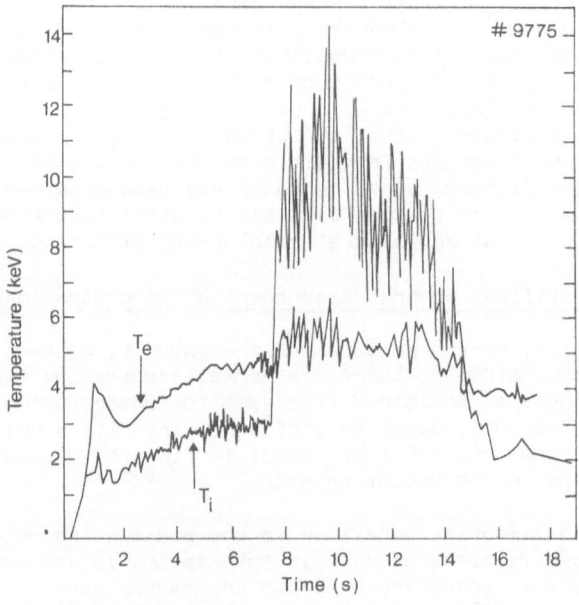

Fig.8b Ion (T_i) and electron (T_e)
temperature as a function of
time during a JET discharge.

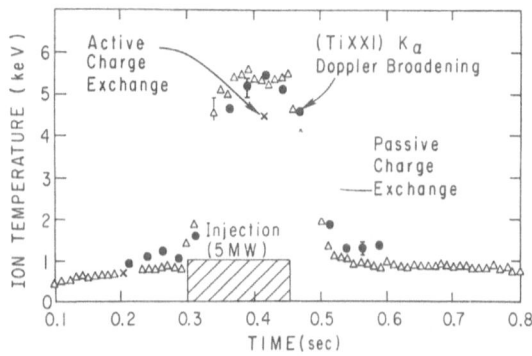

Fig.8c Ion temperature as a function of time during
a discharge on the PLT tokamak [27].

presentation of radially resolved measurements, performed with a flat
crystal instrument of van Hamos type [31,32] scanning spatially on a
shot-to-shot basis, showed the importance of such measurements and also the
possibility to achieve both spectrally and spatially resolved X-ray spectra
with a not too cumbersome instrumentation.

The instruments should also be calibrated to make it possible to
measure absolute concentrations. Thus, a calibrated X-ray source should be
connected to the spectrometer during normal tokamak operations to be able to
obtain the absolute efficiency of the instruments. This is usually not done
at the tokamak machines and therefore there are also several questions
relating to absolute intensities and calibrations [33].

c) Impurity Confinement Time

The confinement time of particles can ideally be measured if the decay
time of the emitted radiation is observed. A technique developed to measure
impurity confinement and transport properties is to purposely inject
impurity elements and to follow the decay time of the emission from different ionization stages [34,35]. An example of an impurity decay time trace
from JET is shown in Fig.9 for Li-like Ni (Ni^{25+}). The experimental data
are shown as points and the predictions from a plasma impurity transport
model with various diffusion coefficients (D = 0.5, 1.0 and 1.5 m²/s) are
shown with solid lines.

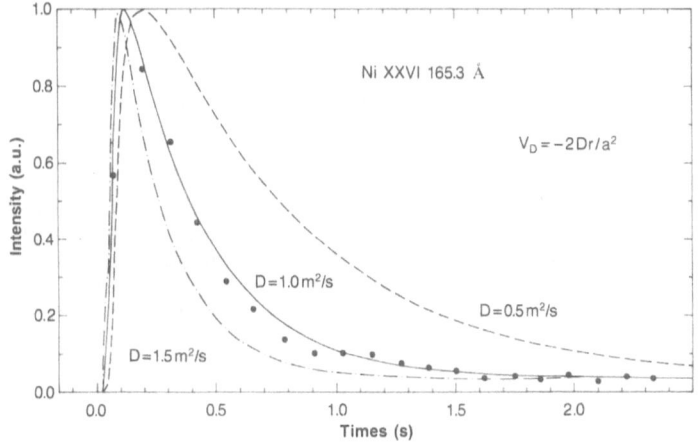

Fig.9 Time evolution of Ni XXVI during a JET discharge. The three lines
represent predictions from an impurity transport code [36].

5. ATOMIC PROCESSES AND X-RAY SPECTROSCOPY FROM THE CENTRAL PLASMA REGION

The central part of the plasma is described as a homogeneous plasma with no gradients in temperature and density (see Fig.3). An ionization stage equilibrium can be predicted based on electron impact ionization and recombination (as shown in Fig.4) assuming any time dependent variations of plasma parameters are slow compared to atomic collision times. For example, at an electron density of 10^{13} cm^{-3} it will take ~ 5 ms to reach Ni^{26+} through successive electron collisions and the "flat top" lasts for several seconds. Thus, during the quiescent phase of the plasma discharge (the flat top of Fig.2) spectrosocopic observations can favourably be done to study details of highly ionized atoms. Here, we will give a few examples of such recent studies.

In the center of the hot tokamak plasma, low Z elements such as C and O are fully stripped and medium Z elements, such as iron and nickel, are mainly present in the H- and He-like ionization stages. For these ions studies now exist up to Z = 28, Ni, [37,38,4], while for lower ionization stages also heavier elements such as Kr and Mo have been studied [39-41].

a) <u>Spectroscopy from Highly Ionized Atoms</u>

The H- and He-like spectra have been studied in detail at several tokamaks and here we will just mention a few results still receiving attention. Figure 10 shows the H- and He-like spectra observed from the JET tokamak [37]. The solid line represents the best fit of the theory to the experimental data. Several points can be made from such a detailed comparison. The absolute wavelengths predicted from the calculations must be adjusted to agree with the experiment; in Fig.10 a systematic shift of all calculated wavelengths of 2.9 mÅ (0.2%) has been done for the best fit. Shifts in wavelengths for different lines in the isoelectronic sequence of He-like spectra have also been studied [42] and agreement to within 1 mÅ between theory and experiment has been found. However, no absolute measurements of these X-ray wavelengths from He-like lines observed in tokamak plasmas have been done; absolute measurements with an accuracy of < 20 ppm have only been presented for H-like Cl and Ar observed from the Alcator C tokamak [43,44]. It can also be observed from Fig.10 that there is a disagreement between experiment and theory in the position of the lines of the H-like Ni spectrum. No shift of the theoretically predicted lines has been done in this case. Thus, there is still a need for accurate, absolute measurements of wavelengths for these few electron systems.

b) <u>Ionization State Balance</u>

A detailed comparison of theoretical predictions with experimental spectra as shown in Fig.10 gives not only line positions but also relative intensity results. To arrive at the solid line, however, requires not only calculations of excitation rates for all the different atomic processes populating the levels involved. It also requires calculations of ionization state balance to obtain relative abundances of the ionization stages involved in the observed transitions. As has been pointed out previously in this paper, an ionization state balance has been assumed valid for the central part of the plasma (coronal equilibrium). However, in order to arrive at the best fit shown in Fig.10 it has been necessary to change the abundance ratios considerably from those predicted from a corona equilibrium calculation. For example, the ratios Ni^{25+}, Ni$^{24\pm}$, Ni^{23+} to Ni^{26+} are deduced to be 0.7, 0.2 and 0.06, respectively, while predicted values are 0.33, 0.09 and 0.017, respectively. Similar results have also been found from the TFTR tokamak at a higher electron temperature (4 keV) than for the spectrum from the JET tokamak (3 keV) [38]. The relative abundance ratios

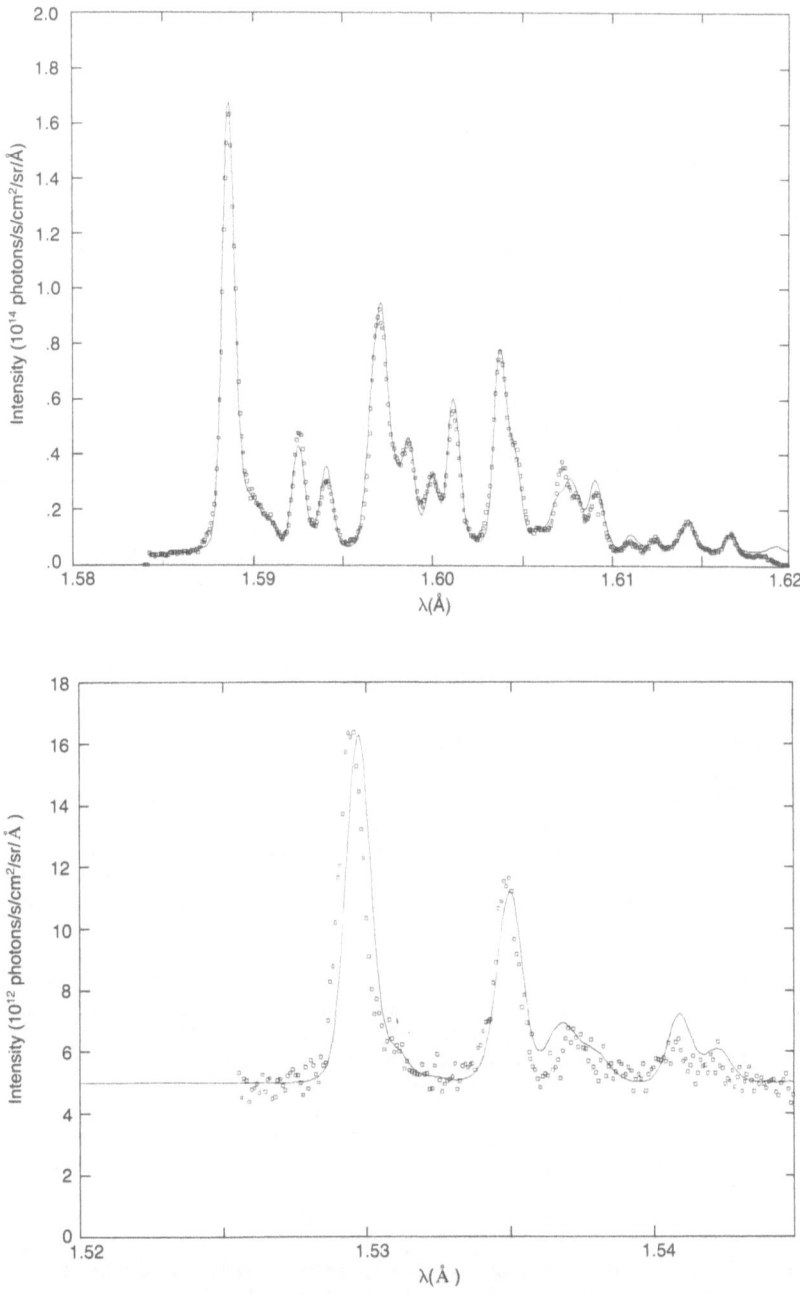

Fig.10 Spectra from He-like (top) and H-like Ni (bottom) observed for the
        JET tokamak. The solid lines are best fit predictions from
        theoretical calculations [37].

differ by up to a factor of two between different calculations so there is a need for more detailed ionization equilibrium calculations and evaluation of accuracy of the calculations as well. Further comparisons with experimental spectra should be done, preferably over a range of electron temperatures in order to understand the difference in abundance ratios deduced from the best fit of experimental data and those predicted from a pure ionization state balance calculation for a homogeneous plasma.

The experimental spectra are observed along a horizontal line-of-sight and modelling is included to take into account effects of radial profiles of electron temperature and density. Of course, it would be preferable to have radially resolved measurements in order to properly describe the ionization stage profiles, although any deviations from values predicted from a transport code are assumed to be small for the ionization stages from the central part of the plasma.

c) Temperature and Density Dependence

The He-like spectra have been commonly observed because of the diagnostic possibilities which exist from the different line ratios in the spectra [45,46]. Observations of line ratios from a well diagnosed plasma, like from a tokamak plasma, are important, since they serve directly as test of the atomic theory, i.e., rate coefficients for the atomic processes involved are measured. In general, good agreement (to within 10%) has been found for line intensity ratios representing dielectronic recombination [47,48] and electron impact excitation. However, there are still uncertainties as to the main diagram lines in the He-like spectra. For example, the ratio G (representing the sum $(x + y + z)/w$ in the notation of [46]) tends to be higher than predicted from a purely ionization balance calculation. Since the observations are done along a chord (integrated line-of-sight) the question of radial profiles of the impurity ionization stages always linger as an uncertainty.

Recently, an analysis has been made on data from JET for line ratios of the He-like Ni spectra with electron temperatures up to 15 keV. Deviations in the main diagram line ratio x/w have been observed as well as in the satellite ratio t/w [49]. This is the first time the theory has been tested at these high electron temperatures and there is a need for better cross section data at high energies.

From the analysis of spectra observed from plasmas we cannot directly measure a cross section since the whole electron velocity distribution is present and contributes to the excitation process. Thus, we use rate coefficients in the modelling of the plasma emission. Therefore, when we observe deviations from the predicted theoretical line ratios this could be signatures of e.g., non-Maxwellian distributions and not only errors in the used cross sections. For example, there are clear evidences of non-Maxwellian distributions during electron cyclotron resonance heating [50] causing changing line ratios in the He-like spectrum. We could, however, also discuss the possibility of e.g., truncation of the Maxwellian distributions during various plasma operation scenarios. These questions are presently most interesting for the plasma physics and there is a need for more experimental data even at the high electron energies in order to separate effects of errors in the cross sections from any other effects such as non-Maxwellian distributions.

6. ATOMIC PROCESSES AND X-RAY SPECTROSCOPY IN PLASMA REGIONS WITH GRADIENTS

In the plasma regions of rapidly varying electron temperature and density there will also be a strong influence of particle transport and the

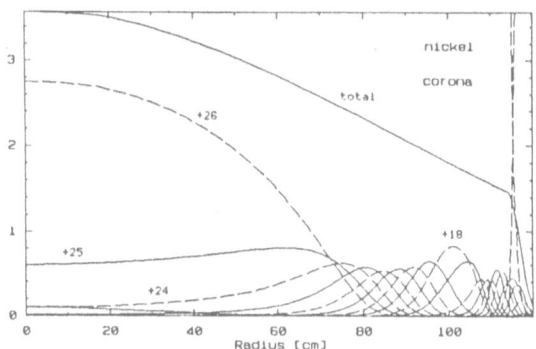

Fig.11a Ion densities
(arb. units)
of Ni calculated
with a corona
approximation
(no transport).

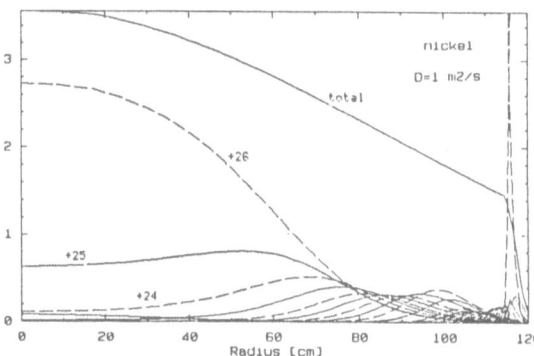

Fig.11b Ion densities
(arb. units)
calculated with
a transport code
using a constant
diffusions of
$1 \, m^2/s$.

ionization stage distributions will no longer be described by a simple atomic model. Figure 11 shows the nickel density distribution both for a model using an ionization balance (corona) and a model including particle transport [51]. Thus, measurements of radial distributions become crucial since the issue of particle transport in tokamak plasma is still not understood and only described with phenomenological concepts.

Not only the ionization stage balance is influenced at the outer radii but also the spectral features are remarkably changed showing that the atomic processes producing the spectra are changing. A clear example of this comes from the first radial scan observations of X-ray spectra from a tokamak [32,52,53]. Figure 12 shows the spectra from He-like argon emitted by a dense tokamak plasma with a central electron temperature of ~ 1.5 keV. The top spectrum is from the central part and the two lower spectra are from the outer regions with temperatures of ~ 800 eV and ~ 350 eV, respectively. The dielectronic satellite lines (K in Fig.12b) show first the expected behaviour with increasing intensity when the temperature falls, while at about 2/3 of the plasma radius the intensity drops suddenly. These results showed for the first time clear evidence in characteristic X-ray spectra of charge exchange recombination with thermal hydrogen. It should be emphasized, that in addition to charge exchange becoming the dominant excitation process instead of electron impact excitation to produce the spectra from He-like argon there must also be sufficiently fast transport of H-like argon to the edge (since the charge exchange process involves the thermal hydrogen and the hydrogen-like argon).

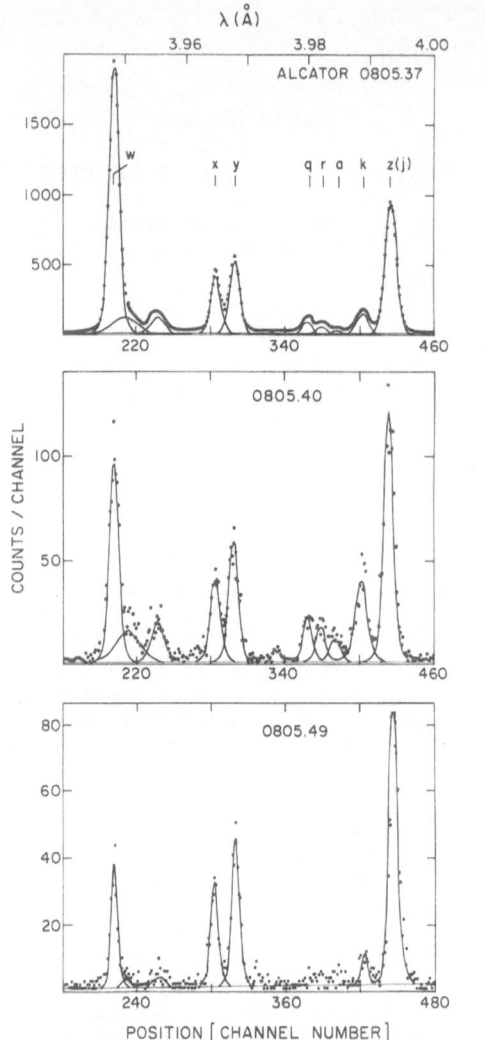

Fig.12a Spectra from Ar^{16+} at three
different radii from the
Alcator tokamak plasma [52]

Fig.12b Line ratios (S=x+y/w,
$G = \frac{x+y+z}{w}$, K = k/w and
Q = q/w) as a function of
plasma radius.

The interpretation of the origin of the spectral changes observed
in the n=2 to n=1 spectrum, as shown in Fig. 12, could be verified with
observations of transitions from the high n states to the ground state
[53] as shown in Fig. 13 [32].   The resonant charge exchange process

$$Ar^{17+} + H \Rightarrow Ar^{16+}(n) + H^{+}$$

at the low velocities present in this experiment (0.2-1.5 keV) will populate only low ℓ quantum numbers of the resonantly populated n=8, 9 shell
[54, 55] and the deexcitation occurs through the Y-rast chain to the
ground state.   It is interesting to note that the characteristics of this
process can be well described by a classical model [55] and the analysis
of this data could yield valuable infromation of the neutral density
distribution as discussed in [53].

Observations of characteristic x-ray spectra from these high n states from a hot plasma source complement nicely direct experiments on ion-atom collisions. So, for example, it could be possible to observe and characterize various collision processes, e.g., by using different fill gas in the plasma (such as H and He from Fig. 13).

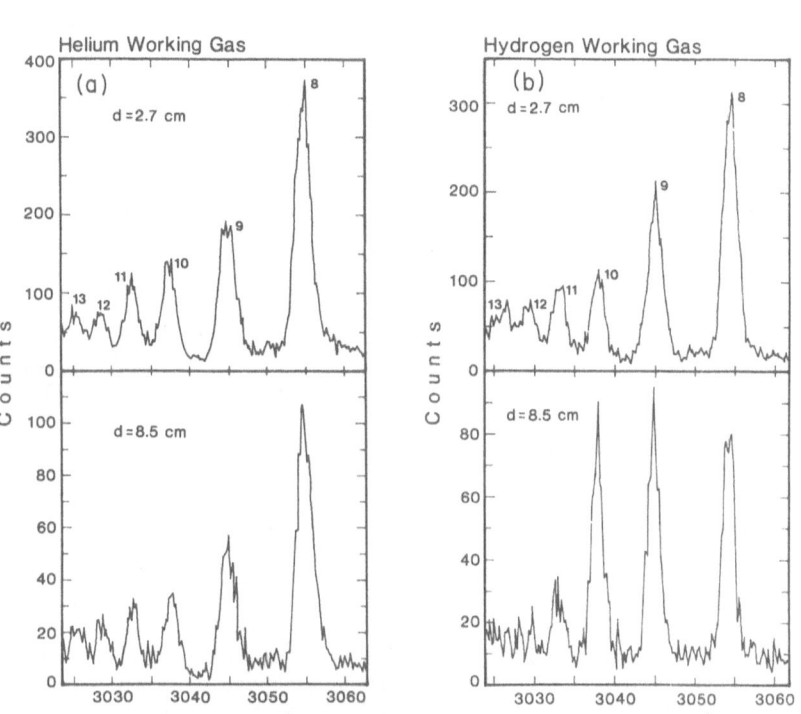

Fig.13. Spectra from 1s8p- to 1s13p - $1s^2$ transitions from two different chords obtained in a) helium working gas and b) hydrogen working gas [32].

Here the resonant charge exchange was probably not observed because it would be outside of the spectrometer range (should occur at n = 7, 8 for the He case of Fig. 13a). Not only the main diagram lines but also inner shell and dielectronic satellite lines have been observed from the hot tokamak plasma [56]. Identification of the satellite structure could be done by comparisons with theoretical calculations and a systematic structure of satellite lines originating from the higher n shells was identified.

Observations of radial profiles of the lines in the spectra from He-like ions become crucial to start to entangle the questions of importance of atomic processes, of radial transport and their relative importance in different temperature regions of the plasma. So, for example observations of the different diagram lines from He-like argon spectra [32] showed that the usually employed model to describe the radial emissivity from the plasma had to be modified by using a much stronger term of recombination (by a factor of five) to get agreement between the observed profiles and those predicted by the model. Further information to entangle these questions must be obtained from simultaneous radially resolved x-ray spectroscopy measurements. On the other hand we have here

shown examples from X-ray spectroscopy from hot plasmas where a switch
of the atomic processes producing the lines in the He-like spectrum could
clearly be observed.

7. ATOMIC PROCESSES AND X-RAY SPECTROSCOPY IN TIME TRANSIENT REGIONS

The discussion so far has been concentrated on observations from the
plasma during the quiescent period of the plasma discharge. However, as
was pointed out earlier, and seen from Fig. 2, the plasma discharge
consists of a current ramp-up phase, a quiescent flat top and a current
ramp-down phase in a simple Ohmically heated discharge. A more thorough
understanding of the atomic physics of the highly charged ions in the
hot plasma can only be obtained in the theoretical predictions can describe even the time-varying phases (which still are of long duration
compared to atomic collision times). A First attempt to introduce the
time dependence to the ionization-recombination model [57] showed that
during the various phases of the discharge different assumptions of the
excitation mechanisms has to be introduced. Here we can merely just
introduce this interesting issue, much more time-and space-resolved observations must be made from the hot plasma source to be able to conclusively describe the spectral features and the influence of the different
atomic processes.

In a similar way, although here much more obvious, to the interpretation of the results from the central stationary part of the plasma,
the ionization balance cannot be simply described by an ionization-
recombination model (corona). Different excitation mechanisms seem to
be important in the beginning of the discharge than during the quiescent
phase of the discharge. The model described in [57] proposed as a solution to introduce excitation by high energy electrons to explain the
relative line ratios in the He-like spectra observed from different tokamak
machines. On the other hand, particle transport could be much more important during the start-up phase as well as the different plasma-wall
interactions being responsible for the original release of the ions into
the plasma.

From Fig. 2 we also observe that during additional heating large
periodic oscillations occur in the plasma parameters. Thus, here we may
observe time varying phenomena on time scales comparable to the atomic
time scales and it becomes absolutely essential with temporally resolved
x-ray spectra to even commence some modelling on the atomic physics processes producing the x-ray spectra. Variation of central plasma parameters, such as electron temperature and density as shown in Fig. 2, might
also be followed by variations of radial distribution functions as discussed in Sec. 6. Indeed, by changing the plasma parameters in a systematic and controlled fashion it might be possible not only to understand
and predict the details of the X-ray spectra but also the collective
phenomena responsible for the periodic oscillations. This, however,
requires extensive systematic experiments with the aim of understanding
the observations. Such experiments still remain to be done at the hot
plasma sources. Here, the new ion sources [58] might be more suitable
sources than the large and often cumbersome tokamak machines.

As a final remark in these lectures we will highlight some earlier
observations from tokamak plasmas still very intriguing but unexplained
phenomena. Fig. 14 shows observations of line intensity variation of the
doublet structure from H-like sulphur observed in the Alcator C tokamak
[59].

262

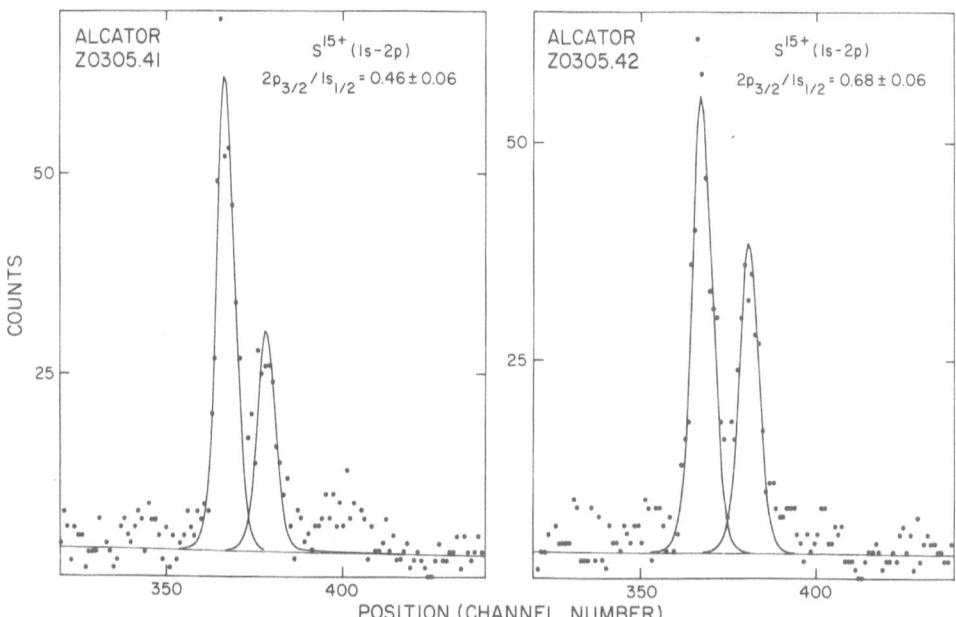

Fig.14 X-ray spectra from H-like sulphur (1s-2p transistions)
from two different plasma dischrages [59]. The solid
lines are best fits to the experimental data.

The large fluctuations observed in the relative intensity ratios from
these lines were not related to any obvious change in plasma parameters.
Nor could any systematic variation in line intensity ratios be observed
for these plasma discharges. Processes such as proton excitation (instead
of the normally dominating electron excitation) could change these line
ratios but not sufficiently to explain the experimental observations for
these plasma conditions. Thus, even in the central, quiescent phase of
the plasma discharge (where these observations were made) there might
be processes influencing the excitation of these relatively simple spectra.
It is clear that a wealth of atomic physics of highly charged ions could
be learnt from systematic and controlled experiments aimed at observing
with high spectral, temporal and spatial resolution X-ray spectra.

REFERENCES

1. M. Bitter, et al., Phys. Rev. Lett. 43,129(1979)
2. E. Källne and J. Källne, in Atomic Physics 10, eds. H. Narumi and
   I. Shimamura, Elsevier Science Publ. 1987, p.395.
3. H. P. Summers, Comments on Atomic and Molecular Physics 21, 277 (1988)
   JET-P(88)02.
4. E. Källne and J. Källne, Physica Scripta T17, 152, 1987.
5. E. Hinnov in Atomic Physics of Highly Ionized Atoms, ed. R. Marrus, NATO
   ASI Series B, Physics Vol.B96, Plenum Press, 1983, p.49.
6. B. Denne, In Proceeding of EGAS, Graz July 1988, Physica Scripta; to be
   published.
7. C de Michelis, in Basic and Advanced Diagnostic Techniques for Fusion
   Plasmas, EUR 10797 EN, Varenna 1986, Vol.I, 1987, p.83.

8. R. Bartiromo, in Ref.7, p.227.
9. W. Engelhardt, in Diagnostics for Fusion Reactor Conditions, Varenna 1982, EUR 8351-1 EN, Vol.I, 1983, p.11.
10. C. de Michelis and M. Mattioli, Rep. Progr. Phys. 47, 1233(1984).
11. "Physics of Plasma Wall Interactions in Controlled Fusion", Val Morin, Canada 1984, eds. D. Post and R. Behrisch, NATO ASI Series B Physics, Vol.13, Plenum Press 1985.
12. M. Keilhacker, et al., Plasma Physics and Controlled Fusion 26, 49(1984) and 29, 1401(1987).
    B. Denne, K. Behringer, A. Boileau, G. Fussmann, M. von Hellermann, L. Horton J. Ramette, B. Saoutic, M. F. Stamp, and G. Tallents, Controlled Fusion and Plasma Physics, Vol.II, Madrid 1987, p.109.
13. J. Jacquinot, et al., in Proc. Roy. Society, 'The JET Project and the Prospects for Controlled Nuclear Fusion', London, 1985, p.95.
14. G. Duesing, P. Lomas, A. Stäbler, P. Thomas, and E. Thompson, in ref.13, p.109.
15. D. J. Campbell, et al., Phys. Rev. Lett. 60, 2148(1988).
16. European Conference on Controlled Fusion and Plasma Physics, Dubrovnik 1988, Madrid 1987, Plasma Physics and Controlled Fusion, vol.29, No.10A, 1987.
17. Plasma Physics and Controlled Nuclear Fusion Research, IAEA, London 1984, Kyoto 1986, Nice 1988.
18. American Physical Society Annual Meeting on Plasma Physics, Bull. Am. Phys. Soc., Vol.31, No.9, 1986 (Baltimore), Bull. Am. Phys. Soc., Vol.32, No.9, 1987 (San Diego).
19. R. Giannella, Journal de Physique C1, Suppl. No.3, 283, 1988.
20. K. W. Hill, et al., in Proc. International School on Plasma Physics, Varenna, 1986, EUR 10797 EN, p. 169 and 201
21. E. Källne, J. Källne, L. G. Atencio, J. Chmielevski, G. Idzorek, and C. L. Morris, Rev. Sci. Instr. 58, 1077 (1987).
22. F. Sauli, Cern Report 77-09, 1977.
23. G. Rupprecht, Soft X-ray detectors for spectral and spatial scanning double crystal monochromators KS1 and KS2 at JET, JET Report 1988.
24. F. Bombarda, R. Giannella, E. Källne, and G. Tallents, to appear J. Quant. Spectr. Rad. Transf. 1988.
25. M. Mattioli, J. Ramette, B. Saoutic, B. Denne, E. Källne, F. Bombarda, and R. Giannella, to appear J. Appl. Phys. 1988.
26. Proceedings of Workshop on Ion Temperature Measurements, Princeton, Nov.1987.
27. M. Bitter et al., Princeton Report PPPL-1891, April 1982.
28. R. Barnsley, K. D. Evans, N. J. Peacock, J. Dunn, and N. C. Hawkes, in International School on Plasma Physics, Varenna 1986, EUR 10797 EN, p.287.
29. U. Schumacher, E. Källne, H. Morsi, and G. Rupprecht, to appear J. Appl. Phys. 1988.
30. W. Engelhardt, J. Fink, G. Fussman, H. Krause, H. B. Schilling, and U. Schumacher, MPI für Plasma Physik Report IPP 1/212, IPP III/81, Garching (March 1982).
31. E. Källne, J. Källne, E. S. Marmar, and J. E. Rice, Physica Scripta 31, 551 (1985).
32. J. E. Rice, E. S. Marmar, E. Källne, and J. Källne, Phys. Rev. A35, 3033(1987).
33. F. Bombarda, R. Giannella, E. Källne, D. Muir, M. Shaw, and G. Tallents, JET Report 1988, A. Helte, MSc Thesis, Royal Inst. of Tech., Aug.1988.
34. E. S. Marmar, J. Cecchi, S. Cohen, Rev. Sci. Instr. 46, 114(1975).
35. E. S. Marmar, J. E. Rice, J. L. Terry, and F. H. Sequin, Nucl. Fusion 22, 1567 (1982).
36. K. Behringer, Description of the Impurity Transport code "STRAHL", JET-R(87)08.

37. F. Bombarda, R. Giannella, E. Källne, G. J. Tallents, F. Bely-Dubau, P. Faucher, M. Cornille, J. Dubau, and A. H. Gabriel, Phys. Rev. A37, 505(1988).

38. M. Hsuan, M. Bitter, K. W. Hill, S. von Goeler, B. Grek, D. Johnson, L. C. Johnson, S. Sesnic, C. D. Bhalla, K. R. Karim, F. Bely-Dubau, and P. Faucher, Phys. Rev. A35, 4280(1987).

39. B. Denne, E. Hinnov, J. Ramette, and B. Saoutic, Bull. Am. Phys. Soc. 33, 945(1988), and to be published.

40. E. Källne, J. Källne and Robert D. Cowen, Phys. Rev. A27, 2682(1983).

41. P. Beiersdorfer, Thesis, Princeton University, Feb.1988, "High Resolution Studies of the X-ray Transitions in Highly Charged Neon-like Ions on the PLT Tokamak", and Phys. Rev. A, 1988.

42. TFR Group, M. Cornille, J. Dubau, M. Loulergue, Phys. Rev. A32, 3000 (1985).

43. E. Källne, J. Källne, P. Richard, M. Stöckli, J. Phys. B17, L115(1984).

44. E. S. Marmar, J. E. Rice, E. Källne, J. Källne, and R. E. LaVilla, Phys. Rev. A33, 774(1986).

45. A. K. Pradhan, Astroph.J. 249,821(1981), ibid, 246, 1031, ibid 263, 477(1982).

46. A. H. Gabriel, Mon. Not. R. Astr. Soc. 160, 99(1972).

47. M. Bitter, from Measurements of Temperatures, Washington, D.C. 1982.

48. D. Belic and A. K. Pradhan, Comments on Atomic & Molecular Physics, Vol.XX, 317, 1987.

49. K-D Zastrow, Diplomarbeit, Technische Universitat München, Aug.1988 "Beobachtung der Intensitätsverhältnisse der Resonanzlinie und Ihrer Satelliten in Röntgenspektrum von Helium Ähnlichem Nickel am JET-Tokamak".

50. P. Lee, A. Lieber, R. Chase, A. Pradhan, Phys. Rev. Lett. 55, 386(1985).

51. A. Weller, D. Pasini, A. W. Edwards, R. D. Gill, and R. Granetz, JET Report JET-IR(87)10.

52. E. Källne, J. Källne, A. Dalgarno, E. S. Marmar, J. E. Rice, and A. K Pradhan, Phys. Rev. Lett. 52, 2245 (1984).

53. J. E. Rice, E. S. Marmar, J. L. Terry, E. Källne, and J. Källne, Phys. Rev. Lett. 56, 50 (1986).

54. R. K. Janev, D. S. Belic and B. H. Bransden, Phys. Rev. A28, 1293 (1983).

55. A. Niehaus, These Proceedings.

56. E. Källne, J. Källne, J. Dubau, E.S. Marmar, and J. E. Rice, Phys. Rev. A38, 2056(1988).

57. T. Kato and K. Masai, Journale de Physique, C1-349, Suppl. No. 3, 1988.

58. M. Levine, These Proceedings.

59. K. Källne and J. Källne in AIP Proc. No. 94, X-Ray and Atomic Inner Shell Physics, ed. B. Crasemann, American Institute of Physics 1982, p. 463.

THE PHYSICS OF SLOW, HIGHLY CHARGED ION ATOM COLLISIONS

A. Niehaus

Buys Ballot Laboratory
Princetonplein 5
3584 CC Utrecht, The Netherlands

1. A Short Review

This review, and also the following discussions, will be limited to
collisions in a velocity range where the typical effects caused by the
large amount of "potential energy", carried into a collision system in
the form of the recombination energy of the highly charged ion A^{q+}, are
not too strongly perturbed by effects caused by the collision velocity.
This velocity region may be characterized by the condition $v_{coll} \lesssim 1$ a.u.
for capture of the outer-shell electrons of neutral target atoms B.
Further, we will limit ourselves to processes involving more than one
"active electron". Single capture processes have been recently reviewed
[1], and extensive information on the recent development of the field of
multiple capture can best be obtained from the proceedings of three
conferences [2-4] devoted to this subject. Here we only summarize the
most important steps in the development during the last years in order to
have a basis for the following somewhat more detailed outline of our
present understanding of the physics of multiple capture processes.

Multiple capture processes may be indicated as:

$$A^{q+} + B \rightarrow A^{r+} + B^{s+} + (r+s-q)e^- \quad , \tag{1}$$

whereby, in the low velocity region, the (r+s-q) electrons may be thought
of as being "spontaneously" emitted, either by one of the collision
partners after the collision, or by the quasimolecule during the

collision. For this kind of ionization the term "transfer ionization (TI)" is commonly used [5]. Except for results from some rather early work [6,7] on reactions of type (1) for ions with low q-values, as available from normal ion sources, nothing was known, as recent as ca 1975, for the more highly charged ions, where the potential energy of A^{q+} is high enough to lead, in principle, to the ionization of several target electrons.

The first step towards a better understanding for the case of high q-values was made by carrying out systematic measurements of absolute capture cross sections, $\sigma_{q,r}$, for a large range of q- and r-values and for different collision velocities [8,9]. It was found that these cross sections are very large ($10^{-16} - 10^{-14}$ cm^2) and - for given values of q and r - are virtually independent of collision velocity. The $\sigma_{q,r}$ were further found to increase nearly linearly with q, and to decrease systematically with the number (q-r) of captured electrons. An interpretation of these data was hampered by the fact that processes characterized by the value-pair (q,r) are not well defined: the same final projectile state (r) can arise for processes leading to different target charge states (s), and, even for a final state defined by the triple of values (q,r,s), the final state can, in general, be reached via different intermediate states of the collision system. In this situation, the logical second step was to carry out measurements of cross sections for the formation of defined charge states of target _and_ projectile. Such coincidence measurements of cross sections $\sigma_{q,r}^{s}$ were performed in several laboratories [10,11,12]. It turned out that, in general, to each (r)-value there belongs a distribution of s values (s > q-r), proving that one or more electrons are emitted in the process. The s-distribution was further found to increase in width with q and with the number (q-r) of captured electrons, indicating that the potential energy set free by binding (q-r) electrons in the highly charged ion is somehow responsible for the emission of the (r+s-q) electrons. These results led to the formulation of a statistical model [13] which was based on the assumption that the potential energy available in a collision system is equipartitioned among the target electrons. On the one hand, this model explained the measured charge state distributions for larger q- and (q-r)-values rather well, on the other hand, however, it did of course not explain 'where the electrons come from' - projectile, target, or quasi-molecule - and, also, was apparently in contradiction with other experimental data that proved the processes to occur in a very "unstatistical" way. Such data were obtained by the methods of "translation spectroscopy (TS)" [14,15,16] and electron spectroscopy (ES) of the autoionization electrons [17,18,19]. By(TS) it was shown that a final projectile charge state r=q-1 arises from two or three different processes, characterized by distinct energy gains (Q) in the energy of relative motion. These processes are, single capture, double capture followed by projectile autoionization and - in some cases - three electron capture followed by the spontaneous emission of two electrons by the projectile. From the values of the corresponding energy gains (Q) in the case of double capture it was further evident that the electrons were very selectively captured into states lying in a rather narrow region of binding energies. The position and the width of this "population window" was further found to depend in a systematic way on the charge state q and on the binding energies of the target electrons. (ES)-measurements, which became feasible with the advent of efficient sources for highly charged ions [20,21] confirmed these conclusions for two electron capture and proved that, quite generally, also in case of multiple electron capture from multielectron target atoms, electron emission by the projectile is dominant [22,23]. In addition, these (ES)-measurements yielded a wealth of new spectroscopic information on multiply excited few electron ions. This information will be further outlined in the last section of this paper. Regarding the capture process itself, it thus became obvious that, instead of scheme (1), a more detailed scheme may be used, namely,

$$A^{q+} + B \rightarrow A^{(q-s)+} + B^{s+}$$

$$A^{r+} + (r - (q-s))e^- \ . \tag{2}$$

This decomposition of the overall process into two successive steps simplifies its description considerably. The first step is a simple (multiple) charge exchange process which, in principle, can be described by established methods such as the Landau-Zener approximation [24]. In practice, however, such a description, which requires the evaluation of transition probabilities at all crossings of the potential curves of the quasimolecule, can only be applied in the most simple cases, because there exist, in general, infinitely many such crossings. It was therefore desirable to find a simpler model that would be able to describe the processes at least approximately, and to evaluate the large amount of available experimental data in terms of physical quantities. For single electron capture such a simple model had already previously been formulated [25], and had been found to be very successfull. This model is based on the so called "overbarrier criterion" - first applied to the description of resonant charge exchange [26] - which says that an electron will be transferred from B to A^{q+} at a critical distance R_c at which the top of the potential barrier separating the Coulomb wells, caused by B^+ and by A^{q+}, has an energy equal to the level occupied by the electron when bound to B. Recently, Bárány et al [27] have formulated a model for the description of multiple electron capture which is based on the same criterion. This extended model allows one to calculate upper limits of multiple capture cross sections without the use of any free parameter. However, it does not include in its formulation the possiblity to make predictions regarding the final states the electrons occupy on A and B after the collision. A formulation not having this limitation was later proposed by the present author [28]. This model will be used for the discussion and the analysis of experimental data, and is therefore outlined below in some detail.

2. The model

The main idea - in contrast to the earlier overbarrier models - is that capture does not occur "on the way in" when the collision partners approach, but rather "on the way out", when they separate. On the way in, a certain number of electrons originally bound to the target become "molecular", because the Coulomb barrier between target and projectile successively ceases to be effective for these electrons at critical distances R_1^i, R_2^i, ... R_t^i ... On the way out then, these molecular electrons may be captured with certain probabilities by projectile or target. This capturing is caused by the rising Coulomb barrier which separates the space accessible to the electrons at critical distances R_t^0. By applying the overbarrier criterion for each target electron, it follows that the molecular binding energies E_1, E_2, ... E_t ..., are ordered in the same way as the target ionization energies I_1, I_2, ... I_t ..., and that $R_t^0 > R_{t+1}^0$, etc. To calculate a certain critical distance R_t^0 from the criterion that the Coulomb barrier height is equal to the binding energy E_t, it is necessary to specify for the number of electrons $r_{t'}$ ($t' > t$) that have been captured by the projectile at smaller distances $R_{t'}^0$. By neglecting the screening of "outer electrons" throughout the collision, the distances R_t^i and $R_t^0(r_t)$ can be calculated for a charge changing collision specified by the numbers r_t, i.e. by defining where each electron finally is captured. To characterize such a well defined process, we use a string (j) in which the position indicates the index (t) and a "1" indicates capture by the projectile while a "0" indicates recapture by the target. A process with string

$$(j) \equiv (10100000) \tag{3}$$

for instance, would indicate a process where the target electrons with ionization potentials I_1 and I_3 are captured by the projectile, while the others of the 8 electrons considered are recaptured or remain "atomic" throughout the collision.

In order to contribute to the example process (3), the turning point of a collision (R_{TP}) has to be smaller than R_3^i, the distance at which the third electron becomes molecular on the "way in". For collisions with turning points $R_{TP} < R_t^i$ ($t > t_0 = 3$) the probability for process (3) to occur depends on t. Calling this probability $P_t^{(j)}$, we can express the cross section in the straight line approximation generally as

$$\sigma_{q,q-r}^{(j)} = \sum_{t=t_0}^{x} \pi\{(R_t^i)^2 - (R_{t+1}^i)^2\} P_t^{(j)} \tag{4}$$

where x is the maximum number of electrons that can become molecular. The probabilities $P_t^{(j)}$ are products of the appropriate single electron probabilities. These single electron probabilities are the crucial quantities of the model. Quantum mechanically one would estimate these probabilities from a projection of the molecular orbital at $R_t^0(r_t)$ onto atomic orbitals centered around projectile and target. Since it was the aim to keep the simplicity of the model, in the original version the degree of degeneracy of the possible Coulomb-orbitals on projectile and target was used to estimate the capture probability. Calling $n(t)$ and $m(t)$ the principal quantum numbers of the corresponding orbitals — i.e. of those orbitals into which the electron would be captured after separation — the single electron capture probabilities are simply given by

$$w_t^{(j)} = n(t)^2/(n(t)^2 + m(t)^2) \qquad (5)$$

where the quantum numbers are assumed to be continuous. In Coulomb approximation these numbers are obtained from the atomic binding energies, which, within the model, are given by

$$EP_t^{(j)} = I_t + q/R_t^i - \frac{t + r_t}{R_t^0(r_t)}$$

$$\qquad (6)$$

$$ET_t^{(j)} = I_t + q/R_t^i - \frac{q - r_t}{R_t^0(r_t)}$$

for capture by the projectile, and by the target, respectively. In addition to the cross sections $\sigma_{q,q-r}^{(j)}$, also the kinetic energy gain can be calculated from the binding energies (6). Denoting the actual binding energy of the (t)-th electron in a process (j) by $\varepsilon_t^{(j)}$, where

$$\varepsilon_t^{(j)} = \begin{cases} EP_t^{(j)} & \text{for capture by projectile} \\ \\ ET_t^{(j)} & \text{for capture by target} \end{cases} \qquad (7)$$

The energy gain is simply

272

$$Q^{(j)} = \sum_{t=1}^{t_0} (\varepsilon_t^{(j)} - I_t) \qquad (8)$$

3. Absolute Cross Section

The model predicts absolute cross sections for capture processes. When the predictions are to be compared to experimental data, one has to consider possible decay processes of the initially formed atoms, and further, one has to sum over all theoretical cross sections which are not distinguished in the experiment. Tawara et al [29] have studied collisions of I^{q+} with He at low collision energies for q = 10-41. They determined absolute cross sections for the fomation of $I^{(q-1)+}$. Since double capture is followed by autoionization for higher q, the measured cross section $\sigma_{q,q-1}$ contains all three theoretical cross sections $\sigma_q^{(10)}$, $\sigma_q^{(01)}$ and $\sigma_q^{(11)}$. In fig. 1 these cross sections are shown together with their sum, which should be compared to the experimental points taken from the work of Tawara et al [29]. The agreement is rather good. This is also true for the average principal quantum number predicted by the model.

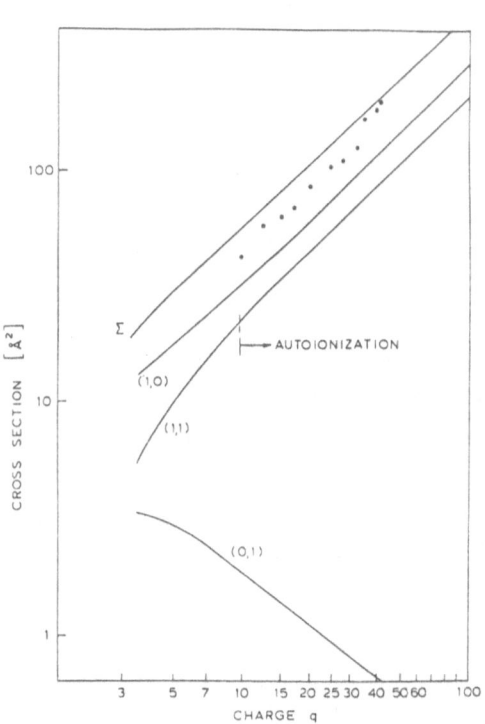

Fig. 1. Model cross sections for single capture (1,0), (0,1), and double capture (1,1) in I^{q+}/He-collisions. The sum (Σ) of these cross sections is compared to the experimental cross section for the formation of $I^{(q-1)+}$ ($\bullet$) [29].

The advantage of the simplicity of the model is that predictions can be made also for complicated systems. Although we do not expect the predictions to be always as accurate as in the case of I^{q+}/He, we believe that the relative importance of the vast number of different charge changing processes is probably rather realistically estimated. In fig. 2 we show in which way for the system Ar^{q+}/Ar the "geometrical cross

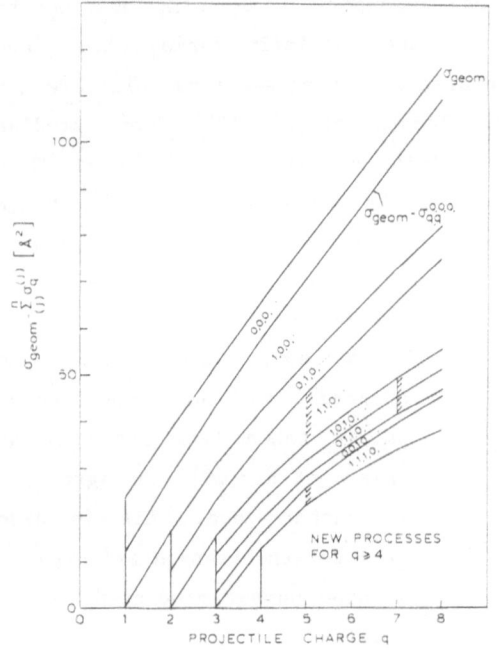

Fig. 2. Calculated composition of the geometrical cross section $\sigma_{geom} = \pi(R_1^i)^2$ by the individual cross sections $\sigma_q^{(j)}$ indicated by the corresponding strings, for the systems Ar^{q+}/Ar. Predicted limits for autoionization are also indicated.

section" $\sigma_{geom} = \pi R_1^2$ is divided up into the partial cross sections $\sigma_q^{(j)}$. For the capture of (r) electrons there are $\binom{q}{r}$ partial cross sections $\sigma_q^{(j)}$ as indicated by the strings (j) in fig. 2. For each individual process the binding energy of each electron is predicted. This allows one to predict whether the process leads to decay by electron emission. Since the binding energy of the electrons generally decreases when q increases, there arise limits on the q-scale beyond which a certain process is followed by autoionization. A few such limits for the case of projectile autoionization are indicated in fig. 2. Target autoionization is expected for $q \geqslant 5$ following the capture of two or three "inner" electrons by the projectile. Generally, the prediction is that projectile autoionization by far predominates.

For collisions of Ar^{q+} with Ar Astner et al [30] have obtained cross sections for the formation of projectiles of charge (k) and target ions of charge (s), using a coincidence technique. These cross sections $\sigma_{q,k}^s$ are composed of several partial cross sections $\sigma_q^{(j)}$. In table 1 we show the relation between the $\sigma_q^{(j)}$ and the $\sigma_{q,k}^s$ with the experimental values. From the rather good agreement of all the numbers we conclude that even in complicated systems the main mechanism of charge exchange is well described by the model.

Table 1. Comparison of experimental cross sections (last column [30]) with model cross sections (last but one column) for the phenomenological processes q → k,s in Ar^{5+} + Ar → Ar^{k+} + Ar^{s+}. In the first column the string (j) characterizing the different processes distinguished in the model is given. The second column gives the absolute model cross section, and in the third column it is indicated whether or not autoionization is predicted for target or projectile. In the fourth column the phenomenological process to which a process of defined string (j) contributes is indicated.

Process (j) $t = 1\ 2\ 3\ 4\ 5$	$\sigma^{(j)}$ [Å²]	Auto-ionization	Phenomenological processes $q \to k, s$	$q \to k, s$	Model cross section σ [Å²]	Experimental cross section σ [Å²]
1 0 0 0 0	17.6	–	5 → 4, 1			
0 1 0 0 0	6.9	–	5 → 4, 1	5 → 4, 1	28	26
0 0 1 0 0	2.4	–	5 → 4, 1			
0 0 0 1 0	0.8	–	5 → 4, 1			
0 0 0 0 1	0.6	–	5 → 4, 1			
1 1 0 0 0	10.9	proj.	5 → 4, 2	5 → 4, 2	10.9	7.8
1 0 1 0 0	3.6	–	5 → 3, 2			
1 0 0 1 0	1.3	–	5 → 3, 2			
1 0 0 0 1	0.9	–	5 → 3, 2			
0 1 1 0 0	3.6	–	5 → 3, 2	5 → 3, 2	11.7	9.2
0 1 0 1 0	1.3	–	5 → 3, 2			
0 1 0 0 1	1.0	–	5 → 3, 2			
0 0 1 1 0	1.3	target	5 → 3, 3			
0 0 1 0 1	1.0	target	5 → 3, 3			
0 0 0 1 1	1.0	target	5 → 3, 3			
				5 → 3, 3	9.2	8.4
1 1 1 0 0	3.6	proj.	5 → 3, 3			
1 1 0 1 0	1.3	proj.	5 → 3, 3			
1 1 0 0 1	1.0	proj.	5 → 3, 3			
1 0 1 1 0	1.3	–	5 → 2, 3			
1 0 1 0 1	1.0	–	5 → 2, 3	5 → 2, 3	5.6	4.6
1 0 0 1 1	1.0	–	5 → 2, 3			
0 1 1 1 0	1.3	–	5 → 2, 3			
0 1 1 0 1	1.0	–	5 → 2, 3			
0 1 0 1 1	1.0	–	5 → 2, 3			
0 0 1 1 1	1.0	target	5 → 2, 4	5 → 2, 4	1.0	0.9
1 1 1 1 0	0.8	–	5 → 1, 4			
1 1 1 0 1	0.6	–	5 → 1, 4			
1 1 0 1 1	0.6	–	5 → 1, 4	5 → 1, 4	3.0	–
1 0 1 1 1	8.7	–	5 → 1, 4			
0 1 1 1 1	8.3	–	5 → 1, 4			

4. Population windows

From measurements of the gain of kinetic energy Q in charge changing collisions [31,32], as well as from measurements of autoionization electrons [33] and of photons [34] it is well known, that states lying within an energy window of a certain width are populated in a capture process, while the model described so far predicts a sharp binding energy. This sharp value probably has to be interpreted as an average value. By comparing the widths of the windows found experimentally, we noticed that they tend to be larger at larger collision velocity. This indicates a dynamic origin. Some dynamic behaviour can be incorporated into the "static model" described so far, by taking into account that the overbarrier criterion is uncertain due to the time variation of the barrier height. We introduced a "minimum uncertainty" of the barrier height which is given by

$$\Delta V_b = \sqrt{v_{rad} \cdot V'_b} \, , \qquad\qquad (9)$$

with v_{rad} the radial velocity and V'_b the derivative of the barrier height with respect to the distance between the collision partners. In this way the binding energies $E_t^{(j)}$ defined in (6) are replaced by a Gaussian distribution of binding energies, and, correspondingly, the energy gain Q by a distribution of energy gains. The widths of the distributions corresponding to different indices (t) are assumed to add up quadratically.

In figures 3 and 4 the Gaussian distributions on the Q-scale calculated for the processes j ≡ (10), (01), (11) in collisions of Ar^{6+} and Ne^{7+} with D_2, are compared with experimental projectile energy gain spectra for single capture. For the very small scattering angles arising in these processes, and for the collision energy which corresponds to a relative energy that is large compared to the relevant Q-values, the projectile energy gain-value differs from the corresponding Q-value by an insignificant amount. The experimental spectra contain contributions from pure single capture, and from double capture followed by autoionization

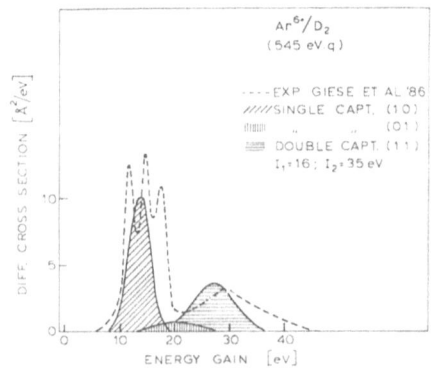

Fig. 3. Experimental energy gain spectrum for Ar^{5+} formed in Ar^{6+}/D_2 collisions (---) [32], and calculated "Gaussians" for the processes indicated by the strings. The absolute cross section scale relates to the Gaussians.

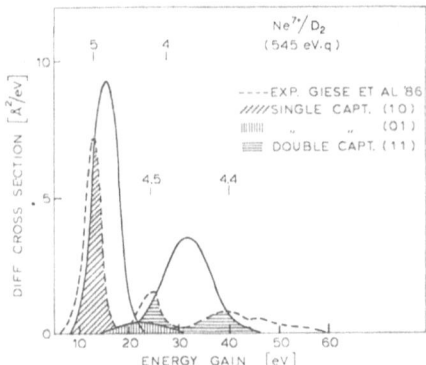

Fig. 4. Experimental energy gain spectrum for Ne^{6+} formed in Ne^{7+}/D_2 collisions (---) [32], and calculated "Gaussians" for the processes indicated by the strings. The absolute cross section scale relates to the Gaussians.

[32]. Width and position of the theoretical spectra agree qualitatively with the experimental ones. The two examples indicate in which way the theoretical "Q-window" should be interpreted: the area under the Gaussians corresponds to the maximum possible cross section, which is realized only if states are available in the relevant energy range. Following this interpretation the predicted cross sections are lower for Ne^{7+}/D_2 than for Ar^{6+}/D_2, as indicated.

The Gaussian distribution of the individual binding energies of the captured electrons lead to the prediction of a "population window" of excited states, and, therefore, to a corresponding "predicted window" for the electrons ejected in an autoionization process.

For a large number of collision systems $A^{q+}/He, H_2$ the position and the width of this "predicted window" was compared to the actual probability distribution of the binding energies of the doubly excited states formed by two electron capture [18,35,36]. This comparison was made on the basis of the measured autoionization electron spectra. In case of autoionization into one final state the intensity distribution of the observed lines directly reflects the probability distribution. Good qualitative agreement was found in almost all cases.

In fig. 5 electron spectra arizing from autoionization following double capture are shown. These spectra are mainly due to population of $(3\ell n'\ell')$-states which decay to the $(2p\epsilon)$-continuum. The electron energy "window" corresponding to this decay and to the calculated population

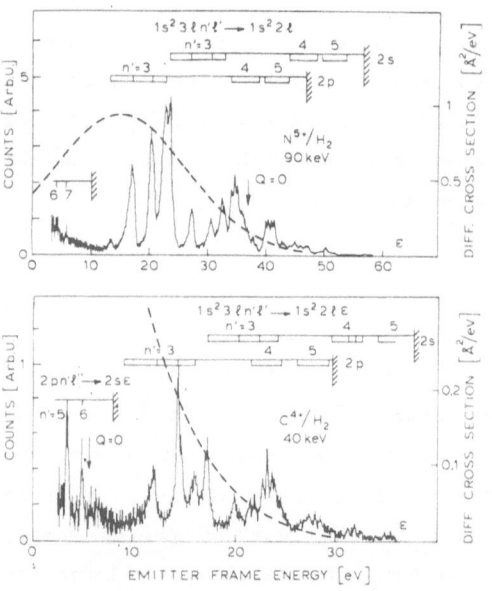

Fig. 5. Autoionization electron energy spectra for N^{5+}/H_2- and C^{4+}/H_2-collisions. Also indicated the predicted "Gaussians". The absolute cross section scale relates to the Gaussians. The count rate scales are the same for the two experimental spectra.

window is indicated. The intensity scales for the two experimental spectra are comparable, and normalized to the absolute differential cross section scale by comparison with the theoretical Gaussians. This normalization is somewhat arbitrary, however, the low cross section observed for C^{4+}/H_2 is well explained by the fact that states are only available in the wings of the population window.

The variation of the spectra with collision energy may be characterized as a broadening of the "population window" whose energy position does not change. This is clearly seen in fig. 6 for the system O^{6+}/H_2.

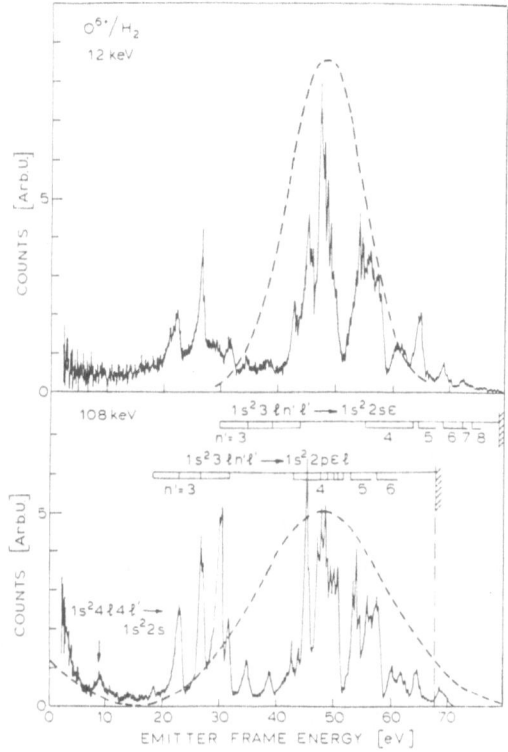

Fig. 6. O^{6+}/H_2-electron spectra at two different collision energies. Otherwise see caption of fig. 5.

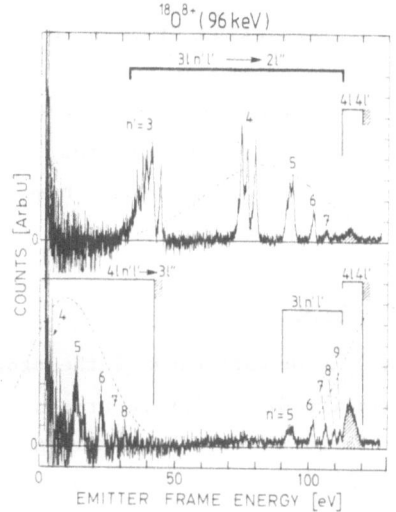

Fig. 7. Electron spectra arising from the autoionization of doubly excited states populated by double capture from the two electron targets He (upper panel) and H_2 (lower panel) into the bare ion O^{8+}.

For electron spectra arising from collisions of the bare ions O^{8+} with He and H_2 the comparison is shown in fig. 7. In both cases states belonging to configurations (3ln'l') and (4ln'l') are populated. For He the population window in electron energy is shown for decay of the populated states into the (21'εl''')-continuum. For H_2 the calculated window is shown for decay into (21''εl''')- and (31''εl''')-continuum. It is seen that the rather different population distributions observed in the experiment, are qualitatively well reproduced by the model. This qualitative agreement also includes the "height" of the calculated windows, i.e., the same normalization of the Gaussian relative to the spectra leads to similar heights of the measured peaks relative to the Gaussian.

The spectrum for H_2 exhibits an interesting feature which is also found in other systems: although, according to the calculated window, the intensity of the lines belonging to the states (3ln'l') should increase with n', actually a decrease of the intensity is found. This suggests that there is some constraint regarding the energy exchange between the two captured electrons, favouring those two-electron states which arise from the one electron states predicted for sequential single capture. For the present case, the most probable one electron states are predicted to have quantum numbers $n_1 = 4.6$ and $n_2 = 3.9$, and the half widths of both n-distributions is approximately only $\Delta n = 1$. One therefore might say that, according to the model an extra mechanism is required to realize the population of states like ($n_1 = 3$, $n_2 = 9$). The situation just discussed is probably the same as the one encountered in several other systems - for instance N^{5+}/H_2 - where the population of Rydberg series 2pn'l' is observed [37,38]. Also in those cases the height of the corresponding autoionization lines, due to decay to the 2s-continuum, are considerably lower than predicted by the model. As an explanation for the occurrence of these lines electron-electron correlation effects have been proposed [37].

5. Angular momentum states populated by capture

The model, so far, does not account for an influence of the orbital angular momentum of a state on its population in a capture process. For the case of single capture recently a first attempt has been made to introduce such an influence by assuming that the electronic angular momentum, caused by the relative motion of the heavy particles, is conserved and appears as orbital angular momentum in the state populated by capture [39]. In this way only states with angular momenta $L < V \cdot R_t^i(L)$ can be populated, with V the relative collision velocity and $R_t^i(L)$ the critical overbarrier distance whose dependence on L is also accounted for. While this approach is certainly much too simple to describe the observed [40] strong and seemingly irregular variations of the subshell cross sections $\sigma_{n\ell}$, it might be appropriate to describe the average value of the angular momentum $\langle L \rangle$ within a shell. For several systems A^{q+}/H, H_2 a comparison of the velocity dependence of the calculated $\langle L \rangle$ with experimental data of Dijkkamp et al [40] has been made [39]. As an example we show in fig. 8 the result for the population of (n = 4)-states

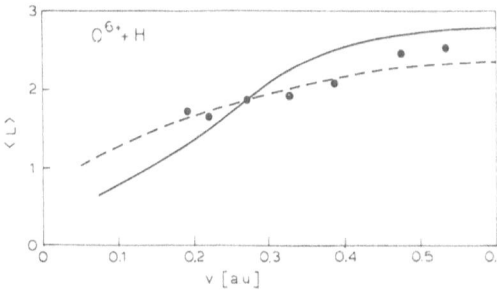

Fig. 8: Variation of the average angular momentum of the $O^{5+}(1s^2 4L)$ states formed in O^{6+}/H_2-collisions. (••) experiment [40]; (steep curve) model [39]; (other curve) modified model [41].

by single capture in O^{6+}/H collisions. Also shown is the result of a somewhat improved description [41] which allows for a certain width ΔL of the angular momentum carried into the electronic state via the heavy particle motion. In all cases studied, the smooth increase of $\langle L \rangle$ with collision velocity is well described by the model.

6. Differential cross sections

Recently it has been shown [42] that the model as described in section 2 above can easily be extended to allow one to calculate differential cross sections.
Eq. (3) implies the assumption of straight line trajectories and simply

means that the cross section is composed of ring shaped areas that contribute with probabilities $P_t^{(j)}$ to the process. The repulsive Coulomb forces at each point of a trajectory are known within the model. These forces change in strength each time an electron becomes molecular at R_t^i, and each time an electron becomes atomic again at $R^{o(j)}_t$. The scattering angle for a trajectory with impact parameter (b) may therefore be calculated as a sum of deflection angles, caused by the forces between two critical distances, R_k and R_k' say. If we use the small angle approximation, which is certainly appropriate here, the corresponding deflection angle is simply

$$\theta_k = \frac{C_k}{2Eb}\left\{\sqrt{1 - \frac{b^2}{R_k^2}} - \sqrt{1 - \frac{b^2}{R_k'^2}}\right\} \quad , \tag{10}$$

when the Coulomb potential is given by C_k/R in the region $R_k < R < R_k'$. The scattering angle for the whole trajectory is then

$$\theta(b) = \sum_k \theta_k \tag{11}$$

where the sum is taken over the appropriate pieces of the trajectory. An expression for the differential cross section at the same "level of crudeness" as expression (3) for the cross section, is now obtained by projecting the contributions $A_t \cdot P_t^{(j)}$ onto the angular range between scattering angles $\theta_t^{(j)}(b=R_t)$ and $\theta_{t+1}^{(j)}(b=R_{t+1})$:

$$W_t^{(j)}(\theta) = A_t \cdot P_t^{(j)} / (2\pi \cdot \sin\theta \cdot |\theta_t^{(j)} - \theta_{t+1}^{(j)}|) \tag{12}$$

The differential cross section for the process (j) thus becomes

$$W^{(j)}(\theta) = \sum_{t=t_o}^{N} W_t^{(j)}(\theta) \tag{13}$$

Expressions (10-13) are used, together with the relevant expressions given in ref. [28], where experimental data are compared with predictions of the model.

Recently the first measurements of differential cross sections for well defined final states were reported [43-45]. Due to the fact that large impact parameter collisions contribute dominantly to the capture processes, differential cross sections peak at rather low values of E.θ (Collison energy × scattering angle ~ 1-10 keV.deg). But because of the same reason, these measurements are very sensitive to details regarding the changes the "electron cloud" undergoes during the collision. For the system O^{8+}/He it was found that the peak in the differential cross section at E.θ ~ 4 [keV.deg] for capture of two electrons into states (n_1 = 3, n_2 = 4) is consistent with the assumption of "sequential capture" – the mechanism also assumed in the model as described in section 2 – and inconsistent with a one-step two-electron capture mechanism [43]. In addition it was found that, in the case of multiple capture studied for N^{7+}/Ar, the assumptions made in the present model [28] regarding the screening of the projectile charge by the "molecular" electrons is more realistic than the assumptions made in the model of Bárány et al [27]. In cases of multiple capture followed by multiple ionziation, a comparison of measured differential cross sections with predictions of the model are complicated because it is not always clear which intermediate states – characterized by the above defined strings (j) – contribute to the observed final state. In fig. 9 we show such a comparison for the differential cross section $\sigma_{9,7}^{4}$ of the system Ar^{9+}/Ar. The experimental curve is adapted from ref [45]. The theoretical curve is obtained using the extended model as described above. The contributions of the following

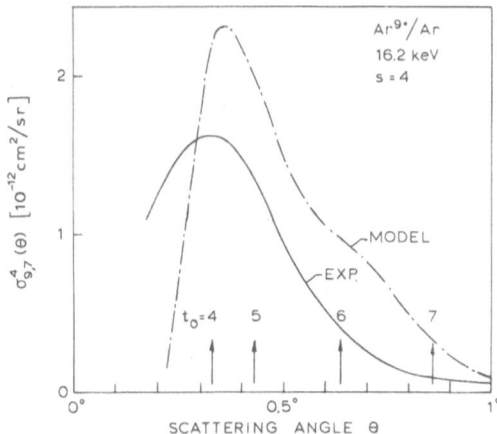

Fig. 9. Comparison of the experimental differential cross section $\sigma_{9,7}^{4}(\theta)$ for the process Ar^{9+} + Ar → Ar^{7+} + Ar^{4+} + $2e^-$ [ref. 45], with the prediction based on the model (see text).

strings are taken into account: one string (111100000) with the innermost "1" at position t_o = 4, four strings (....10000) with t_o = 5, ten strings with t_o = 6, and twenty strings with t_o = 7. The sum of all these contributions is finally smoothed with a Gaussian of .1° width. In the figure it is indicated where the contributions belonging to a certain t_o peak. It should be emphasized that both, experiment and calculation are absolute. The quoted angular resolution of the experiment is ∿.15° . The overall agreement regarding, absolute value, angle position, and width of the two differential cross sections is very good. The absence of the shoulder in the experimental curve indicates that the contributions belonging to t_o = 6 probably do not autoionize twice, and therefore do not contribute to $\sigma_{9,7}^4$, but rather to $\sigma_{9,6}^4$. The processes belonging to strings with t_o = 4,5 lead to the highest excitation of the captured electron and therefore should be responsible for the measured cross section, because emission of two electrons is most probable. This seems to be confirmed by the measurement.

7. Spectroscopic information

In connection with atomic spectroscopy, one interesting aspect is that electron capture into highly charged ions is a rather selective process, which in addition is well enough understood for predictions to be made as to which states are mainly populated in certain collision systems A^{q+}/B. In this way it is not only possible to choose a suitable collision system in order to study certain states, but also to control the population of states to a certain extent – for instance by using different target atoms with a given A^{q+}, or by using $A^{(q-1)+}$ and A^{q+} to populate the same states after capture. Such methods may yield additional information needed to identify states observed. In this connection it is important to emphasize that double electron capture has been shown to conserve electron spin for light ions and two electron targets [46]. Therefore the possible multiplicity of the states populated is known, and can be controlled by choosing appropriate systems.

Since multiple electron capture leads in general to multiply excited states that usually autoionize faster than radiate, the method of optical spectroscopy of states formed by capture processes has been limited to single capture [47,48]. The main spectroscopic information that can, in principle, be obtained using multiple electron capture, will certainly have to come from electron spectroscopy. In the case of states formed in two-electron capture from two-electron targets the application of this method is straight forward and can presently be performed with an accuracy comparable to the accuracy achieved for optical spectroscopy in the X-ray region ($\Delta\varepsilon \backsim .1eV$) . In the case of multiply excited states formed by capture from multielectron targets electron spectroscopy will have to be combined with the coincident detection of – at least – the charge state of the target, in order separate out the dominant contribution to the electron spectra from decay of states formed by double capture. Such measurements have not yet been performed. We will summarize here some recent results concerning states formed by double capture from two-electron targets.

One rule which has been confirmed quite generally [17,18,19] concerns the decay of doubly excited states by electron emission: the decay occurs with very strong preference into the nearest continuum. An example for this behaviour is also the O^{8+}/H_2 spectrum shown in fig.7. The states $O^{6+}(4ln'l')$ that lie above the $(31\infty l')$-limit decay completely to $(31''\varepsilon l''')$, and not to $(21''\varepsilon l''')$ as they do below this limit. No exception to this rule has been found for continua that differ in principle quantum number. For the branching to continua $(..2s\varepsilon l)$ and $(..2p\varepsilon l)$ from states such as $(..31n'l')$ one finds typically a ratio of $\backsim 5$ in favour of the $(2p\varepsilon l)$-continuum [18].

Another rule, which seems to hold for the light ion-systems studied so far, concerns the fluorescence yield for Li-like ^{4}L-configurations of the type $(1s2ln'l')$, which are forbidden to autoionize because of the spin selection rule: the fluorescence yield for the allowed, optical transitions within the quartet system is close to 100% [36]. Only the lowest quartets, $^4P^o$ and $^4P^e$, which are not allowed to decay radiatively, at least partially decay by (forbidden) autoionization, as evidenced by the existence of the corresponding K-LL-Auger lines [36].
The energy positions of many doubly- and triply excited levels in He-like, Li-like, and Be-like ions have been measured by electron spectroscopy with an accuracy of ca .1eV [18,19,38,49]. In some cases

they have been assigned and compared to theoretical calculations. An example for such a comparison is shown in Fig. 10 for Li-like OVI ions of the configuration (1s2121'). Instead of level-energies we show wavelengths corresponding to optical transitions from the doublet states to the ground state $(1s^22s)^2S$, and first excited state $(1s^22p)^2P^o$. These so called "dielectronic satellite lines" have also been measured by optical spectroscopy [50]. From our electron energies, arising from autoionizing transitions to $(1s^2)^1S$, we have obtained the wavelength using the center of gravity energies for the final states [51]. The resulting wavelengths are listed in Table 2.

In Fig. 10 the widths of the heavy vertical lines indicate the quoted uncertainty, which for our data ranges from 0.005 to 0.001 Å. The comparison with recent calculations [52 - 54] demonstrates that accurate experimental data for doubly excited Li-like ions are still needed.

Table 2. Wavelengths of the dielectronic satellite lines as obtained from electron spectra due to autoionization of the upper states indicated.

Transition		Wavelength [Å]
lower state	upper state	
$(1s^22p)^2P^o$	$(1s2p^2)^2S$	21.789
$(1s^22s)^2S$	$(1s[2s2p]^3P^o)^2P^o$	21.848
"	$(1s[2s2p]^1P^o)^2P^o$	22.025
$(1s^22p)^2P^o$	$(1s2p^2)^2D$	22.124
"	$(1s2p^2)^4P$	22.328
$(1s^22s)^2S$	$(1s2s2p)^4P^o$	22.372
$(1s^22p)^2P$	$(1s2s^2)^2S$	23.012

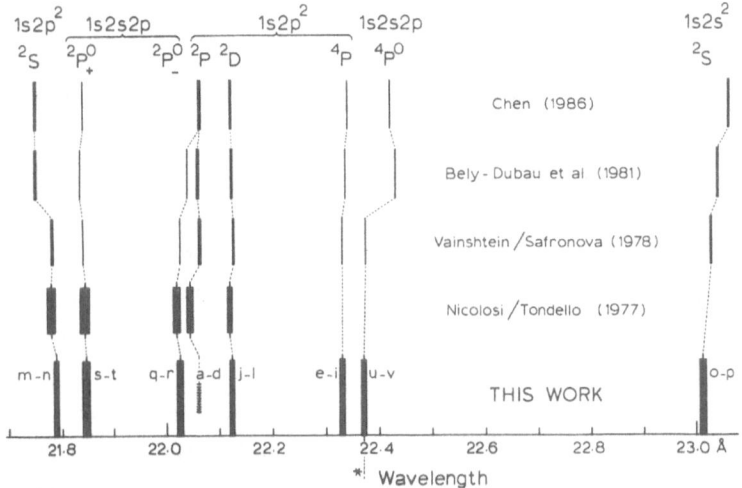

Fig.10. Comparison of wavelengths of the dielectronic satellite lines evaluated from electron spectra [ref. 49], with results from optical spectroscopy (Nicolosi/Tondello (1977): ref. 40) and with theoretical results (Vainshtein/Safranova (1978): ref. 52; Bely-Dubau et al. (1981): ref. 53; Chen (1986): ref. 54).

A case of special spectroscopic interest are the doubly excited He-like ions in states of configurations with two equivalent electrons, because for these states electron correlation is very important [e.g. 55]. We have measured electron spectra arising from autoionization of states of the (3131')-configurations of C^{4+} and O^{6+} [56]. Autoionization occurs into (2s,p). The measured spectrum is shown in Fig. 11. It consists of a superposition of lines which are partially broadened because, due to a rather short lifetime, autoionization occurs when the doubly charged target is still at a distance of v.τ (v ≡ relative velocity, τ ≡ lifetime), which leads to a (PCI)-shift and -broadening of the line. The PCI-shift is given by [57]

$$\Delta \varepsilon = \frac{q_T}{2 \cdot v \cdot \tau} \left\{ 1 - \frac{v}{|\vec{v} - \vec{v}_e|} \right\} \qquad \begin{matrix} q_T \equiv \text{target charge} \\[2mm] \vec{v}_e \equiv \text{electron velocity} \end{matrix} \qquad (14)$$

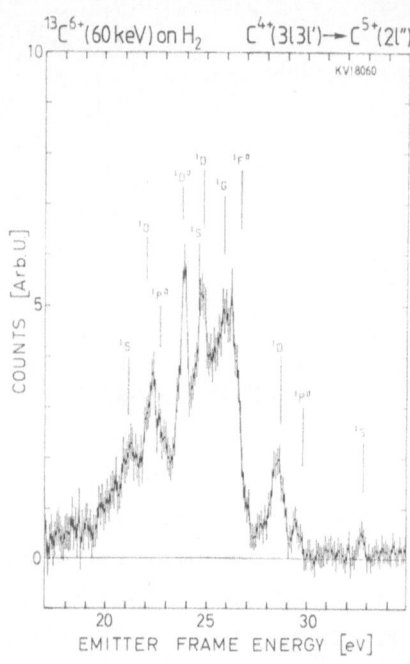

Fig.11. PCI-broadened and -shifted electron spectra arising from collisions of the bare ions C^{6+} with H_2. The part corresponding to decay of the doubly excited states (3l3l') → (2l''εl''') is shown. The vertical bars indicate theoretical positions shifted by the PCI-shift that corresponds to the theoretical lifetime of the respective states (see text).

The level energies and lifetimes τ for the (3l3l') states of C^{4+} have also been calculated [58-60]. In Fig. 11 we have indicated, by vertical lines, the positions obtained when the theoretical energy positions are shifted according to relation (14) using the theoretical τ-values. One notices that, within the accuracy of the measurement the theoretical peak positions agree very well with the experimental ones. This suggests that both, theoretical energy positions and lifetimes of the states are correct. A more detailed evaluaton of the measured spectra in terms of the known line-shapes [57] leads to a more stringent test [61].

Finally, we would like to point out that, in principle, spectroscopic information of the types outlined above for states populated by capture of two electrons, can also be obtained for states populated by capture of three and more electrons. Of especial interest would be states belonging to configurations with three equivalent electrons. Spectra obtained for C^{6+}/Xe show features which indicate that such states are in fact populated [62]. To study such spectra in detail it is necessary to carry out the measurements in coincidence with the detection of Xe^{3+}.

REFERENCES

1   R.K. Janev and H. Winter, Phys. Reports <u>117</u> (1985) 265

2   Production and Physics of Highly Charged Ions; ed. L.Liljeby, Phys. Scripta T3 (1983) 1

3   The Physics of Highly Ionized Atoms; eds. J.D. Silver and N.J. Peacock, Nucl. Instr. and Meth. <u>B9</u> (1985) 359

4   The Physics of Multiply Charged Ions; eds. R. Morgenstern, A. Niehaus, F.J. de Heer, A.G. Drentje, Nucl. Instr. and Meth. <u>B23</u> (1987) 1

5   A. Niehaus, Comm. Atom. Molec. Phys. <u>9(5)</u> (1980) 153

6   I.P. Flaks, G.N. Ogurtsov, N.V. Federenko, Sov. Phys. - JETP <u>14</u> (1962) 1027

7   A. Niehaus and M.W. Ruf, J. Phys. <u>B9</u> (1976) 1401 (and ref. cited therein)

8   See E. Salzborn and A. Müller in "Electronic and Atomic Collisions"; N. Oda and K. Takayanagy, eds.; (1980), North Holland; pp. 407

9   E. Justiniano, C.L. Cocke, T.J. Gray, R. Dubois, C. Can, W. Waggoner and R. Schuch, H. Schmidt-Böcking, H. Jugwersen, Phys. Rev. <u>A29</u> (1984) 1088

10   C.L Cocke, R. Dubois, T.J. Gray, E. Justiniano, C. Can, Phys. Rev. Lett. <u>26</u> (1981) 1671

11   W. Groh, A. Müller, A.S. Schlachter, E. Salzborn, J. Phys. <u>B16</u> (1983) 1997

12   A. Bárány, G.Astner, H. Cederquist, H. Danared, S. Huldt P. Hvelplund, A. Johnson, H. Knudsen, L. Liljeby, K.-G. Rensfelt, Nucl. Instr. and Meth. <u>B9</u> (1985) 397

13   A. Müller, W. Groh, E. Salzborn, Phys. Rev. Lett. <u>51</u> (1983) 107

14   S. Ohtani, "Electronic and Atomic. Collisions"; eds. J. Eichler, I.V. Hertel, N. Stolterfolt; North Holland (1983); pp. 353

15   C.L. Cocke, "Electronic and Atomic Collisions"; eds. D.C. Lorents, W.E. Meyerhof, J.R. Peterson; North Holland (1986) pp. 453 (and refs. cited therein)

16   A. Bárány, P. Hvelplund, Nucl Instr. and Meth. <u>B.26</u> (1987) 40

17   A. Bordenave-Montesquieu, P. Benoit-Cattin, A. Gleizes, A.I.

Marrakchi, S. Dousson, D. Hitz, J. Phys. B17 (1984) L223

18 M. Mack, Nucl. Instrum. and Meth. B23 (1987) p. 74 (and refs. cited therein)

19 R. Mann, Phys. Rev. A35 (1987) 4988

20 R. Geller, Phys. Scripta T3 (1983) 19

21 R. Phaneuf, IEEE Trans. Nucl. Sci. NS-28 (1981) 1181

22 M. Mack, A.G. Drentje, A. Niehaus, "Abstr. Contr. paper XIV ICPEAC, Palo Alto (1985); p. 466

23 A. Bordenave-Montesquieu, P. Benoit-Cattin, M. Bondjema, A. Gleizes, S. Dousson, Nucl. Instr. and Meth. B23 (1987) 94

24 L. Landau, Phys. Z. Sowjetunion 2 (1932) 46
 C. Zener, Proc. R. Soc. London, Series A 137 (1932) 646

25 H. Ryufuku, K. Sasaki, T. Watanabe, Phys. Rev. A 21 (1980) 745

26 D.R. Bates, R.A. Mapleton, Proc. Phys. Soc. 87 (1966) 657

27 A. Bárány, G. Astner, H. Cederquist, H. Danared, S. Huldt, P. Hvelplund, A. Johnson, H, Knudsen, L. Liljeby, K.-G. Rensfelt, Nucl. Instr. and Meth. B9 (1985) 397

28 A. Niehaus, J.Phys. B 19 (1986) 2925

29. H. Tawara, T. Iwai, Y. Kaneko, M. Kimura, N. Kobayashi, A. Matsumoto, S. Ohtani, K. Okuno, S. Tagaki, S. Tsurubuchi, J. Phys. B18 (1985) 337

30. G. Astner, A. Bárány, H. Cederquist, H. Danared, S. Huldt, P. Hvelplund, A. Johnson, H. Knudsen, L. Liljeby, K-G. Rensfelt, J. Phys. B17 (1984) L877

31. E. H. Nielsen, L.H. Andersen, A. Bárány, H. Cederquist, P. Hvelplund, H. Knudsen, K.B. McAdam, J. S rensen, J. Phys. B17 (1984) L139

32. J.P. Giese, C.L. Cocke, W. Waggoner, L.N. Tunnell, S.L. Varghese (1986) (manuscript privately comm. by C.L. Cocke)

33. A. Bordenave-Montesquieu, P. Benoit-Cattin, A. Gleizes, A.I. Marakchi, S. Dousson, D. Hitz, J. Phys. B17 (1984) L223

34. D. Dijkkamp, A. Brazuk, A.G. Drentje, F.J. de Heer, H. Winter, J. Phys. B17 (1984) 4371

35. A. Niehaus, Nucl. Instr. and Meth. B23 (1987) 17

36. M. Mack and A. Niehaus, Nucl. Instr. and Meth. B23 (1987) 116

37. N. Stolterfolt, C.C. Havener, R.A. Phaneuf, J.K. Swenson, S.M. Shafroth and F.W. Meyer, Phys. Rev. Lett. 57 (1986) 74

38. A. Bordenave-Montesquieu, P. Benoit-Cattin, M. Boudjema, A. Gleizes, S. Dousson and D. Hitz, Abstract Contrib. papers 15th Int. Conf. Physics of Electronic and Atomic Collisions Brighton (1987) p. 551

39. J. Burgdörfer, R. Morgenstern and A. Niehaus, J. Phys. B19 (1986) L507

40. D. Dijkkamp, A. Brazuk, A.G. Drentje, F.J. de Heer, H. Winter, J. Phys. B17 (1984) 4371

41. J. Burgdörfer, R. Morgenstern and A. Niehaus, Nucl. Instr. Meth. B 23 (1987) 104

42. A. Niehaus, Nucl. Instr. Meth. B 31 (1988) 359

43. H. Laurent, P. Roncin, M.N. Gaboriaud and M. Barat, Nucl. Instr. and Meth. B 23 (1987) 45

44. H. Danared, H. Andersson, G. Astner, A. Bárány, P. Defrance and S. Rachafi, J. Phys. B 20 (1987)

45. H. Danared, Ph.D. Thesis, Stockholm (1987)

46. M. Mack, A. Niehaus; Nucl. Instr. and Meth. B23 (1987) 109

47. D. Dijkamp, D. Ciric, E. Vlieg, A. de Boer, F.J. de Heer; J. Phys. B18 (1985)4767

48. T. LuDac, D. Hitz, M. Mayo, S. Bliman; Nicl. Instr. and Meth. B23 (1987) 86

49. M. Mack, A. Niehaus; "Abstr. Contr. papers XV ICPEAC, Brighton (1987) 550

50. P. Nicolosi, G. Tondello; J. Opt. Soc. Am. 67 (1977) 1033

51. S. Bashkin, J.O. Stoner jr.; "Atomic Energy Levels and Grotrian Diagrams" (Vol. I) (North Holland, 1978)

52 L.A. Vainshtein, V.I. Safranova; At. Data Nucl. Data Tables 21, 49-68 (1978)

53 F.J. Bely-Dubau, J. Duban, F. Faucher, L. Steenman-Clark; J.Phys. B 14 (1981) 3313

54 M.H. Chen; At. Data Nucl. Data Tables 34 (1986) 301-356

55 C.D. Lin, in: "Adv. At. Mol. Phys. Vol. XXII)" (Acad. Press. 1986)

56 M. Mack, A. Niehaus; Abstr. Contr. papers XV ICPEAC, Brighton (1987) p. 567

57 P. v.d. Straaten, R. Morgenstern; J. Phys. B $\underline{19}$ (1986) 1361

58 Y.K. Ho; J. Phys. B $\underline{12}$ (1979) 387.

     Phys. Lett. $\underline{79A}$ (1980) 44.

     Phys. Rev. $\underline{A35}$ (1987) 2035

59 H. Bachau, J. Phys. B$\underline{17}$ (1984) 1771

60 D.H. Oza; J. Phys. B $\underline{20}$ (1987) L13

61 M. Mack, J.H. Nijland, P. v.d. Straaten, A. Niehaus, R. Morgenstern
     (to be published)

62 M. Mack, Ph.D. Thesis, Utrecht 1987)

ELECTRON COOLING RINGS

D. Liesen

GSI Darmstadt
P.O. Box 11 05 52
D-6100 Darmstadt, West-Germany

INTRODUCTION

Currently, there is a worldwide boom of plans and proposals for heavy-ion storage rings with beam cooling. This boom has three main reasons:

1. Many heavy-ion accelerators are operating successfully all over the world; thus, a lot of experience in design, construction, and operation of such accelerators is available;

2. The development of new ion sources and injectors (EBIS, EBIT, CRYEBIS, RFQ, etc.) yielded beams of highly-charged ions near the space-charge limit;

3. The invention of methods to 'cool the beams' and to achieve extremely high phase-space densities enabled experiments which would never have been performed before.

In this framework, R. Marrus, the director of the 'NATO Advanced Study Institute on Atomic Physics of Highly Ionized Atoms', asked me to present a series of lectures on 'Electron Cooling Rings'. I gladly accepted this invitation – again for three main reasons:

1. Cargèse is a very pleasant place with an stimulating atmosphere – also for physics;

2. To be able to present some of the basic physical concepts of storage rings and electron cooling to physicists who want to use cooled heavy-ion beams for experiments. In most of the 'conventional' accelerator-based experiments, the accelerator merely is a source for the appropriate ions at some energy, emittance, and momentum spread. In contrast, in most of the experiments at a cooler ring the ring itself is part of the experiment, because one uses internal targets, performs experiments in the cooler, and sometimes utilizes the beam itself as target. Thus, it is inevitable that users of those beams know about the basic ingredients of accelerator and cooling physics in order to be able to design and perform experiments within the limits imposed for a safe operation of the machine.

3. To demonstrate that the quality of a stored, cooled ion beam is a challenge for new experiments which could not have been performed with ion beams from a conventional accelerator. Such experiments will become possible because stable ions and exotic, radioactive isotopes in any charge state and with high phase-space densities will become available.

Based upon these points, the lecture will be divided into three parts: In the first chapter, basic principles of 'accelerator phycis for user' will be presented (my excuses to all the colleagues who know everything about accelerator physics !). After that we will discuss the method of electron cooling of an ion beam; this naturally leads to the third chapter which deals with cooling electron-ion recombination processes. It is my hope that especially the subject of the third chapter may indicate the large variety of experiments on atomic and plasma physical questions which can be performed in the electron cooler.

Some words about the references concerning chapter I, are in order. The first chapter deals with general features of storage rings which are discussed in a variety of books and papers. In order to avoid not to give appropriate credit to all of the authors, the Refs. [1,2,3,4] contain a (by far not complete) subjective survey of the latest publications on this subject. In these publications almost all of the relevant papers are quoted. Numerical examples presented in the following always rely on the design parameters of the ESR which currently is constructed at the GSI Darmstadt. Note that 'ESR' means 'Experimentier-Speicher-Ring', and not 'Experimenteller Speicher-Ring.

ACCELERATOR PHYSCIS FOR USERS

The storage of a beam of particles in a ring is a rather ambitious task. This is easily seen by a simple comparison: The earth performed up to now about $3 \cdot 10^9$ revolutions around the sun – on an astronomical time scale. A beam of particles stored for 1 h in a ring with a revolution frequency of 1 MHz performs almost the same number of revolutions – but on a daily-life time scale. In space, the boundary conditions for a stable orbit of the earth around the sun are almost ideal: One particle bound by a well-known force and 'moving in a very good vacuum'. In a ring, the situation is much worse: Many particles which interact with external forces and among themselves, are travelling in a vacuum (typically $10^{-10} - 10^{-11}$ Torr) which is by far not as good as in space. It is obvious that stability of the trajectories of the particles is an important constraint for a successful operation of a storage ring.

The stability of the earth's motion on its equilibrium orbit around the sun is guaranteed by the gravitational force. In a storage ring, a particle is kept on its closed equilibrium orbit by the bending forces of magnetic fields. Supposed, this particle moves around the ring on a trajectory which lies in the center of the beam pipe; in German one would call it a 'Sollteilchen'. However, in a ring not only the 'Sollteilchen' but a beam of particles is to be stored. These particles are produced in a source and transferred into the ring by an appropriate injector. Due to the production mechanisms in the source and the finite acceptance of the injector, the beam enters the ring with a distribution of lateral displacements, angles, and momenta around the 'Sollteilchen'. Since every particle in the beam travels around the ring on its own closed orbit. The beam is spread around the central orbit.

In order to achieve a stable motion also after the many revolutions the particles perform in the ring, the beam must be focused in the two transverse directions. The combination of bending and focusing elements which in a ring are closed in themselves, is called the lattice of a ring.

The basic ingredients of a lattice are dipole magnets which bend the beam around, and quadrupole magnets which focus the beam towards the central orbit. The magnetic field of a quadrupole brings a particle which in one transverse plane moves away from the central orbit, back to the axis, while a particle in the orthogonal plane is deflected further from the axis (see Fig. 1). If the planes of focusing and defocusing are periodically switched, an overall transverse focusing is the net effect – just as for series of converging and diverging optical lenses. this method is called 'strong focusing' or 'alternating-gradient focusing'.

The transverse focusing leads to oscillations of the particles around the central orbit, the so-called betatron oscillations because they were first studied for particle trajectories in a betatron. It can be shown that these transverse oscillations are to first order described by

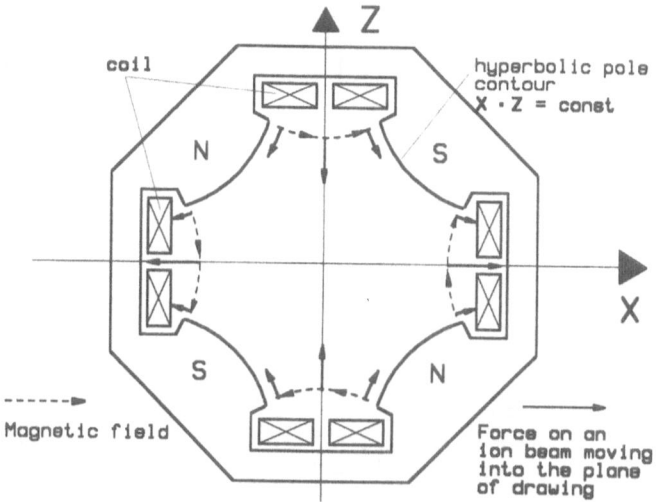

Figure 1. Cross section of a quadrupole magnet.

an equation of Hill's type:

$$\frac{d^2y}{ds^2} + K(s)y = \frac{1}{\rho}\Delta p/p_0 \tag{1}$$

where y is the displacement from the central orbit ('Sollbahn') in any of the two transverse planes, s is the longitudinal coordinate, ρ is the bending radius in a dipole, $K(s)$ is the focusing strength, and $\Delta p/p_0$ is the relative deviation from the design momentum p_0 (see Fig. 2). For a ring of circumference L, the periodicity of the lattice implies

$$K(s+L) = K(s) \tag{2}$$

The general solution of this equation reads:

$$
\begin{aligned}
y(s) &= C(s)y_0 + S(s)y_0' + D(s)\Delta p/p_0 \\
y'(s) &= C'(s)y_0 + S'(s)y_0' + D'(s)\Delta p/p_0
\end{aligned}
\tag{3}
$$

where $C(s)$ and $S(s)$ are two independent solutions of the homogeneous equation, and $D(s)$ is a particular solution of the inhomogeneous equation. C, S, and D are called the principal trajectories (Cosinelike, Sinelike, and Dispersion).

Ignoring the momentum deviation (thus setting $\Delta p/p_0 = 0$), two real solutions of Hill's equation may be written in the form

$$y = a_y \cdot \sqrt{\beta(s)}\cos(\psi(s) + \delta)$$
and
$$y = a_y \cdot \sqrt{\beta(s)}\sin(\psi(s) + \delta) \tag{4}$$

These solutions demonstrate that the particles perform pseudo-harmonic oscillations around the central orbit with an amplitude

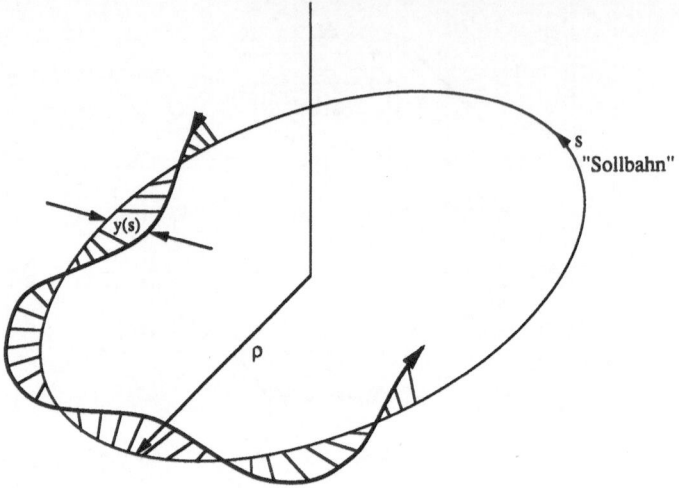

Figure 2. Sketch of a particle trajectory in a ring; $y(s)$ is the displacement from the 'Sollbahn' with radius ρ in any of the two transverse planes, and s is the longitudinal coordinate.

$$A_y \;=\; a_y \sqrt{\beta(s)} \tag{5}$$

and a 'betatron phase'

$$\psi(s) \;=\; \int \frac{ds}{\beta(s)} \tag{6}$$

which both vary as function of the longitudinal coordinate s; δ is an arbitrary constant phase.

The function $\beta(s)$ is called the 'betratron function' which is an intrinsic property of the lattice. More precisely, one deals with two betatron functions which may be different for the two transverse planes.

Let us consider the cosinelike trajectory in more detail; from Eq. (4) one obtains:

$$
\begin{aligned}
y(s) &= a_y \sqrt{\beta(s)}\,\cos(\psi(s) + \delta) \\
y'(s) &= \frac{dy}{ds} = -\frac{a_y}{\sqrt{\beta(s)}}\left\{ \sin(\psi(s) + \delta) - \frac{1}{2}\beta'(s)\cos(\psi(s) + \delta) \right\}
\end{aligned}
\tag{7}
$$

which is a parametric representation of an ellipse in the (y, y') phase plane (Fig. 3). One may look to this ellipse in two alternative ways: Either one considers an ensemble of particles with the same amplitude at a given point s of the trajectory but with different phases $0 \le \delta \le 2\pi$, or one selects one particle with a fixed δ and follows the evolution of the betatron phase $\psi(s)$ for many revolutions at a fixed point s. In both cases, the point $(\,y(s), y'(s))$ moves around an ellipse centered at (0,0) (the 'Sollbahn'). The variety of amplitudes and phases for all the particle leads to a corresponding number of smaller or larger ellipses in the phase plane.

The area of the phase space ellipse is given by

$$F \;=\; \pi a_y \sqrt{\beta(s)} \cdot \frac{a_y}{\sqrt{\beta(s)}} \;=\; \pi a_y^2. \tag{8}$$

It can be shown that the area of the ellipse stays constant and that only its shape and orientation vary if the particle trajectories are transformed through the accelerator. Thus, one obtains the important result that the phase space density is invariant (Liouville's theorem). This is true if the particle energy stays constant and if processes like intra-beam-scattering, residual gas scattering, and beam cooling are neglected.

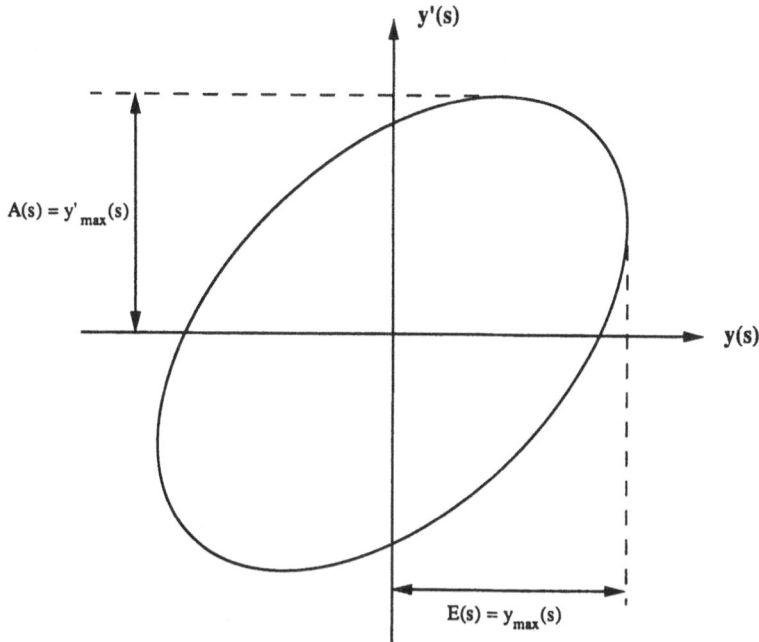

Figure 3. Emittance ellipse in the $(y(s), y'(s))$ plane with $E(s)$ being the beam envelope and $A(s)$ being the maximum divergence.

The important theorem on the phase space density leads to a special notation in Eq. (7), the emittance ϵ, by

$$F = \pi\epsilon \qquad (9)$$

With this expression, the particle trajectory on the outer of all the ellipses can be rewritten as

$$y(s) = \sqrt{\epsilon}\sqrt{\beta(s)}\cos(\psi(s) + \delta), \qquad (10)$$

from which the 'beam envelope'

$$E(s) = y_{max}(s) = \sqrt{\epsilon}\sqrt{\beta(s)} \qquad (11)$$

and the beam divergence

$$A(s) = y'_{max}(s) = \sqrt{\epsilon} \cdot \sqrt{\frac{1 + \left(\frac{1}{2}\beta'(s)\right)^2}{\beta(s)}} \qquad (12)$$

are obtained. $E(s)$ and $A(s)$ are shown in Fig. 3. Thus, knowing the beam emittance ϵ and the betafunction of the lattice, the size and the divergence of the beam can be calculated from Eqs. (10) and (11).

Finally, we note that varying the energy of the particles, the emittance of the beam varies by virtue of Liouville's theorem as $\epsilon \sim 1/p$. Thus, acceleration decreases and deceleration increases the emittance.

The number Q of betatron oscillations per revolution can be calculated from Eq. (6)

$$Q = \frac{1}{2\pi}\oint \frac{ds}{\beta(s)}. \qquad (13)$$

This number is called 'the tune' of the accelerator. Generally speaking: The stronger the focusing the larger is the tune (in 'weakly focusing' machines the tune is smaller than 1, thus there is less than 1 betatron oscillation per revolution). Certain values of the tune have to be avoided in order to prevent resonances between the betatron oscillations and the structure of the machine. Unavoidable imperfections of the elements of the structure may cause a disturbance of the trajectory which (depending on the tune) builds up with every passage and finally leads to beam loss. In principle, one should avoid the following values of the horizontal tune Q_h and the vertical tune Q_v (corresponding to the two transverse planes).

$$kQ_h + \ell Q_v = m$$

where

$$k, \ell, m = 0, \pm 1, \pm 2, \pm 3... \tag{14}$$

In practice, it is extremely difficult to avoid all the resonances given by Eq. (13); fortunately, only the lower resonances cause serious trouble.

Up to now we have considered only the trjactories of particles with zero relative momentum deviation from the design momentum. A particle with momentum deviation will follow a 'dispersed orbit' described by Eq. (3). For particles with momentum $p \neq p_0$ a closed orbit emerges with a displacement $D(s) \, \Delta p/p_0$ around which the particles perform betatron oscillation. The function $D(s)$ which is an intrinsic property of the lattice, is called the 'dispersion function'.

A particle with an off-momentum Δp will leave a bending magnet with an angle which is somewhat different from the angle of the 'Sollteilchen'. If the number of betatron oscillations per revolution is an integer one, the subsequent angular deviations will build up and lead to beam loss. Thus, a finite dispersion in a circular accelerator is possible only if Q is different from an integer.

A simple consequence for the size of the beam pipes of a ring can be drawn from the above considerations: The size of the beam pipe should be large enough to accept a particle beam with a horizontal width of

$$A_h = 2\sqrt{\epsilon_h \beta_h} + D\Delta p/p_0$$

and a vertical width of

$$A_v = 2\sqrt{\epsilon_v \beta_v}.$$

It is of economical and practical interest, to keep A_h and A_v which determine the 'aperture' of the machine, as small as possible. Thus, an ambitious aim in accelerator physics is finding a way to reduce the amplitude of the betatron oscillations and the momentum spread — in other words: To reduce the phase space of the beam, keeping in mind Liouville's theorem. The advantages of such a reduction of transverse and longitudinal phase space also for experiments, especially for high-precision experiments, are obvious.

Since changes of the momentum of the particles will lead to different paths around the ring, also different revolution frequencies result. The change of the revolution frequency ω with momentum is calculated as follows:

From

$$\omega = \frac{v}{R} = \frac{\beta c}{R}$$

with $2\pi R$ = length of the orbit, one finds

$$\frac{d\omega}{\omega} = \frac{d\beta}{\beta} - \frac{dR}{R}.$$

Since from $p = mv\gamma = m\beta c\gamma$

$$\frac{d\beta}{\beta} = \frac{1}{\gamma^2}\frac{dp}{p}, \tag{15}$$

and from $E = mc^2\gamma = p/\beta c$

$$\frac{dE}{E} = \beta^2\frac{dp}{p},$$

one obtains with

$$\frac{dp}{p} = \frac{1}{\alpha}\frac{dR}{R}$$

$$\frac{d\omega}{\omega} = \left(\frac{1}{\gamma^2} - \alpha\right)\frac{dp}{p}. \tag{16}$$

In Eq. (15) we introduced the 'momentum compaction α' which relates the change of the orbit length dR of a particle to the momentum change dp. From Eq. (15) follows that the revolution frequency increases for $\alpha < 1/\gamma^2$ and decreases for $\alpha > 1/\gamma^2$ with increasing momentum, while for $\alpha = 1/\gamma^2$ the revolution frequency is independent of the momentum. The point $\alpha = 1/\gamma_t^2$ is called the 'transition gamma' and $mc^2\gamma_t$ is called the 'transition energy' of the machine. In most cases, γ_t is approximately equal to the horizontal tune Q_h, so that

$$\gamma_t = \sqrt{1/\alpha} \approx Q_h \tag{17}$$

The transition point is critical for the longitudinal (momentum) focusing of a beam in a ring. Longitudinal focusing is achieved if a bunched beam is injected into the ring, and if this beam is passed through a rf cavity in the ring. If a voltage of the form

$$V(t) = V_0 \sin\{n\omega(t - t_0)\}$$

is applied to the cavity, then the energy of the particles is changed across the cavity by

$$\Delta E = qV(t) = qV_0 \sin\{n\omega(t - t_0)\}. \tag{18}$$

In Eq. (17) q denotes the charge state of the ion and n is the harmonic number of the particle revolution frequency ω. If the phase ωt_0 is adjusted so that the 'Sollteilchen' passes the cavity at $t = t_0$, the 'Sollteilchen' will suffer no energy change. Particles which are faster than the 'Sollteilchen' ($\Delta p = p - p_0 > 0$), pass the cavity earlier ($t_1 < t_0$) and get a kick which slows them down; vice versa, lagging particles ($t_2 > t_0$) are accelerated (Fig. 4). As a net result, the longitudinal rf focusing leads to phase oscillations of the particles in a bunch around the phase of the 'Sollteilchen', thus keeping them in the bunch. These oscillations are called 'synchrotron oscillations'. They are stable for energies below and above the transition energy of the machine, if the phase of the rf 'system' is properly adjusted. However, at transition energy the revolution frequency is independent of the momentum. Therefore, the rf system cannot focus the particles longitudinally at the transition point, and one usually avoids operating a ring near the transition energy.

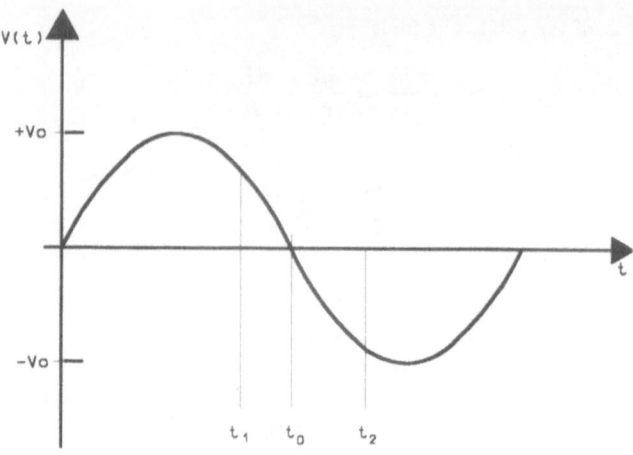

Figure 4. Time dependence of a voltage applied to a rf cavity for momentum focusing; the 'Sollteilchen' is the one that passes the cavity at $t = t_0$ with $V(t_0) = 0$. Particles travelling too fast arrive at t_1 and lagging particles arrive at t_2. For $\gamma \lessgtr \gamma_t$ the rf field leads with proper phase adjustment to acceleration of lagging particles and deceleration of preceding particles.

A finite momentum spread $\Delta p/p_0$ of the beam leads to an effect which is called "chromaticity", namely a shift of the tune of the machine as function of $\Delta p/p_0$. The bending and focusing strengths which determine the tune, depend on the momentum of the particle. A deviation Δp from the design momentum p_0 causes a first order tune-shift (or a betatron-wavelength shift) of

$$\Delta Q = \xi \Delta p/p_0 \qquad (19)$$

which is proportional to the relative momentum spread $\Delta p/p_0$. The factor ξ is called the chromaticity of the machine; its counterpart in optics is the chromatic aberration of an optical lens system. Since one has to avoid the many dangerous, closely spaced Q resonances (given by Eq. (13), a tune spread $Q \pm \Delta Q$ should be kept as small as possible for a safe operation of the machine. Thus, the chromaticity ξ which depends on the betatron function and the focusing strength, has to be compensated. This can be achieved by inserting higher-order magnetic multipole fields — such as sextupoles — into the ring.

Based on these considerations one should be able to understand the function of most of the components of the lattice of a storage ring, an example of which gives the ESR in Fig. 5. The maximum bending power of the magnets ("the magnetic rigidity") is 10 Tm leading to a maximum energy of 2.2 GeV for protons, 834 MeV/u, and 556 MeV/u for fully stripped Ne and U, respectively. The ring circumference is 108.36 m at a mean radius of 17.246 m. This ring is currently built at GSI-Darmstadt.

However, an essential part is still missing in our simplified description of a storage ring, namely the process of particle accumulation. This can be achieved by a method called "rf-stacking", the idea of which is as follows: A number of particles, called "bucket", is trapped after the injection into the ring on a special orbit, the "injection orbit". Then this bucket is accelerated by an rf-field at constant magnetic bending fields towards an outer orbit in the vacuum chamber. After the rf-field is switched off, the particles in the bucket will debunch on their orbit. Next, the rf is switched on again to take care of the next injected bucket,; this bucket is again accelerated to another stacking orbit close to the previous one. This procedure is repeated until the available phase space of the ring is filled with particles. The slight disturbance of a bucket on a stacking orbit by the rf-field applied in order to stack the

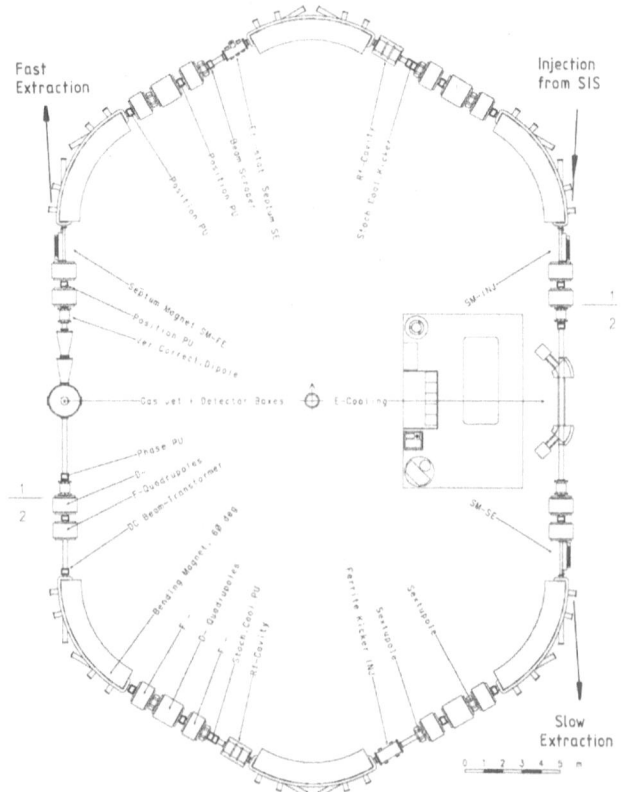

Figure 5. Design of the ESR (Experimentier-Speicher-Ring) lattice with a circumference of 108.36 m. The ion beam will be injected from SIS (Schwer-Ionen-Synchrotron) and extracted either by fast or slow extraction. The cooling device and the set-ups for experiments will be installed in the two straight sections. The abbreviations used for ion-optical and diagnostic elements are: SM = Septum Magnet, PU = Pick Up, SE = Slow Extraction, FS = Fast Extraction, F = Focusing (in the plane of drawing), D = Defocusing. Smaller UHV components are not shown.

following bucket, can be minimized by an artful choice of the rf-voltage applied to successive buckets.

In most cases, the transverse and longitudinal phase space accepted by the ring is not the decisive factor which limits the number N of stored particles. This number is rather determined by effects introduced by the collective interactions of particles among themselves and with the surrounding, for example the vacuum pipe.

The trajectory of a single particle in a storage ring is determined by external forces which are caused by focusing, bending, and accelerating fields. However, in an intense beam the behaviour of the particles differs from that of a single particle, because the beam represents a considerable electric charge and current. Both of them generate electromagnetic fields which are determined by the charge distribution and the boundary conditions imposed by the surroundings of the beam (impedance and geometry), and which act back on the particles ("self forces") [5]. The two most important consequences of the self forces for the ESR beam will shortly be described: The space-charge limit due to a spread of the betatron tune caused by the space-charge field and the longitudinal microwave instability limit imposed by electromagnetic wake fields reflected from the metalic walls of the vacuum pipes.

The space-charge of a beam which strongly depends on charge and mass of the ions stored, leads to a shift ΔQ of the tune of the machine. The maximum tolerable shift for a safe operation of the machine limits the number of stored particles — assuming $\Delta p/p_0 = 0$ — to [6]

$$N_{sc,max} = \frac{\pi A}{rq^2} \Delta Q \frac{B_f}{g} \beta^2 \gamma^3 \epsilon_- \left(1 + \sqrt{\frac{\epsilon_+ Q_-}{\epsilon_- Q_+}}\right)$$ (20)

In Eq. (19) A and q give the mass number and the charge state of the ion, $r = 1.54587 \cdot 10^{-18}$ m, B_f is the bunching factor, i.e. the peak to average ion current, g is a form factor between 1 and 1.5, β and γ are the usual relativistic variables $\beta = v/c$, $\gamma = (1-\beta^2)^{-1/2}$, ϵ_+ and ϵ_- are the larger and smaller of both the transverse emittances, and Q_+ and Q_- the corresponding betatron tunes. Note that $B_f = 1$ for a coasting beam. From the number of stored particles one finds the circulating particle current

$$I = N \cdot \beta c / L = N \cdot f$$ (21)

where L is the circumference of the ring and f is the revolution frequency. β is connected to the particle energy E (in units MeV/u) via

$$\gamma = \left(1 - \beta^2\right)^{-1/2} = 1 + E/E_0$$ (22)

with $E_0 = 931.48$ MeV/u the rest energy of atomic mass unit u.

Table I [6] gives in the ESR for different particle energies E the maximum number of $^{238}U^{+92}$ ions which can be stored according to Eq. (20) under the following assumptions: $\Delta Q = 0.05$, $\epsilon_- = \epsilon_+ = 1\pi\mu m$, $B_f = 1$ (coasting beam), $g = 1.5$, and $Q_-/Q_+ = 1$. The correspondent particle current is calculated from Eq. (21). The last column gives lower limits for the momentum spread $\Delta p/p_0$ of the beam. Below these values, the intenity is mainly limited by microwave instabilities rather than by space charge effects.

Table 1.

E MeV/u	$\beta^2 \gamma^3$	$N_{sc,max}$	f MHz	I pmA	$(\Delta p/p_0)_{min}$ $\times 10^{-4}$
556	2.4750	$9.3 \cdot 10^9$	2.160	295.00	0.78
200	0.5780	$2.2 \cdot 10^9$	1.570	51.00	0.82
100	0.2510	$9.4 \cdot 10^8$	1.190	16.40	0.88
50	0.1160	$4.4 \cdot 10^8$	0.872	5.65	1.06
20	0.0443	$1.7 \cdot 10^8$	0.565	1.41	1.32
10	0.0218	$8.2 \cdot 10^7$	0.402	0.49	1.57
5	0.0108	$4.1 \cdot 10^7$	0.286	0.17	1.85
3	0.0064	$2.4 \cdot 10^7$	0.222	0.08	2.1

As mentioned before, electromagnetic fields ("wake fields") created by the beam itself are reflected from the metalic surface of the surrounding beam pipe, and interact with the beam. A slight perturbation of the density of the beam can grow exponentially and lead to a "microwave instability" by the coupling of the beam current to the corresponding perturbing wake-field. This coupling usually is described by coupling impedances: $Z_\|$ for longitudinal and $Z_\perp$ for transverse fields. A detailed presentation of the rather complicated calculation of coupling impedances is found in [5].

For simplicity, we will give here only the threshold number $N_\|$ of stored ions at which a longitudinal microwave instability (LMI) will occur for a coasting beam with a relative momentum spread $\Delta p/p_0$ (the so-called Keil-Schnell stability criterion) [7]:

$$N_\| \leq F \left|\frac{n}{Z_\|}\right| \frac{A E_0 L}{q^2 e^2 c} \beta \gamma \left|\frac{1}{\gamma^2} - \frac{1}{\gamma_t^2}\right| \left(\frac{\Delta p}{p_0}\right)^2$$ (23)

F is a form factor in the order of unity, and n is the mode number of the wake-field, in general some multiple of the revolution frequency f. One may consider Eq. (22) alternatively as a criterion for the minimum momentum spread required for stability at a given number of stored particles. Above this minimum momentum spread a fast longitudinal mixing in the beam prevents a growth of the instability. This stabilizing effect connceted with the spread is called "Landau damping". A discussion of Landau damping is beyond the scope of this article, but it can be found in many textbooks on plasma physics. Note that according to Eq. (22) a beam with zero momentum spread is always unstable !

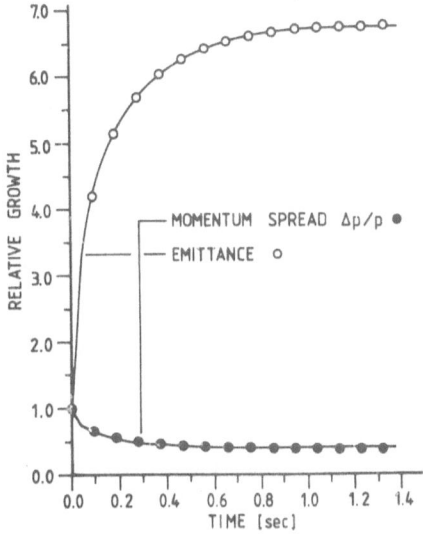

Figure 6. Approach of equilibrium of emittance and momentum spread due to Intra-Beam-Scattering as a function of time for 10^{10} U^{+92} ions stored in the ESR [9]; the initial values were $\epsilon = 0.78\pi$ mm mrad and $\Delta p/p_0 = 1.2 \cdot 10^{-3}$.

The main problem in evaluating Eq. (22) is the rather poor knowledge of the coupling impedance $Z_{||}$ of a ring. A reasonable guess for $Z_{||}$ of the ESR is $|Z_{||}| = 30\Omega$ which has been used with $\gamma_t = 2.66$ to calculate the minimum values of $\Delta p/p_0$ in the last column of Table 1.

Finally, we mention a third intensity-limiting process which is of special importance in heavy-ion storage rings, namele the "intra-beam-scattering (IBS)". With growing phase-space density the number of elastic Coulomb collisions between ions within the beam increases. In every binary collision, kinetic energy and momentum is exchanged between the partners (in all of the three planes). These microscopic interactions which have to be distinguished from the macroscopic space-charge effects, may lead to an excitation or a damping of betatron oscillations and momentum spread. The IBS-theory of Piwinski [8] predicts that the IBS rate increases with Nq^4A^{-2}, thus for fully stripped ions ($q = Z$) with NZ^2. Figure 6 shows the result of numerical calculations based on Piwinski's theory for the temporary development of emittance and momentum spread of a beam of $N = 10^{10}$ stored U^{+82} ions [9]. Values of 1.01 and 10 m were used for the variable γ and for the betafunction in both transverse directions, respectively, and the dispersion function was $D = 0$. The initial values of emittance (0.78π mm mrad) and momentum spread ($\Delta p/p = 1.2 \cdot 10^{-3}$) were chosen to be far from equilibrium. After a time in the order of 100 msec an equilibrium is reached due to the net exchange of energy and momentum between the different degrees of freedom. The equilibrium state is

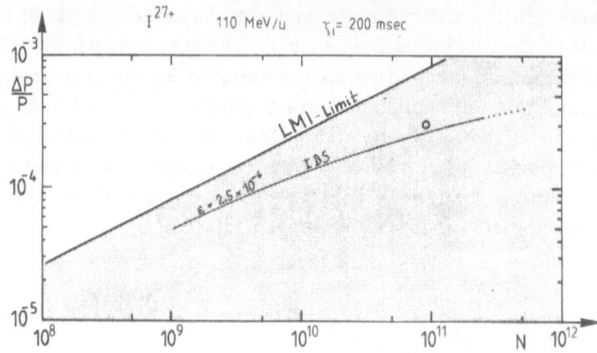

Figure 7. 'Stability diagram' for J^{27+} ions stored in the ESR at 110 MeV/u with respect to Longitudinal Microwave Instability (LMI) and Intra-Beam-Scattering (IBS) [10]. Note that only the white area is stable against LMI.

characterized by an increase of the emittance (about a factor of 6.5) and a reduction of the momentum spread (about a factor of 0.6).

A general plot of the LMI- and IBS-intensity limits for a beam of J^{27+} at 110 MeV/u is presented in Fig. 7 [9]. Note that only the points $(N, \Delta p/p_0)$ lying in the white area are stable according to the Keil-Schnell criterion. The line "IBS" represents for an emittance of $2.5\,\pi$ mm mrad the number of points $(N, \Delta p/p_0)$ with a longitudinal intra-beam scattering rate of 5 sec^{-1} which corresponds to an IBS-time of 200 msec. Thus, 10^{11} stored J^{+27} ions at $\Delta p/p_0 = 3 \cdot 10^{-4}$ and $\epsilon = 2.5\pi$ mm mrad (black circle) is on the safe side with respect to IBS, but not on the safe side with respect to LMI.

ELECTRON COOLING OF HEAVY-ION BEAMS

The process of intra-beam-scattering lead via an exchange of energy and momentum between the three degrees of freedom of the beam particles from an initial non-equilibrium to a final equilibrium state. This suggests a description of the corresponding parameters in thermodynamical terms, i.e. in form of a temperature equilibration. In fact, in a coordinate system moving with the average velocity of the particles, the transverse and the longitudinal motions of all the individual particles due to the betatron oscillations and the finite momentum spread appear as a sort of "thermic motion". Therefore, a characterization of this motion by a "beam temperature" T_i is reasonable.

In the co-moving system, this temperature is defined by

$$3/2kT_i^* = \frac{M}{2}\left\{ \left(\Delta^* v_{\|}\right)^2 + \left(\Delta^* v_h\right)^2 + \left(\Delta^* v_v\right) \right\}, \tag{24}$$

where M is the rest mass and $\Delta v_{\|}$, Δv_h, and Δv_v are the longitudinal, horizontal, and vertical velocity spreads of the particles. The symbol "$*$" indicates that the corresponding variables are measured in the co-moving system.

An expression for the beam temperature in the lab-system is obtained as follows: From the relativistic momentum $p = mv\gamma$, one finds $\Delta v_{\|} = (\beta c/\gamma^2)\Delta p/p_0$, and with $\Delta v_{\|} = \gamma^{-2}\Delta v_{\|}^*$ one gets for the longitudinal temperature

$$\frac{1}{2}kT_{i\|} = \frac{M}{2}\beta^2 c^2 (\Delta p/p_0)^2 \tag{25}$$

The transverse temperature $T_{i\perp}$ follows by noting that for the transverse velocity spreads ($\perp$ stands for both transverse planes with coordinate y)

$$\Delta v_{\perp}^* = dy^*/dt^* = dy/dt^* = \gamma \, dy/dt = \gamma\beta c \, dy/ds$$

(only for the time a Lorentz-transformation has to be applied). Combining this expression with the equations for the beam envelope and divergence, Eq. (10), one finds

$$kT_{i\perp} = \frac{M}{2}\gamma^2\beta^2 c^2 \left\{ \frac{\epsilon_h}{\beta(s)_h} + \frac{\epsilon_v}{\beta(s)_v} + \frac{1}{\gamma^2}\left(\frac{\Delta p}{p_0}\right)^2 \right\}, \tag{26}$$

so that finally

$$\frac{3}{2}kT_i = \frac{M}{2}\gamma^2\beta^2 c^2 \left\{ \frac{\epsilon_h}{\beta(s)_h} + \frac{\epsilon_v}{\beta(s)_v} + \frac{1}{\gamma^2}\left(\frac{\Delta p}{p_0}\right)^2 \right\} \tag{27}$$

Equation (26) demonstrates that the beam temperature is given by the transverse emittances due to the betatron oscillations and the momentum spread. It is clearly very exciting to reduce the temperature of a beam when it is stored in a ring, and to achieve thereby an extremely "brilliant" beam with very small amplitudes of the betatron oscillations and a very sharp momentum distribution around the design momentum.

Such a cooling of the beam offers many experimental advantages and opens possibilities for new experiments:

- high particle densities for fixed target and for colliding and / or merging beam experiments,

- storage and accumulation of 'hot' exotic nuclear reaction products even at low production rates,

- highly charged ions (at the ESR bare, H- and He-like ions up to U) also at small projectile energies (note that without cooling the beam emittance $\epsilon \sim 1/p$, so that deceleration blows up the emittance),

- compensation of "beam-heating" from intra-beam and residual-gas scattering and from gradual growth of the amplitude of the betatron oscillations due to machine errors,

- compensation of energy-loss and small-angle scattering in an internal target.

The two latter points especially demonstrate the importance of cooling: Cooling extends the lifetime of a stored beam. Therefore, the beam particles can interact many times with the target atoms which finally leads to an increase of the event-rate in an experiment.

Now the question for an appropriate refrigerant arises. Remember that in the sense of Eq. (26) cooling of a beam means the reduction of the phase space of a given number of particles. In order to satisfy Liouville's theorem, a coupling of the beam to an outer system by a non-conservative force, for instance a frictional force, is needed. G. Budker [11] was the first to point out that a "cold" beam of electrons moving at the same velocity parallel to the "hot" ion beam on a straight section of the ring could provide a refrigerant. The Coulomb forces between electrons and ions cause a friction between both the beams leading to an equilibration of their temperatures, because the circulating ion beam passes many times a beam of continuously supplied, fresh, core electrons.

In a series of experiments, Budker and his coworkers demonstrated the successful working of electron cooling of proton beams [12]. Additionally, this group elaborated the theoretical framework of electron cooling [13]. Up to now, electron cooling has been applied only to proton beams. But based on the experience gained, an increasing number of projects for electron cooling of heavy ion beams has been pushed forward during the last few years. A comprehensive review of the proton experiments and of the new projects can be found in Ref. 14.

In this context we mention that also other cooling mechanisms are known: Radiation cooling of electron beams by the emission of synchrotron-radiation, stochastic cooling and laser cooling. These mechanisms will not be treated in the present paper which concentrates only on electron cooling.

Let us estimate in a simple way the characteristic cooling time τ_c in which a heavy ion beam can be cooled by an electron beam. We will consider the collisions between ions and electrons in a binary model which clearly illustrates the physics of cooling. A plasma-physical treatment which certainly is more adequate to the problem but much less instructive, can be found in a paper by Sorensen and Bonderup [15]. In Coulomb collisions between ions and electrons kinetic energy and momentum is exchanged between both the partners. The net result of this exchange can be described as a collisional energy loss per unit distance dE^*/ds^* and is given by [16].

$$dE^*/ds^* = 4\pi n_e^* \frac{Z^2 e^4}{m(v^*)^2} \cdot L, \tag{28}$$

where n^* is the density of electrons and m their rest mass, v^* is the relative velocity between ions and electrons. L is the "Coulomb-logarithm",

$$L = ln(b_{max}/b_{min}),$$

where b_{max} and b_{min} are the upper and lower bounds of the impact parameters of the collision. Noting that the upper bound b_{max} is approximately given by the Debye shielding length λ_D

$$b_{max} = \lambda_D = \sqrt{\frac{kT}{4\pi n_e^* e^2}} = 7.4 \cdot 10^2 \, (T/eV)^{1/2} \left(n_e^*/cm^{-3}\right)^{-1/2} [cm]$$

and the lower bound by the distance of closest approach in a head-on collision

$$b_{min} \approx \frac{Ze^2}{\frac{m}{2}(v^*)^2} = 2Z\frac{e^2}{mc^2}\left(\frac{v^*}{c}\right)^{-2},$$

one finds

$$L \approx ln\left\{\lambda_D \left(\frac{v^*}{c}\right)^2 /2Z\frac{e^2}{mc^2}\right\} \approx ln\left(3 \cdot 10^4/Z\right).$$

for typical values of $T = 0.5$ eV, $n_e^* = 10^9$ cm^{-3} and $(v^*/c) = 10^{-3}$. Thus, to a good approximation $L \approx 10$ for all nuclear charges Z.

The equation for the collisional energy loss which originally has been derived by Bohr, may be interpreted as a friction force F^* in the rest frame with a momentum change per unit time

$$\frac{dp^*}{dt^*} = -F^* = \frac{dE^*}{ds^*}.$$

From this momentum change a cooling time τ_c^* can be derived.

$$\tau_c^* = \frac{p^*}{dp^*/dt^*} = \frac{m}{4\pi n_e^* e^4 L}\frac{M}{Z^2}(v^*)^3 \tag{29}$$

With $\tau_{c_i} = \gamma\tau_c^*$ and $n_e = \gamma n_e^*$, Eq. (28) can be transformed into the lab-system; additionally, a factor η which gives the ratio of the length of the cooling section to the circumference of the ring, has to be introduced. With

$$r_e = \frac{e^2}{mc^2}, \; r_i = \frac{e^2}{E_0}, \; M = AE_0, \text{ and } j = en_e\beta c$$

where j is the current density of the beam, one finally finds the cooling time τ_c in the lab-system

$$T_c = \frac{\gamma^2}{\eta} \frac{1}{4\pi L} \frac{e}{r_e r_i} \frac{1}{j} \frac{A}{Z^2} \beta \left(\frac{v^*}{c}\right)^3 \tag{30}$$

The cooling time increases linearly with the ion mass and with the third power of the relative velocity and decreases linearly with the electron current density, and quadratically with the nuclear charge.

As a numerical example we consider a beam of 200 MeV/u U^{+92} ions which corresponds to $\gamma = 1.22$ and $\beta = 0.57$. For $v^*/c \sim 10^{-3}$, $j = 1A/cm^2$, and $\eta = 2 \cdot 10^{-2}$ one obtains from Eq. (29) a cooling time $T_c = 32$ msec.

This simple model has obviously to be extended since it ignores the fact that the relative velocity between electrons and ions is not a singly-valued function but a distribution whose width is determined by the temperatures of the ion beam (Eq. (26) and of the electron beam. Also the latter beam has a finite temperature T_e which is composed of different parts as the temperature of the cathode, ripple of the high-voltage power supplies used to accelerate the electrons up to the energy of the ion beam, imperfections of the beam transport system, etc.

If $f(\vec{v}_e^*)$ denotes the velocity distribution of the cooling electrons and $\vec{v}_i^*$ the velocity of an individual ion, then (in analogy to Eq. (28) the friction force is given by [17].

$$\vec{F}^* = -\frac{4\pi n_e^* L Z^2 e^4}{m} \int d^3 v_e^* f(\vec{v}_e^*) \frac{\vec{v}_i^* - \vec{v}_e^*}{|\vec{v}_i^* - \vec{v}_e^*|^3}. \tag{31}$$

From this equation one obtains the cooling time in the lab-system

$$T_c = \frac{k}{\eta} \frac{e}{L r_e r_i j} \frac{A}{Z^2} \beta^4 \gamma^5 \left(\Theta_e^2 + \Theta_i^2\right)^{3/2} \tag{32}$$

with

$$\Theta_e = v_{e\perp}/\beta c \ (\text{the divergence of the electron beam})$$

and

$$\Theta_i = \left\{\frac{\epsilon_h}{\beta(s)_h} + \frac{\epsilon_v}{\beta(s)_v} + \left(\frac{1}{\gamma}\frac{\Delta p}{p_0}\right)^2\right\}^{1/2}. \tag{33}$$

k is a form factor whose numerical value depends on the velocity distribution of the cooling electrons: for a Maxwellian distribution, $k = 3/\left(2\sqrt{2\pi}\right) = 0.6$.

In the years 1974 – 1976 the first cooling experiments in Novosibirsk gave cooling times smaller than expected [18]. A short time later, this effect of 'fast cooling' was explained by two reasons [13]: The flattened electron velocity distribution and the magnetized electron beam.

The flattening of the original velocity distribution near the cathode in longitudinal direction is a pure kinematical effect. If near the cathode the electrons have some effectie temperature $T_{eff} = m(\Delta v_e)^2/2$, then their energy after acceleration in an electric field with voltage V_0 will be

$$E_e = eV_0 + T_{eff} = \frac{m}{2} (v_0 + \Delta v_e)^2$$

(relativistic: $e\acute{V}_0 + T_{eff} = (\gamma - 1) mc^2$).

For $\Delta v_e \ll v_0$, the longitundinal velocity spread in the lab-system is therefore

$$\Delta v_{e\parallel} = \frac{T_{eff}}{mv_0} \left(\text{relativistic: } \Delta v_{e\parallel} = \frac{T_{eff}}{\beta\gamma^3 mc}\right).$$

With $\Delta v_{e\parallel} = \Delta v_{e\parallel}^*$ (relativistic: $\Delta v_{e\parallel} = \gamma^{-2}\Delta v_{e\parallel}^*$) one obtains for the longitudinal electron temperature in the co-moving system

$$T_{e\parallel} = \frac{m}{2}\left(\Delta v_{e\parallel}\right)^2 = \frac{T_{eff}^2}{4E_e}\left(\text{relativistic: } T_\parallel = \frac{T_{eff}^2}{2\beta^2\gamma^2mc^2}\right).$$

Since $T_{eff} \sim T_{e\perp} \sim T_{cathode}$, the electron temperature is longitudinally flattened with a ratio

$$T_{e\parallel}/T_{e\perp} \approx T_{cathode}/4E_e \ll 1.$$

The important consequence of this flattening for cooling is the reduction of the form factor k for the cooling time (Eq. (31) [17]: For a flattened distribution, $k = 1/2\pi = 0.16$; this leads to a cooling time which is about a factor of 4 smaller than for a Maxwellian velocity distribution ($k = 0.6$).

Nature was very kind to cooling, because besides the flattening it made a second present in form of the "magnetized electron beam". From the equations for the cooling time τ_c follows that $\tau_c \sim j^{-1}$. So a high current density of the electron beam is necessary for short cooling times. However, high current densities cause large space-charge forces, both in longitudinal and in transverse direction [19]. Let us estimate the transverse force in order to get an idea about the transverse blowing-up of the electron beam.

Using the Gaussian law one finds for the radial force in the co-moving system

$$\vec{F}_y^* = \frac{e^2 n_e^*}{2\epsilon_0}y,$$

where y stands for both transverse components. The resulting "transverse acceleration" in the lab.-system is easily obtained from the force.

$$\frac{dy'}{ds} = \frac{ej}{2\epsilon_0 mc^2\gamma^2\beta^3c}y \tag{34}$$

With the following design parameters of the ESR: $j = 1 A/\text{cm}^2$ and electron-beam radius $y = 2.5$ cm, one obtains for $\gamma = 1.22$, $\beta = 0.57$ (corresponding to 200 MeV/u kinetic energy of the ions)

$$\frac{dy'}{ds} \approx 3\,\frac{mrad}{cm}.$$

Thus, the beam divergence grows about 3 mrad per cm. This blowing-up of the electron beam is prevented by a longitudinal magnetic guiding field of typically $B \leq 1.5$ kg. In this field, the transverse acceleration of the electrons is transformed into a macroscopic rotation of the whole electron beam around its longitudinal axis with a frequency [19]

$$\omega_y = \frac{1}{e}\frac{dF_y^*/dy}{B} = \frac{en_e^*}{2\epsilon_0 B} \approx 9 MHz$$

for $n_e^* = 10^8 cm^{-3}$ and $B = 1kGauss$.

Additionally, all the individual electrons rotate around the longitudinal magnetic field lines with the cyclotron frequency

$$\omega_c = \frac{eB}{\gamma m} = 17.6 MHz\frac{B/Gauss}{\gamma}$$

and a gyration radius

$$r_c = \frac{v_{e\perp}}{\omega_c}$$

which is about 15 μm for $B = 1kGauss$, $T = 0.2$ eV, and $\gamma = 1$. Note that in a plane perpendicular to the beam direction the overall motion of the electrons consists of the superposition of both the frequencies ω_y and ω_c resulting in an epicyclic trajectory.

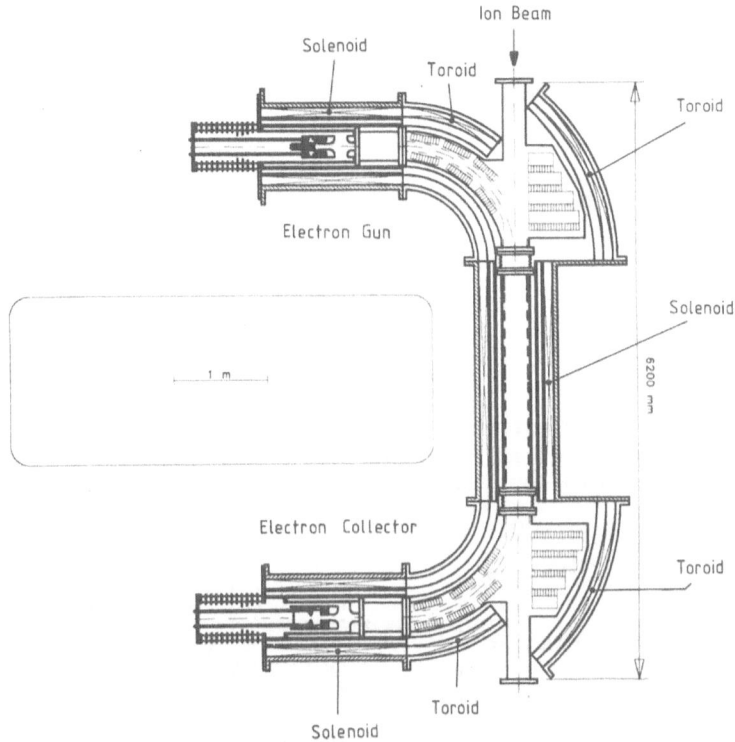

Figure 8. Design of the ESR Cooler. Electrons from the gun are directed into the 2.5 m long cooling section and then bent into the collector by toroidal magnetic fields. Solenoidal fields guide the beam in the straight sections.

Concerning cooling it turned out [15,17,19] that the high-frequency part of this trajectory which is in practical cases given by the cyclotron frequency ω_c, is the important one. The corresponding gyration radius divides the number of Coulomb collisions between electrons and ions into two domains: Into 'adiabatic collisions' with impact parameter $b \gg r_c$ and into 'fast collisions' with impact parameter $b \ll r_c$. The time it takes for an electron for one Lamor revolution is $\tau = \omega_c^{-1} = 5.7 \cdot 10^{-11}$ sec. Collisions with $b \gg r_c$ last for times in the order of $\lambda_D/v^* = 5.2 \cdot 10^{-9}$ sec for typical values $T_{e\perp} = 0.5$ eV and $n_e^* = 10^8$ cm^3. Therefore, during an adiabatic collision the electrons appear to the ions as "Lamor disks" moving on parallel lines rather than free electrons with some transverse and longitudinal velocity distribution. As a consequence, the exchange of transverse temperature between electron and ion beam becomes very small and the effective "cooling temperature" seen by the ions is the longitudinal electron temperature $T_{e\parallel}$. This temperature is due to the flattening effect very small, so that for these collisions the effective Θ_e in Eq. (32) becomes negligible and the cooling time τ_c decreases correspondingly.

In fast collisions with $b \ll r_c$ the electrons appear more or less as free electrons and Θ_e is determined by the transverse velocity $v_{e\perp}$ of the electron beam. Since it is not easy to derive analytical expressions for the cooling time for fast and adiabatic collisions, we only give approximative, but instructive formula for the cooling force which can be derived from Eq. (Hi31). With the abbreviation

$$K = \left(\frac{4\pi n_e^* L Z^2 e^4}{m} \right)^{-1}$$

one finds

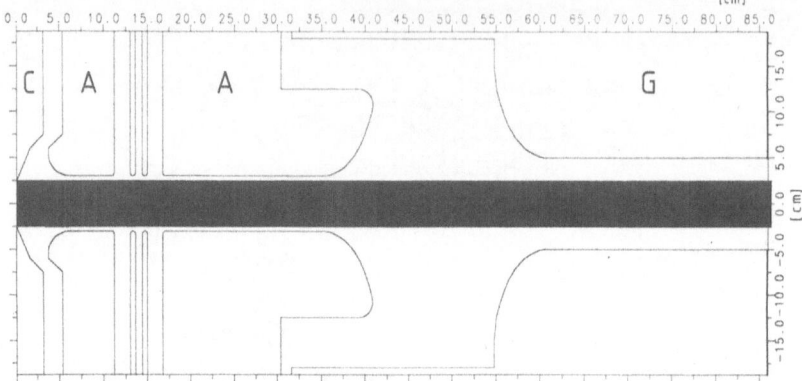

Figure 9. Design of the electron gun. C, A, and G denote cathode, anode, and ground potential, respectively. The black stripe along the axis results from a superposition of many individual electron trajectories at 330 keV.

$$\tau_c \approx K(v^*)^3 \text{ with } v^* = \left\{ \begin{array}{ll} \left. \begin{array}{llll} v_i^* & \text{if } v_i^* & > & v_e^* \\ v_e^* & \text{if } v_e^* & > & v_i^* \end{array} \right\} & \text{fast } (b \ll r_c) \\[2ex] \left. \begin{array}{llll} v_i^* & \text{if } v_i^* & > & v_{e\parallel}^* \\ v_{e\parallel}^* & \text{if } v_{e\parallel}^* & > & v_i^* \end{array} \right\} & \text{adiabatic } (b \gg r_c) \end{array} \right.$$

It is interesting to note that up to now no theory of the cooling forces in collisions with $b \approx r_c$ is available — but it is also to note that the theoretical results for fast and adiabatic cooling converge in the limit $b \rightarrow r_c$ [15]. Notwithstanding of a theory available or not: Nature granted cooling times shorter than originally expected by the happy coincidence of flattening and magnetization of the cooling electron beam [18].

As an example for the design of an electron cooler, Fig. 8 shows the cooler planned for the ESR [20]. The electrons are generated and accelerated in a gun (upper part) which is shown in detail in Fig. 9. Then they are directed into the cooling section (250 cm long) by a toroid and finally bent to the collector (lower part of Fig. 8). The layout is planned for electron energies between 2 and 330 keV and is supposed to deliver electron currents of 10 A at a beam diameter of 5 cm. Computer simulations of the performance suggest longitudinal magnetic fields between 0.1 and 3 kGauss [19]. Concerning Fig. 9 it is to note that the broad black band along the axis of symmetry which represents the shape of the electron beam, corresponds to very low temperatures T_e, not only on the macroscopic scale of Fig. 9 but also on a microscopic scale.

Let us finally mention that besides all the requirements implied by the physcis of electron cooling, large technical problems are encountered in the construction of an electron cooler. The main technical problem stems from the necessity of velocity matching between electrons and ions. This condition leads to a relation between the kinetic energy of the electrons E_e and of the ions E of the form

$$E_e = \frac{mc^2}{E_0} E,$$

from which the acceleration voltage V_0 for the electron beam is obtained.

$$V_0 = mc^2(\gamma - 1)[V] \tag{35}$$

Equation (34) shows that for ion energies above several MeV/u a high voltage system is needed in order to accelerate the electrons to the corresponding energy. Especially at high ion energies ($\geq$ 200 MeV/u) the construction of an appropriate gun ($V_0 \geq$ 110 kV) and an appropriate collector may cause considerable technical problems.

310

It is quite natural that dealing with electron cooling of heavy ions, a simple, but fundamental question arises: What about those processes in which ions capture a cooling electron ? Recombination changes the charge state of the stored ions and will therefore lead to beam loss — except in cases where a ring is operated in a special multi-charge mode.

Thinking about recombination processes, one finds three important aspects:

1. Recombination reduces the lifetime of a stored beam;

2. Recombination provides a tool for diagnostics of both the ion and the electron beam;

3. The photons (and/or the electrons) emitted with the recombination can be used for spectroscopic experiments.

In the following we will discuss these three aspects in more detail. For simplicity we will consider only recombination processes between bare ions and electrons.

Collisional Radiative Recombination

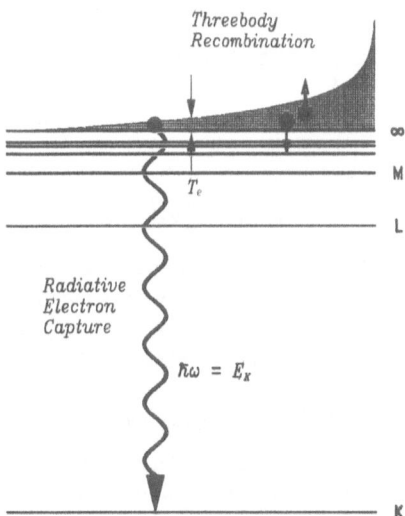

Figure 10. Schematic plot of the Radiative-Electron-Capture and of the Three-Body-Recombination.

Cooling electrons can be captured by bare ions only if a third particle is involved which takes care of energy and momentum conservation. The third particle may be a photon or an electron:

$$A^{+Z} + e^- \rightarrow A^{+(Z-1)} + \hbar\omega \qquad (36)$$

$$A^{+Z} + e^- + e^- \rightarrow A^{+(Z-1)} + e^-. \qquad (37)$$

The process described by Eq. (35) is known as 'Radiative Electron Capture (REC)', and the process given by Eq. (36) as 'Three-body recombination'. A schematic representation of both the processes is shown in Fig. 10.

Let us first deal with the first of the three aspects mentioned above, which obviously is very important for accelerator physics. The REC is the time-reversed process of photoionization, the cross section of which has been calculated already in the year 1930 [21]. With the principle of detailed balance the REC cross section for capture into a shell with principle quantum number n is obtained from the photoionization cross section [22].

$$\sigma_n\left(E_e^*\right) = 2.1 \cdot 10^{-22} \frac{E_{1s}^2}{n E_e^* \left(E_{1s} + n^2 E_e^*\right)} [cm^2] \tag{38}$$

where E_{1s} is the $1s$ binding energy, and E_e^* is the relative kientic energy between ion and electron in the co-moving frame of reference. Introducing the electron velocity distribution $f(v_e^*)$, one can calculate the total REC-rate per ion and sec in the ion rest frame from

$$r^* = \sum_n n_e^* < v_e^* \sigma_n(v_e^*) > = n_e^* \alpha \tag{39}$$

with α being the recombination coefficient

$$\alpha = < v_e^* \sigma(v_e^*) > = \int d^3 \vec{v}_e^* v_e^* \sigma(v_e^*) f(\vec{v}_e^*) \tag{40}$$

Inserting the expression (38) and summing up over all n one finds for a completely flattened electron velocity distribution [23]

$$\alpha_{REC}^{fl} = \frac{3.02 Z^2}{\sqrt{kT_e}} \left\{ ln \frac{11.32 Z}{\sqrt{kT_e}} + \left(\frac{kT_e}{Z^2}\right)^{1/3} \right\} \cdot 10^{-13} [cm^3/sec] \tag{41}$$

where T_e is the transverse temperature of the electron beam. For a Maxwellian velocity distribution one finds:

$$\alpha_{REC}^{Max} \approx \frac{2}{\pi} \alpha_{REC}^{fl}.$$

It follows from Eq. (40) that α_{REC} roughly scales as $Z^{2.2}$, thus with a somewhat larger power than the cooling rate ($\sim Z^2$). Figure 11 shows the recombination coefficients for REC as a function of the nuclear charge at a temperature of 0.2 eV for both velocity distributions.

The total REC-rate in the lab. system is obtained from Eqs. (38) and (40).

$$r_{REC} = n_e \alpha_{REC}^{fl} \cdot \eta \gamma^{-2} \tag{42}$$

where the factor γ^2 stems from the Lorentz-transformation of time and electron density. As an example: For 500 MeV/u projectile energy one finds from Fig. 11 for the flattened velocity distribution

$$r_{REC} \approx \begin{cases} 5 \cdot 10^{-2} sec^{-1} & \text{for } U^{+92} \\ 1.5 \cdot 10^{-2} sec^{-1} & \text{for } Xe^{+54} \end{cases}$$

Thus, the time-constants $\tau_{REC} = r_{REC}^{-1}$ are of the order of 20 sec and above which is considerably larger than the cooling times τ_c.

Concerning the three-body recombination the situation is much more complicated because many open channels are open in this case:
After a three-body recombination into a shell with principal quantum number n

$$A^{+Z} + e^- + e_- \rightarrow A^{+(Z-1)}(n) + e^-, \tag{43}$$

the capture electron may be collisionally re-ionized:

$$A^{+(Z-1)}(n) + e^- \rightarrow A^{+Z} + e^- + e^-, \tag{44}$$

excited:

312

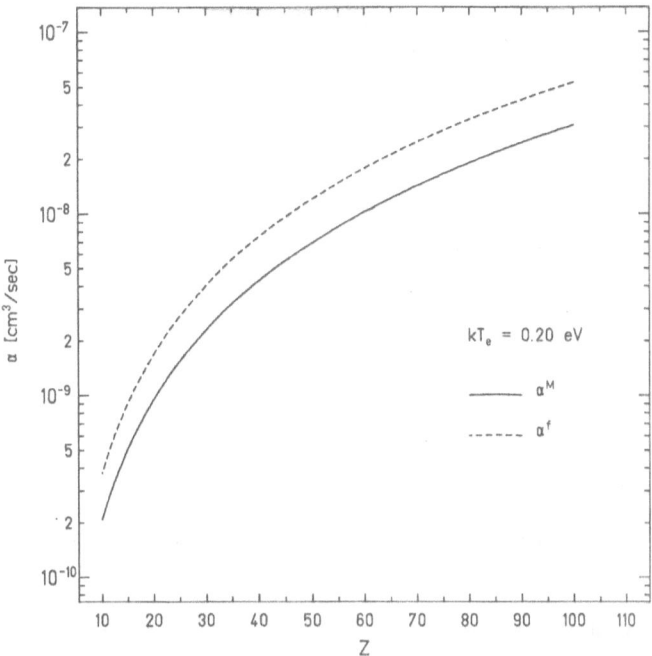

Figure 11. REC recombination coefficients as function of the nuclear charge Z for an electron temperature of 0.2 eV. α^M and α^t are calculated from a Maxwellian and a completely flattened $(T_{e\parallel} = 0)$ electron velocity distribution, respectively.

$$A^{+(Z-1)}(n) + e^- \rightarrow A^{+(Z-1)}(n') + e^- \ (n' > n), \tag{45}$$

de-excited:

$$A^{+(Z-1)}(n) + e^- \rightarrow A^{+(Z-1)}(n') + e^- \ (n' < n), \tag{46}$$

or radiatively de-excited:

$$A^{+(Z-1)}(n) \rightarrow A^{+(Z-1)}(n') + \hbar\omega \ (n' < n). \tag{47}$$

The problem now consists in the calculation of the 'net rate capture' in this compli-cated sequence of events which with the REC included has been called 'collisional radiative recombination' [24]. Numerical calculations have been performed for hydrogen-ion plasmas first by Bates et al. [24] and later in a more refined way by Stevefelt et al. [25]. Stevefelt and coworkers succeeded in fitting their numerical results by a simple analytical formula. It is however not clear, how this formula should be scaled to higher values of the nuclear charge Z. Bates and coworkers gave a prescription how to scale their results to larger Z; unfortunately, these scaling prescriptions are only of limited profit for the parameters which are typical for the cooling of heavy ions ($n_e \sim 10^8 cm^{-3}, kT_e \leq 1$ eV, and $Z > 10$).

Therefore, we developed a model [26] based on the simplifying considerations by Byron et al. [27] rather than performing extended numerical coupled-state calculations. Byron and coworkers realized that under equilibrium conditions a pronounced minimum in the total rate of de-excitation of atoms as a function of the principal quantum number n of the excited states exists. This minimum occurs, because on one hand the collisional de-excitation rate strongly increases with the principal quantum number, whereas on the other hand the radiative transition probability rapidly decreases with n, and the equilibrium population of excited states passes through a minimum. If n^* is the principal quantum number of the excited states at which the minimum of the sum of the collisional and radiative rate appears, then the net rate of recombination is limited to the rate of de-excitation of the level n^* or of a bottleneck of levels around n^* [25].

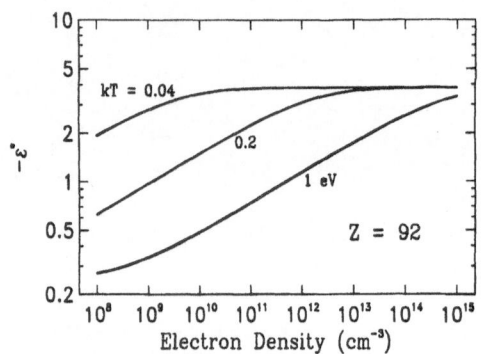

Figure 12. Reduced binding energy $(-\epsilon^* = RZ^2/(n^{*2}kT_e)$ for U ions as a function of the electron density for three different electron temperatures.

It can be shown [26] that the recombination coefficient α_{coll} for the collisional processes (42 – 45) is given by

$$\alpha_{coll} = 2.0 \cdot 10^{-27} \frac{n_e^* Z^3}{(kT_e)^{4.5}} [cm^3/sec] \tag{48}$$

and that the recombination coefficient α_{rad} for the radiative processes (46) is given by

$$\alpha_{rad} = 2.1 \cdot 10^{-13} \frac{Z^{1.5}}{(kT_e)^{0.25}} (-\epsilon^*)^{1.25} e^{-\epsilon^*}$$
$$+ 9.6 \cdot 10^{-14} \frac{Z^2}{(kT_e)^{0.5}} \int_{-\epsilon^*}^{0} d\epsilon \, e^{-\epsilon} \ell n \left\{ \frac{(n^*-1)^2 \, (n(\epsilon)^2 - 1)}{n(\epsilon)^2 - (n^*-1)^2} \right\}. \tag{49}$$

In Eq. (48) we used the reduced energy

$$\epsilon = -\frac{RZ^2}{n^2 kT_e} \tag{50}$$

with R being the Rydberg constant. The dependence of α_{coll} on Z and kT_e can easily be verified using simple classical [23] or dimensional [28] arguments. The first term on the rhs of Eq. (48) stems from the total radiative decay of the state n^* and the second term from the radiative decay of all states n above n^* to all states below n^*. ϵ^*, the reduced energy of the state with principal quantum number n^* (the state at which the minimum occurs) can be calculated from the solution of

$$- (\epsilon^* + 3.83) + 1.87 \cdot 10^{13} \frac{(kT_e)^{3.75}}{n_e^* Z^{0.5}} (-\epsilon^*)^{3.58} (-\epsilon^* - 0.25) = 0 \tag{51}$$

Figure 12 shows $-\epsilon^*$ for $Z = 92$ as function of the electron density for three different values of the electron temperature. In the collisional limit $n_e^* \to \infty$, where radiative transitions

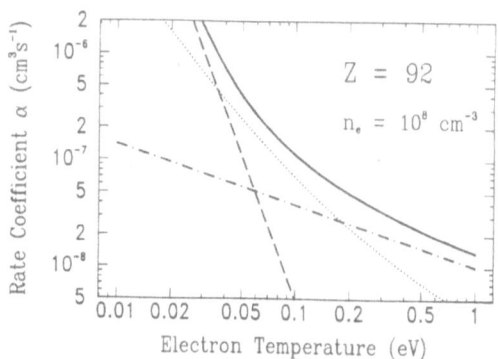

Figure 13. Total recombination coefficient for U ions at an electron density of 10^8 cm^{-3} as a function of the electron temperature. The dashed line represents the collisional contribution α_{coll}, the dotted curve the radiative contribution α_{rad}, and the dashed-dotted line the contribution α_{REC} from radiative electron capture, whereas the solid line gives the sum.

are of minor importance only (note that $r_{coll} \sim n_e^2$, while $r_{rad} \sim n_e$), $-\epsilon^*$ approaches 3.83, independet of Z (cf. Eq. (50)). This corresponds to the well-known fact that the time-reversed process of the three-body-recombination, the electron impact ionization, reaches a maximum cross section at electron impact energies of about 3 – 4 times the binding energies of the electron to be ionized.

With the REC, the collisional and the radiative part the total recombination coefficient can be written as

$$\alpha = \alpha_{REC} + \alpha_{coll} + \alpha_{rad}, \tag{52}$$

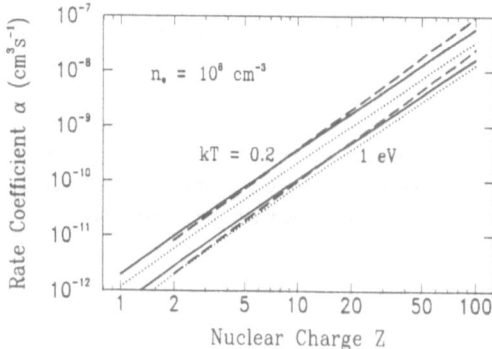

Figure 14. Comarison of the present results (solid curves) for the total recombination coefficient with extrapolations of the numerical results of Bates et al. (dashed curves) and Stevefelt et al. (dotted curves) for a H-plasma to high Z.

315

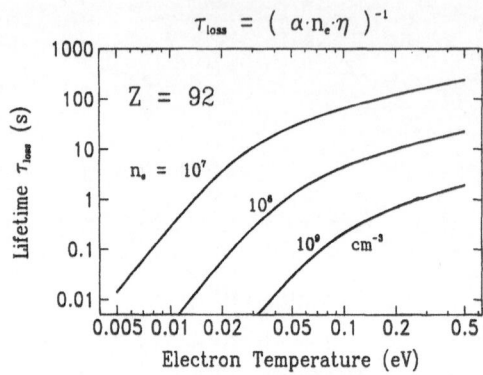

Figure 15. Time constant τ_{loss} of a U^{+92} beam as determined by recombination losses as a function of the electron temperature. At high ion-beam velocities the data shown have to be multiplied with the relativistic variable γ^2.

where α_{REC} is given by Eq. (40), α_{coll} by Eq. (47), and α_{rad} by Eqs. (48 – 50). Figure 13 shows the total recombination coefficient for U^{+92} at an electron density of 10^8 cm^{-3} as function of the electron temperature. The dotted line represents the radiative part, the dashed line the collisional part, and the dashed-dotted line the REC. The sum of all the three contributions is given by the solid line. It is clearly seen that for cold beams ($T_e < 0.1$ eV) the collisional part dominates, while for temperatures above 0.2 eV α_{coll} is completely negligible and the REC governs the total recombination.

The dependence of α on Z at the same electron density is shown in Fig. 14 for two different electron temperatures (solid lines). Figure 14 also presents a comparison to extrapolations of the rather sophisticated numerical calculations of α in H and H-like plasmas by Bates et al. [24] (dashed lines) and Stevefelt et al. [25] (dotted lines). The extrapolations to high Z were made using strictly the scaling prescription given in Ref. 24 ($\sim Z^{2.4}$) although this prescription is supposed to apply only to low Z and to temperatures far beyond those plotted in Fig. 14. The rather good agreement between the three different sets of data is quite encouraging.

In analogy to Eq. (41) we can calculate a time constant $\tau_{loss} = (n_e \alpha \eta \gamma^{-2})^{-1}$ in the lab. system with α given by Eq. (52). Since γ is in the order of 1 for present heavy ion storage rings, we plot in Fig. 15 $\tau_{loss} = (n_e \alpha \eta)^{-1}$ as function of the electron temperature for U^{+92} at different electron densities. Note that for typical ESR parameters ($n_e = 10^8$ cm^{-3} and $T_e = 0.2$ eV) the time constant for total recombination of a U^{+92} beam is about 10 sec and is thus larger than the expected cooling time. However, it is obvious from Fig. 15 that recombination will become a severe problem when electron cooling of heavy ions is envisaged at substantially smaller electron temperatures.

Let us now come to the second and the third aspect of electron-ion recombination mentioned above which are in some respect connected to each other. A rather simple, but for daily operation useful diagnostic tool consists of a measurement of the total REC photon emission rate as function of the relative velocity between electron and ion beams. Since the recombination coefficient strongly increases with decreasing relative velocity, the REC photon

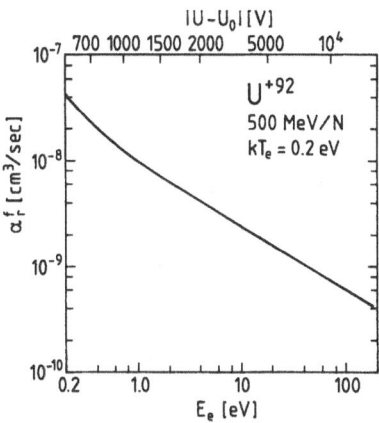

Figure 16. REC recombination coefficient for U^{92+} as function of the relative kinetic energy between electrons and ions (lower scale). The upper scale gives the acceleration voltage detuning $|U - U_0|$ of the electron beam for an energy of the ions of 500 MeV/u. The electrons are described by a completely flattened velocity distribution with a temperature of 0.2 eV.

emission shows a maximum if both beams are completely tuned. This is shown in in Fig. 16 which presents the REC recombination coefficient for a flattened velocity distribution as function of the relative energy in the co-moving frame (lower scale). The connection between the relative kinetic energy E_e in the co-moving frame and the acceleration voltage $|U - U_0|$ in the lab. system (U_0 is the acceleration voltage of the electron beam for complete tuning) is found by a Lorentz-transformation:

$$|U - U_0| = U_0 \frac{\beta}{\beta_0^2} \frac{\gamma + 1}{\gamma_0} \sqrt{\frac{2E_e}{mc^2}} \approx U_0 \frac{\gamma_0 + 1}{\beta_0 \gamma_0} \sqrt{\frac{2E_0}{mc^2}}. \tag{53}$$

Note that even for small E_e the detuning $U - U_0$ in the lab. system becomes rather large. This 'scale-spreading' can be very helpful also for experiments in which transitions of electrons (bound-bound or bound-free) are studied.

A much more detailed diagnosis of both the beams can be performed by a energy-resolved measurement of the REC photon spectrum. The energy- and angle-differential REC recombination coefficient has been calculated by us for different distributions of the electron velocity [29]. Instead of presenting here the somewhat lengthy analytical expressions which can be found in Ref. 29, we show in Fig. 17 the photon-energy and emission-angle differential REC recombination coefficient as function of the photon energy for Ar^{+18} (REC into the ground state). A fixed transverse electron temperature of 0.2 eV and the two longitudinal temperatures of $2 \cdot 10^{-3}$ eV and $2 \cdot 10^{-5}$ eV, respectively, were used for the calculation. The emission angle considered is $\Theta = 0^0$ which corresponds to photon emission parallel to the direction of the beams. Note that the energy scale still has to be transformed into the lab. system (which results in a blue shift for $\Theta = 0^0$) and that E_0 denotes the 1s binding energy for Ar^{+17}. Obviously, the REC photon spectrum associated with this recombination coefficient

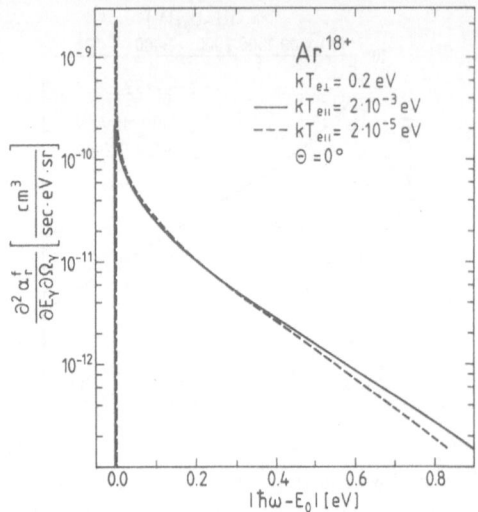

Figure 17. Photon energy- and emission-angle differential REC recombination coefficient for Ar^{+18} at a transverse electron temperature of 0.2 eV and at longitudinal temperatures of $2 \cdot 10^{-3}$ and $2 \cdot 10^{-5}$ eV as a function of the photon energy $|\hbar\omega - E_0|$. E_0 denotes the $1s$ binding energy of Ar^{+17}. The photon detector is placed at $\Theta = 0^0$ in the reference frame of the emitting ion.

shows a very interesting dependence on the photon energy. A sharp edge just at the binding energy E_0 and a fast decrease with increasing photon energy leads to the spike at E_0 whose height and width are determined by $T_{e\parallel}$ (the width is about equal to $T_{e\parallel}$). At higher photon energies the spectrum decays almost exponentially with a decay constant which is determined by the transverse electron temperature.

Clearly, from a precise measurement of such a REC spectrum not only the binding energy of the state into which the electron has been captured, but also the transverse electron beam temperature can be determined. The latter point which is important for the diagnosis of the beams, should be attacked predominantly at low Z ions, because the width of the spectrum is (independent of Z) $\sim kT_{e\perp}$ and the resolution of the spectrometer needed to observe the small asymmetry of the spectrum might be high enough only at small REC energies. Let us finally note that the finite emittance and velocity spread of the ion beam (its temperature T_i) can be taken into account by replacing $T_{e\parallel}$ by $T_{e\parallel} + \frac{m}{M}T_{i\parallel}$ and $T_{e\perp}$ by $T_{e\perp} + \frac{m}{M}T_{i\perp}$, respectively [29]. Thus, the ion temperature broadens the spectrum, so that a time and energy resolved measurement of the REC spectrum enables a determination of the time dependence of the ion beam temperature and thereby of the cooling time. Measurements of such spectra will be an intriguing task for the next future and may bring new data which are of importance for atomic, plasma, and accelerator physics.

References

[1] H. Bruck, Accélérateurs Circulaires de Particules, Presses Universitaires de France, Paris, 1966

[2] W. Scharf, Particle Accelerators and Their Uses, Part 1 and Part 2, Harwood Academic Publishers, Chur, Switzerland, 1986

[3] K. Steffen, in: CAS, CERN Accelerator School, Gif-sur Yvette, Paris, France, September 3 – 14, 1984, ed.: P. Bryant, S. Turner, CERN-Report CERN 85-19

[4] P. Schmüser, in: CAS CERN Accelerator School, Scanticon Conference Center, Aarhus, Denmark, September 15 – 26, 1986, ed.: S. Turner, CERN-Report CERN 87-10

[5] R.L. Glückstern, in: Frontiers of Particle Beams, Proceedings, South Padre Island, Texas, 1986, ed.: M. Mouth, S. Turner, Springer Verlag Heidelberg, 1986, p. 87; A. Hofmann, ibid., p. 99

[6] B. Franzke, Information about ESR Parameters, GSI-ESR-TN/86-01 (1986), and private communication

[7] E. Keil, W. Schnell, CERN Report ISR-TH-RF/69-48 (1969)

[8] Proc. Int. Conf. on High Energy Accelerators, Stanford, 1974, p. 405

[9] R. Mayer-Prüßner, I. Hofmann, GSI Scientific Report 1985, GSI 86-1, 1986

[10] I. Hofmann, private communication

[11] G.I. Budker, Proc. Int. Symposium on Electron and Positron Storage Rings, Saclay, France, 1966 (Presses Universitaires de France, Paris, 1967)

[12] G.I. Budker, N.S. Dikansky, V.I. Kudelainen, I.N. Meshkov, V.V. Parkhomchuk, D.V. Pestrikov, A.N. Skrinsky, B.N. Sukhina, Part. Accel. $\underline{7}$, 197 (1976)

[13] Ya. Derbenev and A.N. Skrinsky, Part. Accel. $\underline{8}$, 1 (1977); G.I. Budker and A.N. Skinsky, Sov. Phys. Usp. $\underline{21}$, 277 (1978)

[14] ECOOL 1984, Proc. of the Workshop on Electron Cooling and Related Applications, ed.: H. Poth, Kernforschungszentrum Karlsruhe, KfK 3846, Juli 1984

[15] A.H. Sørensen and E. Bonderup, Nucl. Instr. Meth. $\underline{215}$, 27 (1983)

[16] J.D. Jackson, Classical Electrodynamics, J. Wiley & Sons, New York (1967)

[17] Ya. Debenev and I.N. Meshkov, Studies on Electron Cooling of Heavy Particle Beams Made by the VAPP-NAP Group at the Nuclear Physcis Institute of the Siberian Branch of the USSR Academy of Sciences at Novosibirsk, CERN 77-08, April 1977

[18] G.I. Budker, A.F. Buchulev, N.S. Dikansky, V.I. Kononov, V.I. Kudelainen, I.N. Meshkov, V.V. Parkhomchuk, D.V. Pestrikov, A.N. Skrinsky, and B.N. Sukhina, New Experimental Results on Electron Cooling, Report on the Vth USSR National Conference on Particle Accelerators, Dubna (1976), Preprint INP 76-32 (translated at CERN by O. Barbalat, PS/DL Note 76-25)

[19] F. Nolden, Optimizing Parameters for Electron Cooling in a Wide Energy Range, GSI-ESR/86-02 (1986)

[20] N. Angert, H.F. Beyer, B. Franzke, B. Langenbeck, D. Liesen, F. Nolden, H. Schulte, P. Spädtke, and B. Wolf, GSI Scientific Report (1985) GSI 86-1

[21] M. Stobbe, Ann. d. Physik $\underline{7}$, 661 (1930)

[22] H.A. Bethe and E.E. Salpeter, Quantum Mechanics of One- and Two-Electron Atoms, Springer Verlag, 1957

[23] V.M. Katkov and V.M. Strakhovenko, Sov. Phys. JETP $\underline{48}$, 639 (1978); M. Bell and J.S. Bell, Part. Accel. $\underline{12}$, 49 (1982); D. Möhl and K. Kilian, Phase Space Cooling of Ion Beams, CERN-EP/82-214, 1982

[24] D.R. Bates, A.E. Kingston, and R.W.P. McWhirter, Proc. Roy. Soc. $\underline{267A}$, 297 (1962)

[25] J. Stevefelt, J. Boulmer, and J.F. Delpech, Phys. Rev. $\underline{A12}$, 1246 (1975)

[26] H.F. Beyer, D. Liesen, and O. Guzmann, On the total Recombination between Cooling Electrons and Heavy Ions, GSI-ESR-88-01, January 1988

[27] St. Byron, R.C. Stabler, and P.J. Bortz, Phys. Rev. Lett. $\underline{8}$, 376 (1962)

[28] Ya. B. Zel'dovich and Yu. P. Raizer, Physics of Shock Waves and High-Temperature Hydrodynamic Phenomena, Vol. I, Academic Press, NY, 1966

[29] D. Liesen and H.F. Beyer, On the Radiative Capture of Free Cooling Electrons: Total and Differential Recombination Coefficients, GSI-ESR/86-04, October 1986

PHYSICS OF THE EBIS AND ITS IONS

Robert W. Schmieder

Physical Science Department
Sandia National Laboratories
Livermore, CA 94551

INTRODUCTION

Low-energy/high-charge-state ions are extraordinary objects. Chemically, they are the most reactive species in the universe; they will locally destroy any piece of matter they come close to. Interaction of one such ion with any molecule will result in the nearly instant explosion of the molecule. They present almost a "black electrostatic hole" to electrons regardless of where they are found. A single bare Xe ion, Xe^{54+}, could in principle extract more than 10,000 electrons from a solid surface, more than 200 times the charge of the ion.

The atomic and interaction physics of these ions are intrinsically interesting, but the ions also are important in technology. They are constituents of hot plasma devices such as X-ray lasers and fusion reactors. They are sought for injection in heavy ion accelerators that must have the highest charge-to-mass ratio ions available. Trapped in confined spaces, they will readily undergo phase transitions to ordered states, showing a whole range of mechanical properties that could be important to designs of machines to study them. They are useful for diagnostics of plasmas and have potential for extending high-precision spectroscopy to higher energies.

In spite of their importance, it's not easy to produce these ions. Various laboratory sources have been, and are being, developed. Among these, the Electron Beam Ion Source (EBIS) has the potential to produce ions of higher charge state (while still keeping the ion energy low) than any other laboratory source. It accomplishes this feat by electrostatically trapping the ions for long times and subjecting them to repeated impacts by an electron beam. Recent developments in EBIS technology, including a "breakthrough" due to the introduction of ion cooling that helps keep the ions in the trap, suggest that heliumlike uranium, U^{90+}, at energies around 1 keV/Q will be obtained eventually, and that reaching bare uranium, U^{92+}, is feasible. However, the technical challenge is sufficiently great that so far most effort has gone into understanding the physics of the sources; we are just now beginning to actually use the sources to study atomic processes.

Because the physics of the sources and the physics of the ions that can be studied using these sources are so tightly linked, we tend to think of "EBIS Physics" as meaning both the physics of the source and the physics of the ions produced by the source. More importantly, in the absence of a friend with a source, anyone contemplating experiments with these ions has to face the daunting task of developing one. Therefore, we will review both the sources and their applications.

Our purpose here is both to summarize the current state of the EBIS art and to stimulate new ideas for the development and application of the EBIS. Since several reviews of the EBIS are available, we shall not attempt to be comprehensive. And since this is meant to be a working paper rather than an archival one, we will take the liberty, where appropriate, and with appropriate disclaimers, to be slightly speculative or suggestive. Time will tell whether any of these ideas will bear fruit.

Historical Background and Related Devices

The EBIS was developed by Donets and co-workers in the Soviet Union beginning in the late 1960's [1]. After two decades of development, Donets apparently has succeeded in producing Xe^{54+} and possibly higher charge states of heavier elements. The outstanding success of the Dubna group has stimulated development of these devices in many laboratories. Donets has provided a series of reviews of the EBIS [2, 3, 4, 5, 6].

Pioneering work was done by Arianer and co-workers in developing the CRYEBIS at Orsay [7, 8, 9].

Every few years a workshop is held that brings together the people actually building EBISs. The proceedings of these workshops have been informally published [10, 11, 12, 13].

The idea of trapping ions to obtain high charge states is not new. One of the early schemes was the HIPAC [14], in which an electron cloud was confined by external magnetic fields. Ions would be trapped in the potential well of the electron cloud, and would reach high charge states after a long "cooking" time. This device proved unworkable due to vacuum problems. A device very closely related to the EBIS was the Electron Ring Ion Trap (ERIT), developed principally as a spectroscopic source [15, 16, 17, 18, 19]. In this device, ions were trapped in the potential well of a toroidal ring of electrons confined by a transverse magnetic field. Extensive modelling and preliminary experiments showed that the ring would generate ions up to Xe^{54+}, but the scheme proved unworkable because instabilities destroyed the ring in times much less than the times necessary to reach the high charge states.

Another device closely related to the EBIS is the Electron Beam Ion Trap (EBIT). developed at Lawrence Livermore Laboratory (LLNL) by Levine and Marrs [20]. This device is essentially a short EBIS: it traps the ions in a very short, very intense electron beam. Already it has shown remarkable success in producing spectra of ions up to Au^{69+}. The importance of this new device can hardly be overstated, since it is a small and relatively inexpensive means for studying very high charge state ions. It is, however, principally a spectroscopic source rather than an ion source, hence will be limited in its ability to provide extracted ions for beam-target experiments. The EBIT and EBIS are really complementary: the former emits photons, the latter emits ions. In this review, we shall include remarks about the EBIT when they are germain to the EBIS.

Electron Cyclotron Resonance Ion Sources (ECRIS) have proven to be reliable and to produce large amounts of high charge state ions [21]. However, because the electron cloud is not as hot as the electron beam in the EBIS, the ECRIS cannot reach as high charge states as the EBIS. Furthermore, although the extracted currents are large, the beam quality is poor compared with the EBIS (in which essentially all the ions are extracted in a narrow pencil beam). Thus, although the ECRIS is being widely used and is certainly useful for certain atomic physics experiments, it cannot compete with the EBIS when the highest charge state and beam quality are crucial.

Another important recent development is the production of low-energy/high-charge-state ions by recoil from heavy ion impact [22]. This source potentially is capable of competing with the EBIS: typical is ions of iodine up to I^{40+} by impact of 2 GeV U^{75+} on HI gas [23]. The energies of the recoil ions are very low; they are emitted at 90° from the beam, and can be extracted and trapped [24, 25]. The charge state spectrum is quite similar to an EBIS operated in DC mode and therein lies the difference: with the EBIS one has the ability to confine the ions for longer times, and the distribution peaks at higher charge for longer time. And of course, to generate recoil ions one needs a high energy heavy ion accelerator ahead of the target, typically a 10-20 person operation; the EBIS is a 2-3 person device and is potentially capable of producing more ions.

PHYSICAL PRINCIPLES OF THE EBIS

The EBIS, shown schematically in Fig. 1, makes use of a magnetically confined electron beam with energy typically several times the ionization energy of the ions, i.e., from several keV to several hundred keV. Ions are electrostatically trapped in and around the beam, radially by the electron beam itself and axially by a set of electrodes ("drift tubes"). The trapped ions are subjected to repeated impacts by beam electrons, and proceed in time to higher charge states. Eventually some process such as recombination or ion loss prevents further ionization, and an equilibrium is established.

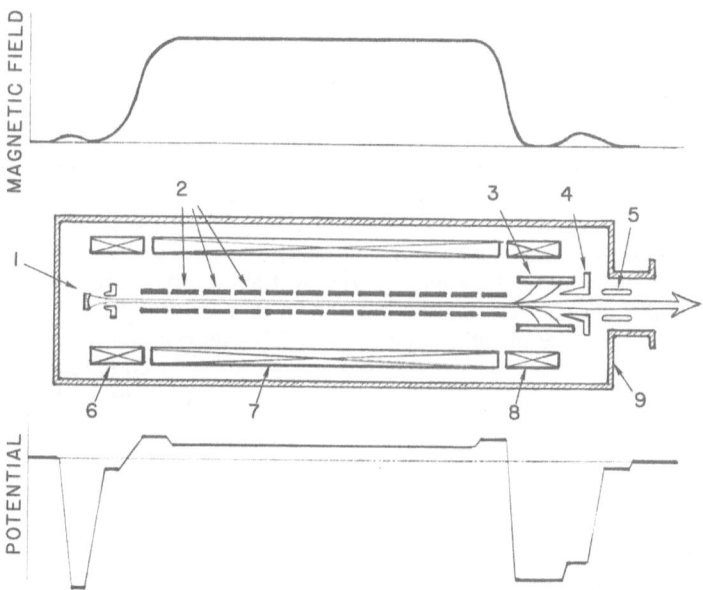

Fig. 1.   Schematic of an EBIS. (1) electron gun; (2) drift tubes; (3) beam collector; (4) ion extractor; (5) ion transport optics; (6) gun bucking magnet; (7) main solenoid; (8) collector bucking magnet; (9) ultrahigh vacuum chamber. (Above) The axial magnetic field; the small side lobes are due to the bucking coils. (Below) The voltages applied to the various elements. The potential experienced by the ions on the axis varies smoothly between these limits, except that the presence of the electron beam makes the potential slightly more negative.

Table 1. Levels of Approximation to the EBIS

Zeroth	PERFECT EBIS	Single isolated ions in uniform electron beam flux
First	CLASSICAL EBIS	Equilibrium between ionization and neutralization by residual gas; static uniform distributions of ions of fixed energy
Second	DISTRIBUTED EBIS	Equilibrium nonuniform spatial distributions of various particles with fixed energies
Third	PLASMA EBIS	Dynamics; collective effects of many particles; time dependent abundances and distributions particle heating and cooling; beam instabilities; plasma effects
Last	REAL EBIS	Experimental device

This device, as with all plasma devices, is extremely complex, and complete models of real devices are not yet available. We can, however, identify several levels of approximation that provide simplified models; these are summarized in Table 1. Depending what the question is, one model may be sufficient, or it may be necessary to invoke a more complicated one.

The model of the Perfect EBIS is useful as a guide to the fundamental lower limit of trapping time. Most past modelling has been based on the equilibrium model of the Classical EBIS. It is only very recently that the spatial distributions of ions, and ion heating and cooling, have been considered, i.e., the Distributed EBIS. Following preliminary experiments at Lawrence Berkeley Laboratory (LBL), work at LLNL and Sandia National Laboratories (SNLL) has shown that the trapping of heavy ions can be enhanced by the presence of light impurity ions. This description of the beam/trapped-ion system as a plasma (rather than as independent particles in an external field) is leading to new understanding and significantly better performance of the EBIS. The net result is clearly a "breakthrough" in EBIS development.

The Perfect EBIS ("0th approx.")

Neglecting all other effects except

* Trapping of the ions, and
* Ionization of ions by impact of beam electrons,

all the electrons will be stripped from the trapped ions, mostly one at a time. Given enough time, all the ions will be reduced to bare nuclei. This device could be called a Perfect EBIS, and could be considered a "zeroth" approximation to a Real EBIS. This approximation actually will be a reasonable model for times very much less than the time of electron pickup from the residual gas.

The product of the mean time $\tau(Q)$ for an ion to reach charge state Q when irradiated with a constant flux J of electrons of energy E producing single-step ionization events is

$$J\tau(Q) = \sum_{q=0}^{Q-1} \frac{1}{\sigma_i(q,q+1)}$$

This relation is useful for getting a rough idea of the engineering requirements for generating any desired ions. The ionization cross sections $\sigma_i(q,q+1)$ can be estimated from any of several semi-empirical formulas [26]; that of Lotz is most often used. The energy of the electron beam is most logically selected at the peak of the ionization cross section, typically about twice the threshold energy. The beam flux J depends on the gun cathode and the profile of the magnetic field into which the beam is injected. Given J, the above relation gives the times necessary to trap the ion to reach any desired charge state.

Table 2.   Ions Attainable with a Perfect EBIS

	Ne Z=10	Ar Z=18	Kr Z=36	Xe Z=54	Au Z=79	U Z=92
0 Bare	Ne^{10+} 2×10^{20} 3 keV 7 ms	Ar^{18+} 2×10^{21} 9 keV 67 ms	Kr^{36+} 3×10^{22} 40 keV 1 s	Xe^{54+} 2×10^{23} 80 keV 7 s	Au^{79+} 6×10^{23} 180 keV 20 s	U^{92+} 2×10^{24} 300 keV 67 s
2 He-like	Ne^{8+} 8×10^{18} 0.6 keV 0.3 ms	Ar^{16+} 1×10^{20} 2 keV 3 ms	Kr^{34+} 2×10^{21} 7 keV 67 ms	Xe^{52+} 2×10^{22} 20 keV 0.7 s	Au^{77+} 6×10^{22} 45 keV 2 s	U^{90+} 2×10^{23} 70 keV 7 s
10 Ne-like		Ar^{8+} 3×10^{18} 0.3 keV 0.1 ms	Kr^{28+} 3×10^{20} 4 keV 10 ms	Xe^{44+} 2×10^{21} 8 keV 67 ms	Au^{69+} 6×10^{21} 17 keV 200 ms	U^{82+} 3×10^{22} 30 keV 1 s
18 Ar-like			Kr^{18+} 1×10^{19} 1.5 keV 0.3 ms	Xe^{36+} 2×10^{20} 5 keV 7 ms	Au^{61+} 1×10^{21} 12 keV 33 ms	U^{74+} 5×10^{21} 20 keV 167 ms
36 Kr-like				Xe^{18+} 6×10^{18} 1 keV 0.4 ms	Au^{43+} 1×10^{20} 4 keV 3 ms	U^{56+} 7×10^{20} 7 keV 23 ms
54 Xe-like					Au^{25+} 2×10^{19} 1.5 keV 0.7 ms	U^{38+} 6×10^{19} 4 keV 2 ms

Key: Ion
    $J\tau$ [e^-/cm^2], requirement from ionization cross sections
    E beam energy, about 2x ionization threshold
    τ confinement time, assuming $J=3\times10^{22}$ e^-/cm^2
    Ref. [6], except Au, which is interpolated.

Typical values of $J\tau(Q)$, the peak beam energy E, and ionization times are listed in Table 2. To get the ionization times, we assumed a beam with about 5000 A/cm^2, a rather high but not unobtainable value. From this table, we can see that confinement times of some seconds are necessary to reach high charge states of heavy ions. This fundamental requirement will run headlong into the limit imposed by residual gas in the trap.

The Classical EBIS ("1st approx.")

The Perfect EBIS approximation fails due to the presence of residual gas in the vacuum system, which causes:

* Neutralization of the ions by electron pickup from residual gas
* Ion generation from residual gas
* Ion charge changes due to charge exchange with other ions or free electron capture.

The result is an equilibrium in which the average charge state is determined by the competition between the ion ionization rate and the ion neutralization rate. This device could be called a Classical EBIS. It represents a "first" approximation to a Real EBIS, and is the most widely used model. The challenge in building such a source is to maximize the ionization rates while minimizing the neutralization rates. In practice, this means generating extremely intense, high energy electron beams and implementing an extraordinarily good vacuum (well below 10^{-10} Torr). This approximation still neglects loss of the ions from the trap, which can be due to several causes.

The equilibrium charge distribution is easily determined from the equilibrium rate equation:

$$\frac{n(Q+1)}{n(Q)} = \frac{n_e \, \sigma_i \, (Q,Q+1) \, v_e}{n_0 \, \sigma_c(Q+1,Q) v_{Q+1}}$$

where $n(Q)$, n_e, n_0 is the number density of ions of charge Q, of the beam electrons, and of the residual gas, respectively; $\sigma_c(Q+1,Q)$, $\sigma_i(Q,Q+1)$ is the cross section for electron pickup from neutral gas and for electron impact ionization, respectively; and v_{Q+1}, v_e is the velocity of ions of charge Q+1 and of electrons, respectively. The equilibrium charge state can be estimated by setting the ratio $n(Q+1)/n(Q) = 1$, and calculating the right-hand-side of the above relation for increasing Q until it exceeds 1. Alternatively, for any desired charge state, we can calculate n_0 as the upper limit to the residual gas density (i.e., pressure) which can be tolerated.

The residual gas causes another problem: it contributes light ions that are trapped in the beam, neutralizing it and ruining the electrostatic trapping after a finite time, called the "compensation time":

$$t_c = [\sigma_{i0}v_e n_0]^{-1},$$

where σ_{i0} is the cross section for electron impact ionization of the residual gas. If we set t_c equal to the ionization time $\tau(Q)$ to produce charge state Q, we get another limit on the residual density n_0. The lower value of n_0 is taken as the requirement for the EBIS.

The Distributed EBIS ("2nd approx.")

The Classical EBIS approximation fails because the spatial distributions of the ions and beam electrons are nonuniform. The ions oscillate within the potential well, both radially and axially. As they execute their orbits, they see a varying potential and varying electron beam current density. Thus, the evolution of the charge states is not simply calculable from the average densities. For instance, it may be necessary to account for trajectories that take the ions outside the beam; trajectories are possible in which the ion is bound to the beam but rarely enters it. Thus, we must account for

* Nonuniform spatial distribution of ions, including halo trajectories, with consequent changes in ionization and neutralization rates.

In addition, other effects can occur:

* Ion loss from the trap due to wall losses and extraction.
* Ion gain due to injection.
* Modification of the electrostatic potential by local accumulation of secondary electrons.
* Compensation of the self-repulsion of the electron beam by accumulating ion charge, causing beam shrinkage.
* Concentration of ions and low energy electrons nonuniformly along the beam due to changes in potential.
* Radial focussing of the ions by the magnetic field.
* Reflux of neutral gas from structures due to X-ray and ultraviolet photons and pulsed beams.
* Ion loss due to drift after neutralization.

These various effects, mostly associated with spatial distributions of particles of fixed energies, could be considered a "second" approximation to equilibrium EBIS behavior. We will call this device the Distributed EBIS. Understanding the EBIS in this approximation can suggest techniques for enhancing its performance, such as controlling the axial drift tube potential to compensate for nonuniform magnetic field and ion space charge.

Ion Trajectories. In general, the motion of an ion bound to the beam is a rapid precessing transverse oscillation around the beam center, plus a relatively independent bouncing between the end barriers. The radial frequency is of the order $\omega_1 = 300 \sqrt{n_e/M}$, where n_e is the beam electron density and M is the mass (in AMU) of the ion. For a typical EBIS, this is about 10-100 MHz. The ions rotate about the beam axis with the Larmour frequency $\omega_2 = QeB/2M$, which is of the order of 1-10 MHz. The axial oscillation frequency is about $\omega_3 = v_z/2L$, where v_z is the ion axial velocity and L is the trap length. This frequency will range up to perhaps 1 MHz. Since the ions typically are confined for seconds, it is obvious that they undergo many oscillations during their confinement, and their phases are essentially random.

We assume that the trap in the EBIS is very long and very thin, and approximate the electrostatic potential as the sum of a radial and an axial part, which are independent:

$$V(r,z) = \Phi(r) + \Psi(z).$$

The hámiltonian is then

$$H = (1/2M) \left[P_r^2 + (1/r^2)\left(P_\theta^2 - \frac{1}{2}QeBr^2\right)^2 + P_z^2 \right] + Qe\Phi(r) + Qe\Psi(z)$$

The trajectories $r=(r,\theta,z)$ are easily found using Hamilton-Jacobi theory in classical mechanics.

$$\phi_1 = \omega_1 t = M \omega_1 \int dr \ 1/\sqrt{S_{12}(r)}$$

$$\phi_2 = \omega_2 t = \theta + J \int dr \ 1/r^2 \sqrt{S_{12}(r)}$$

$$\phi_3 = \omega_3 t = M \omega_3 \int dz \ 1/\sqrt{S_3(z)}$$

where

$$S_{12}(r) = 2M[W - Qe\Phi(r)] - (1/r^2)[J - (1/2)QeBr^2]^2$$

$$S_3(z) = 2M[K - Qe\Psi(z)]$$

These trajectories are functional relationships between r and the constants of the motion, which we write as (E,ϕ): $E=(W,J,K)$, where W is the transverse kinetic energy, J is the angular momentum around the axis, and K is the axial kinetic energy; $\phi =(\phi_1,\phi_2,\phi_3)$ $= (\omega_1 t, \omega_2 t, \omega_3 t)$, where $\omega_1, \omega_2, \omega_3$ are the radial, azimuthal, and axial oscillation frequencies. The velocities v are found from the trajectories by differentiation with respect to time.

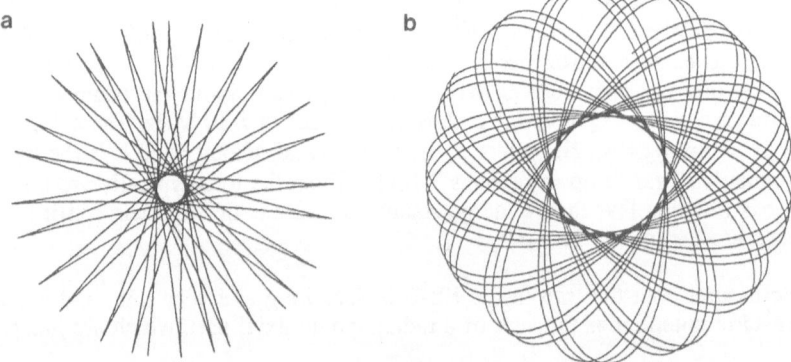

Fig. 2.   Typical radial trajectories of ions trapped in an electron beam. It is assumed that the ions are created inside the beam, which has uniform current density. (a) A $^{131}Xe^{8+}$ ion created with zero initial kinetic energy. Beam energy 50 keV; beam density 6×10^{11} e⁻/cm³; magnetic field 5 T; the radial amplitude of the motion is 0.01 cm. (b) A $^{131}Xe^{1+}$ ion created with 50 eV kinetic energy and large angular momentum. The radial amplitude of the motion is 0.025 cm. In both cases, the ion attempts to fall radially to the center of the beam, but is deflected away from the axis by the axial magnetic field.

Fig. 2 shows typical trajectories of ions trapped on an electron beam of uniform current density. The motion is a combination of the radial oscillation in the electrostatic field of the beam, on which is superimposed the azimuthal cyclotron motion of the ion around the magnetic field. Several points are noteworthy:

(1) The ions are trapped to the beam even when their trajectories take them outside it;

(2) Ions created at rest with no angular momentum never pass through the axis. They begin falling toward the axis, but the magnetic field deflects them and they miss it. The result is a quasi-periodic multi-cusped rosette trajectory;

(3) Ions with angular momentum parallel to the magnetic field will be deflected more away from the axis, while ions with angular momentum antiparallel to the magnetic field will be deflected less away from the axis. Ions with just the right antiparallel angular momentum pass through the axis, while those with more pass outside it, enclosing it within the trajectory.

Distributions. The calculation of the ion distributions can be formulated in a very general and powerful way, using classical hamiltonian mechanics [27]. The technique is to assume they are uniformly distributed in the phase of their periodic motion and use a series of transformations to convert this distribution to distributions over space, velocity, energy, speed, etc. The transformations are generated by calculating the trajectories of individual ions in the electrostatic and magnetic fields.

Let $dn = G(r,v)drdv$ be the number of ions of given charge Q with position r and velocity v in the phase space volume $drdv$. Since the distributions can be parametrized by (E,ϕ) just as well as by (r,v), we define a distribution $F(E,\phi)$ such that the number of ions $F(E,\phi)dEd\phi$ with E,ϕ in $dEd\phi$ is equal to $G(r,v)drdv$. The two distributions are then related by

$$F(E,\phi) = J(rv/E\phi)G(r,v)$$

where $J(.../...)$ is the Jacobian

$$J(rv/E\phi) = \begin{vmatrix} dr/dE & dr/d\phi \\ dv/dE & dv/d\phi \end{vmatrix}$$

which can be calculated directly from the trajectories $r(E,\phi)$, $v(E,\phi)$. Note that J is a 6x6 matrix. Now we transform the phase ϕ to another variable s:

$$F(E,s) = J(E\phi/Es)F(E,\phi) = |d\phi/ds|_{Es}\ F(E,\phi)$$

where the subscript Es emphasizes that J is written explicitly in terms of E and s. Now we define a reduced distribution over the variable s alone:

$$f(s) = \int F(E,s)\ dE\ .$$

Finally we impose the requirement that the ions undergo many oscillations, i.e., that the trajectory phases are uniformly distributed:

$$F(E,\phi) = F(E)\quad .$$

Using this we find

$$f(s) = \int \left| \frac{d\phi}{ds} \right|_{E_s} F(E) \, dE$$

This result allows us to calculate the distribution of the ions in any variable s by making an assumption about the distribution in energy. The assumption of random phases is built into this result.

Because of the separability of the transverse and axial motions, the distribution $F(E)$ separates into transverse and axial parts $F_{12}(E_1, E_2) F_3(E_3)$; the reduced distributions separate similarly:

$$f(s) = f_{12}(s_1, s_2) f_3(s_3)$$

where

$$f_{12}(s_1, s_2) = \int \int \left| \frac{d\phi_1}{ds_1} \frac{d\phi_2}{ds_2} - \frac{d\phi_1}{ds_2} \frac{d\phi_2}{ds_1} \right|_{WJs_1s_2} F_{12}(W, J) \, dW dJ$$

$$f_3(s_3) = \int \left| \frac{d\phi_3}{ds_3} \right|_{Ks_3} F_3(K) \, dK$$

These relations can be used to find other distributions starting from an assumed distribution in energy and angular momentum. For instance, we can calculate the distribution in energy $s_3 = K'$ resulting from collisions of various types, and distributions in space and velocity alone.

A simple but interesting specific case is that of ions trapped inside a beam of uniform current density. If we neglect the angular momentum ($J=0$) and the cyclotron motion ($B=0$), the radial potential well is harmonic, $\Phi(r) = (1/2) ar^2$, and the ion motion is pure harmonic radial oscillation with frequency $\omega_Q = \sqrt{Qea/M}$. In this case, the double integral $f_{12}(s_1, s_2)$ reduces to a single integral, which can be evaluated for various cases. First, given a collection of ions with a distribution $F_Q(R)$ of amplitudes R, the radial spatial distribution will be

$$G_Q(r) = \frac{2}{\pi} \int \frac{1}{\sqrt{R^2 - r^2}} F_Q(R) \, dR$$

and the radial velocity distribution will be

$$H_Q(v) = \frac{2}{\pi} \int \frac{1}{\sqrt{\omega_Q^2 R^2 - v^2}} F_Q(R) \, dR$$

If we start with a distribution $F_Q(R)$ and some process (such as electron impact ionization) changes the ions to charge state Q', the two radial amplitude distributions will be related by

330

$$F_{Q'}(R') = \frac{2}{\pi} \int \frac{R'}{\sqrt{[R^2 - R'^2][R'^2 - (Q/Q')R^2]}} F_Q(R)\,dR$$

This relation allows us to infer what happens to the radial distributions as the ionization progresses toward higher charge states. Indeed, it can be shown that the mean-square-radius of the distribution is

$$\langle R_Q^2 \rangle = [(2Q)\,!/\,2^{2Q}(Q!)^2]\,\langle R_0^2 \rangle \rightarrow 1/\sqrt{\pi Q}\,\langle R_0^2 \rangle$$

This clearly shows that, in the absence of other heating, successive ionization causes the ions to be concentrated more toward the axis. The same analysis shows that the mean radial velocity of the ions increases in proportion to $\sqrt{Q}$.

Some examples of radial distributions for xenon ions created at rest at one radius as the charge increases are shown in Fig. 3. The general trend is obvious: initially, the ions spend most of their time near the turning points, producing a distribution peaked at the amplitude. As Q increases, this peak is pulled inward and broadened.

The Plasma EBIS ("3rd approx.")

The Distributed EBIS approximation fails due to a variety of processes that alter the energies and/or abundances of the ions and electrons; for instance:

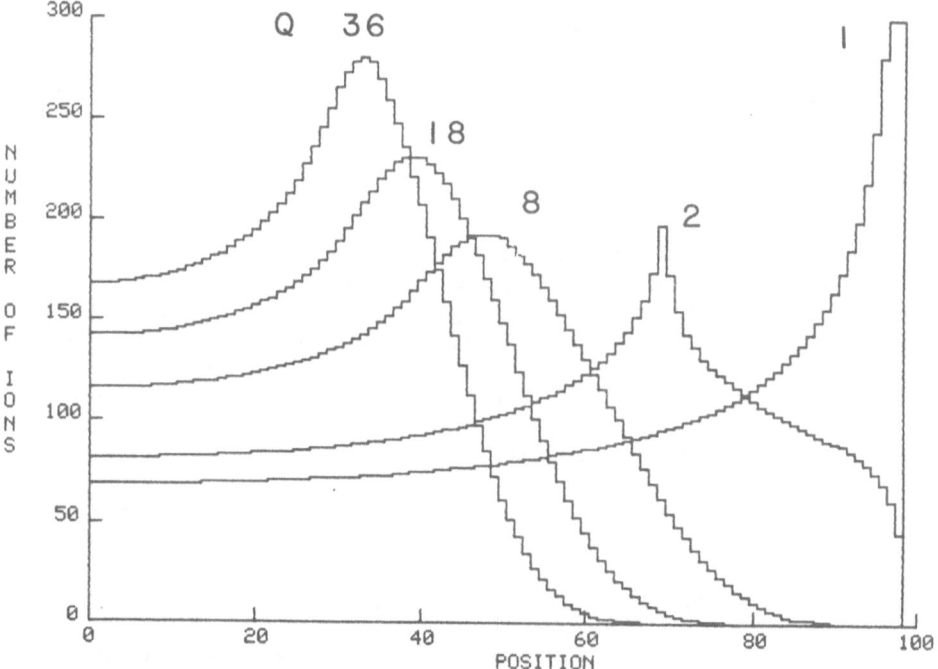

Fig. 3.   Radial distributions of xenon ions trapped in an electron beam, as the charge is increased by ionization. The Q=1 ions all have the same energy, hence the same radial amplitude; this produces the distribution peaked near the turning point. As the charge increased, the distribution becomes concentrated more toward the axis, and the spike at the peak amplitude is smoothed and broadened.

* Axial ion acceleration due to focusing of the electron beam in a nonuniform magnetic field.
* Ion heating and diffusion by electron impact, ion-ion collisions, collective processes such as beam instabilities.
* Ion cooling by similar processes.
* Extraction of residual gas due to ionization in axial field gradients.
* Change of the potential within the beam due to interaction with conducting electrodes such as drift tubes.
* Fluid instabilities.
* Microwave generation.

All these effects, most of which involve change of particle energies and collective fields, are described within classical plasma physics, and could be considered a "third approximation" to EBIS behavior. This device is called the Plasma EBIS.

During 1987-88, experiments at LLNL [20] and SNLL [28] indicated that heavy ions could be trapped in the EBIS far longer than expected based on the classical models. Since then, much data has been obtained that show that the enhanced trapping occurs when impurities are present; it is due to heating of light ions and cooling of the heavy ions. Apparently a dirty EBIS is better than a clean one! This wonderful circumstance is a kind of miracle that may well the allow these devices to live up to expectations developed from more naive models. In retrospect, it should have been öbvious that impurities would enhance the ion yield: it has been common knowledge in the ECRIS community that "magic" gas mixtures produce significantly enhanced yields of high charge states [29]. And what is good for the ECRIS may be as good for the EBIS!

As described above, a standard assumption of the classical model [3] is that the residual gas will be ionized by the beam, and the ions will accumulate inside the beam until the compensation time, when the ion charge equals the electron beam charge. At that time the beam is neutralized, and there is no longer a radial potential well. After compensation, the ions drift freely away, the steady state being maintained by the constant supply of residual gas. This means that ionization of previously trapped heavy ions will stop, freezing the charge state distributions.

That something is wrong with this picture is seen from the following data, taken from experiments with the LBL/MOD EBIS at SNLL: direct observations of the time dependence of the yield of residual nitrogen ions showed $t_c = 20$ ms, consistent with the known cross section for ionization of nitrogen gas at the measured pressure $P = 10^{-9}$ Torr. But ions of xenon gas bled into the trap were found up to at least Xe^{40+}. The time necessary to reach this charge state is $\tau(Q) = [\sigma (Q-1) v_e n_e]^{-1} = 10$ s for the 100 A/cm^2 beam. The mystery is how the Xe ions could be trapped for times so much longer than the compensation time. Extensive experiments at LLNL [30] have directly and spectacularly demonstrated trapping of heavy ions for very long times, up to several hours.

Ion Heating. It came as a surprise to EBIS workers that the compensation time could be exceeded. First hints that the classical model was inadequate came from experiments on ion heating [31] although several persons had remarked on the plasma properties of the EBIS [9, 32]

Ion heating alone can produce enhanced trapping of heavy ions, without invoking any cooling mechanism. Assume the bare beam produces a radial well Φ_0. In steady state with no heating, the beam is completely compensated by residual gas ions, and there is no radial well. If now the ions are heated, they will be driven out of the beam, leaving it less than fully compensated. This occurs because the residence time of the ions is $t_r = \Phi/(dE/dt)$, where Φ is the radial well depth and dE/dt is the heating rate. The fractional compensation is then $f = t_r/t_c$ and the radial well is $\Phi = (1-f)\Phi_0 = [1+(1/t_c dE/dt)]\Phi_0$. A small number of heavy ions can now be trapped in the reduced radial well for relatively long times, so long as their heating rates are much less than the heating rates of the residual gas ions. Since the heating mechanism could involve nonlinear plasma effects, the rates could be very sensitive to the ion species. Thus, we are led to conclude that complete compensation of the EBIS beam never occurs, and the compensation time limit can be greatly exceeded, so long as the number of trapped heavy ions is kept small.

Eventually, of course, all ions, including heavy ions, are heated and escape from the trap. Thus, we are led to the image that there is not a single time for reaching equilibrium in the EBIS, but several times characteristic of the different ions. Direct evidence for this is shown in Fig. 4, in which the residual gas ions are observed to be constant, but the Xe ions are still evolving to higher charge states after 1 s.

Direct evidence of residual gas ion heating is shown in Fig. 5, obtained from the LBL/MOD EBIS operated in the leaky mode [28]. Here, the energy of the extracted ions is plotted as a function of the barrier voltage. The presence of ions at energies above the drift tube potential, observed by allowing them to spill over the end barrier, shows that they can receive up to 20 eV or more before being lost.

Ion heating will be important not only to the production of high charge states, but also to any experiments that attempt to resolve energies of the order of the ion temperature. For instance, if we wished to decelerate Q=50 ions to study low energy impact phenomena, we would begin to lose part of them as we resolved their energy to less than the ion temperature.

The relative importance of various mechanisms of ion heating in the EBIS are not yet known. Several possible mechanisms are listed in the Table 3.

Certainly direct beam electron impact is one obvious heating mechansim. This is usually formulated as Landau-Spitzer heating. Its rate [33],

$$\frac{dE_i}{dt} = \frac{m_e}{M_i} (kT_e - kT_i) \, m_e \, \sigma_{ei} \, v_e$$

where the cross section

$$\sigma_{ei} = 4 \sqrt{2\pi} \left(\frac{Qe^2}{kT_e} \right)^2 \ln \Lambda$$

is proportional to Q^2. Clearly, this heating mechanism becomes more serious for high charge states. A widely quoted calculation of direct collisional heating by Becker [34] indicated heating would prevent attaining ions above about Q=50. Such calculations led to widespread pessimism in the future of the EBIS: many thought it would never produce really high charge state ions.

Ionization heating occurs when the charge state of an ion in a potential well is increased. The ion oscillates across the well between its turning points. If it suddenly changes from Q to Q'>Q when it is not at the bottom of the well, the position and kinetic energy are unchanged, but the depth of the potential well increases. When the ion again falls into the well, this increase appears as kinetic energy. Thus, as the ion charge rises, so does its mean kinetic energy in the trap. The analysis on distributions given above shows that the mean kinetic energy is proportional to Q. This kind of heating, however, does not lead directly to loss from the beam, since the ions are confined even more tightly as their charge increases. Rather, this mechanism will be important if this input of energy

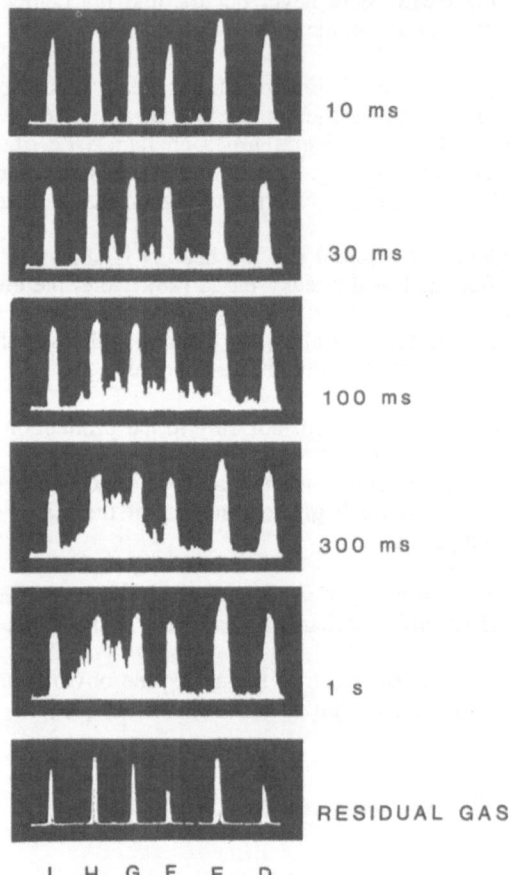

I H G F E D

Fig. 4. Demonstration that the classical compensation time of an EBIS can be exceeded. Residual gas ions reach equilibrium within about 10 ms, but the heavier xenon ions are still evolving toward higher states after 1 s, at which time the distribution peaks near Xe^{30+}. This is a magnetically analyzed spectrum: the residual gas peaks are: D=O^{2+}, E=N^{2+}, F=C^{2+}, G=O^{3+}, H=N^{3+}, I=C^{3+}. These spectra were obtained on the LBL/MOD EBIS, operated as a Gated EBIS by continuously bleeding in xenon gas and confining the ions for the indicated times. At bottom is a residual gas spectrum obtained with the EBIS operated as a Leaky EBIS. The fact that the peaks in the Gated spectra are much wider than in the Leaky spectrum is evidence that the ions are being heated in the trap, hence acquiring a distribution of energies. Data from Ref. [78].

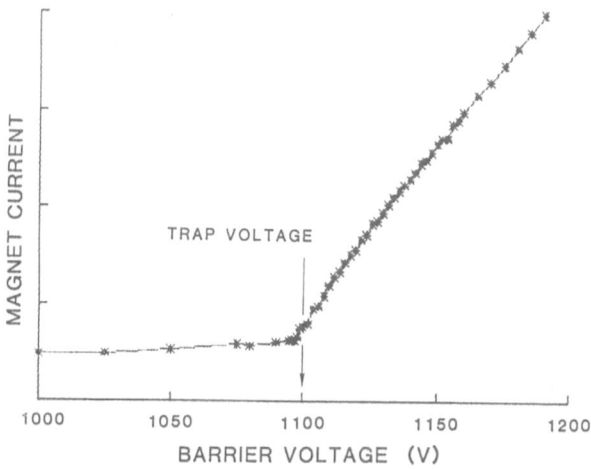

Fig. 5. Demonstration that ions from a Leaky EBIS are monoenergetic. The current in the analyzing magnet (which measures the ion momentum) is plotted as a function of the voltage on the barrier drift tube. When that barrier exceeds the trap voltage, the extracted ions have higher momentum. Ions measured were N^{3+}, peak H in Fig. 4. Data from Ref. [78].

Table 3.   Ion Heating Mechanisms in the EBIS

Landau-Spitzer	Beam electron-ion elastic collisions
Ionization	Increase of charge state of an ion in aa external potential, with conversion to kinetic energy
Plasma ionization	Increase of charge state of an ion in a plasma, with collisional conversion to kinetic energy
Plasma waves	Coupling of ion to coherentcollective fields of the plasma; waves may be generated by instabilities
Ion-ion collisions	Elastic energy transfer from heavy ions to light ions

can be converted to other forms, coupling to other particles, and coupled back on the ions themselves.

For ions in a plasma, there are analogous ionization heating mechanisms. One of these, called "Raman heating" [35] or "ion proximity heating" [36], results from the fact that the ion finds itself in the potential of another ion at a distance about the shielding radius R_s. When the ion charge jumps from Q to Q+1, it gains an energy

$$\Delta E \cong Qe^2 / R_s$$

which is converted to kinetic energy by ion-ion collisions. The heating rate is then

$$dE/dt = \Delta E \; n_e \; \sigma_{ii} \; v_e$$

where σ_{ii} is the cross section for ion-ion collisions. If there is an externally imposed potential Φ, the energy jump is altered:

$$\Delta E = e\Phi \left[1 - \exp \left(\frac{e\Phi}{kT_e} - \frac{e\Phi}{kT_i} \right) \right]$$

This predicts ion heating if $T_i < T_e$, no heating for $T_e = T_i$, and ion cooling for $T_i > T_e$. If $T_i \ll T_e$, the heating rate reduces to the rate for Raman heating.

Another heating mechanism is the coupling of plasma wave energy to the ions. The waves can be generated by instabilities, driven by the input of energy with the electron beam. For this to occur, the diameter of the beam must be larger than the Debye length, i.e., it must act like a plasma. This may well be satisfied for an EBIS beams, especially high-current beams. A variety of instabilities is possible [37], including Penning oscillation, backward-wave oscillators, and various two-stream instabilities. One of these latter, the modified rotational electron-ion two-stream instability, was shown [38] to have growth rates significant for an EBIS. In fact, the fear of such instabilities led Levine to propose making the EBIS very short, eventually developing the concept of the EBIT. Whether instabilities actually pose a significant limit on EBIS performance is still open to question, but it may be significant that the best results from several laboratories were obtained with relatively low currents (typically 15-25 mA), well below the limits for generating instabilities.

Ion Cooling. As shown so beautifully by Levine, Marrs, and others on the EBIT at LLNL [39], ion cooling is the key to making an EBIS work well. The results are nothing short of spectacular: Au^{69+} ions cooled by adding Ti impurities have been trapped for 4 hours! Since the compensation time for that system is of the order of seconds, these results dramatically demonstrate that the classical barrier of beam compensation can be exceeded.

The light ion impurities dramatically increase the intensity of the heavy ion X-rays emitted from the EBIT, as shown in Fig. 6 [30]. This is assumed to result from increased density, which in turn results from lower loss, presumably by heavy ion cooling.

The following simple physical picture emerges: energy enters the plasma with the electron beam. The beam ionizes the ions, which increases their radial potential energies. As the ions oscillate within the radial potential well, their potential energy appears as kinetic energy (ionization heating). Direct beam impact increases their transverse kinetic

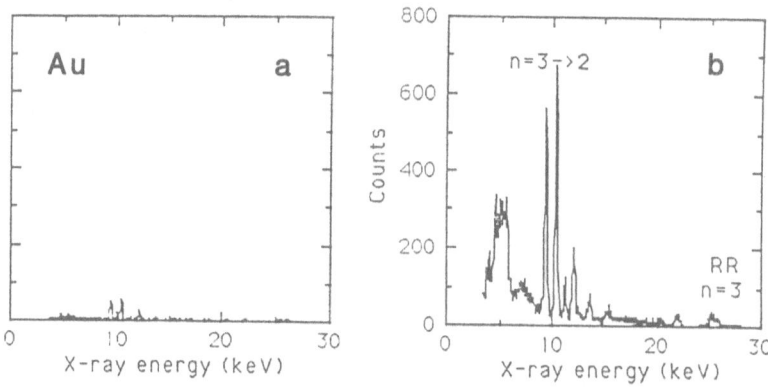

Fig. 6. Increase of X-ray luminosity of trapped ions due to cooling, which increases the
trap time, hence the number of trapped ions. (a) without cooling; (b) with cooling.
Data from Ref. [30].

energy (collisional heating). Ion-ion collisions cause the energetic heavy ions to transfer
energy to the low energy light ions. This cools the heavy ions, and they are confined
deeper in the potential well of the beam. Heating of the light ions drives them out of the
trap to the walls, and new cold ions are created by beam ionization of the residual gas.
The steady state is one of cold gas passing through a hot one, the former carrying away
thermal energy and cooling the latter.

In order to see these processes at work, we have simulated this system with a
Monte Carlo code [40]. The code follows the three-dimensional trajectories of a
numbered collection of ions moving in the collective field of the electron beam and all the
ions. Ions exchange energy in coulomb collisions; ions that gain energy get into
trajectories taking them outside the beam. Ions lost to the walls are replaced by thermal
residual gas ions. Charge states of the ions evolve according to the time spent in the
beam.

Typical results for a model system are shown in Fig. 7. It is seen that the beam
heating tends to drive ions of all types out of the trap; ultimately they would reach the
wall and be lost. Ion-ion collisions cool the high charge state ions, keeping them in the
trap longer.

The equilibrium charge state distributions, shown in Fig. 8, also exhibit the effects
of the ion-ion collisions.

Levine and co-workers [41] have also carried out computer calculations of
evaporative cooling in the EBIT. Their model includes essentially the same physical
effects as those just described (beam heating, ion-ion collisions, and ion wall loss), but is
a based on plasma approximations for the rates. Furthermore, they assume that the
heating is due to the transverse elastic momentum transfer from electron-ion collisions,
rather than ionization heating. This assumption will apply more to the very highest charge
states, for which the ionization proceeds very slowly and there is plenty of time for
collisional heating. They obtain a variety of results, including radial distributions, heating
rates, and ion temperatures. Together with the experimental results, these calculations
are quite convincing that the mechanism is roughly as described above, namely ionization
and collisional heating and collisional cooling of the heavy ions, collisional heating and
evaporation of the light ions.

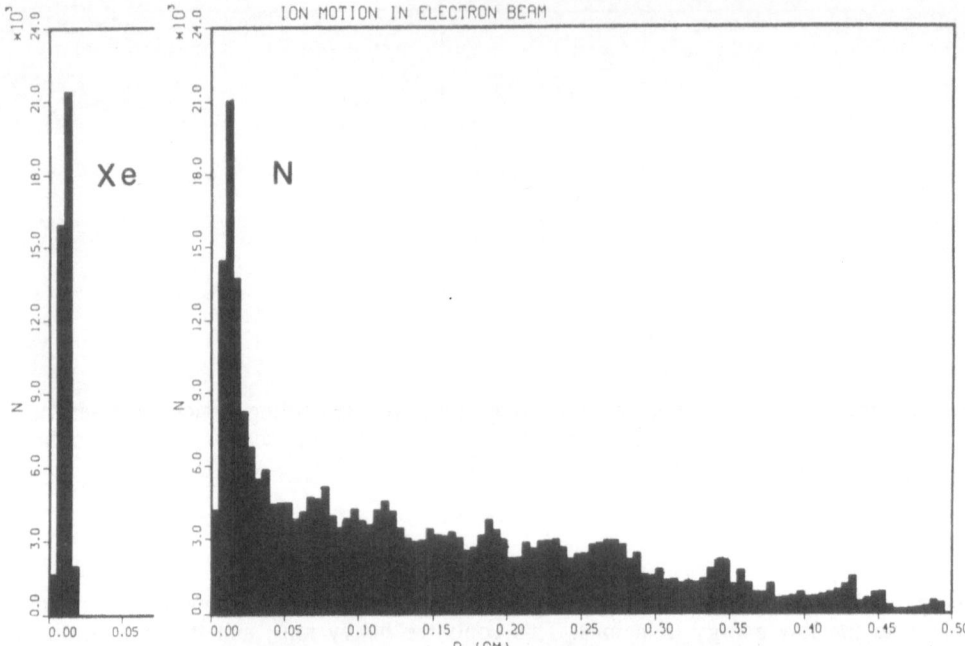

Fig. 7. Radial distributions of ions in a distributed EBIS. The light N^{7+} ions are spread into a thick sheath due to heating, but the Xe^{54+} ions are confined near the axis due to ion-ion collisional cooling. These calculations were done with a Monte Carlo code that assumes 400 N^{7+} and 100 Xe^{54+} ions are trapped in a 2 keV beam with J=25 A/cm^2 and radius a=0.018 drift tube radius is R=0.5 cm and axial barriers are at z=±60 cm. The magnetic field is 3 kG. Initially the ions are randomly distributed across the beam with thermal velocities.

The LLNL plasma code calculation has the advantage of being simpler and faster than the Monte Carlo trajectory code, but it does assume that the plasma rates are applicable to the very tenuous, tiny non-neutral "plasma" of the EBIS. Another advantage of using a plasma description is that various analytical results emerge. For instance, Lyneis [42] quotes the following formula for the confinement time of an ion in the ECRIS:

$$\tau \cong L^2 \sqrt{M} \, Q^2 \, Z_{eff}^2 \, n_i T_i^{-5/2}$$

where L is the trap length, M and Q are the ion mass and charge, Z_{eff} is the effective charge in the plasma, n_i is the ion density, and T_i is the ion temperature. If it really is true that what is good for the ECRIS is good for the EBIS, we might expect the confinement time to rise as $T_i^{-5/2}$, showing the strong advantage of even a little cooling.

Fig. 8. Evolution of xenon ion charge states in an EBIS. The first seven distributions include neutralization from residual gas, but no collisions (Classical EBIS). The next distribution includes beam ion heating, which reduces the trap time and therefore the mean charge state. The bottom distribution includes ion-ion collisions, which cause cooling of the ions, increase of the trap time, and increase of the mean charge. The conditions were chosen to reasonably simulate the LBL/MOD EBIS. From Ref. [40].

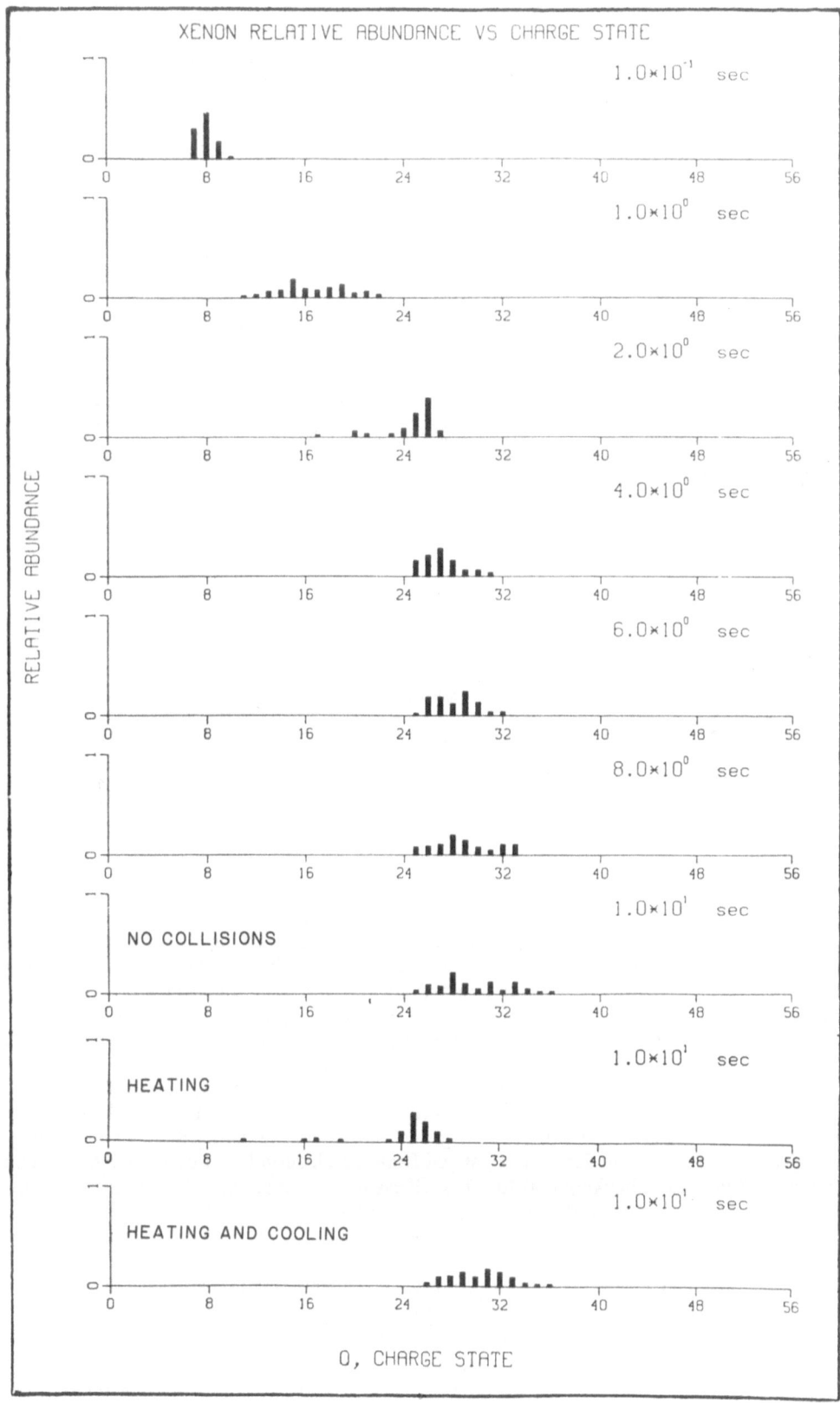

Ion heating and cooling probably ultimately limit performance in a Real EBIS that has good vacuum and good beam quality. The challenge is to minimize the ion heating and implement schemes for cooling the ions. In practice this means controlling the detailed space-time-species distributions in the trap, and possibly implementing clever schemes for selectively extracting energy, such as an external beam of cold particles. Furthermore, it might mean implementing external fields to dynamically control beam instabilities, or selectively extracting undesirable species (for instance, based on their charge-to-mass ratios). One thing seems clear: a slightly dirty gas in an EBIS is better than a very clean one!

The EBIS as a Plasma Device. It has been a major step forward to regard the EBIS as a plasma device. The extraction drift tube is quite similar to the extraction cathode in arc plasma ion sources. The light ions evaporated from the beam form a sheath around it. The similarity of the electron beam to a conducting wire also can be noted: low energy electrons that are either produced in the beam or wander into it are very rapidly heated by the "drag" force (elastic impact). Thus, the beam acts much like a positively biased wire that carries away any electrons that encounter it and also repels excess positive ions to form the sheath. It is perhaps surprising that it took so long to abandon the simple independent-particle models of the EBIS and think of it as a plasma device. But now that that has been done, we can hope that new engineering developments will help the EBIS reach its potential.

PRACTICAL EBIS SOURCERY

There is, of course, a vast difference between the physics of the EBIS as described in the previous section, and actual machines that have been built. The physicist engaged in building an EBIS for use in physics experiments must constantly play the engineering realities against his proposed experiments. An EBIS probably is too complicated to build for a single physics experiment. In this section we describe Real EBIS's, machines that have actually been built, and the kinds of physics experiments that can be done on them. The emphasis is on the interplay between the design/performance and the experimental requirement.

The "Typical EBIS"

A "Typical EBIS" has a solenoidal magnet about 1 m long generating an axial field of a few T which is extraordinarily straight and uniform. In the bore of the magnet is inserted a series of perhaps 20 drift tubes, each a few cm long and internal bore 5 mm. The drift tubes can be independently held at potentals of 1-3 kV. The electron gun generates a beam of energy 3-50 keV, which is compressed by the magnetic field to a diameter of a few hundred microns. The current is 10-100 mA, the current density 100-1000 A/cm^2. The vacuum in the main chamber is around 10^{-9} Torr, but may be less in the drift tubes. Ions are generated in the beam by bleeding a noble gas into one of the drift tubes. The ions are confined longitudinally by biasing the end drift tubes 20-50 V more positive than the trap drift tubes. They are extracted by reducing one end barrier, allowing them to exit axially through the drift tube. The ions are separated from the electron beam by reducing the axial magnetic field with a bucking coil to a few Gauss over several cm, allowing the electrons to expand by coulomb repulsion onto a water-cooled collector. The ions pass into an extraction drift tube and out to the transport optics.

This "Typical EBIS" has taken 3-5 years to design and build, and cost perhaps a million dollars, including overhead. It utilizes much standard accelerator technology such as high voltage and ultrahigh vacuum components. But it also includes some high precision components, such as the solenoid, that have advanced the state of the art. The design of the electron gun and beam collector, and matching them to the solenoid, has been a major effort. The internal components have been carefully designed not only to maintain precise axial symmetry, but also to withstand the rigors of thermal cycling during baking and cryogenic cooling. Because of the critical vacuum, the source is seldom opened; it is designed with sufficient redundancy and backup to live through normal component failures.

The "Typical EBIS" behaves as its designers expected, but much time is required to find the best combination of operating conditions. Much time is spent waiting for the vacuum to settle down after venting, and tuning the various electrical and ion- optical voltages. When tuned up, the xenon ions analyzed by a 90° bending magnet show a regular series falling off toward the highest charge states. The machine is stable over long times, and produces essentially the same results day after day. This machine will last 5 years, to be replaced by one producing greater beam density and better vacuum.

The Leaky EBIS

The simplest way to inject ions into the EBIS is by a continuous bleed of neutral gas through a leak valve. After the gas atoms are ionized, they are extracted continuously by allowing them to "leak" out axially over the end potential barrier. This mode is entirely DC. We call this device a Leaky EBIS [28].

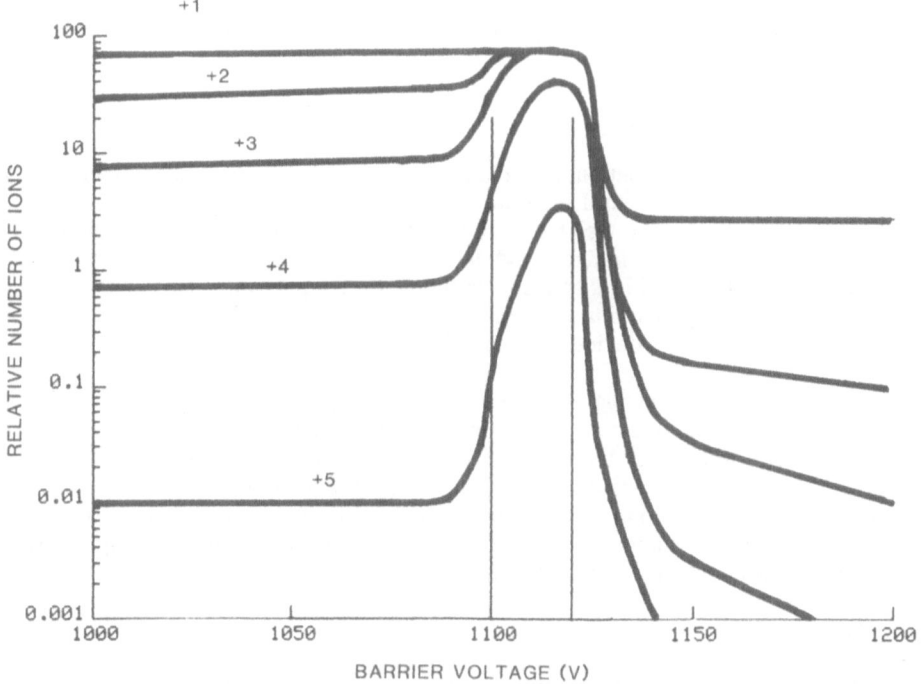

Fig. 9.   Demonstration of a Leaky EBIS. The numbers of carbon ions of different charges were measured as a function of the axial barrier voltage. The Q=1 ions are created from the residual gas, hence show no trapping effect with increasing barrier voltage. For higher charges, the barrier at first increases the trap time and therefore the yield of the ions, then cuts off the axial extraction, forcing the ions to drift radially to the wall. In these measurements, the ions were created in the trap at 1100 V, and the upstream barrier was set at 1120 V; these are indicated by the vertical lines. The trapping enhancement of the C^{5+} ions is about 300x. Data from Ref. [28].

Ions extracted from the Leaky EBIS are monoenergetic. A demonstration of this is shown in Fig. 5, in which the ion energy, measured as the current in the analyzing magnet, is plotted as a function of the exit barrier voltage. This result shows that so long as the barrier is less than the drift tube voltage (1100 V in this case), the ions leaking out are at 1100 eV/Q. But as the barrier is raised, only those ions that have been heated to the barrier voltage can leak out. As soon as the ions reach the barrier voltage, they leak out, with the result that the extracted ions are monoenergetic to within about 1 eV/Q.

The numbers of various residual gas ions extracted from the Leaky LBL/MOD EBIS are shown in Fig. 9. These data show clearly that the yields of ions of charge state Q=1 are not enhanced by the end barrier, but that higher charge state ions are enhanced due to their longer containment time. The greatest enhancement is found for the optimum compromise between increasing the confinement time and increasing the radial losses. After all, if the end barriers exceed the depth of the radial well, all ions *must* go to the walls. The yields of ions falls exponentially as the barrier is increased beyond 20 eV. This exponential tail is characteristic of a quasithermal distribution of ion energies. From the slope we can obtain the ion temperatures: about 6 eV for Q=1 and about 15 eV for ions with Q=3-6.

The Leaky EBIS is a compromise between no confinement and complete confinement/dump. It is an improvement over the time-of-flight TOFEBIS [43] in which the ions make only a single pass through the beam, and the charge states obtainable in this mode are surprisingly high: the LBL/MOD EBIS, a rather modest machine in terms of beam current density, has produced ions up to Xe^{40+}, near the ionization threshold for the beam energy. Much the same has been observed on the Frankfurt EBIS and on CEBIS I (Cornell) operating in the leaky mode.

The Gated EBIS

If the working gas is continuously bled into the EBIS, but the ions are held back by a relatively high axial barrier until they are suddenly released by dropping the barrier, we can enhance the confinement over the Leaky mode. We envision the cycle as starting with cold neutral gas which is ionized and heated by the beam, eventually reaching an equilibrium when the radial loss rate equals the neutral supply rate. This equilibrium will have higher mean charge and ion temperature than the leaky mode, since the confinement time is longer. We call this device a Gated EBIS.

The data in Fig. 4 showing the evolution of Xe charge states continuing for at least 1 s, in spite of the fact that the residual gas had reached equilibrium at 20 ms, were taken in the Gated EBIS mode. At the time these experiments were done, it was slightly mysterious how the gating could enhance the high charge states, since it was assumed that the 20 ms compensation time would stop the Xe evolution. With the realization, supported by this observation and the EBIT results, that heavy ion cooling allows the compensation limit to be exceeded, the Gated EBIS is quite understandable.

Pulsed EBISs

Another significant improvement is realized if the working gas can be injected as a pulse, rather than bled in continuously. Now the low charge states are quickly depleted, reducing the supply of electrons for neutralization. Thus, although the confinement time is not longer than in the Gated EBIS (in which the gate is closed until equilibrium is reached), the ions reach higher charge states before attaining equilibrium. Extraction of the ions using pulsed injection could be either in leaky mode or gated mode.

Pulsing a neutral gas sufficiently fast into an EBIS is problemmatical. Various schemes, including laser vaporization and fast valves have been tried without much success. A better way is to inject the gas as low charge state ions from an external source [44]. This has great advantages in terms of control of the pulse duration and intensity, and maintaining purity.

The pulsed/leaky mode is convenient for studying ion heating in an EBIS, since a fixed leaky barrier automatically releases the ions when they reach the top of the barrier, without fear of transients altering the local potential. Typically one might inject ions for 10 μs and watch them leaking over a 10 V barrier for some seconds, inferring the heating rates from the time profile of the leak rate. This is essentially how the EBIT is operated: a burst of ions from the external source, a MEtal Vapor Vacuum Arc (MEEVA) [45], is injected and a fraction trapped "almost forever" (some hours).

The pulsed/gated mode is the best way to obtain the most ions of the highest charge state, and is the only way to study the evolution of the charge states in time. The strategy is as follows: heavy ions are injected in a pulse and the end barriers are closed, trapping them. Light coolant ions are injected continuously, serving to keep the heavy ions in the beam where they are ionized at the maximum rate. As soon as equilibrium between ionization and neutralization is reached, the barrier is dropped and the ions are extracted. The cycle then repeats. This is the way Donets and co-workers measured the progress of ionization in KRION II [46, 47, 48], demonstrating the expected behavior of the EBIS and obtaining ionization cross sections for high charge state ions.

Mechanical Design

In the absence of a quantitative model of the EBIS, most designers so far have opted for conservatism: the magnetic field had to be extraordinarily straight, the electron beam had to be extraordinarily laminar, the vacuum had to be extraordinarily good. The defense against ignorance was overkill: if a good vacuum works well, a very, very good vacuum will work very, very well. This approach undoubtedly was very costly, and contributed to the reputation that an EBIS is a tricky, complicated, expensive gadget. Seldom was there much more basis for these assumed requirements that an isolated estimate, using models of the Perfect or Classical EBIS. But since the actual conditions in a Real EBIS were relatively poorly understood, one was never certain whether these requirements were essential or just so much added bother.

With the development of the image of the EBIS as a distributed plasma device, and the introduction of ion cooling, some of these requirements can be seen more clearly. The emphasis currently is on understanding and managing the various particle species, and keeping the ions inside the beam. There is also an emphasis on making smaller, simpler, cheaper machines of equivalent performance.

A general criterion can be given for design tolerances in the EBIS: the shift in the potential seen by the trapped ions due to a flaw must be much less than the depth of the potential well inside the beam. Violation of this criterion will lead to ions moving outside the trap, thereby reducing the yield of high charge states.

Certain mechanical aspects appear to be crucial to satisfying the design tolerance criterion: a straight and uniform magnetic field, axial symmetry and good alignment, and a laminar beam. In addition, the ionization balance requires ultrahigh vacuum, and extracting useful ions requires high voltage stability and good ion-optical design. We will offer a few remarks on each of these topics, based on the general experience of the EBIS community.

The Magnetic Field

An EBIS magnet wants to have a field that is both straight and uniform. The straightness requirement arises because the beam interacts with the conducting drift tube wall. Shift of the beam from the tube axis alters the potential inside the beam. Generally it is assumed that the beam must be held in the center of the tube to within a fraction of its diameter, typically 0.1 mm. Based on this criterion, field straightness of 1 part in 10^4 has become a general standard. This means that the magnetic axis can deviate no more than 0.1 mm over the length of a 1 m solenoid. This is clearly a very challenging requirement.

The uniformity requirement arises because the beam density, and therefore the potential well, depends on the value of the field. As the beam enters the magnet it is compressed. If the axial field varies along the axis, so will the potential well seen by the ions, and axial ion pooling will result. Generally, one estimates that a change of 0.1% is tolerable. However, further understanding of the plasma aspects of the EBIS may show that this estimate is too restrictive; it is possible that the general diffusion of charged particles, much more extensive than previously appreciated, will act to smooth the profile, and more reasonable magnets could be used.

The most recent superconducting EBIS magnets are those for the Kansas State University CRYEBIS [49] and the Sandia Super-EBIS [50]. The specifications for both these magnets called for straightness and uniformity roughly as given above. Extensive field mapping by Stockli has shown that the KSU magnet has succeeded in achieving its specifications, and preliminary results on the Sandia magnet indicate the same.

Axial Symmetry

It is easily appreciated that maintaining axial symmetry is essential to EBIS operation. Symmetry can be ruined by several causes, including misalignment, thermal and gravitational distortion, and electrical asymmetry.

One of the most difficult problems in practice is the seemingly trivial job of aligning the EBIS. The fringing field of the magnet is determined by external devices, including bucking coils, possibly iron shields, and structures with nonunity susceptibility. Since the beam starts and ends in the fringing field, the conditions for proper injection into the solenoid are critical. Misalignment will cause reflection of the beam, with consequent vacuum degredation if it hits structures. CRYEBIS II apparently suffered from this problem.

It is not enough to optically align the several components. Brown and Feinberg [51] mapped the LBL EBIS with a rotating Hall probe, a technique that has a long history [52, 53]. Stockli used a similar device to very carefully map the KSU-CRYEBIS magnet and find ways to correct the inevitable sag [54]. The Dubna group has used a similar technique [55]. Kostroun [56] emphasizes that using the beam itself to locate the magnetic axis is the most reliable. The Sandia magnet was designed with a vertical axis to minimize distortions due to gravity. The only other EBIS's with vertical axis are the one in Novosibirsk, a relatively low performance machine, and the EBIT, which is so short that field straightness is not as important.

The designer of an EBIS is well advised to pay careful attention to symmetry. It is not necessarily the best plan to simply include enough screw adjustments to permit multiaxis motion for alignment; a better defense probably is to carefully select materials, maintain axially symmetric geometry, use alignment keys, and hold tolerances.

This last remark applies especially to the problems of thermal distortion. We ask that these machine go through multiple cycles from baking temperatures (say 200 °C), down to LHe temperature, 4 K, and hold certain dimensions within tenths of a mm. Even if all the parts were aligned at room temperature, during the cool-down some will shrink more than others, leaving gaps or exceeding elastic compression strengths. The system may not make it to cryogenic temperatures still aligned even the first time, much less through repeated thermal cycles. Again, the defense is to carefully select the materials for compatible thermal properties and design the parts with symmetry in mind. For instance, in a cryogenic system, one always wants to hold an insulator inside a metal sleeve, so when it cools the sleeve contracts tightly onto the insulator. If the insulator were on the outside, it would pull away from the cold metal, or be broken during the bake as the metal expands.

Laminar Beam

It is believed that a laminar beam is essential in order to prevent ion loss. Very high field gradients associated with nonlaminarity, together with the possibility of instabilities, could, and likely does, cause heating of the ions. This conclusion is supported by general observations that more laminar electron guns have produced better behaved EBIS's.

Designing an electron gun for laminarity is not a job for amateurs. Codes are available that are considered to be both correct and relatively easy to use [57, 58], but considerable experience is needed to design a really good gun.

There is a general trade-off that is useful: more laminar beams can be obtained with low perveance guns. The perveance is defined as $P = I/V^{3/2}$, where I (in A) is the current extracted when the anode voltage is V (in V). A value $P = 2x10^{-6} = 2$ μpervs is typical of guns that are used for high beam compression, but matching the optics of the gun to the solenoidal magnetic field is tricky [59]. Instead, if one uses a gun with $P = 0.01-0.1$ μpervs, the entire optics is more nearly paraxial, and the beam can be made more laminar. Lower total current is the price paid: a $P = 0.1$ μpervs gun operating at $V = 10$ kV will emit 100 mA. In order to get higher current from a cathode of the same size, we could use single crystal LaB_6, which can emit up to 20 A/cm^2 with reasonable lifetime.

Ultrahigh Vacuum

The vacuum requirements for an EBIS at first appear formidable, but they are reasonably within the state of the art. Benvenuti [60] reviews techniques for achieving pressures into the 10^{-14} Torr range. Recognizing that the main source of residual gas in a cryogenic system is dissolved hydrogen diffusing out of the metal parts, the general strategy is to pre-bake these parts in vacuum at sufficiently high temperature to boil off the dissolved gas, or to generate an impermeable layer on the surface that will inhibit diffusion across the surface.

More specifically, it is found that commercially supplied stainless steel and copper outgas at roughly 10^{-12} Torr-liter/s-cm^2. Aluminum is much better, about 10^{-13} Torr-liter/s-cm^2. Vacuum firing each of these metals reduces the rates by at least an order of magnitude. For instance, stainless steel is to be baked at 950 °C in 10^{-6} Torr vacuum for at least 1 hour. A lower temperature bake in air has a similar effect: a 100 °C bake of aluminum in air reduces its outgassing rate below 10^{-14} Torr-liter/s-cm^2, although later baking in vacuum may remove the protective oxide layer. In the absence of these preparations, the system will never reach pressures below the high 10^{-10} Torr range.

Benvenuti remarks that unbaked copper gaskets in Conflat-type flanges may contribute more gas than all the rest of the system.

A variety of pumps must be used on this system if it is to efficiently pump all gasses. The best vacuum is obtained with cooled bakeable getter pumps [60]. Although in general the more pumping area the better, placing the pumping surface strategically to intercept the main sources of gas and prevent them from entering the trap is most important. Systems like this cannot really be treated as lumped; the origin and distributions of the various species must be understood. In this regard, using xenon as the working gas may have an unexpected benefit: it is well-known that frozen layers of noble gases efficiently pump hydrogen. If xenon were bled into the end drift tubes of an EBIS (as it has been in several machines), it would creat a local pump just where needed to prevent incoming hydrogen from passing into the trap.

High Voltage and Ion Optics

In most experiments using extracted ions, the energy of the electron beam need not be controlled precisely; it only needs to be relatively near the peak of the ionization cross section. However, there is an important class of experiments in which a sharp resonance or threshold for excitation or ionization is used to identify the process or to alter the yield of ionic states. Experiments in the EBIT on dielectronic recombination [61] are of this kind: variation of 32 V in the impact energy was enough to move completely off the DR resonance for Ni^{+26}. In order to resolve the details of such resonances, we would require the beam energy to be regulated to within, say 5 V. For power supplies in the range 100 kV, this is a significant challenge. An effective strategy would be to sense the ripple and drift, and feed it back to the trapping drift tube.

Since there will be few ions of the highest charge state, it is crucial that these be extracted and transported to the target with good efficiency. The designer wants to minimize the distance to the target, and select optical elements with minimum aberrations. This problem is straightforward, and there are many texts on the subject [62]; it is just crucial in these devices to do it.

Real EBIS's

EBIS devices have been developed in about fifteen laboratories worldwide. In Table 4, we list the major devices, and provide herewith a few comments.

The KRION II source of Donets has apparently outstripped (!) all others. Donets reports [6, 81] producing bare Xe, Xe^{54+}, in this machine. This implies a performance $J\tau = 2 \times 10^{23}$ e^-/cm^2, sufficient to produce heliumlike uranium, U^{90+}. The required trapping time of 6 s also implies that the vacuum is below 10^{-12} Torr, and that the ions are being cooled in the trap. Donets indicates [3] that cooling is accomplished by periodically injecting fully stripped light ions by a switching technique: the light ions are produced in a drift tube outside the trap, and by dropping the voltage momentarily they are allowed to enter the trap. Although Donets does not say how many Xe^{54+} ions are produced, he has previously quoted 10^9 Ne^{10+} and 10^8 Kr^{35+} ions. As spectacular as this performance is, it is still another order of magnitude in $J\tau$ to reach the goal of bare uranium, U^{92+}. It seems likely that this order of magnitude should be sought not in increasing current density but in increasing trapping time and decreasing neutralization.

Donets has used KRION II in a series of measurements to demonstrate the physics of the EBIS, such as increasing charge state with trap time, and to obtain electron impact ionization cross sections for high charge states of the noble gases [82]. The latter was done by fitting rate equation solutions to the observed charge state distributions. Donets' papers contain a great deal of information about the behavior of the EBIS.

Table 4.   Some Real EBISs

Lab	Person (s)	Name	Purpose	Reference
DUBNA	Donets	IEL	Proof of principle	[1, 63]
		IEL-2	Source development	[64]
		KRION I	Injector	[2, 46]
		* KRION II	EBIS physics	[5, 47
		* KRION III	Source development	[5]
ORSAY	Arianer	SILFEC	Source dev., physics	[8, 65]
		CRYEBIS I	Injector	[7, 8]
		CRYEBIS II	EBIS, ion physics	[9]
STOCKHOLM	Borg	* CRYSIS	Injector	[66]
FRANKFURT	Becker,	TOFEBIS	Source development	[43]
	Kleinod	* Frank. EBIS	Source development	[67]
TEXAS	Kenefick, Hamm	Texas EBIS	Injector	[68]
NOVOSIBIRSK	Abdulmanov	* Novo. EBIS	Injector	[69]
BERKELEY	Brown, Feinberg	BEBIS	Test, instability study	[70]
NAGOYA	Kaneko,	PROTO-NICE	Test	[71]
	Tawara, et.al.	NICE	Coll. physics	[72]
CORNELL	Kostroun	* CEBIS I	Atomic physics	[73, 74]
		* CEBIS II	Atomic physics	[75]
SACLAY	Gastineau, Faure	* DIONE	Injector	[76]
LIVERMORE	Levine, Marrs	* EBIT	Spectroscopy	[20]
SANDIA	Schmieder	* LBL/MOD	EBIS dev., ion physics	[77]
		* Super-EBIS	Atomic, ion physics	[78]
KANSAS	Stockli	* KSU-CRYEBIS	Atomic physics	[79]
TOKYO	Okuno	* MINI-EBIS	Collision physics	[80]

* Operating 1988-89
# The Stockholm machine is a copy of CRYEBIS II and was built at Orsay.

KRION I is being rebuilt at Dubna. This device, shown in Fig. 10, uses iron pole pieces to reduce the fringing field of the superconducting magnet.

DIONE routinely produces large numbers of low-energy/high-charge-state ions: 1×10^9 Ar^{16+} ions in 70 ms [83]. Its superior cryogenic vacuum design apparently is effective in maintaining purity of the ion spectra, but nothing is known yet of the effects of ion cooling in this source. It is used exclusively as an injector.

Several sources, including NICE in Nagoya [72], the Frankfurt EBIS [84], and the LBL/MOD EBIS [77] at Sandia, have been operated mostly in DC mode, as Leaky EBIS's. These machines seem to perform similarly; in Fig. 11 is shown typical charge spectra. Similar results were reported by Kostroun on CEBIS I [85] and by Herrlander on CRYSIS [86]. Taken together, these results reasonably represent the state of the art of EBIS's outside of Dubna. The NICE machine was used in a beautiful series of experiments to measure electron capture cross sections for ions in single collisions with light neutral atoms [87].

The Berkeley EBIS, BEBIS, is an exceptional case [70]. It was built by making a sandwich of four large room temperature copper coils, with spacers between them. The ripple in the axial field was reduced by an ingenious soft iron insert, and iron pole pieces at the ends reduced the field at the gun and the collector. This machine was used to study

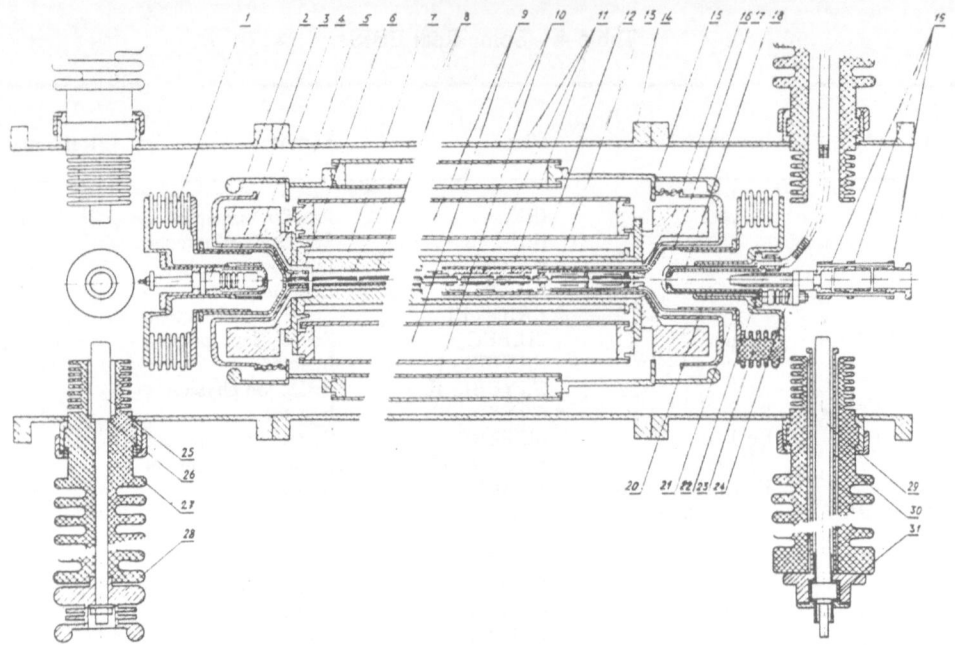

Fig. 10. The KRION-S. This source is a rebuild of KRION I to enable it to operate with higher beam voltage. The pole pieces at the end are iron; there are no bucking coils around the gun or collector. From Ref. [81].

instabilities in the beam and their apparent effects on the ion spatial and charge state distributions [31]. With its warm bore, relatively poor vacuum, and low energy beam, it was not designed to produce very high charge states. When it was decommissioned, it was moved to Sandia Livermore, and rebuilt into LBL/MOD EBIS [77]. The upgrade involved improving the vacuum, replacing the 2.8 μperv gun with one of 0.26 μperv, replacing the drift tube structure, and adding extraction and transport optics and an analyzing magnet. Operated as a Leaky EBIS, it surprised almost everyone by producing ions up to, and possibly beyond, Xe^{40+}. This was the first suggestion, apart from the results on EBIT, that the compensation time limit due to residual gas, could be exceeded. It was quite amazing to see such high charge states produced in a machine with such relatively poor vacuum (approx. $1x10^{-9}$ Torr).

The results on EBIT were far more amazing, of course. Using X-rays emitted from the trapped ions as a measure of the number of ions in the trap, Levine, Marrs, and co-workers [88, 89] were able to confine heavy ions of barium and gold for hours. As described earlier, it is now clear that ion-ion collisional cooling is responsible for this spectacular performance. In the light of this knowledge, the results on LBL/MOD were fortunate: the poor vacuum actually enhanced the ion yield.

Fig. 11. Typical charge state spectra from Leaky EBIS's: (a) Rare gas spectra in the Frankfurt EBIS [94]; (b) Iodine spectrum from NICE (Nagoya) [87]; (c) Xenon spectrum from the LBL/MOD EBIS [78]. The LBL/MOD data were taken using isotopically enriched ^{136}Xe. It may be noted that Xe^{26+} and I^{25+} are slightly peaked in their abundance; these ions have closed n=3 shells.

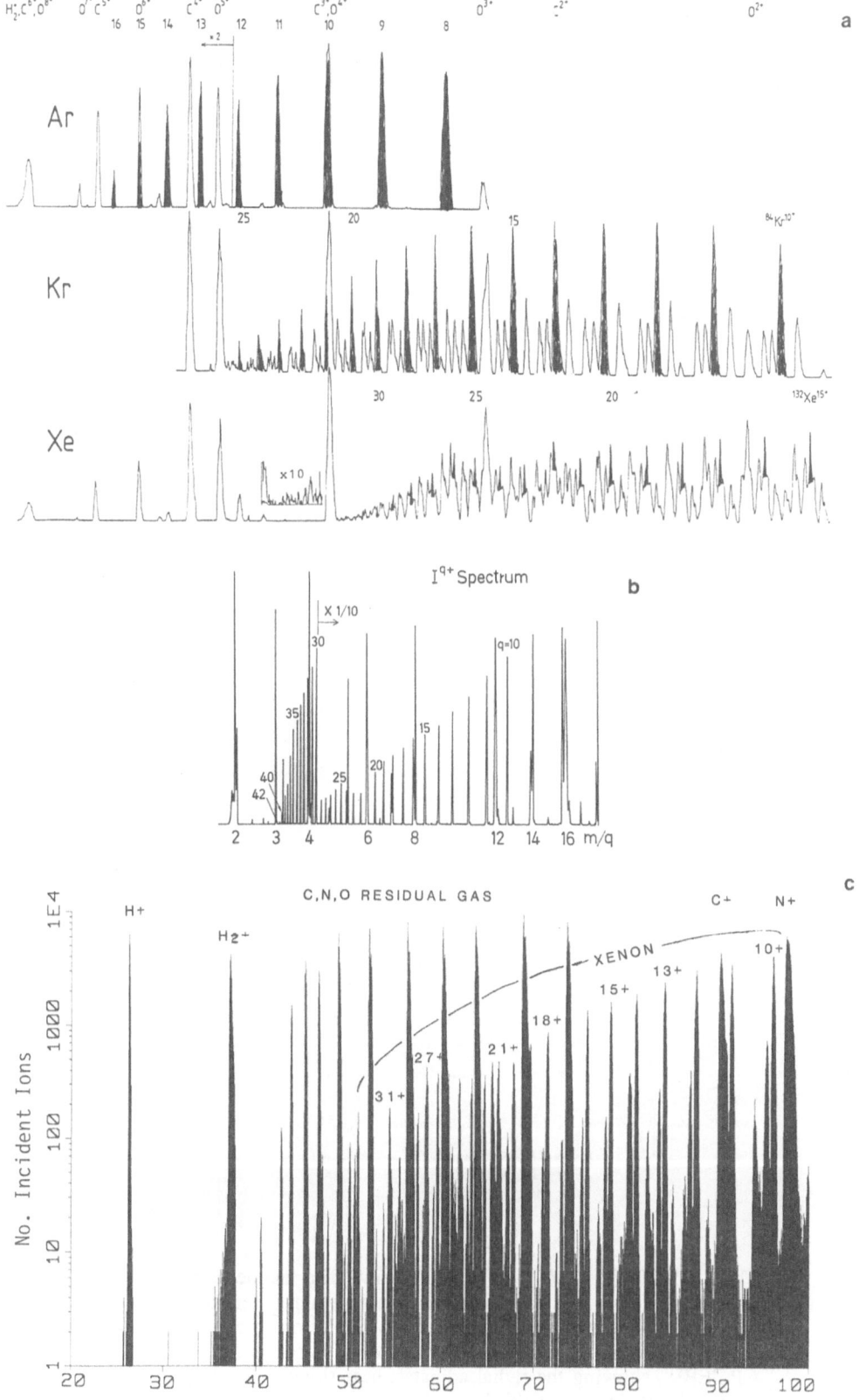

The most recent EBIS's are the KSU-CRYEBIS, built at Kansas State University by Stockli and co-workers [49, 90 91], illustrated in Fig. 12, and the Super-EBIS, built at Sandia National Laboratories by Schmieder and co-workers [78], illustrated in Fig. 13.

The KSU-CRYEBIS was patterned after the Orsay CRYEBIS. It is built around a 1 m, 5 T superconducting solenoid with an 8 cm dia. cold bore. It is surrounded by a steel shield that confines the magnetic field. The bore and magnetic axis were measured coaxial to within 0.1 mm, and the central field line deviates from the optical axis by no more than

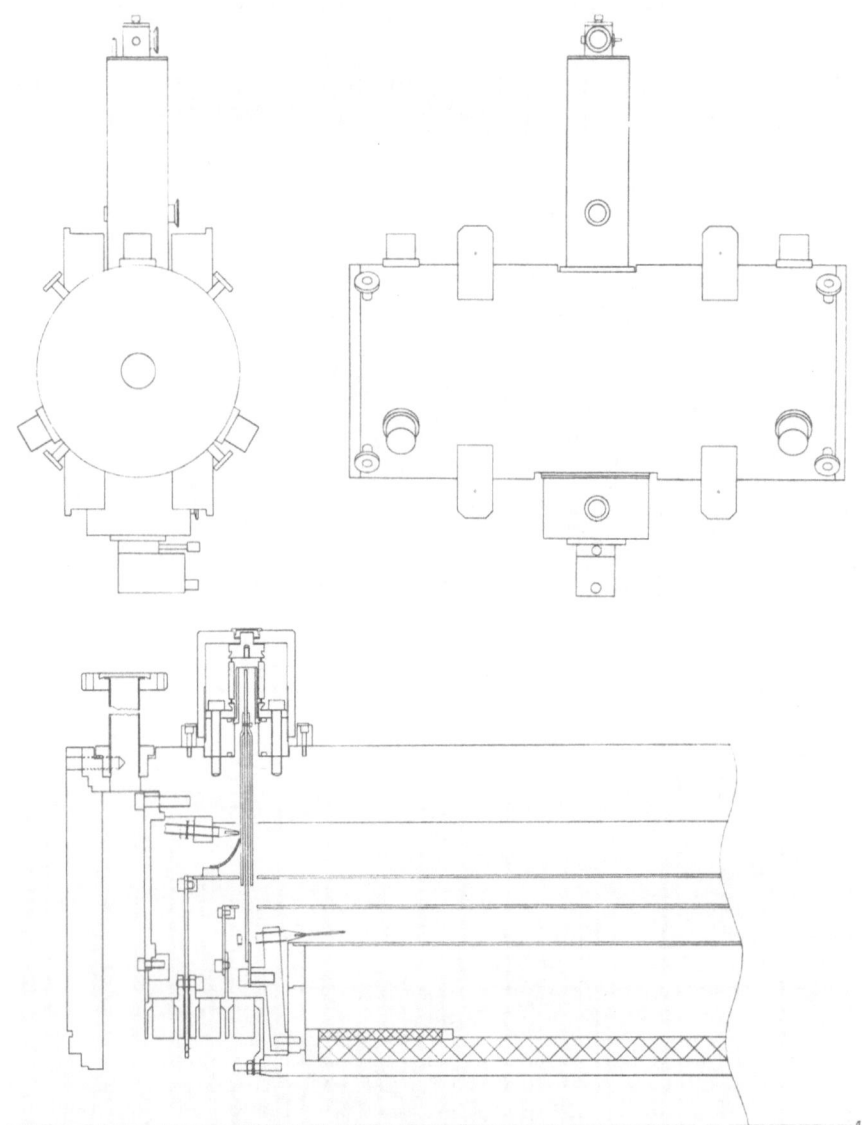

Fig. 12. The KSU-CRYEBIS. (Upper) Outside view. The magnet is held within the cryostat by radial tension adjusters. Liquid helium enters through the stack at the top. The entire magnet is surrounded by an iron shield. The source sits on a 200 kV platform. (Lower) Cross section through one corner of the KSU-CRYEBIS, showing the radial adjuster, heat shields, and the superconducting magnet itself. From Ref. [91].

0.05 mm over the length of the bore. Transverse field is less than 0.04 gauss. Running in persistent mode, the magnet consumes 8 liters of LHe per day. The gun is the same as used on the EBIT; it has a tungsten cathode and a perveance of 0.5 μpervs. The entire source is mounted on a high voltage platform, allowing the energy of the extracted ions to be adjusted from a few keV/Q to 200 keV/Q. The source was completed at the end of 1988 and first operated in early 1989.

The Sandia Super-EBIS (Fig. 13) is constructed on a vertical axis, to minimize deviations from axial symmetry. The gun is at the top, about 4 m above the floor, and the collector is about 2 m below. Ions are extracted through the collector and electrostatically bent through 90° into the horizontal beam pipe. The source is built around a 1.2 m superconducting solenoid producing 5 T in the 15 cm dia. cold bore. There is no external

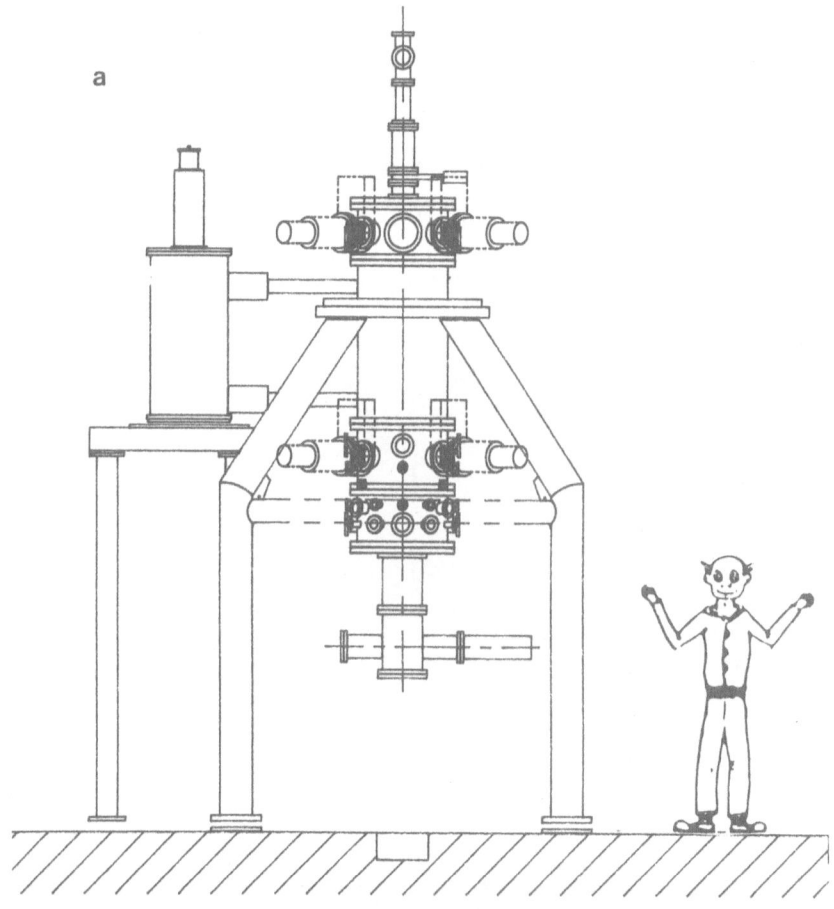

a

Fig. 13. The Sandia Super-EBIS. (a) Elevation, showing the assembly in the non-magnetic stand. The super-conducting magnet is supported on the ring at the top of the stand; the gun is in the spool at the top, and the collector is in the spool just below the magnet. The ions exit the system and are turned into the beam tube by a 90° electrostatic bend. (b) (p. 352) Cross section of the Super-EBIS. The magnet can be put on the gun/collector axis by a set of radial and axial adjusters. From Ref. [78].

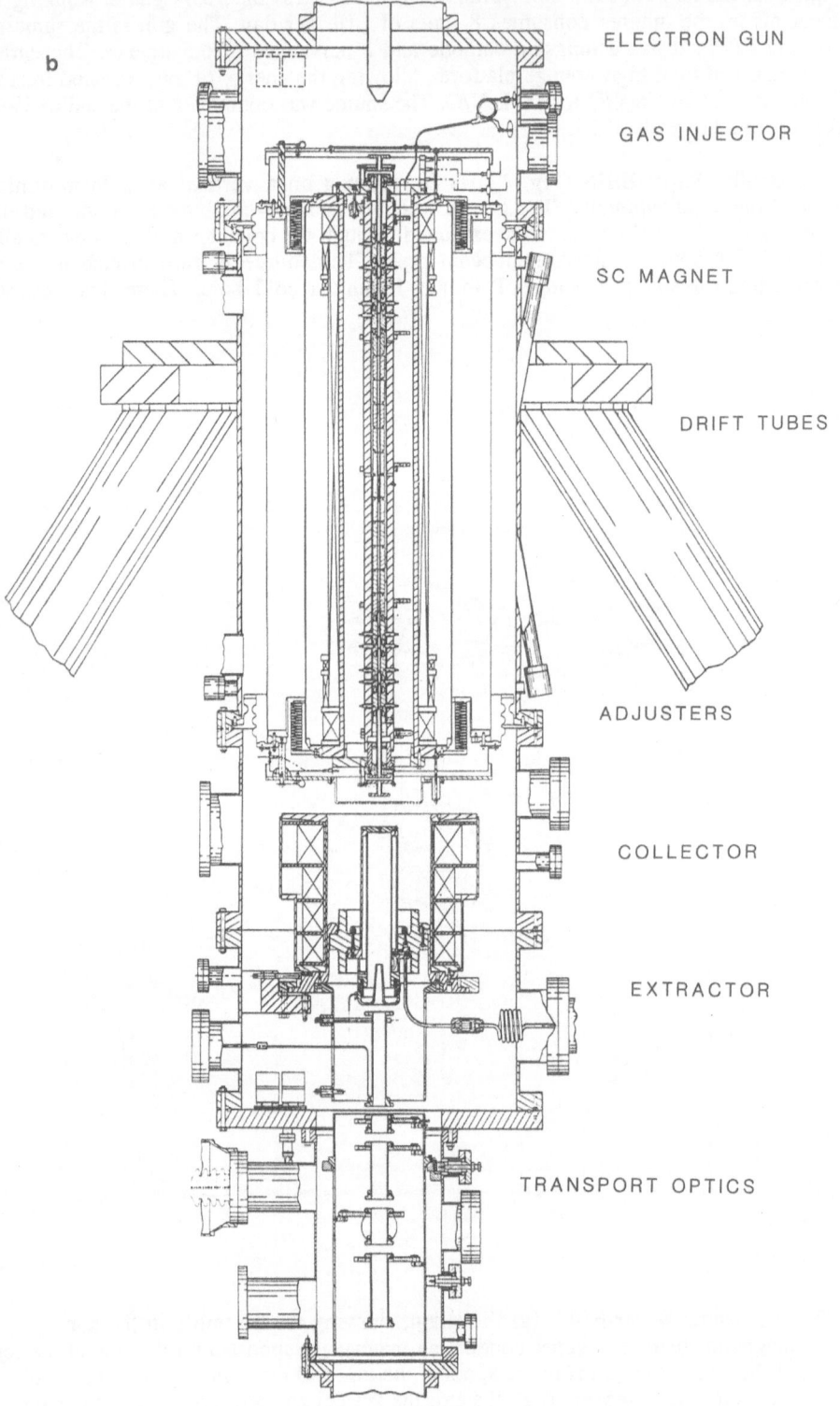

ELECTRON GUN

GAS INJECTOR

SC MAGNET

DRIFT TUBES

ADJUSTERS

COLLECTOR

EXTRACTOR

TRANSPORT OPTICS

b

Fig. 13. continued.

magnetic shield. The stray field between the collector and the electrostatic bend is uncompensated. Rectangular compensating coils are used to cancel the stray field in the beam pipe. The entire stand and supporting structures are non-magnetic stainless steel. Specifications called for straightness and uniformity of the solenoid field comparable to that achieved on the KSU-CRYEBIS. The strategy was to pay the builder [92] to assure coaxiality of the field and bore, thus obviating the need for a magnetic map. A novel feature of the magnet design is that the welded bellows which provide vacuum isolation are pulled down into the annular space between the magnet and the vacuum vessel, allowing the end of the magnet to be open for greater access.

The Super-EBIS is modeled more after KRION II. The gun operates up to 75 kV and is immersed in the fringing magnetic field in order to achieve greater laminarity; the cathode is LaB_6 to obtain the requisite current density. A novel technique was used to support and align the drift tubes: a 6.4 cm dia. longitudinally stretched aluminum rod was gun-drilled with a 3.2 cm dia. bore. The drift tubes are stainless steel, 5 cm long, allowing precision machining; they are slid inside the aluminum tube, together with alumina insulators. At cryogenic temperatures the aluminum clamps firmly onto the drift tubes, aligning them to the bore (and magnetic) axis. This source was scheduled to be completed in early 1989.

Simpler EBIS's

Several people have begun to wonder if a smaller, simpler EBIS could be made with acceptable performance [93]. The Frankfurt group has proposed two ideas [94]: first, a periodic structure of permanent magnets has long been known to be useful for transporting an electron beam, and could be the basis for a simple, inexpensive EBIS. Second, a high voltage gun similar to that used in welding has a cross-over point where the electron density is very high, even without the use of an external field to focus the beam. This high-density section could be exploited for making a magnetic field-free EBIS.

How to Design an EBIS

For the physicist searching for a source of low-energy/high-charge-state ions, the EBIS offers a great deal of performance and versatility. But if he is not fortunate enough to have a friend with one, he may be faced with building his own. This enterprise will take him on a detour from basic physics into the applied physics of the machine — a study fascinating in its own right but not necessarily appreciated at the outset. For such persons we provide here a few remarks and recipes for designing an EBIS.

Suppose we want to build a machine to completely ionize the L-shell of uranium (Z=92), i.e., to produce U^{90+}. The ionization potential I_{89} of the last electron is 32.8 keV, so to maximize the ionization cross section we will want our electron beam to have energy about 90 keV. We estimate $J\tau$ using the simple formula [32]:

$$\sigma(U^{q+} \rightarrow U^{(q+1)+}) = 1.6 \times 10^{-13} \left[\frac{\ln(E/I(U^{q+}))}{EI(U^{q+})} \right] \quad .$$

for q = 82...90, which gives the approximate performance requirement:

$$J\tau(U^{90+}) = 2 \times 10^{23} \text{ e}^-/\text{cm}^2 \quad .$$

Now we assume that we will work hard to design a highly compressed electron beam, getting J = 3000 A/cm^2, or 2×10^{22} e$^-$/s-cm^2. This implies we must confine the ions for

$$\tau(U^{90+}) = 10 \text{ s} \quad .$$

Now we assume the ions pick up electrons from the residual gas (which is hydrogen atoms, ionization potential I(H) = 13.6 eV), with the cross section [95]

$$\sigma_0(U^{90+} \rightarrow U^{89+}) \cong 1.4 \times 10^{-12} \ (Q=89)^{1.17} \ [I(H) = 13.6]^{-2.76}$$

$$= 2 \times 10^{-13} \ cm^2 \quad .$$

Now we further assume that we will implement ion-ion collisional cooling, and succeed in keeping the U^{90+} temperature below about 10 eV/Q (roughly the trapping potential per charge), or about $kT(U^{90+}) \cong 1$ keV. The mean velocity for these ions is $v(U^{90+}) \cong 10^5$ cm/s. Now the ionization equilibrium of the Classical EBIS shows that the residual gas density must be limited to

$$n(H) < [\tau(U^{90+}) \ \sigma_0(U^{90+} \rightarrow U^{89+}) \ v(U^{90+})]^{-1}$$

$$\cong 6 \times 10^6 \ atoms/cm^3 \quad .$$

This requirement drives us to make the drift tubes and much of the rest of the apparatus from carefully baked aluminum, and to protect the trapping region from gas influx by elaborate differential pumping. Every part that can contribute residual gas is selected and processed to reduce outgassing and inhibit transport. The drift tubes and as much of the rest of the apparatus as possible will be held at cryogenic temperatures. Compatible with this, and in order to make the magnetic field as straight and uniform as possible, we decide to use a superconducting solenoid. We add two years to the schedule.

At $E_e = 90$ keV, the beam travels at $v_e = 0.526c$, and has electron density

$$n_e = J/v_e = 1.3 \times 10^{12} \ e^-/cm^2 \quad .$$

We assume that at most we can trap 1% of the electron charge density as heavy ions, or about 10^{10} e^-/cm^3. The ion density is

$$n(U^{90+}) = 10^8 \ ions/cm^3 \quad .$$

Suppose we want to have 10^5 U^{90+} ions incident on an external target in a single dump, but that we can extract and successfully transport only 10% of the ions from the trap. This means we must have

$$N(U^{90+}) = 10^6 \ ions$$

in the trap, and it will have to have a volume AL = 0.01 cm³. We decide to make the trap L = 10 cm long, that is, shorter than previous machines, in order to reduce the likelihood that convective instabilities will disrupt the beam and heat the ions. The beam will have an area $A = 10^{-3}$ cm², or a diameter about $D \doteq 0.02$ cm. The total current in the beam is $I = JA = 3$ A.

In order to achieve the required current density, we assume we will compress the beam coming from the cathode by 300x, based mostly on computer simulations. The cathode will have to emit about 20 A/cm², which means we will have to use LaB_6 rather than the more convenient tungsten. The cathode will have area 3 A/20 A/cm² = 0.15 cm², or a diameter 0.2 cm. In the absence of a better idea, we believe the higher the magnet field, the better, so we opt for the highest field obtainable with a reasonable superconducting solenoid, namely 10 T. The cathode will have to sit in a field of 10 T/300 = 0.333 T. Studying the profile of the fringing field fall-off, we decide we will need to provide additional end solenoids to obtain the right profile. We expect to need an additional solenoid at the collector to reduce the field to zero there.

Having completed the preliminary specifications, we begin to contemplate the overall layout. In order to withstand the high beam voltage, we decide to use gradient columns to bring the beam into and out of the trap. Additional solenoidal magnets are required to keep the beam confined within the columns and match the gun to the central solenoid. We decide to combine the superconducting coils and drift tubes into a single assembly. For simplicity we forgo a series of drift tubes, opting instead for a single trap consisting of a central drift tube and two end electrodes. The bore of the magnet itself will form the trap; there is no need for a separate central drift tube. We decide to keep the trap at high potential, and therefore the ions will fall to ground during extraction, acquiring a kinetic energy of about 8 MeV. The ions will be transported at this energy, and decelerated again at the target. We reserve space to provide radiation shields for parts cooled to LHe temperatures. Our first sketch of what this machine might look like is shown in Fig. 14. We contemplate the challenging engineering requirements, and how to do this within 3 years and a million dollars. Imperiously, we decide to go ahead. In our spare time we begin to think about an EBIS that could produce bare uranium, U^{92+}!

EBIS EXPERIMENTAL TECHNIQUES

We have seen that the EBIS can produce a reasonable number of low-energy/high-charge-state ions, objects that will be of interest to a variety of fields of physics. The physical constraints of the source will most influence what experiments can be done with these ions. In the simplest sense, we can divide the experiments according to: (1) Ions contained within the trap; (2) Ions extracted to a target.

In the rest of this section, we describe some of the practical aspects of doing physics experiments with EBIS ions, within these two broad categories.

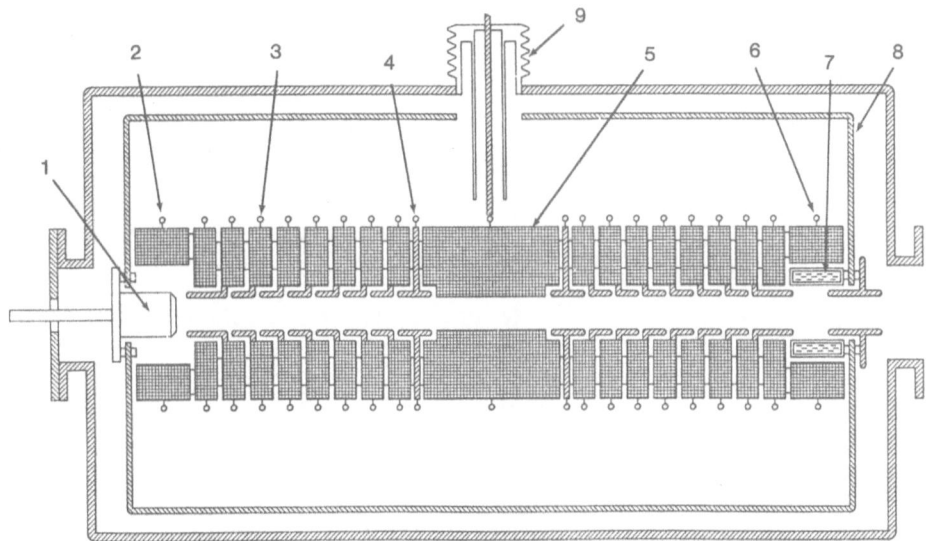

Fig. 14. Conceptual sketch of an EBIS for producing U^{90+}. (1) Electron gun; (2) gun bucking magnet; (3) combination HV gradient tube and superconducting guide field magnet; (4) trap electrodes; (5) central trap and superconducting magnet; (6) collector bucking magnet; (7) collector; (8) radiation shield; (9) HV (90 kV) feedthrough.

Observing Ions Inside the EBIS

Looking into an EBIS is not easy; the intrinsic nature of the trap leaves little room for access ports. In spite of this several successful experiments have been carried out.

Briand et al. [96], looked through a hole in the gun cathode of SILFEC with a Si(Li) X-ray spectrometer and detected X-rays from trapped Ar ions. Varying the beam energy between 2 and 3 keV, they observed a variation in the X-ray yield consistent with KLL dielectronic recombination resonances of Ar^{12+}, Ar^{13+}, Ar^{14+}, and Ar^{15+}. The low-resolution X-ray spectrum included overlapping spectra from unresolved KLM...KXY resonances from the various ions. Furthermore the ±15 eV resolution of the beam energy was too poor to separate the spectra from different charge states.

A similar experiment was carried out on CEBIS I [97], looking up the end of CEBIS I toward the cathode with a Si(Li) detector. A broad spectrum was recorded, tentatively identified as resulting at least partially from bremsstrahlung and characteristic X-rays from metal parts of the structure. Donets has carried out much the same experiment with better collimation, and has observed characteristic X-rays from ions in the trap [81].

Schmieder [98] used a Si(Li) detector looking transversely into the LBL EBIS (after it was moved to Livermore but before it was modified to become LBL/MOD). A hole in one of the drift tubes allowed the X-rays to pass into a tube between the coils and exit through a Be window 80 cm from the beam axis. K X-rays of argon gas leaked into the trap were observed, together with bremsstrahlung, which showed a sharp cutoff at the beam energy. When the beam was switched on suddenly, the bremsstrahlung showed a transient increase generated from reflux of gas from the collector as shown in Fig. 15. The Ar line showed a transient decrease associated with beam pumping of the argon gas out of the drift tubes. This study showed that the X-ray emission could be used as a diagnostic for the gas dynamic processes occuring in a pulsed EBIS.

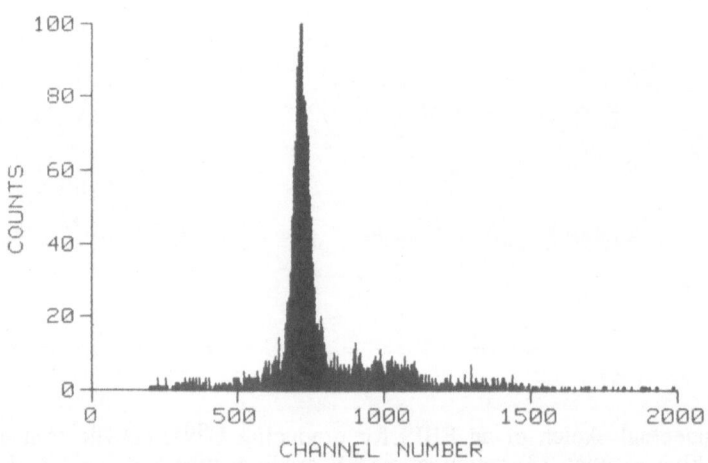

Fig. 15. X-ray spectrum from argon ions trapped in the LBL/MOD EBIS. The spectrum consists of the K_{α} peak at 3 keV and a bremsstrahlung continuum ending near the beam energy (5 keV). From Ref. [98].

Motivated by the great interest in emission spectroscopy of trapped ions, the Sandia group attempted to provide a radial access port in the Super-EBIS [50]. There were two problems: first, the port would necessitate a break in the solenoid windings, resulting in a ripple of about 1% in the axial field, which was considered unacceptable. Second, the port allowed direct access to the bore, which was considered a hazard for maintaining the ultrahigh vacuum. After much effort, the attempt was abandoned.

Many of the difficulties of performing in-situ spectroscopy in an EBIS were solved with the invention of the EBIT [39]. By replacing the solenoid with a Helmholtz pair, a radial port was possible that provided direct access to the trapped ions. A consequence is that the trap is much shorter than a normal EBIS, hence can hold far fewer ions. Thus we see the complementary nature of the EBIS and EBIT: the former is more useful as a source of (extracted) ions, the latter is more useful as a source of (emitted) photons.

It is clear that providing the optical port provides much more than the option for atomic spectroscopy. Any process in which a photon is emitted becomes accessible for study, for instance, electron impact excitation and recombination. In this regard, looking into the EBIT is superior to extracting the ions from an EBIS and attempting a crossed-beam experiment. However, there are also many processes which do not generate photons, in particular processes such as charge exchange and secondary electron emission. For these processes, there is little alternative but to extract the ions.

High Energy Ions Inside the EBIS

Although the primary appeal of the EBIS is the fact that the high charge state ions have very low kinetic energy, once the EBIS is available it can easily produce ions of much higher energy. This happens automatically during extraction: the ions are carried with the electron beam into the collector, which is held at a voltage near that of the electron gun. As the beam is decelerated, the ions are accelerated, and pass through the collector at high energy. For instance, the EBIS in the design example above has a beam energy of 90 keV; the collector probably would be biased at about $V_c = 85$ kV (leaving the electrons with enough energy so their trajectories can be controlled). The U^{90+} ions are accelerated across this potential, reaching 85 keV/Q, or $E\,(U^{90+}) = QeV_c = 7.7$ MeV.

For studies in which the charge state of the ion is not important, such as kinetic damage of solids, nuclear scattering, high energy sputtering and secondary electron emission, the EBIS would be an excellent source of heavy ions in the range 1-10 MeV. The EBIS realizes its advantage over conventional electrostatic accelerators by virtue of the very high ion charge. Furthermore, its small emittance will allow these ions to be focussed into a very small spot (roughly the diameter of the beam). If 10^5 8 MeV U^{90+} ions were produced in a single pulse from the proposed EBIS and focussed onto a solid in an area 10^{-3} cm^2, the resulting energy density would easily exceed the latent heat of vaporization of the solid: a crater would be produced!

Extracting the Ions from the EBIS

As soon as we contemplate extracting the ions from a source, we want to know the properties of this source: how many ions, their energies, their angular spread, and so on. We would like to have some measure for comparing the EBIS with other sources such as the ECRIS and the recoil ion source. There are various figures of merit such as brightness and emittance, which are measures of the number of particles emitted per unit time within a region of parameter space (e.g., per unit area and solid angle). However, low-energy/high-charge-state ions bring with them a qualitatively new function not present in other sources, namely the large amount of potential energy per particle. For most experiments involving slow highly charged ions, it is this enormous potential energy that drives the reactions: the amount of damage to the target and the yield and nature of secondary products will depend strongly on the ion's potential energy. Thus the concepts of emittance and brightness, which keep track only of particles and not their energies, omit an essential aspect of such sources.

The author [99] has proposed a figure of merit particularly relevant to sources of highly charged ions in general, and the EBIS in particular. This quantity is called "Quightness," [100] and has units of energy flux spread: W/cm^2-sr. It is an extension of the well-known quantity "brightness." The Quightness of a source of ions is defined as

$$C = \sum C(Q) \qquad C(Q) = \frac{1}{A\Omega} \frac{dN(Q)}{dt} U(Q)$$

where $dN(Q)/dt$ ions of charge Qe are emitted per unit time from area A into solid angle Ω. The quantity $U(Q)$ is the total potential energy of the ion, equal to the sum of the ionization potentials of all charge states up to (but not including) Q:

$$U(Q) = \sum_{q=0}^{Q-1} I_q$$

Values of $U(Q)$ for selected ions of Ne, Ar, Kr, Xe, Au, and U are tabulated in Table 5 [101].

Table 5.   Total Potential Energies (eV)

Q	Ne Z=10	Ar 18	Kr 36	Xe 54	Au 79	U 92
1	20	14	13	11	8	6
8	934	567	496	421	451	409
9	1130	990	729	601	582	534
10	**3492**	1469	1001	807	731	694
16		5810	3533	2620	2393	2078
17		9930	4121	3026	2783	2458
18		**14356**	4761	3463	3240	2865
34			40581	23403	17117	15763
35			57877	25612	18658	17192
36			**75813**	27914	20255	18687
52				120948	58761	54230
53				161219	63796	57320
54				**202519**	68972	60486
77					332980	211269
78					424495	222495
79					**517750**	233960
90						501560
91		Bold: bare ion				631133
92						**762953**

For relatively large Q, the following formula gives a surprisingly good approximation to U(Q) [99]:

$$U(Q) \cong Q^b \qquad\qquad b = 3 - \frac{Z}{400} + \frac{1}{5}\left(\frac{2Q}{Z} - 1\right)^2$$

where U(Q) is in eV. In fact, the relation $U(Q) = Q^{2.8}$ is not a bad approximation. The range

$$\tfrac{1}{2}Q^3 < U(Q) < Q^3$$

includes essentially all the ions with Q>10, including partially filled K, L, and M shells. These approximations are adequate for estimating the Quightness of various sources of low-energy/high-charge-state ions.

Now we must specify what is meant by the area A and solid angle Ω of the EBIS. Apart from a factor of order unity, $A\Omega$ is the emittance [62]. As shown in Fig. 1, the ions are separated from the electron beam in the collector, after they have been accelerated to the collector voltage V_c. At great distance the ions appear to emerge from an area roughly the size of the electron beam. The solid angle of the beam is roughly the square of the ratio of the transverse ion velocity to its longitudinal velocity, i.e., the ratio of transverse kinetic energy $W = \tfrac{1}{2} M v_t^2$ to longitudinal kinetic energy energy $K = \tfrac{1}{2} M v_z^2$. Now W is roughly Q times half the radial well depth Φ_0, and K is Q times the collector voltage difference V_c. Thus,

$$\Omega = \pi \left(\frac{v_r}{v_z}\right)^2 = \frac{\pi \Phi_0}{2 V_c} \ .$$

The well depth Φ_0 (in V) is

$$\Phi_0 = \frac{1}{4\pi\varepsilon_o} \frac{I_e}{v_e} = 479 \frac{I_e[A]}{\sqrt{V_e[kV]}}$$

where $\varepsilon_o = 8.85 \times 10^{-12}$ MKS and we used (1/2) $m_e v_e^2 = e V_e$. This gives the solid angle (in sterradians, sr)

$$\Omega = \frac{\sqrt{m_e/e}}{\varepsilon_o \, 2^{9/2}} \frac{I_e}{V_c \sqrt{V_e}} = 0.753 \frac{I_e[A]}{V_c[kV]\sqrt{V_e[kV]}}$$

If we assume the collector and gun are at the same potential, $V_c = V_e$. We also recognize $P_e = I_e/V_e^{3/2}$ as the beam perveance. Thus

$$\Omega = 0.024 \, P_e[\mu pervs]$$

where 1 μperv $= 10^{-6}$ A/V$^{3/2}$. For an EBIS with $I_e = 0.1$ A and $V_e = 30$ kV, we find $\Omega = 5 \times 10^{-4}$ sr. This is a half-angle of $\sqrt{\Omega/\pi} = 12$ mrad = 0.7 deg.

An indication of the magnitude of the Quightness for a typical EBIS is obtained by the following: assume that 10^6 ions/s of Q = 50 [U(Q) = $Q^{2.8}$ = 5x10^4 eV] are emitted from an EBIS beam of area 10^{-4} cm^2 into a solid angle of Ω = 5x10^{-4} sr. This would give C(50) = 200 W/cm^2-sr. The yield of lower charge states is higher, but the potential energy is lower. If we assume 10^8 ions/s of charge Q = 30 [U(Q) = $Q^{2.8}$ = 10^4 eV], the same EBIS would give C(30) = 5x10^{20} eV/s-cm^2-sr = 2 kW/cm^2-sr. Since an EBIS typically will emit perhaps 5 nearby charge states, the Quightness would be about C = 10 kW/cm^2-sr. We emphasize that this is due only to the ion potential energy—there is no kinetic energy.

In comparison, typical of the ECRIS [42, 104] is an extracted ion current of 3 nA of Xe^{30+} [dN(Q)/dt = 6x10^8 ions/s, U(Q) = $Q^{2.8}$ = 10^4 eV] ions from an exit area roughly 1 cm^2 into a solid angle 10^{-3} sr. This gives C(Q) = 6x10^{15} eV/s-cm^2-sr = 0.1 W/cm^2-sr, a figure significantly lower than for the EBIS. At the upper limit of the available charge states the yield falls exponentially [29], and is very small above Q=40. The EBIS is seen by a comparison of C(Q) values to be superior for charges above about Q=30. This advantage is achieved mainly by reaching higher charge states in much smaller cross section and with smaller divergence. Of course, the ECRIS emits much larger quantities of lower charge state ions, so its Quightness C summed over these charge states may actually be comparable to the EBIS.

We emphasize the importance of the low emittance of the EBIS: the emittance of a circularly symmetric uniform source of area A is E=$\sqrt{A\Omega}$ / π^2. The area of the EBIS source can be taken to be the area of the beam, typically A=10^{-5} cm^2. Using Ω = 5x10^{-4} sr, we find E=7 mm-mrad. Other estimates [32] and computer calculations [7] gave values near 5 x 10^{-2} mm-mrad, which can be compared with 200 mm-mrad for a typical ECRIS [24]. The much lower emittance of the EBIS translates directly into ability to extract the ions and focus them onto a target, which is just what the experimental physicist wants to do.

Transport of the Ions to a Target

Except for imperfect optics, there should be little loss in transport of the ions to a target. From the cross section for electron pickup from nitrogen by low-energy/high-charge-state ions [105, 106, 107], $\sigma_c(Q,Q') \cong 5 \times 10^{-16} Q^{1.12}$, we can estimate the lifetime in residual gas at 10^{-9} Torr to be 3 ms for Q=50. This is much longer than the few μs needed for 1 keV/Q ions to cover several meters to a target.

Solid Targets

Perhaps the simplest experiment with extracted ions is to let them impinge on a solid target and look at the products. The advantage is that the target is thick (opaque!), so every incident ion creats an event. The difficulty is sorting the events into meaningful pieces. There will, of course, be a multiplicity of secondary particles, including electrons, sputtered atoms and ions, photons, and possibly molecular species. Thus, depending on one's background and interest, a whole range of detectors and spectrometers will be bolted onto the target chamber.

Donets and co-workers [4] observed X-ray spectra generated from argon and krypton ions incident on a solid copper target (Fig. 16). Actually the target was inside the EBIS, a few cm beyond the collector, so the charge states were not separated. However, by varying the confinement time, the distribution of the extracted ions could be peaked at different charges. For instance, 1.8 and 4.8 s confinement times gave significantly different relative amounts of Kr^{35+} and Kr^{36+}. The K_α recombination lines emitted from

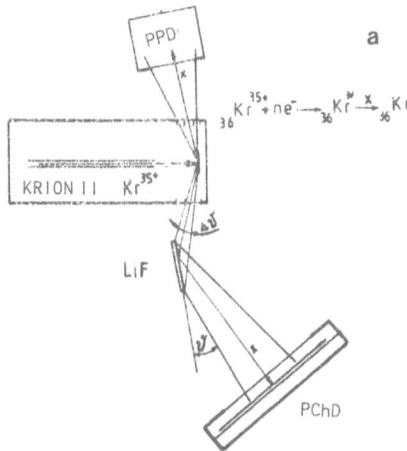

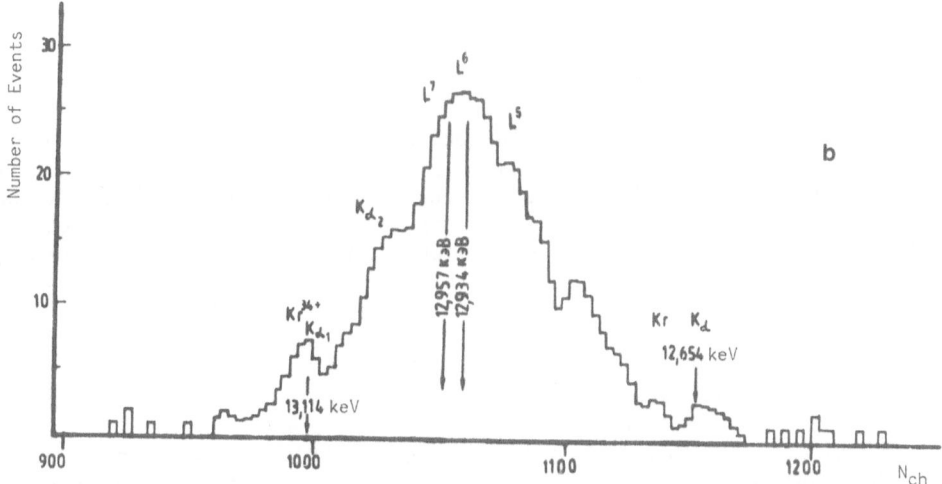

Fig. 16. Medium-resolution Kr³⁴⁺ Kα X-ray spectrum from Kr³⁵⁺ ions incident on a copper target. (a) The experimental technique; (b) the observed spectrum, showing overlapping peaks ascribed to multiple spectator electrons. This spectrum is plotted on a wavelength scale; (c) Part of the predicted line structure for the Kr³⁴⁺ Kα, plotted on an energy scale. From Ref. [109].

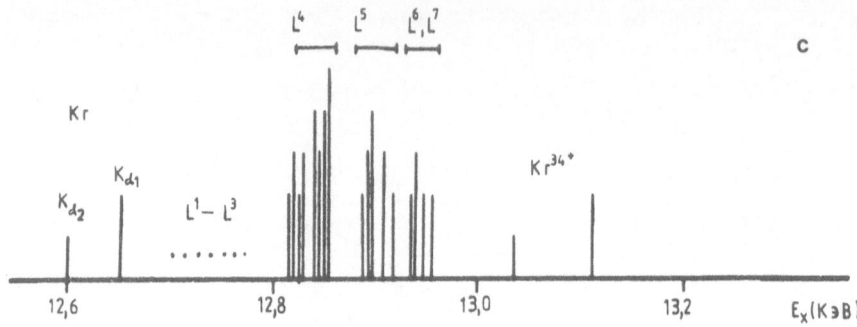

Fig. 16. continued.

the target show satellite structures associated with the different charge states. In low resolution these appear as broadening and shifts [108], but in higher resolution distinct lines identified with specific numbers of L-shell vacancies are seen [109, 110].

One expects that the Auger spectrum resulting from EBIS ions approaching a solid surface will be rich and interesting. The electron spectroscopic techniques are straightforward [111, 112]; only the incident ion is exotic. One quantity of great interest is the yield of secondary electrons as a function of charge state, for ions of very low kinetic energy. The yield should demonstrate the Auger cascade mechanism [113].

Gross surface damage is also possible with extracted EBIS ions incident on a solid. In contrast to the high energy bombardment described above, the damage would result from the potential energy of the ions. The technique is embarrassingly simple: put the target in, irradiate it, take it out and look at it. A variety of analytical tools is available for detecting and characterizing the damage, but since it probably would differ from that done by low charge state ions only on the atomic scale, one would expect to use microscopic imaging. A most interesting instrument for this application is the scanning tunneling microscope (STM), which can image individual atoms on a solid surface. An STM is now available commercially [114, 115], and one could easily fit into a target chamber or be used off-line.

Crossed-beam and Beam-gas Experiments

Experiments with EBIS ions in which charge change occurs are relatively easy. Since individual charge-separated ions are easily detected with electron multipliers, the signature of such events is large and relatively free of interference. Thus, even though the number of ions available from the EBIS will be relatively low, every ion can be used. For instance, the group in Nagoya used iodine ions up to I^{41+} to measure the total charge capture cross sections when these ions are incident on neutral gases [87].

Crossed-beam experiments involving an exotic target, such as an excited atom, an ion, a molecular cluster, or even an electron, are likely to be difficult due to the inability to make a thick target. With so few of the highest charge ions coming from the source, a relatively thick target is essential for a reasonable count rate. For example, one might contemplate crossing the extracted EBIS beam with a dense electron beam in hopes of producing fluorescence due to impact excitation to radiating bound states. We might ask to cause excitation of the $2p^53p\ ^1S_0$ state of neonlike Se^{24+} by passing the ions through a focussed electron beam. Excitation would be registered by detection of a $3p\ ^1S_0 \rightarrow 3s\ ^1P_1$ photon at 183Å. Electron beam targets of 0.3 A/cm^2 have been produced [116], and this could be magnetically compressed to higher densities, but even for a relatively dense electron beam, say 10 A/cm^2, the estimated count rate is only 3/min. To complete a reasonable spectral analysis of the fluorescence would require counting times of weeks or months. The message is that when we think of EBIS ions we are thinking of very exotic species, and there are relatively few of them. Seeing how they interact with common matter is easy, but if we want to see how they interact with another exotic species, probably there just aren't going to be enough of them.

Extracted Ions as Diagnostics for In-beam Processes

Actually, it probably is possible to make a sufficiently dense free electron target for EBIS ions, if extracted ions are used as the indicator of a specific electron-ion charge-changing process occurring inside the EBIS. If we assume that the extracted ions are representative of the trapped ions, we should be able to detect, say a resonant recombination or ionization by monitoring changes in the charge state distributions as the beam energy is changed. To be specific, assume that the EBIS is operated in the Leaky mode, and ions Q leaking over the barrier at energy QeV_b are detected. When the beam voltage is tuned to a dielectronic recombination resonance for the ion Q, some of those ions will have their charge reduced to Q-1, and these will see a lower barrier $(Q-1)eV_b$ through which they will readily pass into the extraction system. The signature will be a decrease in the extracted Q ions and an exactly corresponding increase in the Q-1 ions. Since all other ions remain unchanged, they can be used to normalize the counts; the differential yields are then directly attributable to the recombination process.

External Electron Target

Another technique for studying electron-ion interactions would be to pass the extracted ions through a second EBIS. A 5000 A/cm^2 electron beam presents a flux of 3×10^{22} e$^-$/cm^2-s. A 1 keV/Q Xe^{54+} ion has a velocity 2.8×10^7 cm/s, and would pass through a 1 m beam in 3.5 μs. Thus, any process with cross section larger than 10^{-17} cm^2 will occur during the transit, presumably to be registered by the change in charge of the exiting ion. If we obtain 10^5 ions per EBIS confinement/dump cycle, and want at least one event per cycle, we are sensitive to cross sections of 10^{-22} cm^2. Clearly the limitation is due to the optically thin target; it would be better if we could trap the ions in the target beam for the cycle period, say 10 s, which would increase the event rate to 3×10^4 per cycle. The problem is that processes competing with the one being studied could alter the charge state distributions, ruining the advantage of the long trap time.

Trapping of Extracted EBIS Ions

In the past few years, physicists have become much more aware of the potential of ion traps for carrying out precision experiments [117]. Although the EBIS itself is an ion trap, it probably is such a noisy environment that it would not be suitable for experiments of this kind. Hence it is reasonable to think of extracting the ions from the EBIS and transporting them to another trap. Some thought has been given to the transfer of a DC beam to a trap [118], but trapping EBIS ions is different from trapping low charge state ions. The main difficulty that must be faced is the huge cross section for neutralization by the residual gas. For U^{90+}, residual gas density of 6×10^6 cm^{-3} would limit the trap time to 10 s. If we want to trap such ions for times significantly longer than this, we will have to attain truly spectacular vacuum, the equivalent of less than 10^{-14} Torr. This is not hopeless, however, since the trap could be small and simple. One might make it of very pure beryllium fired in ultrahigh vacuum, with thick walls held at a temperature below 1 K, surrounded outside by a hydrogen getter such as uranium. The ions would be introduced through a very long, small diameter hole that was opened only briefly. Trapping could be done using superconducting coils to form a magnetic well, or electrostatic forces could be applied by segmenting the trap with very small ceramic spacers.

The Limits to the Art and Science

With the introduction of ion cooling, there is every reason to believe that the EBIS can achieve performance predicted when it was thought that the ion heating could be ignored. It is rather easy to foresee the production of U^{90+}. However, ionization into the K shell of very heavy elements requires beam energies above 100 keV, and this involves a different technology. It appears that once ion cooling is fully understood and controlled, the next step will be to increase the beam energy. For long trap times it will also be necessary to understand and improve the ultrahigh vacuum environment. Whether the resources to build such an EBIS could be marshalled in the near future is at present unknown.

PHYSICS WITH EBIS IONS

Much of the interesting physics of EBIS ions is associated with their high charge state and low energy. The criterion for "low kinetic energy" can be written roughly as follows: in the EBIS all ions are created in the trap held at voltage V_0; when extracted they acquire kinetic energy QeV_0. We want this to be less than the total potential energy of the ion:

$$QeV_0 < U(Q)$$

In a previous section we remarked that $U(Q)$ in eV can be approximated by Q^b, where b is slightly less than Q^3. The physical origin of this is easy to see: if we imagine the ion Q is produced as a series of hydrogenic ions of ionization potential Q^2 Ry, then

$$U(Q) = \sum_{q=1}^{Q} q^2 \, Ry = \tfrac{1}{6} Q(Q+1)(2Q+1) \, Ry \cong \tfrac{1}{3} Q^3 \, Ry$$

where we assumed $Q \gg 1$. In the spirit of these approximations, we can write

$$V_0 < \begin{cases} Q^2 & \text{K shell ions} \\ Q^2/2 & \text{otherwise.} \end{cases}$$

According to this criterion, Xe^{36+} are slow ions ("EBIS ions") if $V_0 < 0.65$ kV. For other ions we have V_0 (Xe^{54+}) < 2.9 kV; V_0 (U^{90+}) < 4 kV; V_0 (U^{92+}) < 8.5 kV.

Another aspect of high charge state ions, even those at high energy, is due to the presence or absence of vacancies. Various nuclear processes are sensitive to atomic structure: either a bound electron is necessary to turn the process on, or it turns the process off by filling a needed vacancy. Thus, nuclei can be made "radioactive" or not by simply altering the number of electrons attached to them.

Since the ions are atoms, much of the interest in them lies within the purview of atomic and nuclear physics. However, we will also find interest from the perspective of chemistry, multibody physics, surface physics, plasma physics, precision measurement, and some aspects of applied technology. A complete review of all these interests would take us too far from our present purpose. What we can do is focus on a few selected experiments in which ions of the very highest charge state available are crucial. These are the experiments that would most benefit from a source like the EBIS, and which help motivate those of us who build them.

Spectroscopy of U^{90+}

Heliumlike uranium represents a classic exotic atom, so remote that 20 years ago the mere thought of it generated laughter. But in the past few years, Gould and co-workers [119, 120], using the Berkeley Bevelac, not only succeeded in making it, but also measured transition rates and the Lamb shift. The Lamb shift agreed with QED predictions but the precision was not very high. At 218 MeV/AMU, the ions are flying past the detectors at 0.59c. At this velocity, the transverse Doppler shift is significant, and with only 10^5 atoms/s, the accuracy was limited by statistics. Drake [121] shows that the $1s2p\ ^3P_0$ state radiatively decays to the $1s^2\ ^1S_0$ state not only in the allowed E1 mode, but also in the mixed two-photon E1M1 mode, the latter being 40% of the former. Gould remarks that measuring these spectra under Bevelac conditions may be difficult, and suggests that a source of low energy U^{90+} ions might be better. As shown in the design exercise above, it appears feasible to build an EBIS that could produce U^{90+}, at an energy and cost much below the Bevelac! The n=2 states would be populated by collisional excitation by the electron beam and to a less extent by recombination from U^{91+}.

Mohr [122, 123] reviews QED effects in high Z atoms, and Deslattes [124] discusses the requirements of precision spectroscopy on hydrogenlike and heliumlike ions for meaningful tests of fundamental theory.

Multiply Excited Atomic States

Inner electrons of heavy elements are so tightly bound to the nucleus they act almost independently, interacting relatively weakly with each other . A consequence is that it is exceptionally difficult to get more than one electron at a time in an excited state. But we would like very much to do so, since they would interact relatively more strongly, providing the opportunity to test other terms in the hamiltonian. For instance, the high nuclear charge provides energy to drive Auger transitions, and the electron-electron coupling provides the switch to turn it on. Rau [125, 126] discusses the classification of high-lying 2-electron states, emphasizing that if $n' \cong n$, the atom is best described as an electron-pair bound to the "grandparent" ion. He also emphasizes that these states are relatively long-lived, and are not obtained by photoexcitation.

The good news is that the EBIS provides a way to produce these multiply excited configurations that is both efficient and easy. The mechanism is multiple electron capture.

Janev, Presnyakov, and Shevelko [127, 128] discuss multiple electron capture occuring when a low-energy/high-charge-state ion encounters a target with loosely bound electrons, such as a helium atom. There are two ways the ion can capture two electrons:

(1) in two successive one-electron captures; and (2) in a single simultaneous 2-electron capture. At low impact energy, the first process dominates; at high energy the latter dominates. In either case, the net result is

$$X^{Q+} + A \rightarrow X^{(Q-2)+}(nl;n'l') + A^{2+} \quad .$$

For very large Q, capture of both electrons into excited states dominates capture of one or both electrons into the ion ground state. Experiments by Mann and Schulte[129] confirmed the reality of process (2) for O^{6+} + He, and Muller, Salzborn, and Klinger [130, 131] observed 3e⁻ and 4e⁻ capture processes, the latter presumably sequential.

We thus can expect that allowing EBIS ions to interact with a multielectron target will efficiently produce an ion with two or more excited electrons. At low energies, a semi-empirical cross section for capture of k electrons is [95]

$$\sigma_c(Q,Q-k) = 10^{-12} A_k Q^{B_k} I_0^{-C_k}$$

where I_0 is the first ionization potential (in eV) of the target and A_k, B_k, and C_k are empirical constants [128]. For k=2, $A_k = 1$, $B_k = 0.7$, and $C_k = 2.8$. For large Q and small I, essentially all the captures, including multiple captures, will be to multiply excited states, which then will decay by an Auger transition with probabilty (relative to radiative transition) [127]

$$\frac{W_a}{W_r} = \left(\frac{10}{Q}\right)^4 \left(\frac{n_>}{n_<}\right)^3$$

Although the Auger yield falls quickly with increasing Q, its distinctive signature (an emitted electron) will make detection easy. Matthews [111] reviews ion-induced Auger spectra.

Collision Physics

There is little need to emphasize the value of the EBIS as a source for studying the basic collisional processes involving ions, atoms, and electrons. The intrinsic interest in charge transfer processes for low-energy/high-charge-state ions are reviewed by many authors [26, 132, 133, 134]. In particular, we note that Donets and Osyvannikov [48, 82] used the EBIS to obtain electron impact ionization cross sections for low-energy/high-charge-state ions by fitting solutions to rate equations to the charge distributions of extracted ions as a function of confinement time. The ability of the EBIT to access specific excitation and recombination processes [88, 89] will undoubtedly lead to a sudden rush of new measurements of electron-ion interactions. Fig.17 shows a dielectronic recombination spectrum obtained on the EBIT [61].

Ion-ion charge exchange is notoriously difficult experimentally, since both collision partners are exotic. For resonant exchange, the cross section falls exponentially due to

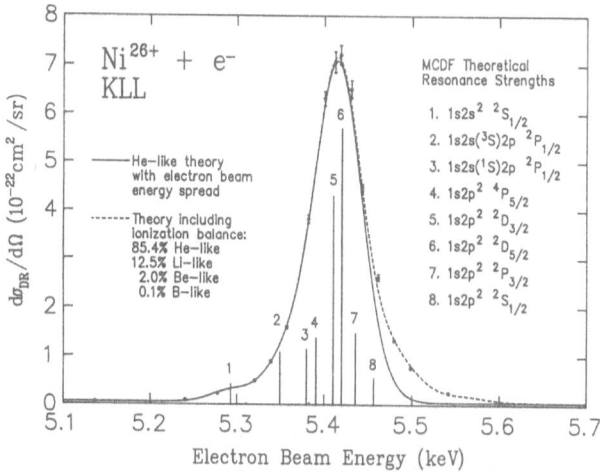

Fig. 17. Cross-section for KLL dielectonic recombination of Ni^{26+}, measured with the LLNL EBIT. The vertical lines are the theoretical predictions. From Ref. [61].

the coulomb repulsion [128]. But it is thought that this process might form the ultimate limit to the ionization balance in the EBIS. Therefore, we might expect that fitting the observed equilibrium (long confinement time) charge distributions to equilibrium rate equation solutions would lead to improved knowledge of these processes.

Ion Gas Crystals

Several theorists recently have used computer simulation of collections of trapped ions to predict that if the ions interact sufficiently, they will spontaneously form ordered spatial structures similar to crystals and shells [135, 136, 137]. These structures exhibit a wide range of mechanical phenomena such as shear and wave propagation. In order to prevent "melting" of these ordered states into a disordered gas, the ion thermal energy must be kept below the interparticle interaction energy. Thus, if the ions have charge Q, interparticle spacing D, and temperature T_i, the parameter

$$\Gamma = Q^2/DkT_i$$

must be large, typically $\Gamma > 100$. Clearly, a defense against being unable to cool the ions sufficiently is to give them high charge. Since Γ increases as Q^2, one gains relatively more from increasing Q than from decreasing T_i.

These ordered states have been observed in ion traps using laser cooling to reduce the ion temperature T_i. We might ask whether the ions in an EBIS would crystallize. That is, will the ion-ion collisional cooling keep the trapped heavy ions cool enough for them to freeze?

Assume that there are 10^6 Xe^{54+} ions in the EBIS. If the ions are all confined within the beam, which has area 10^{-4} cm^2 and is 100 cm long, the ion density is 10^8 cm^{-3}, or a mean interparticle spacing D=0.002 cm. The coulomb interaction potential between ions separated by this distance is $Q^2e^2/D = 0.2$ eV. Now the temperature of the ions is roughly Q times the potential well depth (say 20 eV), or about 1000 eV, from which we find $\Gamma \cong 2\times10^{-4}$. Since the ion temperature is so much larger than their interaction temperature, we conclude that the ions in the EBIS do not crystallize.

Coulomb Explosion

Electrons are the glue that cements nuclei together; in its absence, matter would quickly fly apart due to coulomb repulsion. One expects that if we suddenly removed a significant number of electrons from a small volume of a solid, the remaining material would quickly and violently disassemble. This process is called "coulomb explosion;" it has been discussed for years [138] but never clearly observed.

Earlier we noted that it would be possible to focus EBIS ions on a solid with enough energy to crater it. This is not, however, coulomb explosion; it is kinetic energy driven sputtering. We would seek coulomb explosion at very low ion energies; as the ion approaches the surface, it extracts electrons by field emission. For a limited range of velocities, it is postulated that the ion will capture electrons successively into high-lying quasi-atomic states which then eject the electrons by Auger transition. This process, called "Auger neutralization" [139], has the potential to extract many electrons; the ion acts like an "electron pump," pulling the electrons out of the solid and casting them away in a quasi-continuous process. Since this process costs about 20 eV per electron [113], a large number of electrons could in principle be pumped out of a solid; U^{90+} could extract perhaps 40000 electrons. We emphasize, however, that the process has so far been observed only for relatively low charge state ions; it is unknown whether the extrapolation to low-energy/high-charge-state ions ("EBIS ions") is correct. The proposed experiments are simple in principle: the yield of sputtered particles would approach a finite value at low incident energies, instead of falling to zero.

The processs of coulomb explosion occurs readily in rare gas clusters, where the binding energy is so small that ionization leads quickly to fragmentation, sometimes phrased as evaporation of the superheated cluster [140]. Computer simulations [141, 142, 143] indicate that the charge migrates around before fragmentation, and that the pattern of fragmentation is sensitive to these dynamics. Ionized clusters remain stable only if they have more than a critical number of atoms, e.g., 30, 45, and 72 Pb atoms for 2-, 3-, and 4-times ionized. It appears reasonable to expect that low-energy/high-charge-state ions could provide a means for ionization heating and coulomb explosion of more strongly bound clusters, such as oxides.

Ion-surface Interactions

If the ions do not actually cause disruption of a condensed material such as a solid, but come close enough to be interact with it, the ion can be considered a probe of the long-range interactions. One way to make ion-surface interactions more gentle is to let the ion impact at glancing angles, and observe the changes in the ion trajectories or emission. For instance, Burgdorfer and co-workers [144] develop a detailed model for resonant charge transfer in glancing ion-surface collisions. Andrä and co-workers [145, 146] emphasize the symmetry-breaking aspects of the interaction, which results in strong anisotropy of emission from the scattered ion. These authors show that the effects become much larger for high charge states, hence one could anticipate that an EBIS would be an attractive source for such studies.

High Precision Mass Measurements

Various authors have explored the Penning trap for making precision measurements of ion masses [147]. Van Dyck [148, 149] emphasizes the advantages of using multicharged ions in the trap. He suggests that if it were possible to trap U^{91+} and U^{92+}, the mass difference $\Delta M = M(U^{91+}) - M(U^{92+})$ would show the binding energy of the 1s electron, some 130-140 keV. This result, if sufficiently precise, could impact QED calculations, perhaps yielding more precise values for small corrections.

Faced with no source of low energy U^{91+} and U^{92+}, Van Dyck proceeds to lay out a design for an ion trap to produce them, a device almost identical to the EBIT. He proposes using cyclotron resonance to cool the ions, selectively purging the trap of unwanted ions, and finally extracting them to another trap for weighing. Since the experiment is of value to QED only for very heavy bare and 1- or 2-electron ions, it is obvious that the EBIS, EBIT, or some equivalent device is the only way.

Beta Decay into Bound States

Takahashi and Yokoi [150] describe a mode of nuclear decay not yet observed, namely beta decay into bound atomic states. In normal nuclear beta decay (neutral atoms), the emitted fast electron quickly exits from the atom; the likelihood of being captured is very small. But if there happened to be a K-shell vacancy at the moment of nuclear decay, there would be an appreciable probability of capturing the beta. Normally, of course, there would never be a K vacancy at the moment of decay. But if the atom were highly ionized, either to one electron or the completely bare nucleus, the vacancy would be there waiting. Thus, a normally stable nucleus would be induced to radioactively decay by merely stripping its electron away.

Several specific nuclei, all fully stripped, have been proposed for seeking this process [150, 151]:

$$^{163}Dy^{66+} \rightarrow {}^{163}Ho^{66+} \qquad T_{1/2} = 27 \text{ d}$$

$$^{187}Re^{75+} \rightarrow {}^{187}Os^{75+} \qquad T_{1/2} = 10 \text{ y}$$

$$^{205}Tl^{81+} \rightarrow {}^{205}Pb^{81+} \qquad T_{1/2} = 19 \text{ y}$$

The experiment would involve fully stripping and storing a collection of the parent ions, and looking for the conversion into the daughter. Note, however, that neither the charge nor the mass number of the ion changes; roughly the net effect is to convert a neutron into a proton, and nothing else. Detecting this change in real time will pose a challenge. It was proposed [152] that the ions be stored for varying lengths of time, then extracted, neutralized, and chemically separated to measure the increase in daughter nuclei. But storing such ions for such long times in attainable vacuum appears impossible.

Instead, we propose that the daughter ions be detected in real time by their characteristic K X-ray emission. Electron capture from the residual gas will produce constant source of X-ray emission from the trapped ions. But the spectrum from a hydrogenlike Dy^{65+} (Z=66) will differ form the spectrum of heliumlike Ho^{65+} (Z=67), due to the shielding by the captured beta in the latter. In addition to capture/cascade emission, a relatively small number of the betas could be captured directly in the L shell, producing a prompt K X-ray characteristic of the hydrogenlike ion. Thus a measurement of the emission spectrum should show satellites characteristic of the daughter nucleus, and evidence for the process.

We emphasize that this process does not even occur in ions unless they are high Z and very highly stripped, preferably to bare or one electron, and observing the effect requires trapping them at low energy. The EBIS appears to be the only way to study this mode of nuclear decay.

Monoenergetic Nuclear Pair Production

By now it is common knowledge among physicists that a narrow peak is seen in the energy spectrum of positrons emitted in energetic heavy ion collisions [153, 154]. Identifying the origin of this peak has occupied dozens, perhaps hundreds, of enthusiastic physicists for about five years. The range of proposals put forth to explain the peak is dazzling, and the breadth of experiments showing that it is maddeningly invariant to experimental parameters is impressive. At present the best bet is that it results from the decay of a previously unknown neutral bound state, called only apologetically a "new particle."

Among the plethora of possible processes occuring during the collision, besides the sought-for direct production of positrons in the super-critical vacuum of the combined nuclei, one expects nuclear coulomb excitation [155]. One takes this seriously because the width of the positron peak is consistent with nuclear lifetimes. Given the excited nucleus, it is energetically possible for it to decay by photon emission and internal conversion, which do not produce positrons, and by internal pair production, which does. If the nucleus does produce the e^+e^- pair and the pair escapes from the atom, the positron energy spectrum is a broad continuum, not the observed peak. But if the pair is produced and the electron is captured in a bound atomic state, the positron will be monoenergetic. Could this be the origin of the peak? Concensus is that it is not, because one does not expect any K vacancies to be available to capture the electron.

Whether or not this mechanism is important in the heavy ion collision experiments, it is of interest in its own right. The problem is that K electrons effectively mask the process. The K electrons do two things to the spectrum [156]: they suppress the monoenergetic positron yield by 3 to 5 orders of magnitude by using up the K shell vacancies, and they provide electrons for the internal conversion process, making it 10^2 to 10^3 time as probable as pair conversion, thereby using up excited nuclei uselessly. Lichten and Robatino [157] suggest that the collision experiments should use bare uranium nuclei instead of uranium ions. But the only present source of U^{92+}, high energy accelerators like the Bevalac, produce them at 1 GeV/AMU, a long way from the 8 MeV/AMU needed in the positron experiments. What is needed is a source of low energy bare uranium nuclei. Perhaps in the light of this review, it is not fantasy to suggest that the EBIS is likely to be the only source of these ions.

Applications to Technology

As a final word, we mention briefly that the EBIS is useful for several rather applied problems. For instance, electron-ion streaming plasma instabilities are easily accessible. All one needs to do is increase the beam current and load it with ions; it will rapidly self-destruct due to the instabilities. The effects on the ionization balance are seen by monitoring the charge states of extracted ions.

Of course, one of the major reasons for building EBIS's in the first place was as injectors for accelerators. Presently, DIONE at Saclay, the EBIS at Novosibirsk, and CRYSIS at Stockholm are being used for that purpose.

The EBIS is also an ideal device in which to study collective ion acceleration [103].

Much effort has gone into understanding how clouds of electrons can capture ions and drag them to very high energies. Clearly, the higher the ion charge, the greater the tolerable acceleration. With its high trapping potential and good control of the trap and diagnostics, the EBIS would be an ideal device with which to investigate some of these processes.

Finally, perhaps the most mundane application of the EBIS is to vacuum technology: it is the best vacuum gauge in the world. An expensive one, naturally, but there is no other way to measure the density of residual gas in the 10^{-15} Torr range. If the EBIS could be made simple and inexpensive, perhaps a whole new range of vacuum science would be accessible.

ACKNOWLEDGEMENTS

I would like to thank Prof. R. Marrus not only for the opportunities provided by the 1988 NATO Workshop in Corsica, but also for fruitful and stimulating collaboration and discussion on the subject of highly ionized atoms, dating back to 1969. Several colleagues kindly provided data and information about their EBIS projects in advance of publication, including J. Arianer, R. Becker, J. P. Briand, I. G. Brown, E. D. Donets, J. C. Faure, C. J. Herrlander, V. O. Kostroun, M. Levine, R. E. Marrs, V. P. Ovsyannikov, M. P. Stockli, and H. Tawara. Information on gas mixing in the ECRIS was provided by F. Meyer and C. Lyneis. The author is grateful to his colleagues at Sandia who have contributed to the development of this subject, including C. L. Bisson, S. Haney, N. Toly, A. R. Van Hook, and J. Weeks.

REFERENCES

[1]   E. D. Donets, V. I. Ilyuschenko, and V. A. Alpert, Preprint JINR R7-4124, Dubna (1968).
[2]   E. D. Donets, *IEEE Trans. Nucl. Sci.* **NS-23,** 897 (1976).
[3]   E. D. Donets, *Sov. Phys. Part. Nucl.* **13,** 387 (1982).
[4]   E. D. Donets, *Physica Scripta* **T3,** 11 (1983).
[5]   E. D. Donets, *Nucl. Inst. Meth.* **B9,** 522 (1985).
[6]   E. D. Donets, "Electron Beam Ion Sources," in I. A. Brown, Ed., *Ion Sources* (Wiley, New York, 1989).
[7]   J. Arianer, A. Cabrespine, and C. Goldstein, *Nucl. Inst. Meth.* **193,** 401 (1982).
[8]   J. Arianer, A. Cabrespine, C. Goldstein, T. Junquera, A. Courtois, G. Deschamps, and M. Oliver, *Nucl. Inst. Meth.* **198,** 175 (1982).
[9]   J. Arianer, *et al.*, *Physica Scripta* **T3,** 36 (1983).
[10]   B. Wolf and H. Klein, Eds., *Workshop on EBIS and Related Topics*, Rept. GSI-Bericht P-3-77, Darmstadt (1977).
[11]   J. Arianer and M. Oliver, Eds., *II EBIS Workshop* (May 12-15, 1981).
[12]   V. O. Kostroun and R. W. Schmieder, Eds., *Proc. 3rd Intl. EBIS Workshop* Cornell University, USA (May 20-24, 1985).
[13]   A. Herscovitch, Ed., *Proc. Intl. Symp. EBIS and their Applic.*, Brookhaven (Nov. 14-18, 1988).
[14]   G. S. Janes, R. H. Levy, H. A. Bethe, and B. T. Feld, *Phys. Rev.* **145,** 925 (1966).
[15]   R. W. Schmieder, Rept. LBL-2476, Lawrence Berkeley Laboratory, (1973); also Electron Ring Accelerator Group Rept. ERAN-221 (1973).
[16]   W. W. Chupp, A. Faltens, E. C. Hartwig, D. Keefe, G. R. Lambertson, L. J. Laslett, W. Ott, J. M. Peterson, J. B. Rechen, A. Salop, and R. W. Schmieder, *Proc. IX Intl. Conf. High-Energy Accelerators*, Stanford, USA (1974), p. 235.
[17]   J. M. Hauptman, L. J. Laslett, W. W. Chupp, and D. Keefe, *Proc. IX Intl. Conf. High-Energy Accelerators*, Stanford, USA (1974), p. 240.
[18]   R. W. Schmieder, *Phys. Lett.* **47A,** 415 (1974).

[19]   H.-U. Siebert, *et al.*, "Possibility of Using Spectroscopic Methods in Investigation of Some Beam Parameters of the Collective Heavy Ion Accelerator," JINR Comm. P9-9366, Dubna (1975).

[20]   M. A. Levine, R. E. Marrs, J. R. Henderson, D. A. Knapp, and M. B. Schneider, *Physica Scripta* **T22**, 157 (1980).

[21]   *Proc. Intl. Conf. ECR Ion Sources and Applic.*, Michigan State Univ., East Lansing, MI, USA (Nov. 16-18, 1987).

[22]   I. A. Sellin, J. C. Levin, C.-S. O, H. Cederquist, S. B. Elston, R. T. Short, and H. Schmidt-Bocking, *Physica Scripta* **T22**, 178 (1988).

[23]   R. Mann, *Z. Phys.* **D3**, 85 (1986).

[24]   H. Winter, "Production of Multiply Charged Ions for Experiments in Atomic Physics," in Ref. [25], p. 455..

[25]   R. Marrus, Ed., *Atomic Physics of Highly Ionized Atoms* (Plenum Press, New York, 1983).

[26]   D. H. Crandall, "Electron-Ion Collisions," in Ref. [25], p. 399.

[27]   R. W. Schmieder, *Physica Scripta* **T22**, 312 (1988).

[28]   R. W. Schmieder, "The Leaky EBIS," Bull. Amer. Phys. Soc. March, 1988, and to be published.

[29]   M. Mack, J. Haveman, R. Hoekstra, and A. G. Drentje, "Gas Mixing in ECR Sources," in *Proc. 7th Workshop on ECR Sources*, Rept. Jul-Conf-57, KFA, Julich (1986), p. 152.

[30]   M. B. Schneider, M. A. Levine, C. L. Bennett, J. R. Henderson, D. A. Knapp, and R. E. Marrs, "Evaporative Cooling of Highly Charged Ions in EBIT: an Experimental Realization," in Ref. [13].

[31]   M. A. Levine, R. Marrs, and R. W. Schmieder, *Nucl. Inst. Meth.* **A237**, 429 (1985).

[32]   R. Becker, H. Klein, and W. Schmidt, *IEEE Trans. Nucl. Sci.* **NS-19**, 125 (1972).

[33]   L. Spitzer, *Physics of Fully Ionized Gases* (Interscience, New York, 1956).

[34]   R. Becker, "Acceleration and Heating of Multiply Charged Ions in Dense Electron Beams," in Ref. [11], p. 185.

[35]   R. M. More, *in Laser Program Annual Report*, UCRL-50021-84, Lawrence Livermore National Laboratory, (1984), pp. 3-68.

[36]   P. Hagelstein, "Physics of Short Wavelength Laser Design," UCRL-53100, Lawrence Livermore National Laboratory (1981).

[37]   A. A. Galeev and R. N. Sudan, *Basic Plasma Physics* (North-Holland, Amsterdam, 1984).

[38]   C. Litwin, M. C. Vella, and A. Sessler, *Nucl. Inst. Meth.* **198**, 189 (1982).

[39]   R. E. Marrs, C. L. Bennett, M. H. Chen, T. Cowan, D. Dietrich, J. R. Henderson, D. A. Knapp, M. A. Levine, M. B. Schneider, and J. H. Scofield, "X-ray Spectroscopy of Highly-Ionized Atoms in an Electron Beam Ion Trap (EBIT)," Preprint UCRL-99699 (1988).

[40]   R. W. Schmieder and C. L. Bisson, "Heating and Cooling of Ions in the EBIS: Monte Carlo Model Calculations," in Ref. [13].

[41]   B. M. Penetrante, M. A. Levine,. and J. N. Bardsley, "Computer Predictions of 'Evaporative' Cooling of Highly Charged Ions in EBIT," in Ref. [13].

[42]   C. M. Lyneis, "ECR Ion Sources for Cyclotrons," Lawrence Berkeley Laboratory Rept. LBL-22450 (1986).

[43]   R. Becker and H. Klein, *IEEE Trans. Nucl. Sci.* **NS-23**, 1017 (1976).

[44]   J. Faure, B. Feinberg, A. Courtois, and R. Gobin, *Nucl. Inst. Meth.* **219**, 449 (1984).

[45]   I. G. Brown, J. E. Galvin, J. E. MacGill, and R. T. Wright, *Appl. Phys. Lett.* **49**, 1019 (1986).

[46]   E. D. Donets and A.I. Pikin, *Sov. Phys.* JETP **45**, 2373 (1975).

[47]   E. D. Donets and V. P. Ovsyannikov, JINR Rept. P7-9799, Dubna (1976).

[48]   E. D. Donets and V. P. Ovsyannikov, "Production of Nitrogen, Oxygen, Neon, and

Argon Nuclei in the Cryogenic Electron Beam Ion Source," JINR Rept. P7-10438, Dubna (1977).

[49] M. P. Stockli, J. Arianer, C. L. Cocke, and P. Richard, "The KSU-CRYEBIS," Proc. 10th Intl. Conf. Applic. Accel. in Res. and Ind., Denton, TX (Nov. 7-9, 1988).

[50] R. W. Schmieder, K. Battleson, D. Buchenauer, A. R. Van Hook, J. Vitko, J. Weeks, L. Hansen, R. Wolgast, V. O. Kostroun, and R. Becker, "Specifications for the Superconducting Magnet for the SNLL EBIS," SNLL EBIS Technical Note TN-012 (1986).

[51] B. Feinberg, I. G. Brown, K. Halbach, and W. B. Kunkel, *Nucl. Inst. Meth.* **203**, 81 (1982).

[52] H. Nishihara and M. Tereda, *J. Appl. Phys.* **39**, 4573 (1968).

[53] H. Nishihara and M. Tereda, *J. Appl. Phys.* **41**, 3322 (1970).

[54] M. P. Stockli, C. L. Cocke, J. A. Good, and P. Wilkens, "Magnetic Precision Alignment of a Long Horizontal Ultra-Straight Solenoid," in in Ref. [13].

[55] Yu. V. Kulikov, V. P. Ovsyannikov, and A. Yu. Starikov, "Measurements of Magnetic Field of Superconducting Solenoid," JINR Rept. P9-88-263, Dubna (1988).

[56] V. O. Kostroun, "Electron Beam Alignment in an EBIS," in Ref. [13].

[57] W. Hermansfeldt, Stanford Linear Accelerator Report SLAC 166 (1973).

[58] R. Becker has made substantial improvements to the Hermansfeldt code of particular use to EBIS design.

[59] R. Becker, "Magnetic Compression into Brillouin Flow," in [KS85].

[60] C. Benvenutti, J. Hengevoss and E. A. Trendelenburg, *Vacuum* **17**, 495 (1967).

[61] D. A. Knapp, R. E. Marrs, M. A. Levine, C. L. Bennett, M. H. Chen, J. R. Henderson, M. B. Schneider, and J. H. Scofield, "Dielectronic Recombination of Helium-like Nickel," LLNL Rept. UCRL-99921 (1988).

[62] A. Septier, *Focussing of Charged Particles*, Vol. 2, Ch. 3, 4 (Academic Press, New York, 1967).

[63] E. D. Donets, V. I. Illuschenko, and V. A. Alpert, Proc. 1st Conf. sur les Sources d'Ions INSTM, Saclay, (1969), p. 625.

[64] V. A. Alpert, *et al.*, Preprint JINR D7-5769, Dubna (1971).

[65] J. Arianer and C. Goldstein, *IEEE Trans. Nucl. Sci.* **NS-26**, 3713 (1979).

[66] S. Borg, H. Danared, and L. Liljeby, "CRYSIS - A Status Report," in Ref. [12], p. 47.

[67] R. Becker, H. Klein, and M. Kleinod, Rept. GSI-P-3-77, Darmstadt (1977), p. 3.

[68] R. W. Hamm, L. M. Choate, and R. A. Kenefick, *IEEE Trans. Nucl. Sci.* **NS-23**, 1013 (1976).

[69] W. G. Abdulmanov *et al.*, *Proc. 10th Conf. High Energy Part. Accel.* **1**, 345 (1977).

[70] I. G. Brown and B. Feinberg, *Nucl. Inst. Meth.* **220**, 251 (1984).

[71] H. Imamura, Y. Kaneko, T. Iwai, S. Ohtani, K. Okuno, N. Kobayashi, S. Tsurubuchi, M. Kimura, and H. Tawara, *Nucl. Inst. Meth.* **188**, 233 (1981).

[72] N. Kobayashi, S. Ohtani, Y. Kaneko, T. Iwai, K. Okuno, S. Tsurubuchi, M. Kimura, H. Tawara, and T. Hiro, Rept. IPPJ-DT-84, Inst. Plasma Phys., Nagoya Univ. (1984). In Japanese.

[73] V. O. Kostroun, E. Ghanbari, E. N. Beebe, and S. W. Jansen, in Ref. [11], p. 30.

[74] V. O. Kostroun, E. Ghanbari, E. N. Beebe, and S. W. Jansen, *Physica Scripta* **T3**, 47 (1983).

[75] V. O. Kostroun, in Ref. [12], p. 55.

[76] B. Gastineau, J. Faure, and A. Curtois, *Nucl. Inst. Meth.* **B9**, 538 (1985).

[77] R. W. Schmieder, K. W. Battleson, M. A. Libkind, A. R. Van Hook, and J. Vitko, "The Sandia EBIS Facility," in Ref. [12], p. 69.

[78] R. W. Schmieder, C. L. Bisson, S. Haney, N. Toly, A. R. Van Hook, and J. Weeks, "The Sandia EBIS Program," in Ref. [13].

[79] M. P. Stockli and C. L. Cocke, *Rev. Sci. Inst.* **57**, 751 (1986).

[80] K. Okuno, "Performance of MINI-EBIS Cooled by Liquid Nitrogen," in Ref. [13].

[81]   E. D. Donets, private communication (1988).

[82]   E. D. Donets and V. P. Ovsyannikov, *Sov. Phys.* JETP **80**, 916 (1981).

[83]   J. Faure, "Le Point sur Dione," Lab. Natl. Saturne Rept. LNS/SD Dione 87-133 (1987).

[84]   R. Becker, M. Kleinod, and H. Klein, *Nucl. Inst. Meth.* **B24/25**, 838 (1987).

[85]   V. O. Kostroun, private communication, 1988.

[86]   C. J. Herrlander, private communication, 1988.

[87]   H. Tawara, Ed., "The Collected Papers of NICE Project/IPP, Nagoya," Rept. IPPJ-AM-43, Nagoya Univ., Japan (1985).

[88]   R. E. Marrs, M. A. Levine, D. A. Knapp, and J. R. Henderson, "Measurement of Electron Excitation and Recombination for Ne-like Ba^{46+}," in H. B. Goodbody, W. R. Newell, F.H. Read, and A. C. H. Smith, Eds., *Electronic and Atomic Collisons* (Elsevier, Amsterdam, 1988).

[89]   R. E. Marrs, M. A. Levine, D. A. Knapp, and J. R. Henderson, *Phys. Rev. Lett.* **60**, 1715 (1988).

[90]   M. P. Stockli, K. Carnes, C. L. Cocke, B. Curnutte, J. S. Eck, T. J. Gray, J. C. Legg, and P. Richard, *Nucl. Inst. Meth.* **B10/11**, 763 (1985).

[91]   M. P. Stockli, K. Carnes, C. L. Cocke, B. Curnutte, T. J. Gray, S. Hagman, J. C. Legg, and P. Richard, "The Kansas State Atomic Collision Physics Facility Dedicated to the Studies of Atomic Interactions of Highly Charged, Low and Medium Energy Ions," *Proc. XI Natl. Conf. Particle Accel.*, Dubna, (1988). Also see Ref. [13].

[92]   Cryogenic Consultants, Ltd., London.

[93]   V. O. Kostroun, private communication (1987).

[94]   M. Kleinod, R. Becker, and H. Klein, "Progress Report on the Frankfurt EBIS," in Ref. [13].

[95]   A. Müller and E. Salzborn, *Phys. Lett.* **62A**, 391 (1977).

[96]   J. P. Briand, P. Charles, J. Arianer, H. Laurent, C. Goldstein, J. Dubau, M. Lovlergue, and F. Bely-Dubau, *Phys. Rev. Lett.* **49**, 1325 (1984).

[97]   B. M. Johnson, K. W. Jones, V. O. Kostroun, E. Ghanbari, and S. W. Janson, "X-ray Spectra from the Cornell EBIS (CEBIS-I), in Ref. [12].

[98]   R. W. Schmieder, "Observations of X-rays in the EBIS," EBIS Tech. Note TN-017 (1986).

[99]   R. W. Schmieder, "Quightness: A Proposed Figure-of-Merit for High Charge State Ion Sources," in Ref. [13].

[100] N. Rostoker, in Ref. [99].

[101] J. Scofield, LLNL, private communication, 1989.

[102] C. L. Olson, "Collective Ion Acceleration with Linear Electron Beams," in Ref. [103].

[103] C. L. Olson and U. Schumacher, *Collective Ion Acceleration* (Springer-Verlag, Berlin, 1979).

[104] R. Geller, F. Bourg, P. Briand, J. Debernardi, M. Delaunay, B. Jacquot, P. Ludwig, R. Pauyhenet, M. Pontonnier, and P. Sortais, "The Grenoble ECRIS Status 1987 and Proposals for ECRIS Scalings," in Ref. [21].

[105] H. Ryufuku and T. Watanabe, *Phys. Rev.* **A19**, 1538 (1979).

[106] R. E. Olson, "Electron Capture Between Multiply-Charged Ions and One-Electron Targets," in Ref. [107], p. 391.

[107] N. Oda and K. Takayanagi, Eds., *Electronic and Atomic Collisions* (North Holland, Amsterdam, 1980).

[108] E. D. Donets, S. V. Kartashov, and V. P. Ovsyannikov, "Production, Identification, and Ion-at-Surface X-ray Spectroscopy of Kr^{35+} and Kr^{36+}," JINR Rapid Publ. No. 20-86, Dubna (1986).

[109] W. Wagner, E. D. Donets, V. Danin, and S. V. Kartashov, "Spectrometry of Characteristic X-ray at Hydrogen-like Neutralization Kr^{35+}," JINR Rapid Publ. No. 4(24)-87, Dubna (1987).

[110]   G. Zschornack, G. Musiol, and W. Wagner, "Dirac-Fock-Slater X-ray Energy Shifts and Electron Binding Energy Changes for all Ion Ground States in Elements up to Uranium," Rept. ZfK-574, Akademie der Wissenschaften der DDR, Zentralinstitut fur Kernforschung, Rossendorf bei Dresden (1986).

[111]   D. L. Matthews, "Ion-Induced Auger Electron Spectroscopy," in Ref. [112], p. 433.

[112]   P. Richard, Ed., *Atomic Physics - Accelerators* (Academic Press, New York, 1980).

[113]   U. A. Arifov, E. S. Mulhamadiev, E. S. Parillis, and A.S. Pasyuk, *Sov. Phys. Tech. Phys.* **18**, 240 (1973).

[114]   L. K. Technologies, Bloomington, IN

[115]   Digital Instruments, Santa Barbara, CA

[116]   R. Becker, A. Muller, Ch. Achenbach, K. Tinschert, and E. Salzborn, *Nucl. Inst. Meth.* **B9**, 385 (1985).

[117]   A. Barany, A. Kerek, M. Larson, S. Mannervik, and L.-O. Norlin, Eds., *Proc. Workshop and Symp. Phys. of Low-Energy Stored and Trapped Particles, Physica Scripta* **T22** (1988).

[118]   R. B. Moore and S. Gulick, *Physica Scripta* **T22**, 28 (1988).

[119]   H. Gould, *Nucl. Inst. Meth.* **B9**, 658 (1985).

[120]   C. T. Munger and H. Gould, *Phys. Rev. Lett.* **57**, 2927 (1986).

[121]   G. W. F. Drake, *Nucl. Inst. Meth.* **B9**, 471 (1985).

[122]   P. J. Mohr, "Relativistic and QED Calculations for Many-Electron and Few-Electron Atoms," in Ref. [123], p. 301.

[123]   J.-P. Briand, Ed., *Atoms in Unusual Situations* (Plenum Press, New York, 1986).

[124]   R. D. Deslattes, *Nucl. Inst. Meth.* **B9**, 668 (1985).

[125]   A. R. P. Rau, "High Excitation of Two Electrons," in Ref. [126], p. 491.

[126]   R. S. Van Dyck, Jr. and E. N. Fortson, Eds., *Atomic Physics 9* (World Scientific Publ. Co., Singapore, 1984).

[127]   R. K. Janev, L. P. Presnyakov, *Phys. Rep.* **70**, 1 (1981).

[128]   R. K. Janev, L. P. Presnyakov, and V. P. Shevelko, *Physics of Highly Charged Ions* (Springer-Verlag, Berlin, 1985).

[129]   R. Mann and H. Schulte, *Z. Phys.* **D4**, 343 (1987).

[130]   A. Müller, H. Kinger, and E. Salzborn, *Phys. Lett.* **55A**, 11 (1975).

[131]   H. Kinger, A. Müller, and E. Salzborn, *J. Phys.* **B8**, 235 (1975).

[132]   M. Barat, "Charge Exchange Processes Involving Multicharged Ions: The Quasimolecular Approach," in Ref. [25], p. 365.

[133]   E. Salzborn and A. Müller, "Transfer Ionization in Collisions of Multiply Charged Ions with Atoms," in Ref. [134].

[134]   F. Brouillard, Ed., *Atomic Processes in Electron-Ion and Ion-Ion Collisions* (Plenum Press, New York, 1986).

[135]   J. P. Schiffer and P. Kienle, *Z. Phys.* **A321**, 181 (1985).

[136]   A. Rahman and J.P. Schiffer, *Phys. Rev. Lett.* **57**, 1133 (1986).

[137]   A. Rahman and J. P. Schiffer, *Physcia Scripta* **T22**, 133 (1988).

[138]   I. S. Bitenskii, M.N. Murakhmetov, and E. S. Parillis, *Sov. Phys. Tech. Phys.* **24**, 618 (1979).

[139]   E. S. Parillis, *Dokl Akad. Nauk.* SSSR, Ser. Fiz. **37**, 2565 (1973).

[140]   E. Recknagel, "Small Atomic Clusters," in Ref. [126], p. 153.

[141]   H. Haberland, "What Happens to a Rare Gas Cluster when it is Ionized?" in Ref. [142], p. 597.

[142]   J. Eichler, I. V. Hertel, and N. Stolterfoht, Eds., *Electronic and Ionic Atomic Collisions* (North-Holland, Amsterdam, 1984).

[143]   J. G. Gay and B. J. Berne, *Phys. Rev. Lett.* **49**, 194 (1982).

[144]   J. Burgdorfer, E. Kupfer, and H. Gabriel, *Phys. Rev.* **A35**, 4963 (1987).

[145]   H. J. Andrä, R. Zimmy, H. Winter, and H. Hagedorn, *Nucl. Inst. Meth.* **B9**, 572 (1985).

[146]   H. Hagedorn, H. Winter, R. Zimmy, and H. J. Andrä, *Nucl. Inst. Meth.* **B9**, 637 (1985).

[147] H.-J. Kluge, *Physica Scripta* **T22**, 85 (1988).

[148] R. S. Van Dyck, Jr., *Physica Scripta* **T22**, 228 (1988).

[149] F. L. Moore, D. L. Farnham, P. B. Schwinberg, and R. S. Van Dyck, Jr., *Physcia Scripta* **T22**, 294 (1988).

[150] K. Takahashi and K. Yokoi, *Nucl. Phys.* **A404**, 578 (1983).

[151] P. Kienle, "Studies of Radioactive Decay of Completely Ionized Nuclei in a Heavy Ion Storage Ring," Preprint GSI-86-57, Darmstadt (Dec.,1986).

[152] S. Datz, L. H. Anderson, J. -P. Briand, and D. Liesen, *Physica Scripta* **T22**, 224 (1988).

[153] G. Soff, "Electron Excitation Processes and Quantum Electrodynamics in High-Z Systems," in Ref. [25], p.177.

[154] B. Muller, "Quantum Electrodynamics (QED) in Strong Coulomb Fields: Charged Vacuum, Atomic Clock, and Narrow Positron Lines," in R. Marrus, Ed., *Physics of Highly Ionized Atoms* (Plenum Press, New York, 1989). [This volume].

[155] F. Bosch, "Superheavy Atoms - Electrons in Strong Fields," in Ref. [123].

[156] U. Muller, T. de Reus, J. Reinhardt, B. Muller, W. Greiner, and G. Soff, *Phys. Rev.* **A37**, 1149 (1988).

[157] W. Lichten and A. Robatino, *Phys. Rev. Lett.* **54**, 781 (1985).

ELECTRONIC INTERACTION OF MULTI-CHARGED IONS WITH

METAL SURFACES AT LOW VELOCITIES

H.J. Andrā

LAGRIPPA - CEA-CNRS - C.E.N.-Grenoble

F-38041 Grenoble Cedex, France

1) Definition of topic

The goal of this lecture is to cover the main steps to our present understanding of the interaction of Multi-Charged Ions with Metal Surfaces at Low Velocities starting with the work of Hagstrum on moderately charged ions (q = 5) dating from 1954 [1]. This readily allows to establish the relation between this particular topic and the much wider theme of Electronic Interactions of Ions with Metal-Surfaces which was covered in another lecture recently [2]. The interpretations of Hagstrum were confirmed by Arifov at al. [3] in 1973 via theoretical model calculations and verified by experiments with charge states up to q = 7. With the invention of powerful sources for the efficient production of highly charged ions [4,5,6] a renaissance of these studies with further increasing charge states up to q = 12 has considerably improved the experimental techniques and possibilities in Groningen [7], Oak Ridge [8], Grenoble [5], and Moscow [6]. These experiments have produced a considerable amount of data for which a consistent interpretation is presented in this lecture. It will give new insights into the dynamic interaction of a multi-charged ion _in front_ of a metal surface.

With the underlined specification all interactions of a multi-charged ion with the bulk of the metal are excluded from this lecture. The surface is therefore defined as the vacuum-metal interface in front of the first atomic layer. Only in exceptional situations the interaction of an initially multi-charged ion with the first or second atomic layer will be included in the discussions.

Many aspects of the more general ion-surface interaction have been discussed in ref. [2] to which the reader is referred to including the references therein. Additional citations are given only for the examples discussed here and are therefore far from complete.

2) The trajectory of an ion interacting with a metal surface

The trajectory of an ion close to a metal surface depends on the interaction potential $V(z)$ between the ion and all surface atoms and the potential V_{im} due to the image charge. The potential $V(z)$ is obtained as the superposition of the contributions $V_A(r)$ from all surface atoms. The potential $V_A(r)$ stems from the Coulomb repulsion between the ion and one particular surface atom due to imperfect screening by the electrons at decreasing distances r.

The interaction potential $V_A(r)$ is generally expressed by the product of two terms [9]

$$V_A(r) = (Z_1 \cdot Z_2 / r) \cdot \Theta(r/a) \tag{2.1}$$

where the first describes the Coulomb repulsion and the second the screening with the screening length $a(Z_1, Z_2)$. Various forms of the screening function $\Theta(u = r/a)$ have been described in the literature [9,10] which fit best to different relative velocity regimes of the collision partners. As an example may serve here the Thomas-Fermi-Moliere potential [11]

$$\Theta(u) = 0.35 \cdot \exp(-0.3 \cdot u) + 0.55 \cdot \exp(-1.2 \cdot u) + 0.1 \cdot \exp(-6 \cdot u) \tag{2.2}$$

with $a = 0.8853 \cdot a_0 / (\sqrt{Z_1} + \sqrt{Z_2})^{2/3}$.

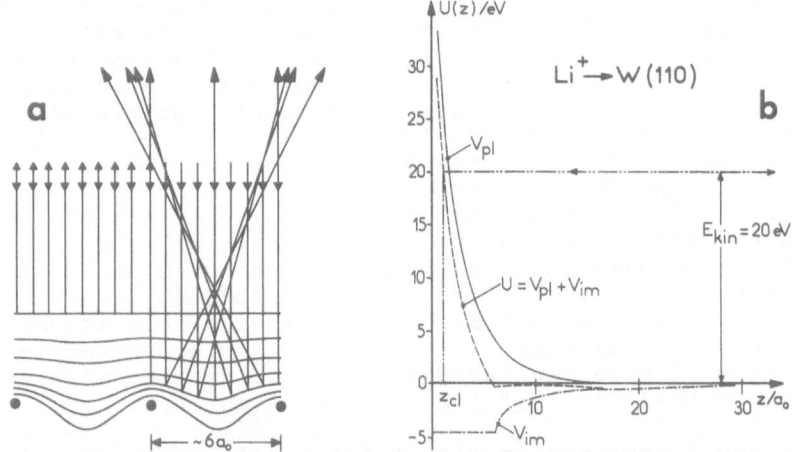

Fig. 1 a) Reflection of ions with two different energies from a single crystal surface. b) Planar potential for Li^+-W(110) interaction including the image potential.

The resulting ion-surface interaction potential $V(z)$ is indicated in fig.1a. It is corrugated at smaller distances and becomes more and more planar at larger distances from the first atomic layer. In this planar region the sum over all surface atoms may be replaced by the so called planar potential [12] which becomes for our example

$$V_{pl}(z) = 2\Pi \cdot Z_1 \cdot Z_2 \cdot \mu \cdot a \cdot f_{pl}(z/a); \quad a = 0.8853 \cdot a_0 / (\sqrt{Z_1} + \sqrt{Z_2})^{2/3} \tag{2.3}$$

$$f_{pl}(u=z/a) = 1.167 \cdot \exp(-0.3 \cdot u) + 0.45 \cdot \exp(-1.2 \cdot u) + 0.017 \cdot \exp(-6 \cdot u).$$

where μ is the density of surface atoms per unit area in a.u. and z is measured from the first atomic layer. The example for a Li⁺-ion interacting with a W(110)-surface is shown in fig.1b.

To this planar potential the image potential V_{im} has to be added in order to obtain the total ion-surface interaction potential $U(z)$. When an ion with charge q is approaching the surface from infinity it is subject to an attractive image force

$$\underline{F} = -q^2 \cdot (\underline{z} - \underline{z}_{im}) / (4 \cdot |\underline{z} - \underline{z}_{im}|^3) \tag{2.4}$$

which yields an image potential

$$V_{im} = -q^2 / (4 \cdot (z - z_{im})). \tag{2.5}$$

According to the so called jellium model of metals the jellium edge is located at $z_j = d/2$ and the plane of the induced surface charge, i.e. the mirror plane, at $z_{im} = d/2 + dz_{im}$ in front of the first atomic layer. d is the lattice constant and $1 < dz_{im} < 2$ a_0 [13,14,15]. With the distribution of the induced surface charge centered at z_{im} and with a half width of about dz_{im} it is obvious that this image potential is valid only at distances $z > d/2 + 2 \cdot dz_{im} \approx 6$ a_0. V_{im} is therefore set constant for $z < 6$ a_0 in order to approximately construct the total ion-surface interaction potential $U(z)$ for a Li⁺-ion interacting with a W(110)-surface in fig.1b.

With the potential $U(z)$ depicted in fig.1b a Li⁺-ion with an initial energy of 20 eV approaches the surface up to the turning point at about 1.4 a_0 and then returns to be nearly totally reflected. The potential of this example clearly shows that only rather low vertical velocity components parallel to the normal of the surface $(0.01 < v_v < 0.05$ a_0 corresponding to $2.5 < E < 50$ eV/u) are admitted in order to limit the interaction to the geometrical⁴ region in front of the surface. An ion arriving with a higher charge close to the first atomic layer sees of course a higher repulsive potential and is thus reflected up to higher v_v. This allows an adiabatic (or quasi-adiabatic) treatment of the electronic interaction between the ion and the metal surface. At these conditions the probability of the ion penetrating the first surface layer of atoms of an ideal metal surface becomes very small (< 0.05). As soon as higher vertical velocities are used fractional penetration into the bulk will occur so that attempts have to be made to subtract the signals stemming from this penetration from the total signal, attributed mainly to the interaction in front of the surface.

In the planar region one expects nearly perfect specular reflection whereas in the corrugated region a wide angular scattering distribution is expected. It is obvious from fig.1a that this angular distribution (including interference effects) carries information on the surface structure and indeed this technique is being widely used to determine the structure of surfaces with hyperthermal He beams (< 100 meV) and in particular the distribution of adsorbed atoms on such surfaces [16,17,18].

Following the above discussion of the potentials it is obvious that impinging ions in higher charge states will experience higher potentials which can, however, only be estimated. Furthermore charge exchange processes will cause a dynamic change of the ion-surface potential during the interaction so that a clear-cut prediction of the ion trajectory becomes impossible for the time being. Taking e.g. the image potential as an example one can estimate the energy gain of an initially multi-charged ion under the assumption of a continuous neutralization while it is approaching the surface from infinity. If the first electron is captured by an ion with $q = 9$ at $z_0 = 59$ a_0 and the next electrons are captured after aequidistant intervals $\delta z = (z_0 - d/2)/q$, in order that the ion be neutral at $z = d/2$, one obtains an energy gain of

$$dE \approx \sum_{i=1}^{q} i^2 / (4 \cdot (\delta z \cdot i - dz_{im})) = 51.76 \text{ eV}. \tag{2.6}$$

This corresponds to a vertical velocity towards the surface of $v = 15684$ m/s for Ar. This velocity then sets a lower limit to the interaction time of the ion with the surface even for the case where its initial velocity at large distance from the surface was close to zero.

In view of these problems of the dynamic behavior of the ion-surface potential one may adopt a simplified approach for the following discussion of the electronic interaction of projectiles in front of the surface. It will be sufficient to assume that the projectile is either reflected from the surface or that one observes quantities which are not very much affected when the projectiles penetrate into the surface.

3) Image potential seen by the atomic electron at intermediate and large distances

The image charge not only creates an attractive potential for the ion charge q as a whole but allows in more detail the description of its influence on an ionic electron bound to an ionic core with an effective charge Z_{core}. The concept of the image charge is an ingenious approximation for the description of the interaction of all electrons of the conduction band with the ionic electrons [19]. In this approximation a valence elec-

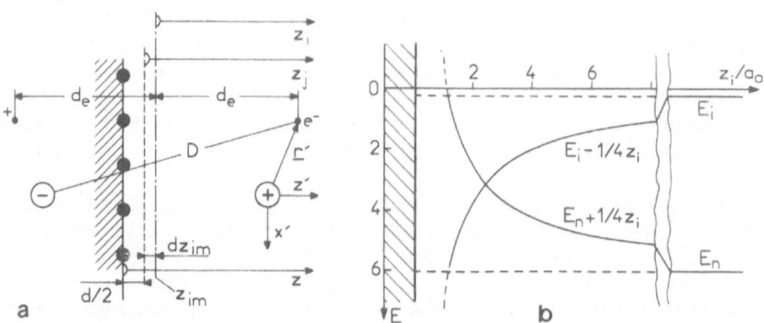

Fig. 2 a) Construction of the image potential of a valence electron. b) Modification of terms of a negative ion and of a neutral atom due to the image potential.

tron in the primed coordinate system of the ionic core sees an additional image potential besides the core-potential

$$V_{im}(\underline{r}',z-z_{im}) = Z_{core}/D - 1/4d_e \; ;$$

$$D = \sqrt{x'^2+y'^2+(z'+2\cdot(z-z_{im}))^2} \; ; \; d_e = z-z_{im}+z' \qquad (3.1)$$

as defined in fig.2a.

The term $1/4d_e$ is obtained when approaching the electron from infinity towards the surface with a fixed ion in front of the surface. When expanding $V_{im}(\underline{r}',z)$ in terms of z' [20] the image potential

$$V_{im}(z',z) = (2Z_{core}-1)/4\cdot(z-z_{im}) - (Z_{core}-1)\cdot z'/4\cdot(z-z_{im})^2 + \ldots \qquad (3.2)$$

in the primed coordinate system of the atom is obtained which has the following consequences on the atomic terms involved, depending on Z_{core} which forms with one valence electron an ion with charge q':

$$Z_{core} = 0, \; q' = -1 : \; V_{im}(z',z) = -1/4\cdot(z-z_{im}) + z'/4\cdot(z-z_{im})^2 \qquad (3.3)$$

With $Z_{core} = 0$ the valence electron forms a negative ion which is exposed to an electric field $1/4\cdot(z-z_{im})^2$ and the eigenstates of which are shifted downwards by $-1/4\cdot(z-z_{im})$ when the ion approaches the surface.

$$Z_{core} = 1, \; q' = 0 : \; V_{im}(z',z) = 1/4\cdot(z-z_{im}) \qquad (3.4)$$

With $Z_{core} = 1$ the valence electron forms a neutral atom, the eigenstates of which are shifted upwards by $1/4\cdot(z-z_{im})$ when the atom is approaching the surface.

$$Z_{core} = 2, \; q' = +1 : \; V_{im}(z',z) = 3/4\cdot(z-z_{im}) - z'/4\cdot(z-z_{im})^2 \qquad (3.5)$$

With $Z_{core} = 2$ the valence electron forms a positive ion which is exposed to an electric field $-1/4\cdot(z-z_{im})^2$ and the eigenstates of which are shifted upwards by $3/4\cdot(z-z_{im})$ when the ion approaches the surface.

$$Z_{core} = q, \; q' = q-1 :$$

$$V_{im}(z',z) = (2\cdot q-1)/4\cdot(z-z_{im}) - (q-1)\cdot z'/4\cdot(z-z_{im})^2 \qquad (3.6)$$

With $Z_{core} = q$ the valence electron forms a positive ion with charge $q-1$ which is exposed to an electric field $-(q-1)/4\cdot(z-z_{im})^2$ and the eigenstates of which are shifted upwards by $(2\cdot q-1)/4\cdot(z-z_{im})$ when the ion approaches the surface. For the model case of $q = 9$ one obtains thus a level shift at $z = z_0 - z_{im} = 57\; a_0$ of $dE = 2.03$ eV. This is a considerable shift which has to be discussed later on.

The unprimed and primed coordinate systems in fig.2a for the surface-(lab-)- and for the ion-frame, respectively, will be used throughout this lecture.

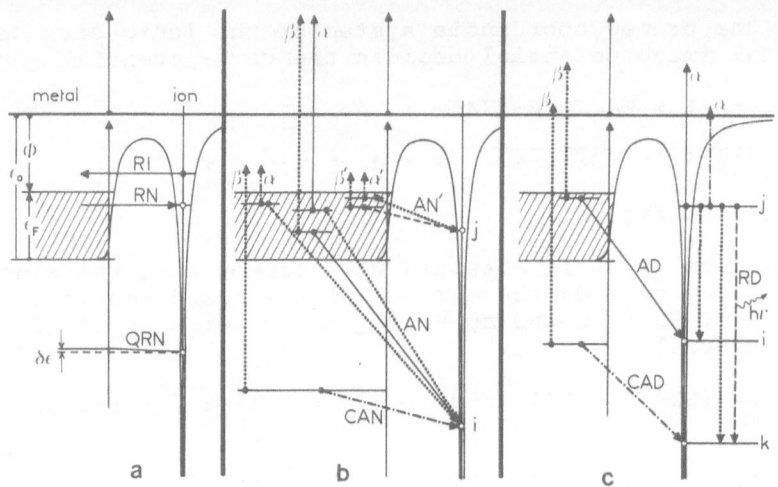

Fig. 3a,b,c Ion(atom)-surface charge exchange processes.

4) Ion-surface charge exchange processes

Charge exchange between ions(atoms) and metal surfaces is a long known phenomenon. The different types of processes possible have been summarized by Hagstrum in an excellent review [21] and are depicted in fig.3. The electronic properties of the metal surface are described by a conduction band with ε_0 being the depth of the potential well for the free electron jellium model (independent free electrons imbedded into a continuous and constant positive charge distribution) and ϕ the workfunction to yield the Fermi energy $\varepsilon_F = -(\varepsilon_0 - \phi)$ as measured from the bottom of the well whereas ε_0 and ϕ are measured from the vacuum level.

If an ion at a given distance in front of the surface has an occupied term above the Fermi energy (or Fermi edge) of the metal this term is facing an empty continuum and can consequently resonantly loose one electron into this continuum via an adiabatic single electron tunneling process which is called Resonance Ionization (RI). If on the other hand the ion has an empty term below the Fermi edge this term is facing a completely filled continuum so that it may gain an electron via Resonance Neutralization (RN). From this simple picture in fig.3a one can deduce at once that the neutralization of a scattered ion beam will strongly depend on the energetic position of the Fermi edge relative to the corresponding atomic ground state. The well known Langmuir-Taylor detector [22] for Alkali atoms at thermal energies may serve as the best example for a long existing and very successful application.

Also indicated in fig.3a is a charge exchange process between a filled core level of the metal and an empty deep lying ion term which are separated accidentally only by a very small energy gap. If the kinetic energy of the ion is sufficient to bridge this energy gap a Quasi Resonance Neutralization (QRN) can take place.

Besides the single electron resonance charge exchange phenomenon the electronic interactions involving two electrons, which we call in general Auger transitions, are playing an important role in the exploration of the ion surface interaction. One may distinguish two classes of Auger processes:

The Auger-Neutralization (AN) in fig.3b and the Auger-Deexcitation (AD) in fig.3c. In an AN process an electron from the conduction band of the metal jumps into some lower lying vacant ionic level (i or j). For reasons of energy conservation this jump is possible only when another electron of the conduction band takes up the released energy and is excited either into the continuum (AN) or into empty states of the metal above the Fermi energy (AN'). It is to be noted that the AN' process to the ionic term (j) occurs in competition to a RN process the relative importance of which has to be discussed later on.

In an AD process an ion excited in a (metastable) term (j) is dexcited in front of the surface via an electron which jumps from the conduction band to some lower lying vacant ion-level (i) forcing the excited ionic electron (j) to leave into the continuum again for reasons of energy conservation. Also indicated in fig.3b is a Core Auger Neutralization (CAN)-process from a low lying core level of the metal to an empty ion term under excitation of a conduction electron into the continuum of the metal. This process occurs in competition to the QRN process in fig.3a. Furthermore a Core Auger Deexcitation (CAD)-process may occur in fig.3c which is, however, in strong competition to AD transitions, except for particular situations.

In all five cases of Auger transitions these so called direct processes (α) cannot be distinguished from the corresponding exchange processes (β) (dotted lines) where the initial electrons are interchanged.

In competition to the AD process one has to consider Radiative Deexcitation (RD) which becomes possible after an RN- or AN'-process if a very low lying empty ionic term exists so that the fluorescence yield may become important.

5) Transition rates of RN and RI processes

Due to the energy shift of the ionic energy terms (induced by the image charge) such a term may be below the Fermi edge and is therefore being neutralized when the ion (atom) is at large distance but may move above the Fermi edge and is being ionized when the ion is approaching the surface. This behavior implies a dynamic treatment of the interaction along the whole ion trajectory in front of the surface.

The coupling matrix element between the electrons of the ion and the metal is governed by the operator Z/r [23], where Z

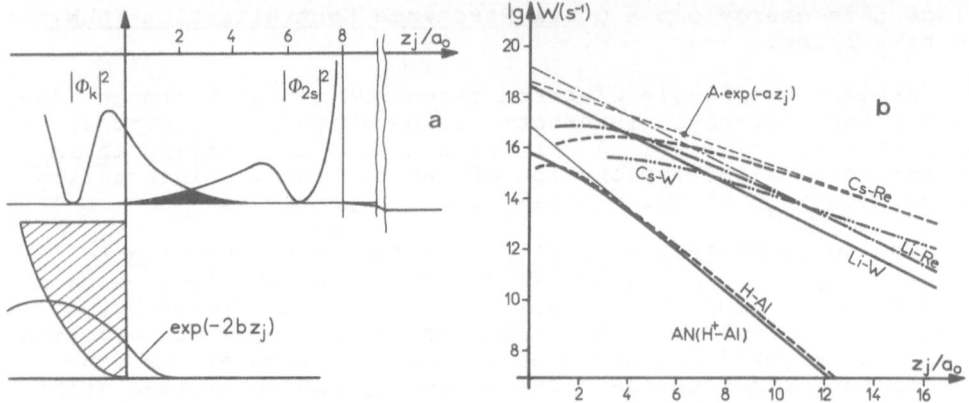

Fig. 4 a) Visualization of the tails of the squares of the
metal wavefunctions and of their overlap with an atomic
wavefunction. b) Transition rates for the charge exchange
between some alkalis and metals.

is the effective charge of the ion "seen" by the electron to be
exchanged. For the case of RI the discrete term of the ion interacts via this transition operator with the empty, near resonant continuum states of the metal. The density of these states
[24] in the free electron model is

$$n(\varepsilon) = (L^3/2\cdot\pi)\cdot\sqrt{8\cdot\varepsilon}. \tag{5.1}$$

It is proportional to $\sqrt{\varepsilon}$ and continues up to the maximum value
$n(\varepsilon_0)$ at the ionization limit. It is only the multiplication of
$n(\varepsilon)$ by the Fermi function which differentiates between occupied
and unoccupied states:

$$n(\varepsilon)' = n(\varepsilon)\cdot\{1+\exp[(\varepsilon-\varepsilon_F)/k\cdot T]\}. \tag{5.2}$$

With this density of the continuum states the Golden rule may be
applied

$$w(s^{-1}) = 2\cdot\pi\cdot|\langle\varepsilon|Z/r|a\rangle|^2\cdot n(\varepsilon) \tag{5.3}$$

to obtain the ionization rate $w(s^{-1})$ as given in fig.4b.

Since this ionization rate depends on the overlap of the
metal and the atomic wave functions as indicated in fig.4a,
which both decay exponentially with distance it can immediately
be deduced that the ionization rate has to depend exponentially
on the distance z_J of the ion from the surface in fig.4b. In
fact for many practical applications [21] the rate can be approximated by

$$w(z_J) = A\cdot\exp(-a\cdot z_J). \tag{5.4}$$

Taking a 2s-atomic wave function, however, as an example in
fig.4a one can deduce in addition that this exponential behavior
is only valid for distances where the non-exponential internal
structure of the atomic wave function is not overlapping yet
with the tail of the metal wave function.

Furthermore the concept of unperturbed atomic wave functions is only valid for a small overlap with the potential well of the metal, i.e. for distances $z_j > \langle r_{n,1} \rangle$. For smaller distances modified atomic wave functions have to be used [25] so that deviations from the exponential behavior of $w(z_j)$ may occur.

Without going into any detailed arguments it is said [23] that the rate of electron exchange is independent of the population of the metal continuum. **The rate for resonance ionization can thus be set equal to the rate of resonance neutralization when the ionic term is facing an occupied continuum with the same $n(\varepsilon)$.**

One last but important feature can furthermore be deduced from the exponential tails of the metal wave functions into the vacuum. These tails decay proportional to $\exp(-2 \cdot b \cdot z_j)$ where $b = h/\sqrt{2 \cdot m \cdot (\varepsilon_0 - \varepsilon)}$ so that the exponential tails of the energetically lowest electrons are considerably shorter than those for electrons near the Fermi energy or those which are even higher lying in energy. The same general statement applies to atomic wave functions the exponential tails of which become shorter and shorter with increasing binding energy as measured from the ionization limit. As an important consequence the rate for resonance charge exchange at a given distance z of the ion from the surface decreases drastically with increasing binding energy of the atomic and metallic electrons involved.

6) Transition rates of AN- and AD- processes

A qualitative understanding of the transition rates of the AN/AD processes can again be derived from the structure of the transition matrix element and from the spatial extent of the metal- and ionic wavefunctions as shown in fig.5a,b for the specific case of H^+-neutralization on aluminium [26]. As for RN we can use the "Golden Rule" to express the transition rate w_{AN} as a function of the density of the final continuum states $n(\varepsilon")$ and of a matrix element. Into this matrix element enters the

$$\langle \text{ overlap I } \rangle = \langle \Phi(1s,r_1) \mid \Phi_m(k,\varepsilon,r_1) \rangle \qquad (6.1)$$

and the

$$\langle \text{ overlap II } \rangle = \langle \Phi_m(k',\varepsilon',r_2) \mid \Phi_c(k",\varepsilon",r_2) \rangle \qquad (6.2)$$

"weighted" by the operator $1/r_{12} = 1/(r_1 - r_2)$ where r_{12} is the distance between the two electrons involved. Due to antisymmetrization direct (full lines) and exchange (dotted lines) matrixelements have to be taken into account. For the ion being at a given distance z_f from the surface the $\langle$overlap I$\rangle$ in fig.5a becomes largest when ε is approaching ε_F due to the increasing exponential tail of the metal wavefunction with increasing ε. At the same time the $\langle$overlap II$\rangle$ stays roughly constant because the energy difference $\varepsilon" - \varepsilon' = \varepsilon - E(1s)$ stays constant independent of ε' which may vary along the double bar which is proportional to the number of possible initial states for this given ε. In fig.5b ε is lowered and $\varepsilon - E_{1s}$ is reduced. Consequently the $\langle$overlap I$\rangle$ is considerably reduced due to the shorter exponential tail of $\Phi_m(k,\varepsilon,r_1)$ but the $\langle$overlap II$\rangle$ is

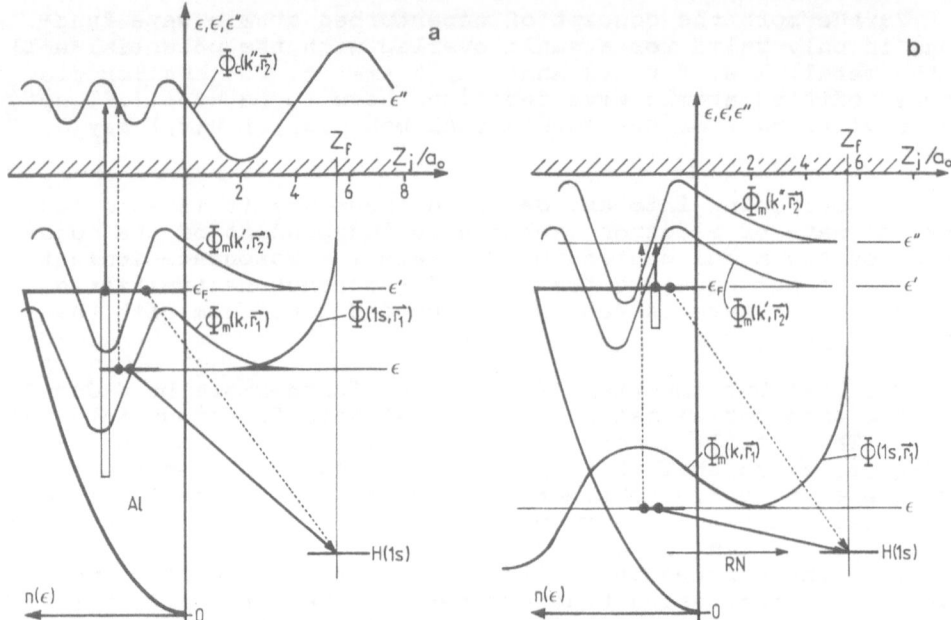

Fig. 5 Derivation of the AN-rates with the help of the shapes
of the wavefunctions; a) for a large energy gap between ε and
H(1s), b) for a small energy gap between ε and H(1s).

strongly increased due to the smaller energy difference ε"-ε'
with respect to the situation in fig.5a. It is to be noted,
however, that the double bar is much shorter.

As for the RN-process the <overlap I> is responsible for
the behavior of the AN/AD-transition rates as a function of
distance from the surface which thus drop off exponentially as
for RN. Without discussing the modifications by the exchange
contributions the AN-probability can now be set proportional to
the square of [(<overlap I>)·(<overlap II>)]² times the number
of initial states times the density of final states. With the
arguments given above on the overlaps, we can thus roughly conclude that all states of the conduction band above the energy of
the H(1s) ground term do contribute to the AN-transition rate
with slightly increasing probability with ε up to a maxi-mum at
ε = ε_F . Although each particular ε,ε',ε"-AN-transition probability is clearly smaller than any RN-transition proba-bility it
is the large number of states involved in AN which may lead to
comparable total transition rates for AN as for RN. This may be
particularly the case when comparing with unfavou-rable RN-cases
as indicated in fig.5b where the <overlap I> (responsible for
RN) at the given distance is very small.

The comparison of RN- and AN-transition rates in fig.4
clearly supports these considerations since the resonance
transition rates are observed to be systematically higher than
the AN-rates except for the unfavorable case of H⁺ on Al just
mentioned which is of the same order as the corresponding AN-

rate. Not included in this figure are favorable AN'-cases when the energies $\varepsilon, E(b)$ becomes small as shown in fig.3b. Here again the AN- and RN-transition rates are expected to be of similar size but this has still to be shown in a proper calculation.

7) Where do RN/RI- and AN/AD- transitions take place ?

This question has been answered in ref.[2] for three different situations of receding and approaching ions which do also well apply to the various stages of the interaction of a multi-charged ion with a metal surface. In contrast to ref.[2], however, it is the multi-charged ion approaching the surface which undergoes the most important electronic transitions. Therefore only the case of an ion approaching the surface will be considered here. The case of a reflected and receding ion will play only a minor role during this lecture so that it is referred to ref.[2], chapter 5.

As will be shown later the most important electronic process for a multi-charged ion approaching a metal surface is Resonance Neutralization (RN) at large distance from the surface. If level-shifts due to the image potential and level widths can be neglected one can assume that RI is not counteracting RN at any time. Under the assumption of an adiabatic interaction the neutralization probability p_n can then be given for a level below the Fermi-edge by a straight forward differential rate equation

$$dp_n(z_J)/dt = w_n(z_J) \cdot [1-p_n(z_J)] = w_n(z_J) \cdot p_+(z_J), \qquad (7.1)$$

where $[1-p_n(z_J)] = p_+(z_J)$ is the probability that the ion has not yet captured an electron on its way from infinity to the distance z_J from the surface. Using $dt = dz_J/v_v$ and $p_+(\infty) = 1$ one obtains

$$p_+(z_J) = \exp\left[-\int_{z_J}^{\infty} w_n(z_J') \cdot (dz_J'/v_v)\right]. \qquad (7.2)$$

The probability of the electron capture to take place at the position z_J in the element dz_J is then

$$p_t(z_J, v_v) \cdot dz_J = w_n(z_J) \cdot p_+(z_J) \cdot dz_J/v_v,$$

which reduces with the approximation $w_n(z_J) = A \cdot \exp(-a \cdot z_J)$ to

$$p_t(z_J, v_v) \cdot dz_J = (A/v_v) \cdot \exp\left[-a \cdot z_J - A \cdot \exp(-a \cdot z_J)/(a \cdot v_v)\right] \cdot dz_J. \quad (7.3)$$

This function $p_t(z_J, v_v)$ gives a range of distances where the electron capture may take place. It is strongly peaked with its maximum at the distance

$$z_{Jf} = (1/a) \cdot \ln[A/(a \cdot v_v)], \qquad (7.4)$$

which is called the "freezing distance" [27] and which is often taken as "the distance" where the electron capture takes place on the average. This capture distance $z_f = d/2 + z_{Jf}$ is largely dominated by the factor A and shifts only slightly towards the surface for higher v_v and away from the surface for lower v_v.

With even less stringent assumptions the same considerations and formulae apply to the AN/AD transitions, except for
W_{AN} replacing W_D. See ref.[2], chapter 7.

8) Auger-Neutralization spectroscopy

Besides the famous calculation [28] and application of the
AD-process as a detection scheme in the early Lamb-shift experiments [29] it was in particular the systematic study of the
AN-process by H.D.Hagstrum [1,21,30,31] which can now be considered as the foundation of the field of Ion Neutralization
Spectroscopy (INS). Hagstrum's goal was to develop an alternative technique for the measurement of the density of states of
metals by observing the electron spectrum following AN. To this
end he used an experimental scheme [30] of which a modern version by M.Delaunay et al. [32] is shown in fig.6 which will be
relevant for the next chapter too.

A polycrystalline W-target is placed at a pressure of 2-
$4 \cdot 10^{-8}$ Pa in the center of three hemispherical grids G1-G3 and a
hemispherical collector C (180°-LEED system) (a simple spherical
collector in Hagstrum's original work). The achievement of this
vacuum was an extremely difficult task in 1953 so that Hagstrum
devotes several pages of reference [31] to vacuum problems !
Singly charged He⁺ ions from an electron impact ion source were
mass and momentum analyzed and then directed via a lens system
onto this target after sputter cleaning with Ar⁺ ions. For the
measurement of energy distributions the grids 1 and 3 were connected to the target (biased if necessary), a retarding potential was applied to grid 2 (to the collector in Hagstrum's original work), and the electron current emitted from the target
surface was measured via the collector as a function of the
retarding potential either biased or simply to ground. For the
measurement of the total yield of electrons emitted per ion the
target was biased or grounded and all three grids were connected

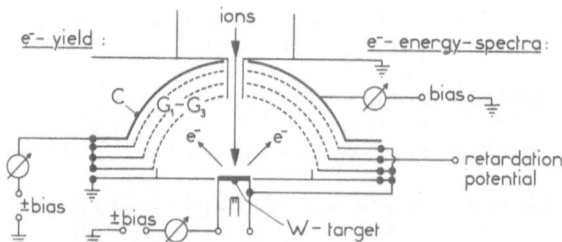

Fig. 6   Experimental apparatus for the measurement
of yield and spectra of secondary electrons emitted
when ions approach a metal surface.

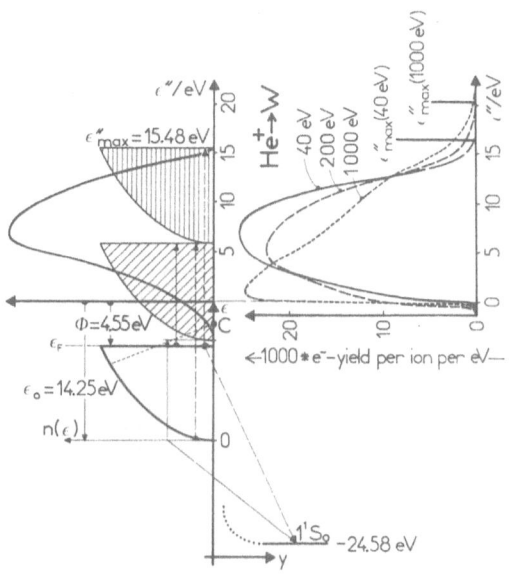

Fig. 7 Electron spectra observed when He⁺ is
approaching a W-surface at three different energies.
Reconstruction of the electron spectrum for v_v = 0.

to the collector to form a simple collector as in Hagstrum's
original work [30]. The determination and subtraction of spurious currents from reflected ions is very well described in
this latter reference.

        The electron spectra and their reconstruction obtained from
He⁺ interacting with a clean, polycrystalline W-surface are
shown on the left side of fig.7. Assuming the free electron jellium model [13-15] for the conduction band of W with ε_0 = 9.7
eV, constant AN-transition rate for all values of ε (see fig.5)
and no effects from level shifts or level broadenings one can
construct the electron spectrum to be expected. An Auger-transition from ε_F to the He ground term yields the replica of the
density of states, hatched vertically, with ε''_{max} = 15.48 eV
while an Auger-transition from ε = 0 to the He ground term
yields the replica of the density of states, hatched diagonally,
with ε''_{min} = -3.92 eV. When allowing now Auger-transitions from
all metal states 0 < ε < ε_F one expects the complete energy
spectrum of the emitted electrons in full line which is a self-
convolution of the density of states of the metal. Since ε''_{min}
is negative the low energy part of the spectrum cannot escape
from the metal.

It was these observations which led Hagstrum to hope that the n(ε)-distribution could possibly be recovered by a deconvolution of the observed spectrum. One has to note, however, that serious problems for such a deconvolution have emerged from his work during the years thereafter. His own critical analysis sheds the best light on the inherent limitations of this method in particular in comparison to similar problems in photoelectron spectroscopy, the well accepted technique for probing the electronic structure of solids and surfaces. In his excellent review on the comparison of both techniques [21] Hagstrum comes to the conclusion that both techniques should corroborate in order to obtain the best possible results.

Another striking feature in the experimental spectra, which represents a serious problem to the initial goal of Hagstrum, is the strong dependence of the spectral shape on the incident energy. The corresponding He velocities for energies between 40 and 1000 eV lie between 0.02 and 0.1 a.u. of which the upper value is somewhat above the limit of the validity of the adiabatic or Born-Oppenheimer approximation. Hagstrum [1,31] therefore attributes the observed modifications of the spectra to time dependent effects to which another aspect was added in ref.[2].

In order to avoid these problems connected to the incident velocity of charged particles Hagstrum has used the lowest possible ion energies of about 5 eV in his systematic study of the INS. [21]. More recently G.Ertl et al. [34-37] have further improved the INS by using neutral rare gas atoms in metastable excited states at thermal velocities. The metastable states lie above the Fermi edge so that the atoms approaching this surface are resonance ionized at large distance so that a singly charged rare gas ion is continuing to approach the surface at thermal energy plus the energy due to the attractive image force. Under such conditions one can expect the velocity dependent effects to be as small as possible and the authors could indeed successfully employ this method for the study of modifications of n(ε) due to the adsorption of atoms or molecules on a surface.

9) Early studies of highly charged ions interacting with metal surfaces

To the great surprise of Hagstrum low energy electrons were dominating the AN-spectrum when the charge state of Kr-projectiles impinging at 200 eV on a clean tungsten surface was increased from 1 through 4 as shown in fig.8 - a behavior which he also found for the other rare gases [1,31]. He further observed:

i) that the electron yield Γ (number of electrons emitted per incident ion) increases with ionic charge and hence with total potential energy available in the ion;

ii) that faster and faster electrons are produced with more highly charged ions but that the mean energy of the ejected electrons is relatively independent of ionic charge;

iii) that Γ drops slowly with increasing ion energy.

From the observations i) and ii) he concluded that the main features can be well interpreted if one assumes that multiply charged ions are neutralized in a series of approximately iso-energetic steps of mean energy $d\varepsilon$". This conclusion was con-firmed by a theoretical model calculation by Arifov et al. [3] in 1973. The authors consider RN and subsequent <u>Auto Ionization (AI)</u> as well as AN processes between a surface and Hydrogen-like states of highly charged ions. The wavefunctions of these states are, however, approximated by an adapted simple exponential only.

$$\Phi_{n\,l} = 1/\sqrt{\pi} \cdot \exp(-r/b) \qquad (9.1)$$

With these functions the authors obtain the average energy of an ejected electron to be $(15-2\cdot\phi)\,eV < d\varepsilon" < (30-2\cdot\phi)\,eV$. On this basis they assume $d\varepsilon"$ to be a constant for all ions so that the electron yield Γ has to become a linear function of W_q — the total potential energy of the ion which is set free when neutra-lizing it in free space or of W_q' when neutralizing it in front of a metal with workfunction ϕ:

$$\Gamma = k\cdot W_q = k\cdot\Sigma_q U_B(q^+) \approx k'\cdot\Sigma_q\,[U_B(q^+)-2\cdot\phi] = k'\cdot W_q'. \qquad (9.2)$$

This relation for the so called <u>Potential Emission (PE)</u> could be experimentally confirmed for highly charged ions of the rare gases interacting with a molybdenum surface up to q = 7 [3].

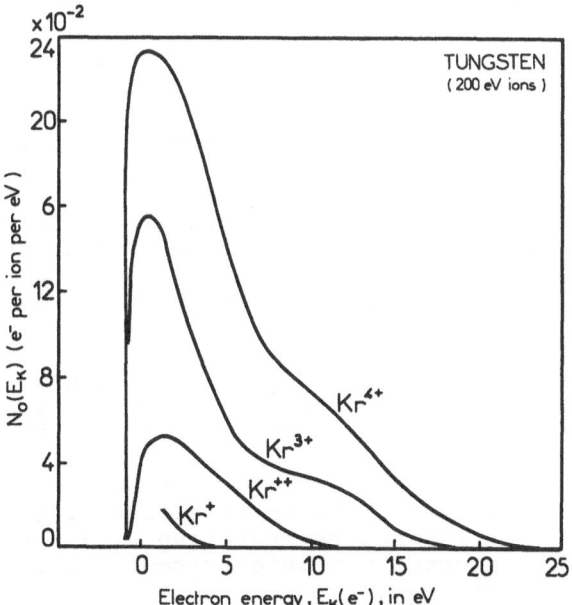

Fig. 8 AN-spectra observed when Kr-ions with various charges interact with a W-surface.

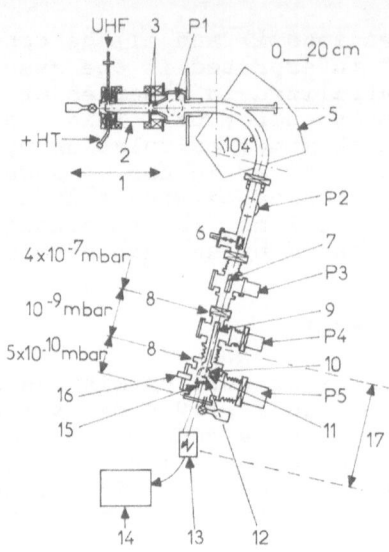

Fig. 9 Apparatus for studies on the
interaction of highly charged ions with
surfaces: (1) ECR-source, (9) retardation
lens, (10) electron-collector, (12) target.
See ref.[39] for more details.

10) <u>Recent investigations of the electron yield per ion</u>

With the invention of powerful sources for the efficient
production of highly charged ions [4,5,6] a renaissance of these
studies became possible with further increasing charge states in
Moscow [6], Groningen [7], Oak Ridge [8], and Grenoble [5]. A
review by P.Varga on this more recent work has just appeared
[38].

The setup in Grenoble is chosen and shown in fig.9 as an
example since its electron detection scheme corresponds closely
to Hagstrum's method and was shown already in fig.6. This detection assembly sits at the end of a dedicated ultra high vacuum
beam line which allows to achieve via two differential pumping
stages a pressure of $5 \cdot 10^{-8}$ Pa in the target region. Ions extracted from an ECR ion source operating at 10 or 14 GHz were
q/m-analyzed, transported via an electrostatic lens system to
the insulated target chamber, decelerated, and focused on a
cleaned, polycrystalline tungsten target. The cleaning was performed by Ar^+-sputtering and flash heating of the target. Tar-
get- and electron collector- currents were measured at the positive deceleration potential so that total electron emission
yields Γ and spectra could be measured down to ion energies of
5 eV/u, corresponding to a vertical velocity $v_v \approx 3 \cdot 10^4$ m/s.

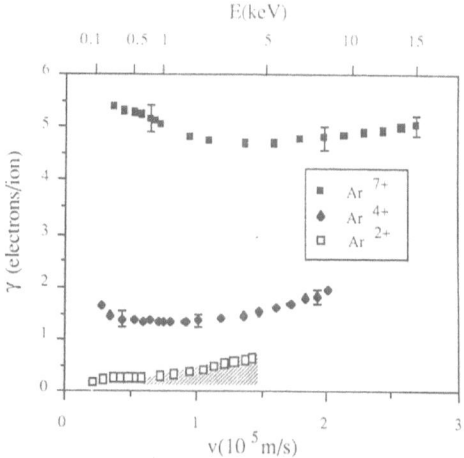

Fig. 10 Electron yield for Ar^{q+}-ions
versus v_v. Shaded area indicates contribution from KE.

A typical result for the electron yield Γ is shown in fig.
10 [39]. From the data for q = 2 the authors deduce that the
<u>Kinetic Emission (KE)</u> of electrons is not negligible so that the
Γ observed is generally composed of $\Gamma = \Gamma(PE) + \Gamma(KE)$. Under the
assumption that PE is negligible for q = 2 the KE can be ob-
tained from the shaded area and can be accounted for in measurements with q > 2 by subtracting this slowly increasing KE-con-
tribution via a linear extrapolation. The data in fig.10 thus
represent dominant PE which after subtraction of KE clearly increases with q and decreases slowly with vertical velocity above
$1 \cdot 10^5$ m/s.

This interpretation is valid, however, only when the KE
does not depend on the charge state q of the ion. This suppo-
sition is questionable since higher charge q implies stronger
time dependent electric fields seen by the collision partner and
hence higher ionization probability or secondary electron yield.
Therefore a q-dependent increase of the KE may be expected and
has been clearly observed in ion-atom collisions at high energies [40]. In the case of ion-surface collisions a systematic
comparison of Γ and in particular of $\Gamma(q)-\Gamma(q')$ as a function of
q and velocity could eventually allow a separation of KE(q,v)
and PE(q,v).

To this end a very promising technique has recently been
employed by H. Winter and his team [41], which allows to study
the detailed statistics of the secondary electron emission when
ions with charge q approach and hit a metal surface. As shown in
fig.11a the ions are directed onto a sputter cleaned polycrystalline Au target at 0 volt which is situated inside a cage at –
20 volts. The design of this assembly is chosen such that an extraction electrode at +3.5 kilovolts allows to direct and post-
accelerate all secondary electrons from the target onto a silicon detector at +30 kilovolts. The trajectories shown are for
electrons emitted at an energy of 20 eV into a solid angle of
2Π. With respect to the time resolution of the silicon detector

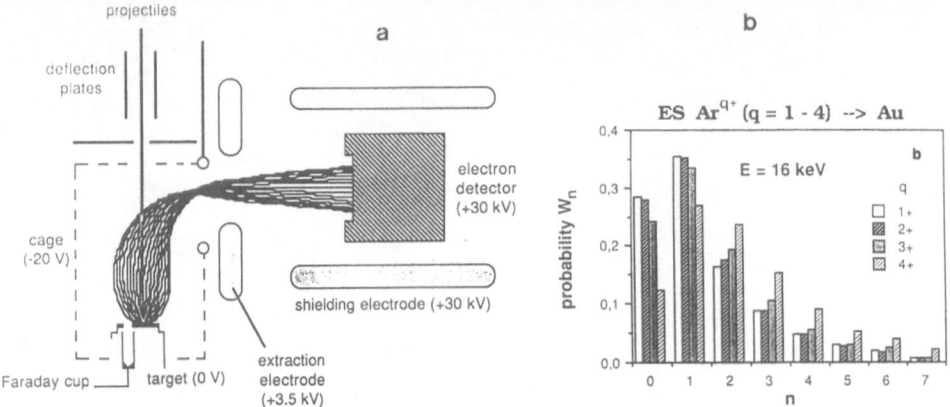

Fig. 11 a) Experimental setup for measuring electron emission statistics. Typical electron trajectories are shown.
b) Statistical distributions of emitted electrons for Ar^{q+} (q = 1-4) approaching a polycrystalline Au-target at 16 keV.

and its electronics the PE and KE are sudden and the difference in travel times of electrons emitted with energies between 1 and 50 eV is negligible. Hence the pulse height registered by the detector corresponds to the number of electrons emitted in a single event of one ion interacting with the surface and the pulse height distribution yields correspondingly the statistics of the electron emission of these events. In order to transform these pulse height distributions into distributions of W_n of probability of emission of n electrons per ion, it is necessary to determine W_0. W_0 is obtained either via extrapolation of a Poisson- or Polya-distribution or via normalization of the distribution to the total yield of electrons Γ using the relations

$$\Gamma = \sum_0^\infty n \cdot W_n , \qquad \sum_0^\infty W_n = 1. \tag{10.1}$$

To this end a measurement of the ion current in a Faraday cup next to the target is carried out. One of the first results is shown in fig.11b for W_n of Ar(q = 1-4) interacting at 16 keV with a polycrytalline Au-target. Obviously the averaged n increases with q as expected from equ. (9.2) but the resulting Γ is somewhat too high -- a systematic observation which the authors attribute to residual gas absorption layers on the target. Once using better cleaned surfaces this technique promises to be particularly well suited to study the extra PE and eventually the extra KE when increasing q over a large variety of velocities.

When plotting the data of fig.10 after the subtraction of the KE at $v_v = 4 \cdot 10^4$ m/s and at $v_v = 2 \cdot 10^5$ m/s as a function of W_q in fig.12a,b, respectively, one observes in fig.12a the linear dependence of Γ on W_q as predicted by the formula (9.2). For fig.12b, however, a systematic deviation from the linear dependence is observed which the authors ascribe to the higher vertical velocity. They argue that the time $T = z_0/v_v$ available for the ion within a given surface distance z_0 becomes shorter and shorter so that the multi-step neutralization becomes less and less complete before the ion hits the surface. This implies that an increasing fraction of W_q cannot be converted into free

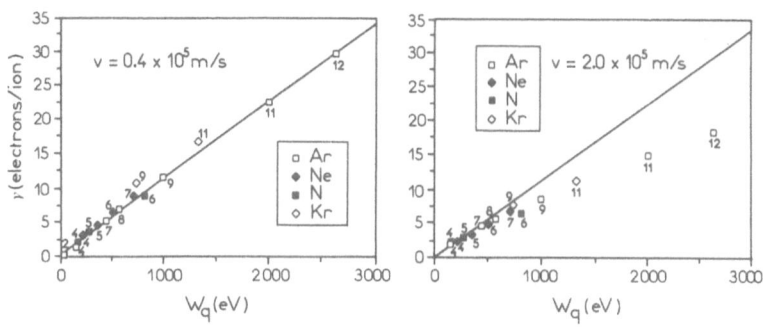

Fig. 12 Electron yield vs W_q, a) at $v_v = 0.4 \cdot 10^5$ m/s,
b) at $v_v = 2.0 \cdot 10^5$ m/s.

electrons so that Γ has to decrease with increasing vertical
velocity. This effect is found to be and has to be more pronounced for high q-values because the number of iso-energetic steps
is increasing with q.

11) Analysis of electron spectra

A more detailed analysis of the electron spectra [42-44,
52,53] sheds, however, more and different light on the data in
Figs. 7 through 12 and on the interpretation above. Hagstrum's
data in fig.8 already imply a very slow but steady increase of
the mean energy dε" of the emitted electrons with increasing q.
This increase of dε" becomes more pronounced when 2p-holes are
existing in the incoming Ar-ions for q ≥ 9 which give rise to
significant Ar-LMX (M < X < ?) Auger emission at around 200 eV
as shown in fig.13 and first observed in ref. [53]. With a LMX-
Auger-contribution of as low as 1% to Γ the mean energy dε" is
nevertheless increased by 13 % from dε" = 14.9 eV to dε" = 16.8
eV. It is thus obvious that the reduction of the $\Gamma(Ar^{9+})$-value
with respect to the straight line in fig.12b is at least partially due to this difference in dε" of 13 %. S.T.de Zwart
[43,44] could indeed show that the deviation from the straight

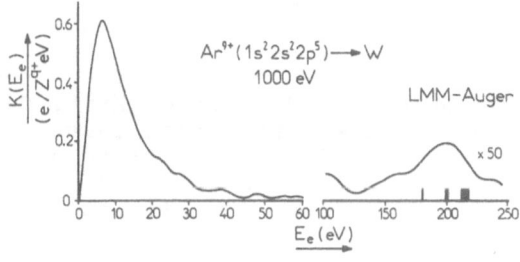

Fig. 13 Electron yield and spectrum per eV for Ar^{9+}
approaching a W-surface.

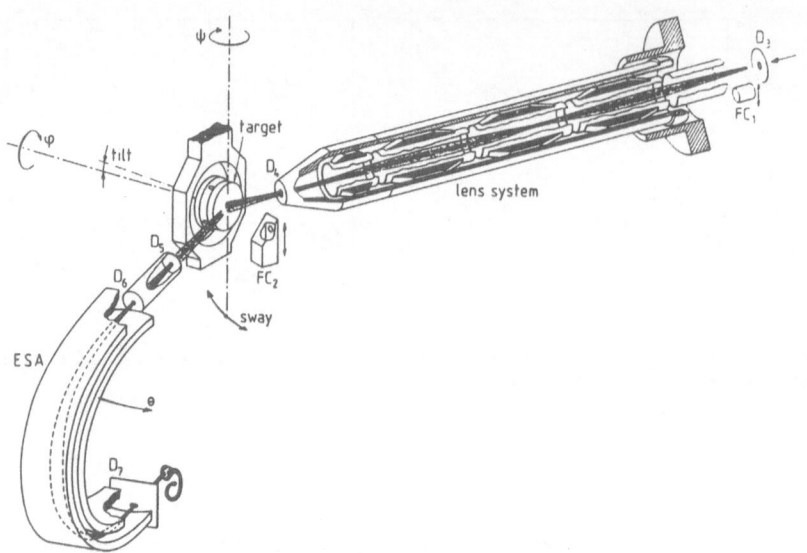

Fig. 14 Differential electrostatic electron
spectrometer (ESA), decelerating lens system
and target manipulation for the work in
refs. [43,44].

line in fig.12b is close to constant for all vertical velocities
from $v_v = 1.3 \cdot 10^4$ m/s to $v_v = 3 \cdot 10^5$ m/s in contrast to figs.12a,
b which show a clear vertical velocity dependence. Whether these
differences are due to the different experimental techniques or
data treatments has still to be shown.

   This latter remark seems necessary since de Zwart's differential electron spectrometer in fig.14, which accepts only a
small solid angle of all electrons emitted, is not ideally suited for measurements of Γ, since it requires at every energy and
incident angle of ions an additional measurement of the angular
distribution of the electrons. This has been done by S.T.de
Zwart for various situations from which he could draw even further conclusions [44]. Nevertheless his spectrometer is better
suited for the observation of spectral features and in particular for the observation of the Doppler effect of the LMX Auger
electrons of Ar^{9+} in fig.15a. With these results he could verify
that the Auger electrons are emitted from freely moving projectiles in front of the surface which start to disappear and wash
out, however, in fig.15b for vertical velocities higher than
$4 \cdot 10^4$ m/s while saturating for vertical velocities below $3 \cdot 10^4$
m/s. This is related to the former noted interpretation that the
time T of interaction with the surface starts becoming too short
for a complete filling of the 2p-hole of Ar^{9+}. One has to recall
in addition that an LMX-Auger transition leaves at least one M-
hole which has to be refilled for a complete neutralization
after the LMX-Auger transition has occurred. The vertical velocity v_v required for a complete neutralization of Ar^{9+} has
therefore to be smaller than $3 \cdot 10^4$ m/s. These results and considerations corroborate very well the findings in Grenoble where a
strong increase of Γ is found consistently in this low vertical
velocity domain for $N(q+)$ (see fig. 24), $Ne(q+)$ [42], and $Ar(q+)$
[39] without reaching the saturation of Γ at $v_v \approx 3 \cdot 10^4$ m/s.

12) <u>Renewed interpretation of the PE</u>

In spite of some discrepancies in the data from the different groups a slightly renewed interpretation of the PE-phenomenon emerges due to an improved comprehension of the RN- and AN- processes and stimulated by the systematic measurements of the electron spectra.

When an ion with $Z_{core} = 9$ is approaching a metal surface its energy levels are shifted upwards due to the image potential according to equ. (3.6). At a distance of 57 a_0 from the image charge in front of the surface this amounts to 2.03 eV ! For initially unperturbed hydrogenic levels with $10 \leq n \leq 14$ and in resonance with the conduction band such a shift of 2.03 eV represents a strong perturbation. For the calculation of any electronic interaction of this ion with the surface the unperturbed wavefunctions would thus be at first sight completely inadequate. For this reason wavefunctions have been constructed by solving the Schrödinger equation using the complete ionic-,

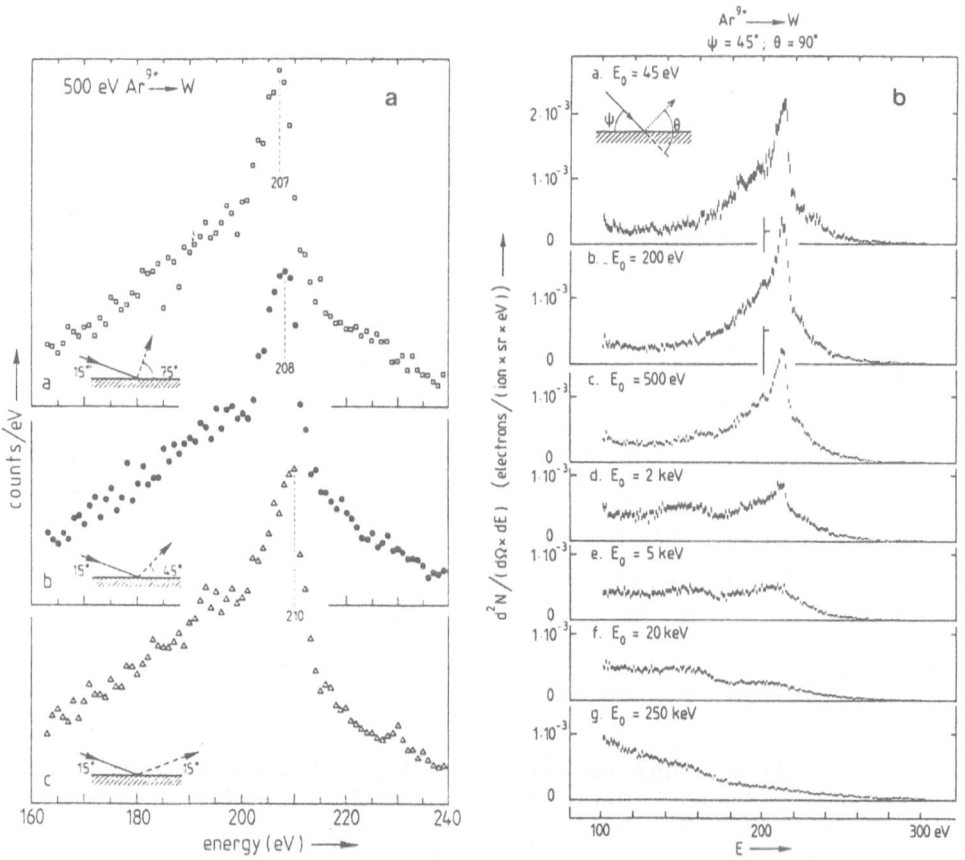

Fig. 15   a) Doppler-shift of the LMX-Auger emission into three different angles with respect to the incoming Ar[9+]-ions.
b) Attenuation of the Ar-LMX-Auger emission for increasing energy of the incoming Ar[9+]-ions.

image-, and metal-potential seen by the electron in order to calculate the resonance charge exchange rate [25,45] between the metal and the ion as a function of the distance. Unfortunately both authors use the same erroneous image potential seen by the electron and find thus reasonable results only for Hydrogen while for increasing Z_{core} the rates turn out to be too high for large distances [45]. One is left therefore with a situation to either repeat these calculations with improved image potentials or to search for approximations which correspond to the basic physics involved.

Choosing the latter approach one can argue that the use of unperturbed wavefunctions without energy shift is not such a bad choice after all: The image potential of an ion with $Z_{core} = 9$ at a distance of 57 a_0 would shift the n = 12 level from 7.65 eV binding energy to 5.62 eV, i.e. very close to the binding energy of the n = 14 level of 5.62 eV. From the point of view of the total potential seen by the electron in such a n = 12 state the image potential reduces the total potential and thus causes a reduction in binding energy and consequently also an increase in the spatial extension of the wavefunction -- at least in the direction towards the surface. Since it is this spatial region close to the surface which, according to chapter 4, is respon- sible for the electronic interaction between the ion and the surface one has to use a spatially expanded n = 12 wavefunction with some Z_{eff} corresponding to the above energy shift. Further- more it was concluded in chapter 4 that only the region of the exponential tails of the wavefunctions are admitted to a pertur- bational calculation. The number of oscillations of a high n wavefunction is thus of no importance at all - only the spatial extension of its exponential tail counts for the treatment of the electronic interaction process by the Golden Rule. One can therefore to a good approximation replace the expanded n = 12 wavefunction by the unperturbed n = 14 wavefunction, the spatial extension of the exponential tail of which corresponds approxi- mately to the one of the n = 12 wavefunction shifted in energy by 2.03 eV. It is of course understood that the n = 14 wave- function is used only for determination of the distance where the capture of electrons from the surface sets in. For the cal- culation of internal properties of the ion, however, like its auto-ionization-rates, the $n_r = n-2 = 12$-wavefunction with the appropriate Z_{eff} should be used. To a first approximation the n_r = 12-wavefunction with Z_{core} will be used in this lecture.

With these arguments in mind the unperturbed energetic situation of an ion with $Z_{core} = 9$ approaching a surface with $\phi = 5$ eV and $\varepsilon_0 = 10$ eV is thus shown in fig.16. The ionic terms with 10 < n < 14 are in resonance with the conduction band and can therefore be populated by RN- and AN'-processes. As an exam- ple the n_s-wavefunctions for n = 14 and n = 9 are shown, the spatial extent of which clearly favour RN-processes to the term energetically closest to the Fermi edge. This argument equally well applies to AN' which is most likely to occur with tran- sitions of type (j) in fig.3b into terms closest to the Fermi edge. It is important to note that such AN'-processes do only produce electrons which heat up the metal but which are not ejected from the metal! AN-processes producing electrons with ε'' > 0 are negligible in comparison to RN as is visible from the negligible overlap of the (n = 9)-wavefunction with the metal wavefunction at the Fermi edge in fig.16. For the ion with Z_{core} = 9 the population of the terms with n = 14,13,12 (n_r =12,11,10)

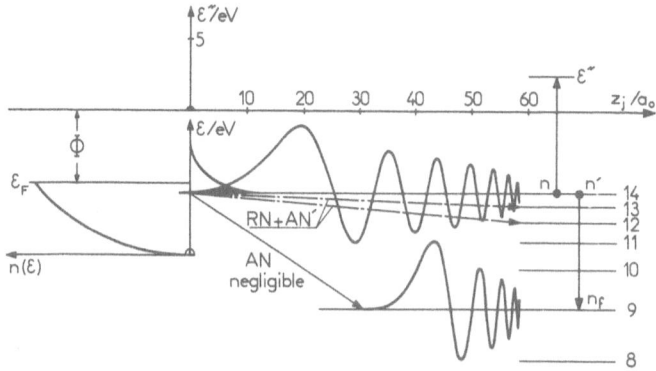

Fig. 16 Rydberg-levels of an ion with Z_{core} = 9 in front
of a metal surface. The overlap of the n = 14,s and of the
k_F-metal-wavefunctions clearly indicates the dominant near
resonant RN- and AN'- processes.

take therefore place dominantly via RN and AN' and sets in at
distances of about z_0 = 60 a_0. For this estimate only geometric
arguments have been used. In a refined
discussion the transition rates and the freezing distances have
to be calculated.

To this end a straight forward numerical calculation of the
RN-rates as a function of the distance z of the ion from the
surface has been carried out (for an analytic approximation see
ref.[44]). In equ.(5.3) hydrogenic Φ_{nlm} wavefunctions with quantization axis along the normal of the surface and standard box-
normalized metal wavefunctions for a potential well of ε_0 = 10
eV (see e.g. ref.[46]) have been used for an integration from z_j
= 0 to z_j = 10b, with b = $h/\sqrt{2 \cdot m \cdot (\varepsilon_0 - \varepsilon)}$ and z_j being counted
from the jellium edge. The resulting RN-rates as a function of

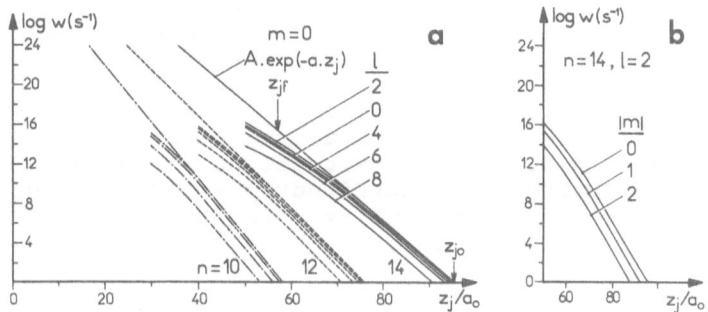

Fig. 17 a) Transition rates for resonant capture into
Z_{core} = 9, n, l, m = 0 levels as function of distance
from the jellium-edge. b) Variation of the transition
rate with |m|.

z_j are compiled in Figs. 17a-c for $Z_{core} = 9$ and a combination
of n,l,m - values. According to the spatial electron density
distribution of the wavefunctions one obtains within the n = 14
manifold the largest rates for l = 0-4, m = 0 in fig.17a which
then systematically decrease with further increasing l-values at
constant m = 0. The same behavior is shown in fig.17a for the n
= 12,10 manifolds at reduced RN-rates corresponding to their reduced spatial extension. For all n,l-levels the wavefunctions of
the |m|-states possess a reduced spatial extension along the z-
axis with increasing |m|-values. Correspondingly one obtains a
reduction of the RN-rate with increasing |m|-value as shown for
n = 14, l = 2, |m| = 0,1,2 in fig.17b. These trends are a rep-
lica of the spatial structure of the wavefunctions "first
touching" the surface. It is therefore obvious that electrons
will be captured preferentially into the lowest l- and m- values
within a n-manifold. For the total capture of an electron into a
n-manifold the capture distance z_f (see equ.(7.4)) has to be
determined with the sum

$$W_n = \sum_{l=0}^{n-1} \sum_{m=0}^{l} w(n,l,m) \qquad (12.1)$$

of the rates which then allow to deduce A and z_{j0} (w=1) in order
to calculate a with equ.(5.4). With this value for a equ.(7.4)
is used with $v_v = 5 \cdot 10^4$ m/s to obtain the freezing distances
where the averaged capture of an electron takes place:

$$z_f (n = 14) = z_{jf} + d/2 \approx z_f (\phi) = 61.5 \ a_0 \qquad (12.2)$$
$$z_f (n = 12) \approx 48.5 \ a_0$$
$$z_f (n = 10) \approx 35.5 \ a_0$$

This table clearly indicates that the n = 14(n_r = 12)-level
which is closest to the Fermi-edge obviously dominates the first
capture of an electron so that one can set $z_0 = z_f (\phi)$. This discussion could be refined by using Stark-eigenstates as basis-
functions in the image field. The relative behavior of the
n,l,m-capture probabilities would, however, not change except
for coherences which are of no importance for the following.

After the capture of the first electron into a 14s-state,
the whole level-scheme in fig.16 has to be shifted slightly
upwards due to screening. For the case of $Z_{core} = 9$ the screening is assumed to yield $Z_{eff} = 8.7$ for the next electron to be
captured. It will thus again be captured into a n = 14(n_r = 12)
-level the energy of which still is below the Fermi-edge. As
soon as this electron is captured strong correlation will cause
autoionization for which one can estimate the rate depending on
the final term n_f of the ion with $Z_{core} = 9$. Only terms with n_f
≤ 9 are considered which allow a free electron to be ejected
with energy $\varepsilon'' > 0$. A model calculation is based on the use of
uncorrelated hydrogenic s-wavefunctions for the bound states and
Coulomb s-wavefunctions for Z" = 8 for the free electron. The
result of this simple estimate is shown in fig.18 and suggests
three significant features:

i) The rate of AI is rather low (of the order of 10^{13} s^{-1}) for
these high lying Rydberg levels.

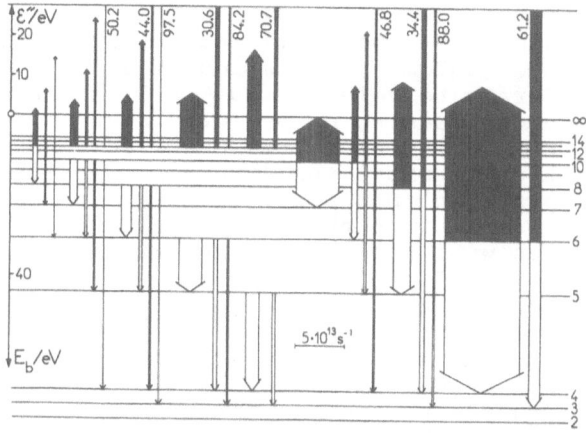

Fig. 18 Autoionization rates for two electrons calculated
with initial and final uncorrelated Hydrogenic s-wavefunctions and s-Coulomb-wavefunctions. The arrows indicate
the direct transitions (black for the ejected electron) and
their widths the transition probability in units given in
the lower center of the figure.

ii) The rate of AI is peaked at the lowest possible energy ε'' of
the free electron and drops rapidly with increasing energy ε''.
For $n = n'$ but both constant and n_f = variable the rate drops
roughly proportional to $c(n,n') \cdot \exp(-c' \cdot \varepsilon'')$ and for $n = n' =$
constant and n_f = variable the rate drops roughly proportional
to $c''(n,n')/\varepsilon''$. This behavior is indicated in fig.18. It depends
on the structure of the wavefunction of the free electron and is
thus very similar to the behavior of photo-ionization probabilities as a function of the energy of the free electron.

iii) The constants $c(n,n')$ and $c''(n,n')$ are increasing with decreasing n' (see left half of fig.18) and also with decreasing n
$= n'$ (see right half of fig.18).

These observations hold for two s-electrons in the hydrogenic approximation for all n,n',n_f ($1 = 1' = 1_f = 0$) terms. The
inclusion of the $1,1',1_f = 0$ terms does not increase the AI-
rates considerably so that the overall behavior corresponds to
fig.18. In the two-electron approximation the high Rydberg
levels (with AI-rates of only 10^{13} s^{-1}) form thus a bottleneck
for the neutralization of a highly charged ion in front of a
surface, since the total time available in front of the surface
is only of the order of $z_f/v_v \approx 10^{-14} \div 10^{-13}$ s.

All presently existing data on the interaction of highly
charged ions with surfaces suggest a significantly higher AI-
rate (see ref.[2]) for these high Rydberg levels. The following
model may therefore be adopted: After the capture of the first

two electrons further electrons will be resonantly captured
still at large distances and still into high Rydberg levels
until the combined Coulomb correlation of all these electrons
increases the AI-rate to values $> 10^{15}$ s^{-1}. It has to be noted
of course that an increasing number of electrons in a given n-
shell will lower the binding energy per electron so that they
may move energetically above the Fermi-energy. This then allows
efficient re-ionization into empty metal-states so that further
resonant capture first into the (n-1)-shell, then into the (n-
2)-shell and so on will become more important until a sufficiently large number of electrons is accumulated in such high
Rydberg levels so that the AI-rates eventually become high
enough. Such a detailed discussion of the capture- and AI-processes becomes thus extremely complicated. These details are
furthermore of not much help since the time-energy uncertainty-
-at the high AI-rates required -- introduces an energy uncertainty of a few eV so that the distinction of specific levels
becomes useless. On the average one therefore better has to
speak of a continuous flow of electrons towards Rydberg levels
with binding energies close to ϕ until the number of electrons
in these levels becomes sufficient to provoke high enough AI-
rates via multiple Coulomb correlation.

This exciting new situation of the de-excitation dynamics
of a multiply (q-times) excited ion can unfortunately not be
treated theoretically for the time being. Nevertheless this
extreme state of nature is new and very stimulating and merits
intensified studies since it may lead the way to questions like
the stability of such excited multi-electron systems. For the
time being, however, one has to assume a model on the basis of
the results shown in fig.18: For a multiple Coulomb correlation
the dependence of the AI-rates on the energy of the free electrons will still dominantly depend on the structure of the wave
function of the free electrons and will thus stay similar to
that shown in fig.18, with the absolute rates increased,
however.

This implies for the first AI-processes that electrons will
preferentially drop from a binding energy of $\approx \phi$ to a binding
energy of only $\approx \geq 2 \cdot \phi$ in order to liberate electrons with the
smallest possible kinetic energy, corresponding to a distribu-
tion shown at the outer left of fig.18. For the first or second
electron at binding energy $\approx \geq 2 \cdot \phi$ the AI-rate will again be too
low for an immediate pursuit of their descend to lower energies.
During continued flow of resonant electrons towards Rydberg
levels of $E_b \approx \phi$ and during continuous AI with liberation of
free electrons with very small kinetic energy a new bottleneck
will thus exist at $E_b \approx \geq 2 \cdot \phi$. It will exist until enough electrons are accumulated in this region of E_b in order to provoke a
sufficient AI-rate for the further descend of electrons to $E_b \approx \geq$
$4 \cdot \phi$ - again liberating preferentially electrons with the smallest possible kinetic energy. This type of liberated electrons
of very small kinetic energy thus readily explains the dominant
maximum observed in the electron spectra at very low energies
which is indeed found to be at lower energies than predicted by
Arifov et al. [3].

During further continued flow of resonant electrons towards
Rydberg levels of $E_b \approx \phi$ and during further continuous AI with
liberation of free electrons with very small kinetic energy a
Distribution Of Electrons in the ion will develop (called DOE

for the following) which corresponds to a convolution of exponentials of the type $\exp(-c \cdot [E_b - \phi])$ and of functions of the type $c'/(E_b - \phi)$. The total number of electrons involved in this dynamics stays of course close to the initial charge q.

While this DOE energetically further drops down into the potential well of the ion the screening for the outer electrons increases such that the quantum number $n(E_b \approx \phi)$ to which the continuous flow of resonant capture proceeds will steadily decrease. Simultaneously the screening for the inner electrons of this DOE will stay very small so that the energy levels and AI-rates of fig.18 are still valid for these electrons in a purely hydrogenic case. (In a case with core electrons, like Ar^{9+}, the internal energetic structure has to be taken into account of course.) Consequently the DOE will energetically stretch out further and further in a peculiar manner such that the binding energy E_{bmax} of its maximum stays close to a few times ϕ while its tail proceeds to higher and higher binding energies. The AI of the outer electrons with $\phi < E_b < \approx 2 \cdot E_{bmax}$ of this DOE will further contribute to the electrons emitted at very low energy while the AI of the inner electrons with $E_b > 2 \cdot E_{bmax}$ of this DOE will produce free electrons with systematically increasing energy -- with small probabilities, however. This readily explains the rather slow increase of the mean energy of the emitted electrons $d\varepsilon''$ as observed for instance in fig.8.

The ion is thus completely filled from the outer shells by a DOE, the maximum of which will stay in the region of binding energy of a few times ϕ. Simultaneously the energetic descend of the tail of this DOE is accelerated due to increasing AI-rates of pairs of inner electrons of the DOE at decreasing n-values. An equilibrium for the filling process (neutralization) of an ion will thus develop which will depend on Z and on the initial charge of the ion, i.e. also on its structure. For small Z (Z<10) and high q (q≈Z) the acceleration of the tail of the DOE versus low n-values may dominate whereas for medium Z (10 < Z < 30) and medium to high q (Z/2 < q ≤ Z) the relative stability of the maximum of the DOE (the bottleneck at higher n-values) may dominate.

The energy uncertainty and the statistics of all low energy AI-transitions (see fig.18) will obscure any structure in the low energy part of the electron spectrum. The quantum structure of the ion will therefore start playing a role only for AI-(Auger)-transitions involving lower lying shells so that discrete structures in the electron spectrum may occur. Such transitions start occurring when the first electron in the tail of the DOE reaches one of the lower lying levels while (q-1) electrons are still in higher n-shells. The high resolution spectroscopy of the resulting Auger-transitions may give a handle to study the time development of the DOE when details of LMX- (see fig.15b) or KLY-transitions (M < X < ? and L < Y < ?) can be resolved and can be measured as a function of the vertical velocity. For ions with initially low charges q < ≈ 4 such structures may be expected in a well resolved electron spectrum at some tenths of eV while for initially higher charges q > ≈ 4 structures above 50 eV may start occurring. For the latter case the well resolved high energy LMX Auger electrons may serve here as example, which are observed by Delaunay et al. [39,42] and de Zwart [44] in Figs. 13 and 15, respectively, when Ar^{9+} with a hole in a low lying 2p-shell is used.

13) <u>Re-analysis of the electron spectra</u>

The scenario of the neutralization dynamics presented in the foregoing chapter implies that inner shell Auger-transitions will take place at some intermediate time of the total neutralization process. When the vertical velocity of an ion approaching the surface is increased it is therefore these inner shell transitions which will disappear first when the time $T = z_0/v_v$ available in front of the surface is reduced. This is in very good agreement with the observations in fig. 15b by de Zwart or by Delaunay who clearly see the Auger electrons emitted from Ar^{9+} disappear much faster than the electrons at lower energy when v_v is increased. The higher v_v becomes the more inner shell holes survive up to very small distances from the surface, and the lower v_v becomes the more inner shell holes are filled before the ion touches the surface.

This feature experimentally allows to establish a time scale for the filling of inner shell holes of various ions when plotting the integral of the intensity of Auger-transitions filling this hole as a function of v_v. To this end S.T.de Zwart [44] has integrated the electron intensities of fig.15b from 175 to 300 eV, corrected the integrals for differences in the angular distribution, which depend on the energy of the ion, and plotted the so obtained total yield of LMX Auger electrons in fig.19. When assuming with chapters 7 and 12, that the distance $z_0 = z_f(\phi) \approx 61.5$ a_0 is rather independent of v_v and when furthermore assuming that the depth of penetration of the ion, from which clearly resolved Auger-structures can still be observed, is small compared to z_0, fig.19 can directly be converted into a time scale for the filling of the 2p-hole of Ar9+ in fig.20. It is striking to see the data follow an exponential very well with a mean time for the filling of the 2p-hole of Ar9+ of $\tau(2p) \approx 5 \cdot 10^{-14}$ s and with a value of saturation of $\Gamma_{sat} \approx 0.35$ Auger electrons per ion. In a further refined discussion the velocity dependence of $z_f(\phi)$ in equ.(7.4) and the penetration depth of the ion still emitting LMX-Auger-electrons in the same energy range should be included.

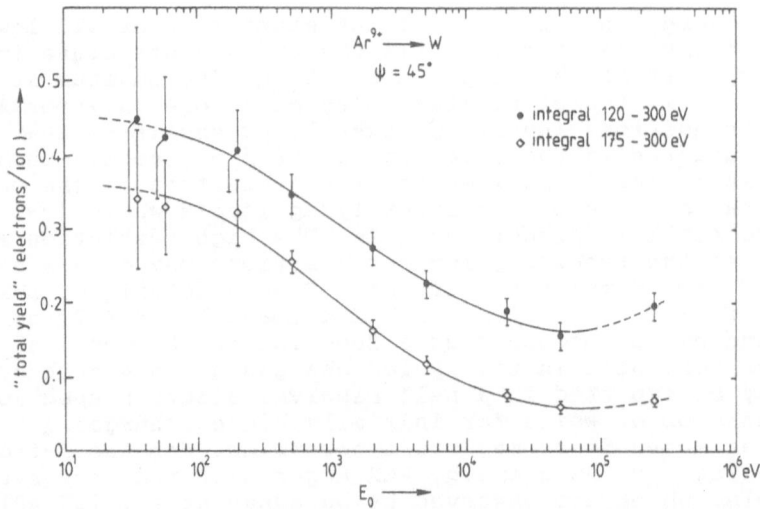

Fig. 19 The total yield of electrons with energies in the range from 175 to 300 eV as derived from fig.15b.

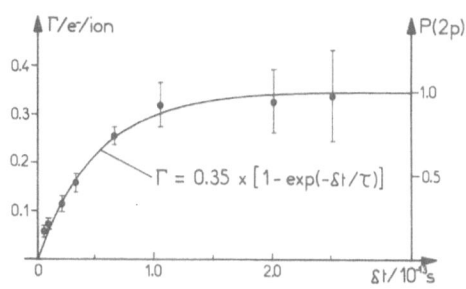

Fig. 20 Time dependence for the filling of
the 2p-hole of Ar⁹⁺ approaching a W-surface.

From $\tau(2p)$ one can first of all deduce that w_{AI} in the
high n-levels has to be of the order $\geq 1 \cdot 10^{15}$ s^{-1}. One may also
roughly estimate a mean time for the neutralization of Ar⁹⁺ of
$\tau(neutral) \approx 2 \cdot \tau(2p)$ when applying the logic of the model in
chapter 12. Such a mean time for the neutralization requires
extremely low v_v-values in order to complete the neutralization
in front of the surface. As pointed out in chapter 2, however,
very low energies of the ion allow to consider a high degree of
reflection so that the time available for the final neutralization grows faster than proportional to $1/v_v$. This readily explains the strong increase of the total electron yield as mentioned at the end of chapter 11 and shown in fig.24, since the
final neutralization yields again large numbers of low energy
electrons. It has to be noted that the ion will come quite close
to the first atomic layer when being reflected at very low energies or when entering into the surface at higher energies. These
are distances where the rates for direct AN- and AD-transitions
or QNR-, CAN-, and CAD-transitions of fig.3a,b, c, respectively,
become non-negligible. All these transitions may therefore contribute during a small fraction of the total time available for
the neutralization process in competition to the RN-AI-transitions which continue to dominate at larger distances from the
surface.

From $\Gamma_{sat} \approx 0.35$ in Figs. 19,20 instead of the expected
value of 1 electron per ion S.T.de Zwart [44] draws important
conclusions on the detection efficiency of LMX-Auger electrons
emitted isotropically from an ion in front of the surface. Since
the reflection probability of electrons directed towards the
surface is very small [47] he expects to detect only 50 % of the
LMX-Auger electrons. This value lies within the error bars of
the 45 % observed when the integration of the electron spectrum
is carried out from 120 to 300 eV (filled circles in fig.19) as
justified by the curves in fig.21 where it becomes evident that
the whole structure from 120 to 300 eV is related to the
presence of the 2p-hole in Ar⁹⁺.

The detailed structures of the electron spectra in figs.15b
and 21 and in particular their significant dependence on the
energy of the ion may serve to indicate some information on the
time development of the DOE during the neutralization of Ar⁹⁺ in
front of a metal surface. S.T.de Zwart [44] has very well recognized that the structures observed belong to LMX-Auger tran-

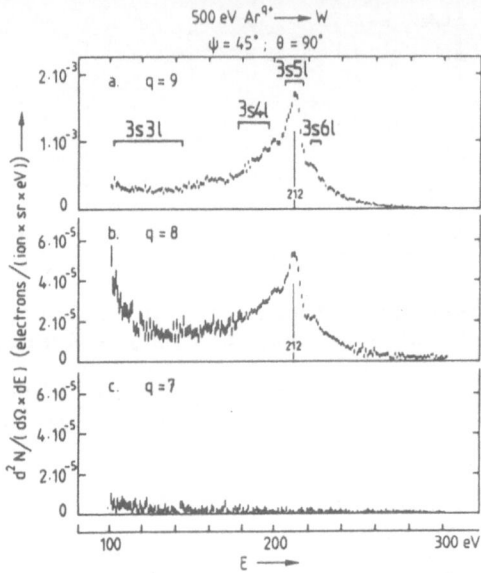

Fig. 21 The LMX-Auger structure as a
function of the initial charge state.
Note the difference in scale between
q = 9 and q = 8. Assignment of the
electron configurations.

sitions where M ≤ X ≤ P. With the help of experimental [48-49]
and theoretical [50] values on the $Ar^{7+}(2p^5 3snl)-Ar^{8+}(2p^6)-AI-$
transitions he could assign the 3s3l-, 3s4l-, 3s5l-, and 3s6l-
fractions of the spectrum. It thus appears that the dominant
peak at 212 eV is due to LMO-Auger-transitions. This is a very
important observation which is fully consistent with the model
in chapter 12 and allows therefore some preliminary
interpretation.

Since the screening for inner shells is small while outer
shells are heavily populated one can assume the validity of the
AI-rates for pairs of electrons in the hydrogenic approximation
of fig.18. One may thus plot the AI-rates for a (n',l'= 3s) ->
(n",l"= 2s) transition under emission of one ns-electron with 4
≤ n ≤ 7 in fig.12 as a function of the realistic energies of the
n-levels derived from fig.8b [31] with E_b (n' = 3 and n" = 2)
fixed at the unscreened values of Ar^{9+} [32]. These rates should
of course be improved by taking into account the structure of
Ar^{9+}, all transitions with non-zero angular momentum, and the
existence of many electrons in the n = 4-7 shells, but for a
first guidance to an interpretation they may nevertheless be
used here. These rates w_{AI} have to be multiplied in a first approximation by the number of electrons in the n-shells in order
to reproduce the intensities of LMX-(X = n)-Auger transitions in
the spectroscopic structures observed in Figs. 15b and 21.
Taking into account the relative widths of the 3snl-contribu-
tions in fig.21 one obtains with the histogram of intensities in
fig.22 the three relative populations in the n = 4,5,6-levels,

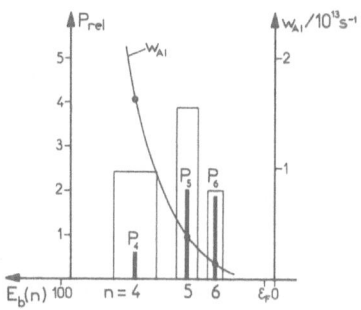

Fig. 22 Reconstruction of the averaged
relative electron distribution in Ar^{9+}
as deduced from the LMX-Auger spectrum.

depicted as black bars with identifications p_4, p_5, p_6, respectively. These p_n-values represent approximately the DOE as expected from the model in chapter 12 at the moment when the LMX-
transitions take place, i.e. when on the average one electron
has arrived in the n = 3 shell. For the completion of the DOE
one has to keep in mind that the total number of electrons in
the DOE should not exceed q and that the screening by the
acquired electrons in n = 3,4,5,6-levels shift the n > 6-levels
considerably upwards (e.g. E_b [n = 7, Z_{eff} = 6]= 10 eV) so that
only one or two levels are left below the Fermi-edge for the
DOE. This problem sheds of course some doubt on the energy position and identification in Figs. 21 and 22, since only unscreened 3snl-configurations have been considered there instead
of taking into account increased screening with increasing n.
Like all the other approximations mentioned this problem has to
be analyzed more carefully in the future when more and more
experimental data exist for comparison.

When one accepts the interpretation in fig.22 the spectroscopic details of the LMX-Auger electrons offer an excellent
tool for the study of the time development of the DOE. The significant differences in the spectra in fig.15b(a.) at 45 eV ion
energy and fig.15b(c.) at 500 eV clearly demonstrate the great
potential in this type of approach. One may e.g. conclude that
the spectrum in fig.15b(a.) shows the DOE at a very late stadium
of the neutralization and consequently has significantly less
3s6l-intensity compared to the spectrum in fig.15b(c.) which is
the result of a DOE at an earlier stage of the neutralization.

Also very exciting is the possibility indicated in fig.21
with the existence of LMX-Auger transitions in Ar^{8+} where normally a 2p-hole does not exist. It is well known, however, that
ECR-sources produce Ar^{8+}-beams with small fractions of Ar^{8+}-
(2p^{5}3s)-metastable ions in this particular case with again one
2p-hole (note the reduction of intensity in fig.21b by a factor
of about 30 with respect to the one in fig.21a). Since the 3s-
electron is existing already in the incoming ion a slightly
earlier stage of the DOE should appear in the LMX-Auger spectrum
of fig.21b compared to the spectrum in fig.21a where the DOE has
to develop further in time for the creation of the 3s-electron.

The statistics of the data in fig.21 is not good enough, however, in order to clearly detect a difference, but a slight shift of the 3s61-contribution to higher energy (less screening !) and a stronger high energy tail in fig.21b may be noted.

One can conclude therefore that the high resolution electron spectroscopy will certainly become a very powerful tool for the future study of astonishing details of the neutralization dynamics of highly charged ions in front of metal surfaces on a time scale of 10^{-13} to 10^{-14} s ! In particular the systematic study of ions with small Z (Z<10) and high q (q≈Z) as well as medium Z (10 < Z < 30) and medium to high q (Z/2 < q ≤ Z) should proof or correct the model of chapter 12.

The electron spectroscopy has already proved to be a very powerful tool in the domain of higher v_v where, according to fig.20, an increasing probability exists for the survival of a 2p-hole in Ar^{9+} or of inner shell holes in any ion up to very short distances from the surface. As mentioned earlier, in such a situation the rates for direct AN- and AD-transitions and in addition for the QNR-, CAN-, and CAD-transitions of fig.3a-c, respectively, become non-negligible. All these transitions may therefore contribute at higher vertical velocities to the neutralization process in competition to the RN-AI-transitions which dominate at vertical velocities $v_v < 5 \cdot 10^4$ m/s. As a consequence more and more moderately resolved structures will occur in the electron spectra with increasing vertical velocity. Some of these structures may, however, also stem from pure intra-metal Auger transitions which occur to refill the core holes of the metal after QNR-, CAN-, or CAD-transitions. Most of the spectral features obtained in Oak Ridge [52,53] can so be explained when velocities $v_v \gtrsim 5 \cdot 10^4$ m/s are assumed. This v_v corresponds to an energy of Ar of 550 eV at which, according to fig.20, ≈ 27 % of the 2p-holes survive up to the first atomic layer. This situation corresponds to the use of 80 keV Ar in a geometry with grazing incidence [2] of 5° on a highly polished Au-target so that the appearance of Au-NNV-Auger transitions in the electron spectrum is no surprise. It is due to a QNR-, or CAN-, or CAD-transition from the Au-N-shell to some lower lying empty level in Ar^{9+} taking place very close to the surface or already below the first atomic layer. In a similar manner all the data obtained in Oakridge with highly charged O- and N-ions can be interpreted [52,53,54].

The high quality spectroscopy of these inner shell transitions taking place in the close vicinity of target atoms could reveal interesting information on the hole-situation of ions at higher velocities. Kinetic emission (KE) may, however, spoil a lot of the overall structures to be expected in these spectra. KE certainly increases with respect to the simple relation derived from fig.10 when ions with inner shell holes (or residual charge q_r) penetrate the surface at velocities $v_v \gtrsim 1 \cdot 10^5$ m/s (see discussion in chapter 10). This increase of KE may be one reason for the insufficient quality of the spectra existing so far with the emphasis on the QNR-, CAN, and CAD-transitions. A quality of the order of that in Figs. 15 or 21 has to be obtained in order to draw further conclusions on the neutralization dynamics of the ion. For the time being the phenomenon of these transitions as such is therefore the major concern of the experimentalists involved.

14) Re-analysis of the measurements of total electron yields

The appearance of discrete high energy structures in the electron spectra destroys the physical basis of the simple relation for the electron yield Γ to be proportional to W_q. Instead the total energy released by free electrons per ion should now be related to W_q. Defining $n(\varepsilon") = d\Gamma/d\varepsilon"$ one obtains for the total energy released by free electrons

$$E_q = \int_0^\infty n(\varepsilon") \cdot \varepsilon" \cdot d\varepsilon" + 2 \cdot \phi \cdot \Gamma \qquad (14.1)$$

where the second term accounts for the fact that RN feeds the ion with electrons approximately from the Fermi edge and that AI has to overcome ϕ in order to produce a free electron. Not included in these definitions are the AN-processes which excite electrons into the continuum of the metal $(\underline{E_{AN}})$, the RD-deexcitations (significant for high q-values) $(\underline{E_{RAD}})$, and a geometric factor which has to be introduced due to the emission of 50 % of the free electrons with velocity vectors pointing into the metal of which an as yet unknown fraction is reflected (see the discussion of fig.19). One can therefore relate E_q to W_q by where

$$E_q = C_{geom} \cdot (W_q - E_{AN} - E_{RAD}), \qquad (14.2)$$

$0 < E_{AN} < 0.4 \cdot \phi \cdot \Gamma$ (roughly estimated from the relative rates), $E_{RAD} \ll E_q$ for ions with q < 13, and $0.5 < C_{geom} < 1$. The expression (14.1) for E_q requires a rather precise absolute knowledge of the whole electron spectrum particularly at high energies. This region has so far been measured only in a few cases, unfortunately, with moderate absolute accuracy and with an intermediate part missing. An attempt with a linearly interpolated intermediate part is nevertheless being made here to integrate the spectra of N^{6+} impinging at $v_v = 1.1 \cdot 10^5$ m/s on W [32] in fig.23 and of Ar^{9+} impinging at $v_v = 7 \cdot 10^4$ m/s on W [56] to yield $E_q (N^{6+}) = 443$ eV $= 0.54 \cdot W_q$ and $E_q (Ar^{9+}) = 324$ eV $= 0.32 \cdot W_q$, respectively, for which uncertainties are difficult to estimate.

As striking result one notes that E_q is a larger fraction of W_q for N^{6+} than for Ar^{9+} although the interaction time T is shorter for N^{6+} because of the higher vertical velocity and because of the smaller z_0. More important radiative energy losses

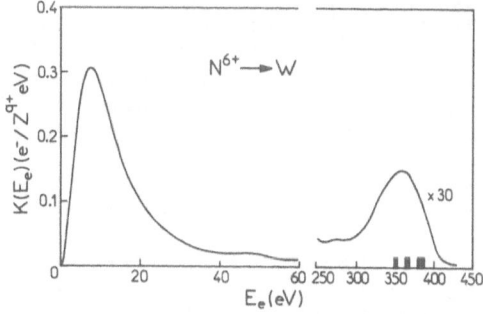

Fig. 23 Yield versus kinetic energy of electrons emitted when N^{6+} is approaching a clean polycrystalline W-target at 900 eV energy.

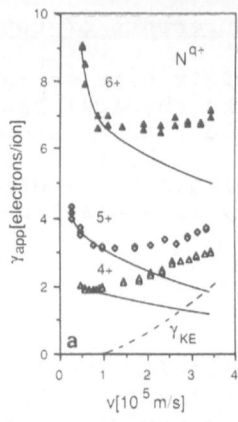

Fig. 24 Total electron yield per ion
as a function of the vertical velocity
of N^{q+} (q=4-6) approaching a W-surface.

for Ar^{9+} are not likely to be responsible for this different
behavior because the radiative transition rates are on the contrary higher to fill the K-hole of N than to fill the L-hole of
Ar. E_{RAD} and E_{AN} can therefore be neglected for the further discussion of these results. The key to a consistent interpretation
may be the observation that for N^{6+} the largest contribution to
E_q is stemming from values of ε'' above 60 eV while for Ar^{9+} more
than 50 % of E_q stems from values of ε'' below 60 eV. This difference suggests that the K-hole of N is filled with a probability
close to one during the interaction time T while the L-hole of
Ar is filled only with a probability of the order of 0.5 in
agreement with fig.20. As a conse-quence the refilling of the L-
holes in N after the KLL-Auger transition will set in at higher
vertical velocities than the refilling of the M-holes in Ar
after the corresponding LMX-Auger transition. Therefore the
steep rise of Γ due to the filling of the L- or M- holes, respectively, under emission of low energy electrons will set in at
higher velocities for N than for Ar in good agreement with the
observations of Γ at the lowest vertical velocities [56]. From
these observations and and their interpretations it is, however,
also obvious that the neutralization is not completed in either
case before the penetration of the surface occurs since otherwise a saturation of Γ would show up at the lowest vertical velocities similar to the saturation of the LMX-Auger yield in
Figs. 19 and 20. In order to better detect such a saturation
effect, Γ and E_q should better be plotted as functions of $1/v_v$
for v_v being still smaller than all present data. It should
however, be noted that still other mechanisms might be responsible for the steep increase of Γ at the very low velocities.

15) <u>X-ray spectra</u>

From the model in chapter 12 and from fig.20 one can im-
mediately deduce that Ar^{17+} or Ar^{18+} at $v_v \geq 5 \cdot 10^4$ m/s will keep
their 1s-hole with very high probability all along from the be-
ginning of the resonant capture of electrons at $z_0 \approx 110$ a_0 into
n = 28 (!) of Ar^{17+} up to the surface layer of atoms. Although
the time T = z_0/v_v available to fill the 1s-hole is

410

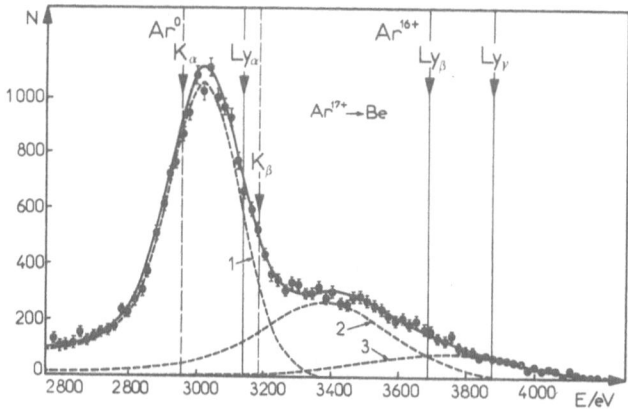

Fig. 25 X-ray spectrum of Ar¹⁷⁺ ions incident on a Be-target with an energy of 420 eV/nucleon.

roughly doubled with respect to Ar⁹⁺ at the same velocity the neutralization dynamics of Ar¹⁷⁺ will last considerably longer than that of Ar⁹⁺. Besides the RN-AI-processes discussed in chapter 12, Ar¹⁷⁺ has sufficient K-fluorescence yield (11.8 %) [57] to allow for the observation of np — 1s radiative decays, indicated as RD-transitions in fig.3c.

E.D. Donets et al. [58-61] were the first to observe these transitions in the interaction of Ar¹⁷⁺ with a beryllium target at an energy of 0.4 keV/u. The Ar¹⁷⁺ ions were produced by the EBIS-source KRION-2 [58,59] and directed onto a Be-target in-clined 84° with respect to the ion beam. The X-rays are observed through a window of 0.1 mm Be with a Ge-detector at 90° with respect to the ion beam. The result is shown in fig.25. The spectrum is decomposed into three contributions [61] which are compared with the K_{α} - and K_{β} - lines of neutral Ar (broken ver-tical lines) and with the Lyman-series of He-like Ar¹⁶⁺ (full vertical lines). The first component is attributed to the 2p → 1s-transition which is shifted by +69 ± 11 eV with respect to the K_{α}-line of neutral Ar. The second component is attributed to the 3p → 1s-transition which is shifted by +215 ± 40 eV with respect to the K_{β}-line of neutral Ar. The third component is at-tributed to np → 1s-transitions with n ≥ 4. Taking into account the absorption in the Be-window, in air and in the detector it-self, it is found that the K_{β}-line is about three times more in-tense with respect to the K_{α}-line than in neutral Ar. Further-more it is noted that the spectrum in fig. 25 does neither change significantly with the target material (cleanness not specified) nor with a reduction of the energy from 420 to 50 eV/u.

An analysis of these observations [61] allows to conclude that the positions of the K_{α} - and K_{β}-lines are consistent on the average with an emission from Ar-atoms with one K-hole, four L-holes, and two or three M-holes. From the halfwidths of the com-ponents 1 and 2 in the spectrum it is derived that the statis-tics of all combinations of L-holes from one to eight do contri-bute.

This analysis is fully consistent with the model in chapter 12 and corresponds to a rather late stage of the development of the DOE in the lower n-levels. This is due to the fact that the K-X-rays are on the average only emitted when the non-radiative transitions have filled already the L- and M-shells with the probabilities mentioned. High resolution X-ray spectra observed with a crystal spectrometer could thus deliver very helpful information on this stage of the development of the DOE in the same way as the Auger-spectra in Figs. 15b and 21 give detailed insight into the neutralization dynamics. The KLY-Auger spectra of Ar^{17+} would be of corresponding importance. One obstacle for high resolution X-ray and Auger studies could, however, be the statistics of the large number of electrons involved which may wash out detailed structures in the spectra. It is, however, known that only the L-electrons have a significant influence on the $2 \rightarrow 1$ transition energies so that all other electrons may be neglected [64]. A first such measurement is thus eagerly awaited for. It has to be added, however, that X-ray measurements suffer somewhat from the fact that the distinction between X-rays emitted in front of the surface and the ones emitted inside the bulk is much less obvious than for Auger electrons. In order to avoid this difficulty the vertical velocity has to be reduced to values where the dominant fraction of the ions is reflectd from the surface. The geometry of grazing incidence [2] offers great potentials for such studies.

The measurement of X-ray spectra could be extended with the same ion-source to Kr^{35+} interacting with a Cu-surface at 0.4 to 8 keV/u [62] in a very similar manner as the Ar^{17+}-measurements. The observed spectrum can again be decomposed into three contributions at 12.88, 15.0, and 16.0 keV, respectively. The analysis suggests on the average an emission from a Kr-ion with 1 K-hole, 6 L-holes, and a filled M-shell. This corresponds to a DOE in the lower n-levels at an earlier stage of development compared to the one of Ar^{17+} which can readily be explained by the increased total $2 \rightarrow 1$ transition probability in Kr^{35+}. These

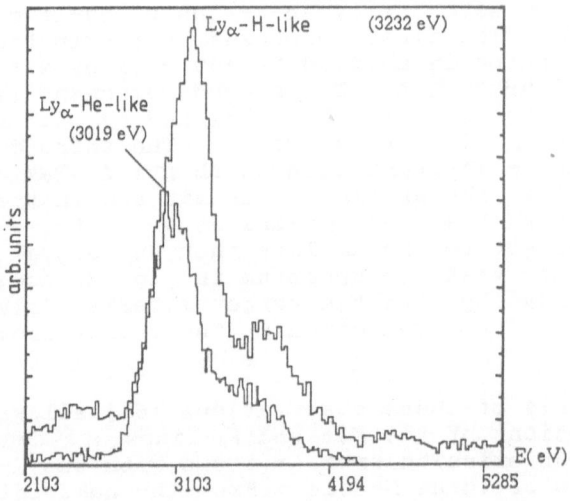

Fig. 26 X-ray spectrum from Ar^{18+} interacting with an Al-surface in comparison to the spectrum obtained with Ar^{17+}.

radiative transitions thus sample the very first electrons of the DOE arriving in the n = 2 shell.

More recently the measurements with Ar^{17+} interacting with surfaces have been repeated with an ECR-source in Grenoble [5] by J.P. Briand and S. Essabaa [63] under very similar conditions as those of Donets et al.. Their results indeed confirm the observations of Donets et al. for all aspects discussed. The same group has also been able to observe the interaction of Ar^{18+} with an Al-surface in fig.26. Although the Si(Li)-detector used did not allow to resolve hypersatellite- and satellite-lines it is noted that the positions of the lines are significantly shifted to higher energies compared to the results with Ar^{17+}. This is obviously due to the hypersatellite component which stems from the transition with initially two K-holes. Conclusions going beyond those drawn from the results with Ar^{17+} are, however, not possible so that again measurements with better resolution are eagerly awaited for.

16) Discussion and Summary

The measurements presented in this lecture clearly show that the whole field of the neutralization of ions in front of surfaces, is still in its infancy due to the small number of adequate ion-sources available in the past. Most experiments (except for one or two) have been of exploratory character and suffer consequently from a lack of good resolution and statistics. Theoretical work on ions or atoms multi-excited in outer shells (more than 2 electrons) is not known, so that models have to be used which go far beyond existing theories.

In recent years a remarkable increase of the number of ECR sources available for low energy physics can be noted which will supposedly boost the field to exciting new horizons. One purpose of this lecture was therefore to suggest interpretations which may stimulate the discussion on future experimental and theoretical work.

The model presented allows a consistent description of all data on Potential Emission (PE) within the adiabatic approximation which excludes high vertical velocities and consequently also most contributions from Kinetic Emission (KE). It accounts for the dominance of the very low energy part of the spectra of secondary electrons as well as for the rather slow increase of the averaged kinetic energy of all electrons with increasing charge of the ion. In essence it suggests high resolution measurements of the very low energy spectrum (taking into account all contact potentials) in order to proof that this spectrum is peaked at $0 \leq \varepsilon \leq 2eV$.

The model also accounts for the LMX- and KLY-Auger transitions observed and suggests the deduction of populations of n-shells from a careful analysis of the fine structure of the Auger spectra. It furthermore allows to conclude that the plotting of the integral LMX- or LMY- intensities (or maybe even of individual Auger-decay channels) as a function of the inverse of the vertical velocity gives access to the pursuit of the time development of the electron distribution in the ion during neutralization. This procedure allows to establish internal clocks for the neutralization dynamics on a time scale of 10^{-14} to 10^{-}

^{13}s. For the shortest times accessible the model suggests to plot structures in the spectrum (whenever resolved) as a function of $1/v_v$ which correspondend to AN-, AD-, or CAN- and CAD-transitions and occur when the ion arrives close to the first atomic layer with a large number ($\approx$q) of inner holes.

All these suggested studies are of course supposed to produce important feed-back for the improvement and refinement of the model so that a clear-cut picture of the physics may evolve in due time.

The experimental work has to be corroborated of course by theoretical investigations of the physics of a large number of simultaneously excited electrons in high n-levels. The interaction of a highly charged ion with a metal surface allows for the first time in physics to populate such high n-levels abundantly by resonance transfer from the conduction band. Unfortunately the resulting multi-excited ion will always be subject to a strong polarisation by the image charge and to a continued electron exchange with the metal so that the ideal multi-excited system in an isotropic environment cannot be studied in this way. An alternative method for the production of less perturbed multi-excited systems may therefore be the resonant capture of a limited number of equivalent electrons into highly charged ions from metal-clusters with a well defined number of atoms. The post collision Coulomb interaction is of course also present in this case as a non-isotropic perturbation of the multi-excited system. It can be assumed, however, that the Coulomb-exploding cluster is a less severe perturbation than a surface. These experimental possibilities may perhaps encourage theoreticiens to attack these extremely difficult but exciting problems.

It is of course evident that inner shells are much less polarized by the image charge. The calculation of inner shell Auger- or X-ray-transitions with a great variety of M-, N-, O-, P- shell populations should therefore be justified in an isotropic approximation. As shown for the X-ray measurements such information would be very helpful for the interpretation of future experimental work.

All data clearly show that the ions in front of the surface passess extremely inverted populations which strongly suggest their exploitation for X-ray lasers. It is of course obvious that densities in slow ion beams are far from being sufficient for superfluorescent emission processes. One can, however, envisage that the expansion of a laser produced plasma against a metal surface could help to better localize an extremely high density of highly charged ions with strong inversion compared to a freely expanding laser produced plasma.

When increasing the charge of the incoming ion, a key question becomes more and more important: How fast can a metal, a semiconductor, and finally an insulator supply the large number of electrons required for the neutralization dynamics. This very interesting problem of the electron-dynamics in a surface and of the charge exchange with such a surface could probably be solved by an inter-disciplinary effort of theoreticiens from the various fields involved - supported of course by corresponding experiments.

17) Acknowledgement

This lecture has been stimulated by lively discussions with M. Delauney, M. Fehringer, S.T. de Zwart, J.J. Geerlings, P. Varga, and H. Winter (Vienna) who initiated the renaissance of this field. I profited particularly from the preprint collection of H. Winter concerning the work of Donets et al. I am also very indebted to J. Hansen who gave the first hints that a serious normalization error existed in the calculation of the AI rates in ref.[2]. Using the "Cowan-code" he supplied AI-rates which served to verify the hydrogenic calculations in this lecture. I have enjoyed the fruitful collaboration over many years with Helmut Winter and his crew in Muenster and in particular the intense exchange of theoretical aspects with R. Zimny and H. Nienhaus. It was a great pleasure to discuss the actual developments in the field with T. Aberg, J.P. Briand, S. Essabaa, W. Heiland, F.W. Meyer, R. Morgenstern, A. Niehaus, K. Snowdon and many other colleagues.

18) References

1) H.D.Hagstrum, Phys.Rev.96, 325 (1954); ibid.96, 336 (1954).

2) H.J.Andrä, NATO summer school on "Fundamental Processes of Atomic Dynamics", ed. J.S. Briggs, Plenum Press 1988.

3) U.A.Arifov, L.M.Kishinevskii, E.S.Mukkamadiev, and E.S.Parilis, Zh.Tekh.Fiz. 43, 181 (1973); Sov.Phys. -Tech.Phys. 18, 118 (1973).

4) R.Geller, IEEE Trans.Nuc.Sci., NS-23 (2) (1972).

5) M.Delaunay, S.Dousson, R.Geller, B.Jacquot, D.Hitz, P.Ludwig, P.Sortais, and S.Bliman, Nucl.Instr.Meth. B23, 177 (1987) and references therein.

6) E.D.Donets, Physica Scripta T3, 11 (1983).

7) A.G.Drentje, Nucl.Instr.Meth. B9, 526 (1985).

8) F.W.Meyer, Nucl.Instr.Meth. B9, 532 (1985).

9) J.F.Ziegler, J.P.Biersack, and U.Littmark, "The Stopping and Range of Ions in Solids", Vol.1, Pergamon Press, New York 1985.

10) E.S.Mashkova and V.A.Molchanov, "Medium Energy Ion Reflection from Solids", North-Holland, Amsterdam 1985.

11) G.Molière, Z.Naturf. A2, 133 (1947).

12) D.S.Gemmel, Rev.Mod.Phys. 46, 129 (1974).

13) N.D.Lang and W.Kohn, Phys.Rev. B1, 4555 (1970).

14) N.D.Lang and W.Kohn, Phys.Rev. B3, 1215 (1971).

15) N.D.Lang and W.Kohn, Phys.Rev. B7, 3541 (1973).

16) G.Comsa and B.Poelsema, Appl.Phys. A38, 153 (1985).

17) J.P.Toennies, Phys.Scr. T19A, 39 (1987).

18) J.Lapujoulade, Surf.Sci. 178, 406 (1986).

19) R.Remy, J.Chem.Phys. 53, 2487 (1970) and references therein.

20) R.Zimny, H.Winter, B.Becker, A.Schirmacher, and H.J. Andrä, Nucl.Instr.Meth. B2, 252 (1984).

21) H.D.Hagstrum, in "Electron and Ion Spectroscopy of Solids", edited by L.Fiermans, J.Vennik, and W.Dekeyser (Plenum, New York, 1978), and references therein.

22) W.Schroen, Z.Phys. 176, 237 (1963).

23) J.W.Gadzuk, Surf.Sci. 6, 133 (1967); Phys. Rev. B1, 2110 (1970).

24) N.W.Ashcroft and N.D.Mermin,"Solid State Physics", Holt-Saunders Intern. Ed., Tokyo, 1981.

25) A.V.Chaplik, Soviet Physics JETP 27, 178 (1968).

26) R.Hentschke, K.J.Snowdon, P.Hertel, and W.Heiland, Surf.Sci. 173, 565 (1986).

27) E.G.Overbosch, B.Rasser, A.D.Tenner and J.Los, Surf. Science 92, 310 (1980).

28) A.Cobas and W.E.Lamb,Jr., Phys.Rev. 65, 327 (1944) and references therein.

29) W.E.Lamb,Jr. and R.C.Retherford, Phys.Rev. 72, 241 (1947); ibid. 79, 549 (1950); ibid. 81, 222 (1951); ibid. 85, 259 (1952); ibid. 86, 1014 (1952).

30) H.D.Hagstrum, Rev.Sci.Instr. 24, 1122 (1953).

31) H.D.Hagstrum and G.E.Becker, Phys.Rev. B8, 107 (1973).

32) M.Delaunay, M.Fehringer, R.Geller, P.Varga, and H.Winter, Europhys. Lett. 4, 377 (1987).

33) J.Burgdörfer, Phys.Rev. A35, 4963 (1987).

34) H.Conrad, G.Ertl, J.Küppers, W.Sesselmann, and H.Haberland, Surf.Sci. 121, 161 (1982).

35) W.Sesselmann, H.Conrad, G.Ertl, J.Küppers, B.Woratschek, and H.Haberland, Phs.Rev.Lett. 50, 446 (1983).

36) B.Woratschek, W.Sesselmann, J.Küpers, G.Ertl, and H.Haberland, Phys.Rev.Lett. 55, 611 (1985); ibid. 55, 1231 (1985).

37) B.Woratschek, W.Sesselmann, J.Küppers, G.Ertl, and H.Haberland, Surf.Sci. 180, 187 (1987).

38) P.Varga, Appl.Phys. A44, 31 (1987).

39) M.Delaunay, M.Fehringer, R.Geller, D.Hitz, P.Varga, H.Winter, Phys.Rev. B35, 4232 (1987).

40) J.Ullrich, H.Schmidt-Böcking, S.Kelbch, C.L.Cocke, S.Hagmann, P.Richard, A.S.Schlachter, and R.Mann, Nucl.Instr.Meth. B23, 131 (1987).

41) G.Lakits, F.Aumayr, and H.Winter, J.de Phys. 50, Coll. C1, 533 (1989).

42) M.Fehringer, M.Delaunay, R.Geller, P.Varga, and H.Winter, Nucl.Instr.Meth. B23, 245 (1987).

43) S.T.de Zwart, Nucl.Instr.Meth. B23, 239 (1987).

44) S.T.de Zwart, Thesis, Groningen (1987).

45) R.K.Janev, J.Phys. B7, 1506 and L359 (1974).

46) B.A.Trubnikov and Y.N.Yavlinskii, Sov.Phys.JETP 25, 1089 (1976).

47) R.W.Strayer, W.Mackie, and L.W.Swanson, Surf.Sci. 34, 225 (1973).

48) M.Mack, Nucl.Instr. Meth. B23, 74 (1987).

49) A.Bordenave-Montesquieu, P.Benoit-Cattin, M.Boudjema, A. Gleizes, and S.Dousson, Nucl.Instr.Meth. B23, 94 (1987).

50) F.Folkman and K.M.Cramon, Phys.Scr. T3, 166 (1983).

51) S.Bashkin and J.O.Stoner,Jr.,"Atomic Energy-Level and Grotrian Diagrams", Vol. II, North-Holland Publ., Amsterdam, 1978.

52) F.W.Meyer, C.C.Havener, S.H.Overbury, K.J.Snowdon, D.M.Zehner, W.Heiland, and H.Hemme, Nucl.Instr.Meth. B23, 234 (1987).

53) D.M.Zehner, S.H.Overbury, C.C.Havener, F.W.Meyer, and W.Heiland, Surf.Sci. 178, 359 (1986).

54) F.W.Meyer, C.C.Havener, S.H.Overbury, K.J.Reed, K.J. Snowdon,and D.M.Zehner, J.de Phys. 50, Coll. C1, 263 (1989).

55) M.Delaunay, C.Benazeth, N.Benazeth, R.Geller, and C.Mayoral, Surf. Sci. 195, 455 (1988).

56) M.Fehringer, Thesis, T.U. Wien (1987).

57) M.O.Krause, J.Phys.Chem.Ref.Data 8, 307 (1979).

58) E.D.Donets, Phys.Scr. T3, 11 (1983).

59) E.D.Donets, Nucl.Instr.Meth. B9, 522 (1985).

60) E.D.Donets, G.A.Tutin, G.Zschornak, et al.,
    preprint nr. P7-83-627 of the Joint Inst. for Nucl.
    Research, Dubna (1983).

61) G.Zschornak, G.Musiol, and W.Wagner, invited talk
    on the Workshop on "Energiereiche atomare Stöße" in
    Oberstdorf/FRG, Jan. 1985.

62) E.D.Donets, S.V.Kartashov, and V.P.Ovsyannikov, Proc.
    Int.Conf.on Phenomena in Ionized Gases, Budapest (1985).

63) S.Essabaa, Thesis, Univ. de Paris VI (1988).

64) C.P.Bhalla and T.W.Tunnell, J.Quant.Spectr.&Rad.Transf.
    32, 141 (1984).

THEORY OF ATOMS IN DENSE PLASMAS

R. M. More

Lawrence Livermore National Laboratory
University of California
Livermore, CA 94550

1. INTRODUCTION

Laser Plasmas and High-Charge Ions

High-charge ions are easily produced in the laboratory by intense
pulsed laser radiation. With laser intensities of 10^{13} to 10^{15} Watt/cm^2,
the surface of a solid target is heated to temperatures of 200 to 1000 eV
and the target material is strongly ionized (Fig. 1). The ions are
identified by their x-ray line emission. Neon-like and nickel-like ions
are easily produced and identified because they have closed-shell
configurations.

For example, a recent study by Tragin et al. (1988) gives the spectra
of ions having Z = 73 to 82 in charge states 41+ to 53+ (Fig. 2). The
spectra show prominent series of equally spaced peaks (line clusters)
resulting from transitions in a sequence of ionization states (Fig. 3).
Such high-Z spectra have been analyzed by Bauche-Arnoult, Bauche, and
Klapisch (1979, 1982, 1985); each line-cluster is identified as a family
of transitions between two electron configurations; the cluster position
and width are calculated by a rigorous moment expansion technique, the
UTA (Unresolved Transition Array) method.

Qualitatively one can predict which ions will be produced by the rule
that the charge state Q increases until the ionization potential I(Z, Q)
$\simeq$ 13.6 eV $(Q/n)^2$ is a few times the temperature kT.

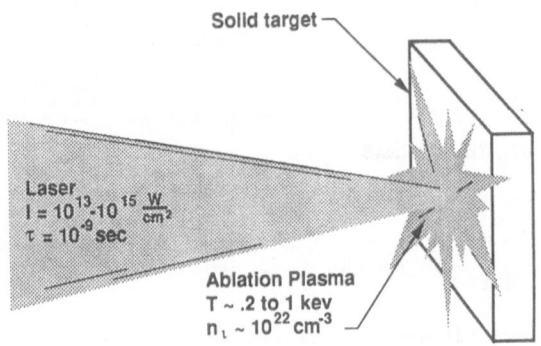

Figure 1 In the hot dense plasma produced by laser irradiation, the atoms
of any target material are strongly ionized.

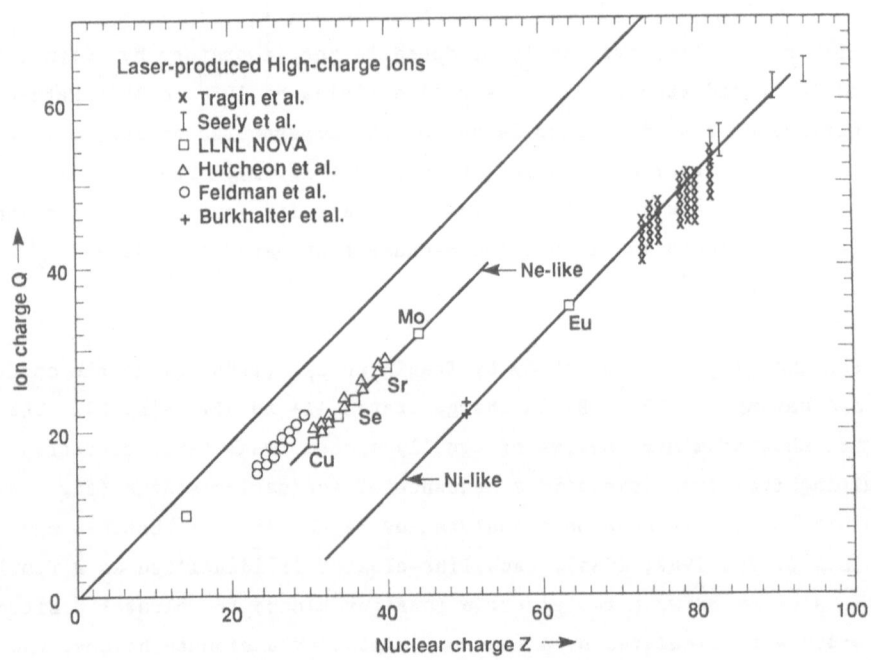

Figure 2 A sampling of ionization states identified in recent
laser-plasma experiments.

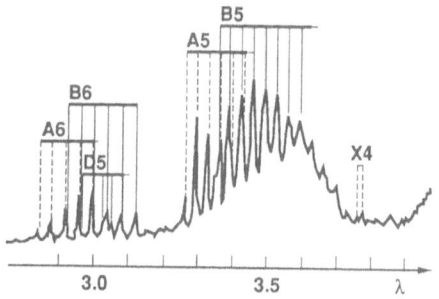

Figure 3 Laser plasma spectra of lead showing spin-orbit split unresolved
transition-arrays (Tragin et al., 1988).

A unique feature of laser plasma spectroscopy is that these ions occur
at high densities $\geq 10^{20}/cm^3$. At such high density the ions are close
together and perturb each other with strong electric fields; due to the
density they also receive frequent impacts by ambient free electrons.
This raises a fundamental question: what does the dense plasma
environment do to the structure and behavior of high-charge ions?

In this paper we examine one popular theoretical description, the
average-atom model, which results from generalizing the usual atomic self-
consistent field theory to finite density and temperature. Our goal is to
identify strengths and weaknesses of this theory and to relate it to
simpler theories which give qualitative insight and/or rapid computational
modelling.

The paper includes discussion of several recent results including a
WKB method for calculation of atomic matrix-elements, a new formulation of
non-equilibrium rate equations for the average-atom model and analysis of
electron population correlations in non-equilibrium plasmas.

Spherical-Cell Model

The spherical-cell model is a basic theoretical picture of high-
density atomic structure. It has a long history (Slater and Krutter,
1935; Morse, 1940; Mayer, 1947; Feynman, Metropolis and Teller, 1948).
The more recent work combines this cell model with quantum self-consistent
field theory (Rozsnyai, 1972; Liberman, 1979; Perrot and Dharma-Wardana,
1982; Cauble et al., 1984).

In the cell model the nucleus is at the center of a spherical cavity
of radius R_o; outside the cavity there is a continuous positive charge

density ρ_+, usually taken to be uniform (Fig. 4). There are Z electrons in the cavity so the cell is neutral as a whole. The external region is neutral point by point because the positive charge density is assumed to be neutralized by a gas of free electrons. The <u>ion–sphere radius</u> R_o is determined by the volume per ion,

$$\frac{4\pi}{3} R_o^3 = 1/n_i \qquad (1\text{--}1)$$

One imagines the plasma to be made up of these ion spheres packed densely together. In the high-density case, the Coulomb repulsion is strong enough to prevent other ions from entering into the ion sphere.

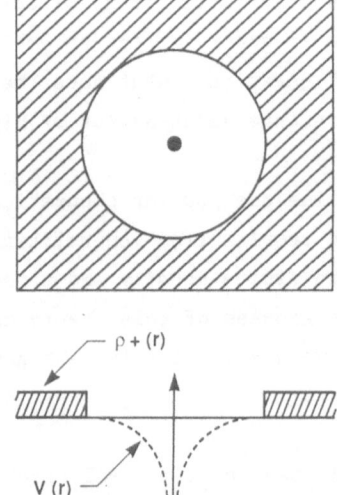

Figure 4 Schematic representation of the spherical–cell model. The cross-hatched region represents the continuous positive charge density, neutralized by an electron gas. The cell itself contains a nonuniform electron gas and one nucleus at the center.

The potential Ze/r of the central nucleus is modified by bound and free electrons to produce a short-range self-consistent electrostatic potential V(r). Four basic equations are used to calculate this potential, the Schroedinger equation,

$$-\frac{\hbar^2}{2m} \nabla^2 \psi_s - eV(r) \psi_s = \varepsilon_s \psi_s \qquad (1\text{--}2)$$

Laplace's equation,

$$\nabla^2 V = 4\pi e\, n(r) \tag{1-3}$$

Fermi-Dirac statistics,

$$P_s = \frac{D_s}{1 + \exp(\varepsilon_s - \mu)/kT} \tag{1-4}$$

and the addition of one-electron charge-densities,

$$n(r) = \sum_s P_s\, |\psi_s(r)|^2 \tag{1-5}$$

In the one-electron Schroedinger equation, s represents the complete set of quantum numbers for a spherically-symmetric system, so that s = (n, ℓ, m, σ) for nonrelativistic bound-states. The shell degeneracy Ds is

$$D_{n\ell} = 2(2\ell + 1)$$

$$D_n = 2n^2 \tag{1-6}$$

Degenerate sublevels of given n, ℓ are assumed to be equally populated so that the total electron density n(r) is spherically symmetric. Both bound and free states must be included in Eqs. (1-2, 4 and 5). The chemical potential μ is determined by the condition of electrical neutrality in the cell.

The potential V(r) is self-consistent if it simultaneously solves Eqs. (1-2, 5). Self-consistency is achieved by iteration.

The plasma density enters through the boundary conditions. Because the ion sphere is assumed to be electrically neutral, the electric field on its surface must vanish. The potential is usually also required to vanish at R_0:

$$\frac{\partial V}{\partial r} = 0 \quad , \quad V = 0 \quad \text{at} \quad r = R_0 \tag{1-7}$$

In the model the exterior environment is electrically neutral so the potential V(r) remains zero for all $r > R_0$; of course, this feature of the model is not very realistic. At small radii the nuclear potential is dominant, so $V(r) \rightarrow Ze/r$.

What is the difference between the isolated ion and the same ion in a dense plasma environment? In the spherical-cell picture the differences come from two sources: the gas of free electrons, and the constraint of

neutrality in a finite cell. The free electrons tend to maintain the ion
in an excited state, especially at high densities. This appears in the
Fermi functions p_s which assign nonzero average populations to excited
states. The boundary condition, Eq. (1-7), transfers the electrostatic
effect of neighbor ions and this will change the ionization potential,
photoelectric absorption edges and atomic matrix-elements. At high
enough compression the electron eigenstates are progressively destroyed
by increasing interaction with the neighbor ions.

A Critique

It is not difficult to make technical improvements to the theory
sketched above. For example, one can introduce a local-density exchange
and correlation potential, or one can use the Dirac equation for one-
electron wave-functions (Liberman, 1979). One can describe the external
plasma using the radial distribution function $g(r)$ obtained from a fluid-
structure theory or Monte-Carlo simulations (Perrot, 1982).

However these changes do not alter the basic framework which consists
of electrostatic self-consistency, the spherically symmetric average
potential $V(r)$, the one electron wave-functions Ψ_s, and the use of average
populations given by Fermi functions in thermal equilibrium. (This last
step is the "average-atom" approximation.)

The self-consistent field theory faces two difficult questions
directed at this basic framework.

First, _is it realistic to represent neighbor ions by a smooth charge
distribution when they are really point charges_? In general, we expect
that the neighbor ions destroy the spherical symmetry of the environment,
generating microfields, quadrupole fields, etc. We expect the neighbor
ions to intrude into the cell, perhaps developing quasi-molecular states
with the electrons already present, perhaps carrying along extra
electrons. The ions surely cause Stark broadening, which is omitted from
the spherical-cell model and must be added by hand. In addition, the ion
environment modifies the free-electron states, as happens in solids where
one has the formation of energy-bands and band-gaps.

Second, _what does the average population p_s mean_? For a plasma which
contains a mixture of hydrogen-like, helium-like and some lithium-like
ions, the "average ion" will have vacancies in its K-shell and fractional
populations in its L-shell. Do average ions with these populations

424

behave anything like the real thing? Section 4 will examine particular nonequilibrium states of a plasma in which the average-atom model predicts far too much or far too little radiative emission. Many of these difficulties can be traced to correlations in the populations of electrons of different shells.

Applications and Experiments

The <u>ionization state</u> $Q(\rho, T)$ is calculated from the spherical-cell model as follows: one assumes the free electron density in the plasma to equal the value $n(R_o)$ of the electron density at the outer edge of the spherical cell. Then the number of free electrons per ion is

$$Q = \frac{4\pi}{3} R_o^3 \, n(R_o) \tag{1-8}$$

This definition can be debated (More, 1985) but it is simple and easily used (Fig. 5).

From calculations of $Q(\rho, T)$ we identify two main ranges in density-temperature space. For low-density and/or high-temperature conditions, the ionization depends mainly on temperature. This is <u>thermal ionization</u>. On the other hand, at low-temperature high-density conditions the free electrons become degenerate and all the bound-states which exist are fully occupied. The ionization rises with compression almost exclusively due to progressive destruction of bound states caused by overlap with neighboring ions. This is the range of <u>pressure ionization</u>, believed to be important in dense cool stars such as the white dwarfs, but also seen in laser-driven shock-wave and implosion experiments.

In the Thomas-Fermi version of the spherical-cell model, the ionization state $Q(\rho, T)$ is a smooth function of density and temperature (Fig. 5). The quantum versions give similar results, rarely different by more than 10% (of Z), but show effects of shell closure, so that Helium-like or Neon-like ions occupy a somewhat larger range of temperatures than in the Thomas-Fermi calculations.

Figure 5 refers to an equilibrium plasma. In optically thin layers of low-density plasma, radiation can escape and the photon population falls far below the black-body distribution. In this case, radiative recombination and line emission advance relative to their inverses, resulting in a less-ionized plasma made up of less-excited ions. In the limit of low densities, one has isolated ground state ions ("coronal"

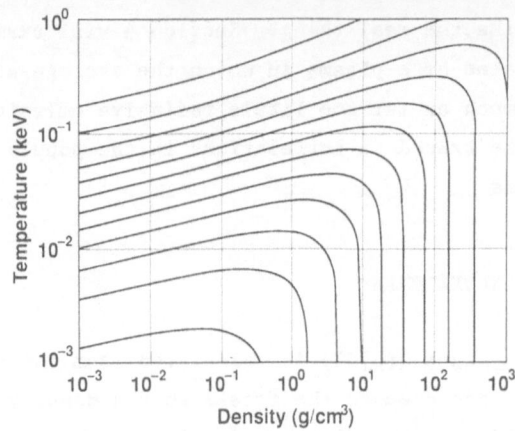

Figure 5 Ionization contours (Q = 1, 2, ..., 12) for aluminum plasma
calculated by the Thomas-Fermi version of the spherical-cell
model. The nearly vertical ionization contours indicate
pressure ionization; at the low densities the ion charge depends
primarily upon temperature.

plasma). For these conditions one must solve NLTE (non-equilibrium) rate
equations described in Section 4.

The pressure can be calculated by a method like Eq. (1-8). One
assumes the electron pressure is essentially the pressure of a free-
electron gas calculated from the values assumed at the outer edge of the
ion sphere,

$$p \simeq n_e(R_o) \, kT \qquad\qquad (1\text{-}9)$$

Equation (1-9) is appropriate for nondegenerate free electrons. The
electron pressure can be corrected for degeneracy, exchange and
correlation effects, and there is also an extra pressure associated with
the ion kinetic energy.

A pressure calculated this way, based on the spherical-cell model and
supplemented by simple representations of chemical and solid-state bonding
effects, is able to reproduce experimental hydrodynamic data for strong
shock-waves and plasma expansion over an extraordinary range of physical
conditions (More, Warren, Young and Zimmerman, 1988).

The pressure and energy determine plasma flow or expansion and therefore represent an inteface between fundamental atomic models and the experimentally observed behavior of matter at extreme conditions.

While the comparisons do not test the detailed spectroscopic features of the average–atom theory, the calculated pressure is very sensitive to the treatment of pressure ionization and the modification of continuum states by density effects; thus the experiments are able to convincingly rule out certain versions of the cell model (More, 1985).

The expansion of laser plasmas is described by hydrodynamic calculations. For example, in one–dimensional plasma motion, the hydrodynamic state of the plasma is defined by the density $\rho(x, t)$, the velocity $v(x, t)$, and an internal energy density $E(x, t)$. The planar hydrodynamic flow is then determined by

$$\frac{\partial \rho}{\partial t} = - \frac{\partial}{\partial x}(\rho v)$$

$$\rho \frac{\partial v}{\partial t} = - \rho v \frac{\partial v}{\partial x} - \frac{\partial p}{\partial x}$$

$$\frac{\partial}{\partial t}\left[\rho\left(E + \frac{v^2}{2}\right)\right] = - \frac{\partial}{\partial x}\left[\rho\left(E + \frac{v^2}{2}\right)v\right] - \frac{\partial}{\partial x}(pv) - \frac{\partial q}{\partial x} \qquad (1\text{--}10)$$

To evaluate the right-hand side of these equations, we need to know the pressure $p(\rho, T)$, where the temperature T is readily extracted from the energy–density, and we need to know the heat–current $q(x, t)$. The heat current can result from electron heat–conduction or from radiation heat–conduction.

Experimental plasma hydrodynamic flow has been compared with theoretical calculations in a number of geometric configurations (Fig. 6). Charatis et al. (1986) report holographic determinations of the plasma density and comparisons with hydrodynamic simulations; this work was aimed at producing a characterized plasma for x–ray laser research. Some recent laser–driven shock–wave research is reported by Ng (1986) and Fabbro et al. (1986).

One of the most important applications of the spherical–cell calculations is the prediction of x–ray absorption and emission cross–

Examples of hydrodynamic flows:

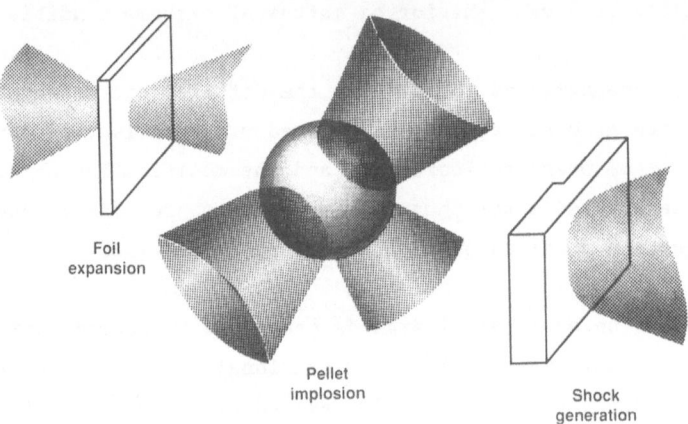

Foil
expansion

Pellet
implosion

Shock
generation

Figure 6 Typical geometries used to test equation of state data resulting
from self-consistent field theories. The foil expansion data is
taken with detailed holographic interferometry. Implosions have
been studied with time-resolved x-ray images. In stepped flat
targets, shock-wave travel-times are measured by streak-camera
recording of the shock breakout through the rear surface.

sections. The cooling by loss of x-ray energy is a significant aspect of
plasma expansion dynamics. The opacity is defined as the (averaged) x-ray
absorption cross-section per gram.

Each part of the average-atom model is used in the calculation of
opacity. The eigenvalues $E_{n\ell}$ determine the photon line spectrum, $h\nu =$
$E_{n\ell} - E_{n'\ell'}$. The populations $p_{n\ell}$ enter because the line emission or
absorption rate is proportional to the number of electrons in the initial
(one-electron) state multiplied by the number of holes in the final state.
The one-electron wave-functions are used to calculate matrix-elements and
oscillator-strengths. Continuum wave-functions are used to calculate the
photoelectric absorption spectrum including the changes caused by high-
density screening phenomena, principally a shift in the continuum boundary
(threshold).

There are many questions about these calculations. What is the
relation between the real spectrum and the theoretical lines calculated

as differences of average-atom eigenvalues? What is the best way to add a description of line-broadening? Are there line shifts, i.e., density-dependent line-spectra? The existence (and sign) of line shifts has been a key question for recent theory and experiment. While the simplest use of the average-atom model predicts substantial red shifts, a somewhat more careful analysis finds little or no line shift (More, 1985). Evidently the actual plasma line-shift is largely a reactive effect associated with electron-impact line-broadening, and rarely larger than an asymmetry of the line (Griem, 1988). This question of line shifts directs our attention toward the electrostatic structure of high-charge ions in dense plasmas and the practical question how to select the screened charges which characterize the motion of radiating electrons.

2. SCREENING IN THE SPHERICAL-CELL THEORY

Electrostatics and Screening

An effective charge Z* is often used to write energy-levels or matrix-elements for a many-electron ion in hydrogenic form in order to take advantage of the known properties of the hydrogen atom.

In order to give a precise definition to this effective charge, one is led to re-examine the textbook ideas of inner and outer screening (Slater 1951, Sommerfeld 1934, Condon and Shortley 1967, Bethe and Salpeter 1957). Here we give a brief development of these ideas which are important for the definition of the effective charge Z*, for the calculation of the electrostatic part of the equation of state, and for the physics of line shifts produced by high density conditions.

Given a self-consistent electrostatic potential V(r), we can define an effective charge or screening function Z(r) by

$$V(r) = Z(r) \ e/r \tag{2-1}$$

Clearly Z(r) equals the nuclear charge at small radii, so $Z(0) = Z$. If the potential vanishes at R_o, then $Z(R_o) = 0$. This representation of the potential is very natural.

However a different effective charge is obtained by writing the electric field inside the atom as

$$\left| \vec{E}(r) \right| = - \ dV/dr = Q(r) \ e/r^2 \tag{2-2}$$

When $Q(r)$ differs from $Z(r)$, it is possible that $Q(r)$ will be more pertinent because the force on electrons is determined by $\vec{E}$.

Comparing Eqs. (2-1, 2) we see

$$Q(r) = Z(r) - r \frac{dZ}{dr} \qquad (2-3)$$

Since $Z(r)$ decreases with radius, $Z'(r) = dZ/dr < 0$ and therefore $Q(r)$ is larger than $Z(r)$. The difference is often substantial, as in Fig. (9). We must ask which effective charge gives the best description of the physics. For example the hydrogenic photoelectric cross-section is proportional to Z^4, a strong sensitivity. For emission or absorption of radiation the acceleration of electrons is the essential process and therefore the radiative cross-sections is probably be governed by $Q(r)$ rather than $Z(r)$. We must still decide which radius is most important for any given transition; this can be done by semiclassical methods (see Section 3).

The significance of these two effective charges becomes clear if we recall the two ways to write Laplace's equation,

$$\frac{1}{r^2} \frac{d}{dr} (r^2 \frac{dV}{dr}) = \frac{1}{r} \frac{d^2}{dr^2} (rV) = 4\pi en(r) \qquad (2-4)$$

Using the first combination of derivatives we have

$$Q(r) = Z - \int_o^r 4\pi r'^2 n(r') \, dr' \qquad (2-5)$$

This is just Gauss' law: the electric field at radius r is determined uniquely by charges inside that radius. Thus $Q(r)$ reflects <u>inner screening</u> or screening interior to r. From the other combination of derivatives we find

$$\frac{dZ}{dr} = \int_r^{R_o} 4\pi r' n(r') \, dr' \qquad (2-6)$$

Thus dZ/dr is entirely determined by electrons outside radius r. Combining these equations, we have:

<u>Theorem 1.</u> The potential $V(r)$ can be uniquely decomposed into the potential of a core of charge $Q(r)$ and the potential produced by concentric outer shells of charge density $n(r')$ at radii $r' > r$,

430

$$V(r) = \frac{Q(r) \ e}{r} - e \int_{r}^{R_o} 4\pi r' \ n(r') \ dr' \qquad (2-7)$$

Atomic processes such as thermal excitation, ionization and production of inner-shell vacancies typically contribute to the first term. Effects of the plasma environment enter through the second term.

The next theorems express the total electrostatic energies in terms of $Z(r)$, $Q(r)$. The proofs are very easy. The results are used to work out the thermodynamic functions and to justify the simplified screened-hydrogenic model of Section 4.

Theorem 2. The electron-nucleus interaction energy is given by

$$U_{en} = Ze^2 \left(\frac{dZ}{dr}\right)_{r=0} \qquad (2-8)$$

Theorem 3. The electron-electron interaction energy is given by

$$U_{ee} = - e^2 \int_{o}^{R_o} [\frac{dZ}{dr}] \ 4\pi r^2 \ n(r) \ dr \qquad (2-9)$$

This formula can be referred to as an outer build-up rule, because it sums the electrostatic interaction of each shell of electrons with other electrons outside that shell.

Theorem 4. The total electrostatic energy $U_{tot} = U_{en} + U_{ee}$ is

$$U_{tot} = - \int_{o}^{R_o} \left[\frac{Q(r) \ e^2}{r}\right] 4\pi r^2 \ n(r) \ dr \qquad (2-10)$$

This equation can be called an inner build-up rule because it calculates the electrostatic interaction of each shell of electrons with a core of charge $Q(r)e$.

The representations given in Eqs. (2-8 to 10) avoid doubly counting the electron-electron interaction because they count in one direction only.

A Generalized Bohr Model

In order to see which effective charge is important, we first consider

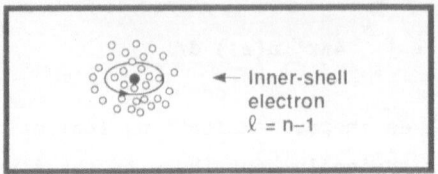

Inner-shell
electron
$\ell = n-1$

Figure 7 Schematic Bohr model for an electron moving among other bound
electrons.

an electron moving in a circular orbit in this self-consistent potential
at radius r_n with velocity v_n, using the Bohr model (Fig. 7). The basic
equations are:

$$m\ v_n^2/r_n = Q(r_n)\ e^2/r_n^2 \qquad (2\text{--}11)$$

$$m\ v_n\ r_n = n\hbar \qquad (2\text{--}12)$$

The first equation represents Newton's law and the second is the Bohr
quantization rule. Solving these equations,

$$r_n = a_o\ n^2/Q(r_n) \qquad (2\text{--}13)$$

$$v_n = Q(r_n)\ e^2/n\hbar \qquad (2\text{--}14)$$

Thus these dynamical quantities depend only upon $Q(r_n)$. However the energy
E_n depends also on the electrostatic potential. Using Eq. (2--3) we find

$$E_n = \frac{m\ v_n^2}{2} - eV(r_n) = -\frac{Q^2(r_n)\ e^2}{2\ a_o\ n^2} - e^2\ Z'(r_n) \qquad (2\text{--}15)$$

In the second expression, the first term is a generalized hydrogenic
formula which contains the electron kinetic energy and bonding to the
core. The second term represents a shift in the energy zero due to the
surrounding (outer) electrons. Thus the energy involves both inner
screening in $Q(r_n)$ and outer screening in $Z'(r_n)$.

Quantum Treatment of a Core Electron

We now give a slightly more elaborate derivation of the same result
by a variational method; the trial function will be taken to be (Fig. 8)

$$\Psi_{n\ell m\sigma}(r) = \frac{1}{r}\ \Psi_{n\ell}(r)\ Y_\ell^m(\theta,\ \phi)\ \chi_\sigma$$

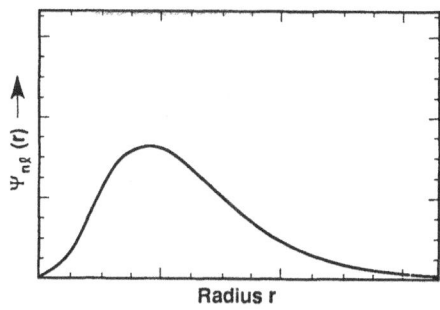

Figure 8 Schematic trial function for a core electron.

$$\Psi_{n\ell}(r) = A r^n e^{-\alpha r} \qquad (\ell = n - 1) \tag{2-16}$$

which is the hydrogenic wave function for angular momentum $\ell = n - 1$. The coefficient A is determined by normalization, and α is the variational parameter.

The expectation value of the radial and angular kinetic energy is then

$$K = \frac{\hbar^2}{2m} \int |\nabla \Psi|^2 \, 4\pi r^2 \, dr = \frac{\hbar^2 \alpha^2}{2m} \tag{2-17}$$

The potential energy is the average

$$U = - \int \frac{Z(r) \, e^2}{r} |\Psi|^2 \, 4\pi r^2 \, dr \tag{2-18}$$

In the integrand we use the Taylor's series expansion of $Z(r)$ around a radius $r_n = n/\alpha$ which locates the maximum of $4\pi r^2 n(r)$. With this expansion

$$U \cong - \frac{Q(r_n) \, e^2 \alpha}{n} - e^2 \, Z'(r_n) \tag{2-19}$$

The total energy is a minimum for $\frac{\partial}{\partial \alpha} (K + U) = 0$ or $\alpha = Q(r_n)/a_o n$. The most likely orbit radius is then $r_n = a_o n^2/Q(r_n)$ in agreement with the Bohr model, and the energy is again

$$E_n = - \frac{Q^2(r_n) \, e^2}{2 \, a_o \, n^2} - e^2 \, Z'(r_n) \tag{2-20}$$

It is useful to go one step further and calculate the screening by a shell containing P_n electrons having the wave-function $\Psi_{n\ell}$. Ignoring other bound electrons, we find

$$Q(r) = Z - P_n\left(1 - e^{-2\alpha r} \sum_{j=0}^{2n} \frac{(2\alpha r)^j}{k!}\right) \qquad (2-21)$$

and

$$Z'(r) = -\frac{P_n}{r_n} e^{-2\alpha r} \sum_{j=o}^{2n-1} \frac{(2\alpha r)^j}{j!} \qquad (2-22)$$

The second formula gives a useful result,

$$eZ'(o) = -\frac{P_n e}{r_n} \qquad (2-23)$$

This is equal to the potential inside a thin spherical shell of total charge $-$ e P_n located at radius r_n. Having constructed $Z'(r)$ and $Q(r)$ in Eqs. (21, 22) it is easy to construct the potential from $Z(r) = Q(r) + r\, Z'(r)$.

In this case, as in general, $Q(r)$ is nearly constant near the nucleus while $Z(r)$ decreases linearly there; this linear decrease reflects the constant potential produced by exterior charges.

The potential constructed by use of Eqs. (2-22, 23) has the functional form introduced by Klapisch (1971) under the name parametric potential. It turns out to be useful that this is an analytic function of radius so it can be used for calculations which require deformation of contour integrals (Section 3).

Effect of an External Electron Gas

Next we consider a nucleus of charge Ze at the center of a uniform spherical distribution of electrons in a sphere of radius R_o. The electrons have constant density $n = Z/(4\pi R_o^3/3)$ and we impose the boundary condition $V(R_o) = 0$. Then

$$V(r) = \frac{Ze}{r} + \frac{1}{2}\frac{Ze}{R_o}\left(\frac{r}{R_o}\right)^2 - \frac{3}{2}\frac{Ze}{R_o} \qquad (2-24)$$

and so

$$Z(r) = Z\left[1 + \frac{1}{2}\left(\frac{r}{R_o}\right)^3 - \frac{3}{2}\left(\frac{r}{R_o}\right)\right] \qquad (2-25)$$

$$Q(r) = Z\left[1 - \left(\frac{r}{R_o}\right)^3\right] \qquad (2-26)$$

434

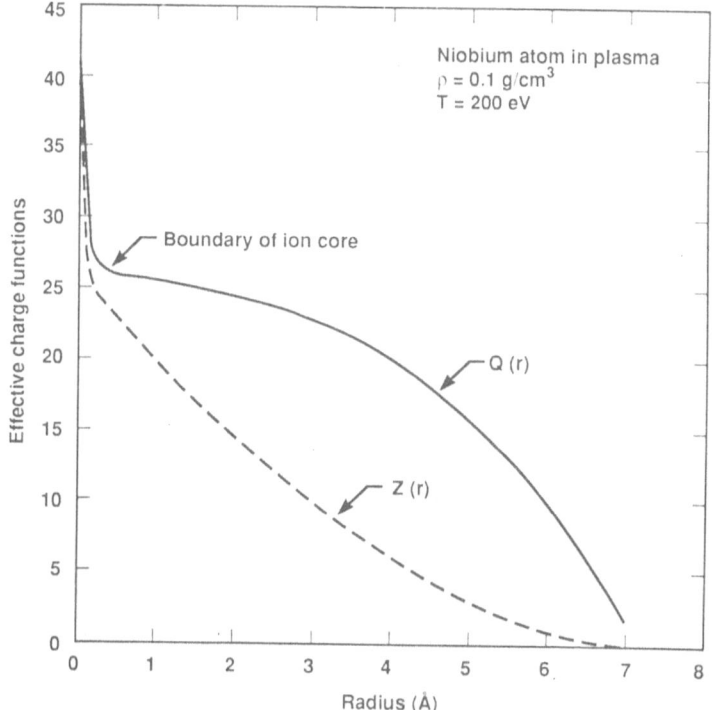

Figure 9 Effective charges Q(r), Z(r) calculated from the self-consistent
potential for a niobium plasma at density 0.1 g/cm^3 and
temperature of 200 eV. In the outer portion of the ion sphere,
Q(r) is substantially larger than Z(r).

Again we see that Q(r) is constant near r = 0 while Z(r) has linear
variation there.

This system — the point charge in a uniform electron gas — has some
similarity to the results of the dense plasma self-consistent field
theory (Fig. 10). For reference we give expressions for various average
properties

$$U_{en} = - \int \frac{Ze^2}{r} \, n(r) \, d^3r = - \frac{3}{2} \frac{Z^2 e^2}{R_o}$$

$$U_{ee} = \frac{e^2}{2} \int \frac{n(r) \, n(r')}{|r - r'|} \, d^3r \, d^3r' = + \frac{3}{5} \frac{Z^2 e^2}{R_o}$$

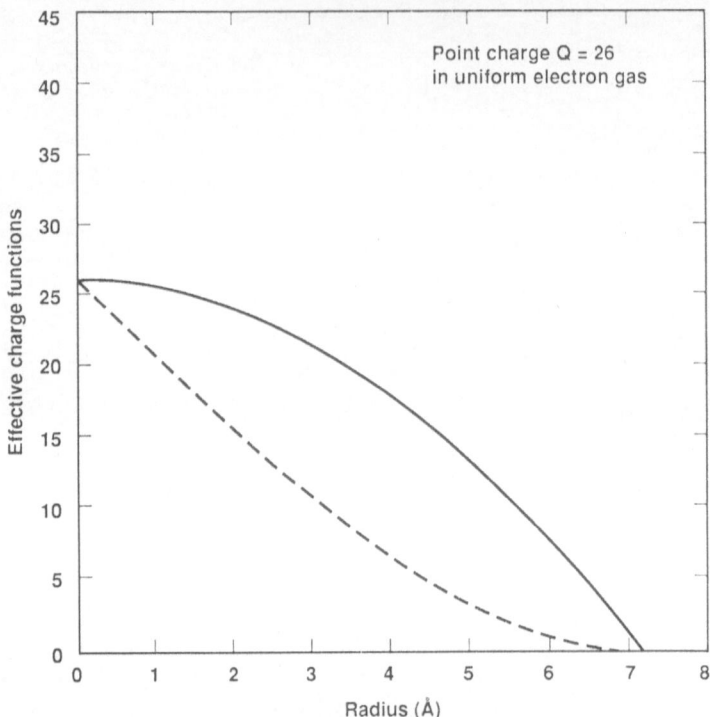

Figure 10 Effective charges Q(r), Z(r) for a point charge in a uniform
electron gas with parameters chosen for comparison to Figure 9.

$$U_{tot} = - \frac{9}{10} \frac{Z^2 e^2}{R_o}$$

$$\langle V(r) \rangle = \frac{3}{10} \frac{Ze^2}{R_o}$$

$$\langle V^2(r) \rangle - \langle V(r) \rangle^2 = \frac{108}{175} \left(\frac{Ze^2}{R_o} \right)^2 \qquad (2-27)$$

We obtain something more useful by a slight extension of this uniform
electron-gas picture. Instead of a point charge Z, we consider a ground
state ion core of charge Q in a uniform free-electron gas of density
$Q/(4\pi R_o^3/3)$. From this approximate picture we obtain approximate formulas
for the total energy and other quantities: we add the energy required to
ionize the ion to the Qth charge state, the energy of the free electrons
(e.g., 3/2 Q kT in the nondegenerate limit) and the interaction energy in
Eq. (2.27).

However there are corrections to this simple picture and the value of the spherical-cell approximation is what it tells us about the corrections. First, at high densities the ions are often excited and the excitation energy represents a significant heat content comparable to the free electron kinetic energy. Second, the free electrons can be (weakly) polarized by the potential of the ion, so they are not exactly uniform, and this increases the Coulomb energy of Eq. (2-27). Third, the electrons can have resonance states in the potential of the ion, so they are not simply free electrons in the ordinary sense.

3. EIGENVALUES AND MATRIX-ELEMENTS (SEMICLASSICAL METHODS)

To explore the physical content of the self-consistent field theory we would like explicit formulas for electron energy-levels and wave-functions, even if they are only approximate. The variational method gave Eqs. (2-16, 20), but they are limited to states of large angular momentum. The semiclassical (WKB) theory gives more general results and also provides a powerful new method for approximate calculation of atomic matrix-elements. With this approach we can see the effects of screening (by bound or free electrons) in a very clear fashion.

WKB Wave-Functions

We consider the radial motion of electrons in a spherically symmetric self-consistent potential $V(r)$. Radial wave-functions can be constructed from WKB travelling waves,

$$\Psi_{n\ell}^{(\pm)}(r) = \frac{\Psi_{n\ell}}{\sqrt{q(r)}} e^{\pm i[\phi_{n\ell}(r) - \pi/4]} \qquad (3-1)$$

The phase integral $\phi_{n\ell}(r)$ and radial wave-vector $q(r)$ are

$$\phi_{n\ell}(r) = \int_{r_1}^{r} q(r')\, dr' \qquad (3-2)$$

$$q(r) = \sqrt{\frac{2m}{\hbar^2}} \left[E_{n\ell} + eV(r) - \frac{\hbar^2}{2m} \frac{(\ell + 1/2)^2}{r^2} \right]^{1/2} \qquad (3-3)$$

The wave-vector $q_{n\ell}(r)$ is $m/\hbar$ times the classical (radial) velocity, and the phase-integral $\phi(r, E, \ell)$ is $1/\hbar$ times the classical action function, so Eq. (3-1) is entirely constructed from classical ingredients.

Bound-electron energy-levels are determined by

$$\int_{r_1}^{r_2} q(r')\, dr' = (n - \ell - 1/2)\, \pi \qquad (3\text{-}4)$$

where r_1, r_2 are inner and outer <u>turning points</u>, defined as zeroes of $q(r)$, and n is the principal quantum number.

A WKB approximation to the quantum wave-function is

$$\Psi_{n\ell}(r) = \frac{1}{2}\, \text{Re}\left[\Psi_{n\ell}^{(+)}(r) + \Psi_{n\ell}^{(-)}(r) \right] \qquad (3\text{-}5)$$

Equation (3-5) gives a satisfactory result in both allowed and forbidden ranges, and no additional connection formula is needed (see Eqs. (3-14, 15) below).

In our work the normalization $c_{n\ell}$ is determined by the following equivalent equations:

$$1 = \frac{1}{2}\, \text{Re} \int_{r_1}^{r_2} \Psi_{n\ell}^{(+)}(r)\, \Psi_{n\ell}^{(-)}(r)\, dr = \frac{1}{2}\, |\, C_{n\ell}\, |^2 \int_{r_1}^{r_2} \frac{dr}{q(r)} \qquad (3\text{-}6)$$

$$|C_{n\ell}|^2 = \frac{2m}{\pi \hbar^2} \frac{d\bar{E}_{n\ell}}{dn} \qquad (3\text{-}7)$$

In Eq. (3-7), $dE_{n\ell}/dn$ is a derivative defined by differentiating Eq. (3-4) rather than a finite difference.

These formulas refer to electrons in an arbitrary spherically symmetric self-consistent potential. For hydrogen everything can be calculated in closed form. The WKB energy-levels are $E_{n\ell} = -\, z^2 e^2/2a_o n^2$ as in quantum mechanics. The wave-vector simplifies to $q(r) = (K_n/r)$ $[(r - r_1)(r_2 - r)]^{1/2}$, where

$$r_{1,2} = (a_o n^2/Z)\, [1 \mp \epsilon] \qquad (3\text{-}8)$$

$$\epsilon = \sqrt{1 - (\ell + 1/2)^2/n^2} \qquad (3\text{-}9)$$

The phase-integral is

$$\phi_{n\ell}(r) = rq(r) + n[\tfrac{\pi}{2} + \sin^{-1} S_1(r)] - (\ell + \tfrac{1}{2})\, [\tfrac{\pi}{2} + \sin^{-1} S_2(r)] \quad (3\text{-}10)$$

where we define

438

$$S_1(r) = \frac{2r - (r_1 + r_2)}{r_2 - r_1} \qquad (3\text{-}11)$$

$$S_2(r) = \frac{r(r_1 + r_2) - 2\,r_1\,r_2}{r(r_2 - r_1)} \qquad (3\text{-}12)$$

The normalization is $|c_{n\ell}|^2 = 2Z^2/a_o^2 n^3$, an equation which paraphrases a well-known classical result (Kepler's third law).

For the Coulomb potential, the classical bound-state orbit is an ellipse given by

$$r = \frac{(1 - \varepsilon^2)\,(a_o\,n^2/Z)}{1 - \varepsilon\cos\theta}$$

This is derived by differentiating $\phi(r, E, \ell)$ with respect to ℓ.

The quantities $q(r)$, $\phi(r)$ and $\Psi^{(\pm)}(r)$ can be extended to complex values of radius r (More and Warren, 1989a,b). For this extension it is necessary to employ the half-integer quantum number exactly as in Eqs. (3-3, 4). The wave-vector $q(r)$ has a pole at $r = 0$ and a branch-cut from r_1 to r_2 along the real axis, and no other singularity. Across the branch-cut, $q(r)$ changes sign, which physically corresponds to the possibility of incoming or outgoing velocities.

$\Psi_{n\ell}^{(\pm)}(r)$ also has a branch-cut from r_1 to r_2, with

$$\Psi_{n\ell}^{(\pm)}(r - i\varepsilon) = \mp\,\Psi_{n\ell}^{(\mp)}(r + i\varepsilon) \text{ for } (r_1 < r < r_2) \qquad (3\text{-}13)$$

Again, the branch-cut corresponds to exchange of incoming and outgoing waves. At $r \to 0$ and $r \to \infty$, the limiting behavior of the Coulomb WKB functions is

$$\Psi_{n\ell}^{(+)} \to \frac{A_{n\ell}}{r^{\ell}} \qquad \Psi_{n\ell}^{(-)} \to B_{n\ell}\,r^{\ell+1} \qquad (r \to 0) \qquad (3\text{-}14)$$

$$\Psi_{n\ell}^{(+)} \to \frac{C_{n\ell}}{r^n}\,e^{Zr/na_o} \qquad \Psi_{n\ell}^{(-)} \to D_{n\ell}\,r^n\,e^{-Zr/na_o} \qquad (r \to \infty) \qquad (3\text{-}15)$$

$\Psi_{n\ell}^{(+)}$ has improper behavior in both limits, but its coefficients $A_{n\ell}$, $C_{n\ell}$ are purely imaginary so this improper (growing) term is automatically removed by Eq. (3-5). $\Psi_{n\ell}^{(-)}$ is decreasing in both limits, and its coefficients $B_{n\ell}$, $D_{n\ell}$ are real. For this reason $\Psi_{n\ell}$ is well-behaved and even accurate to a few percent except near the turning-points. Thus with

Eq. (3-5) there is no need for an additional connection formula. The reality property is apparently quite general (More and Warren, 1989b) but for the general potential q(r) might have complex turning points which would complicate the analytic properties.

WKB Matrix-Elements

The results of this section were reported for the first time at the NATO summer-school on _Atomic Physics of Highly-Ionized Atoms_. The question is how to calculate radiative matrix-elements using WKB wave-functions. This has proven difficult because the matrix-element is the integral of an oscillatory function, so small errors in the wave-functions can give large errors in the integral. Since the WKB wave-function has singularities at the turning points, the direct numerical evaluation of the matrix-element is neither easy nor very accurate.

Our new method consists of writing

$$R_{n\ell}^{n'\ell'} \simeq \frac{1}{2} R_e \int \Psi_{n\ell}^{(-)}(r) \; r \; \Psi_{n'\ell'}^{(+)}(r) \; dr \tag{3-16}$$

This step will be called the _restricted interference approximation_. The integral proves to have a saddle-point at a complex radius r_s which is found by writing

$$\Psi_{n\ell}^{(-)}(r) \; r \; \Psi_{n'\ell'}^{(+)}(r) \equiv e^{ig(r)} \tag{3-17}$$

The saddle-point is the solution of $g'(r_s) = 0$. The WKB result for the matrix-element is then

$$R_{n\ell}^{n'\ell'} \simeq \frac{1}{2} R_e \left[\Psi_{n\ell}^{(-)}(r_s) \; r_s \; \Psi_{n'\ell'}^{(+)}(r_s) \sqrt{\frac{2\pi i}{g''(r_s)}} \right] \tag{3-18}$$

This is a (nearly) closed-form expression for the matrix-element. Table I and Figs. (12-17) give a sample of the results. Equation (3-18) is nearly always within about 10% of the exact answer, for both large and small matrix-elements.

Equation (3-18) proves to have a remarkable physical interpretation which goes a long way to clarify the classical limit of a quantum transition process. The interpretation is based on a geometric construction shown in Fig. (11). We rotate the orientation of the classical orbits for initial and final states until the two orbits have second-order orbit-contact. For the hydrogenic transition n, $\ell \rightarrow$ n', $\ell +$ 1 (n' > n) the orbits intersect at a radius

440

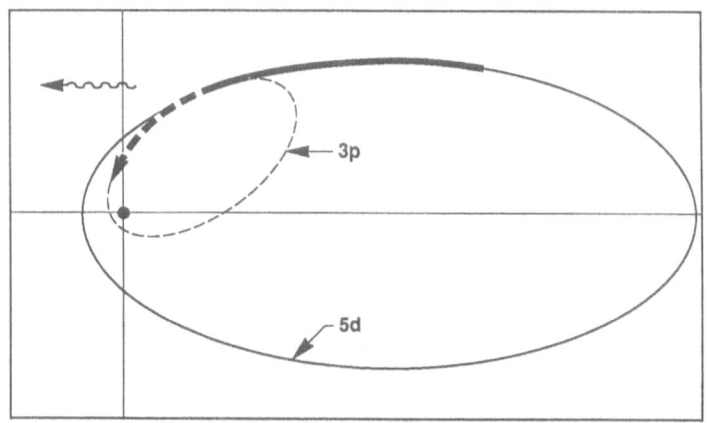

Figure 11 Semiclassical picture for the radiative transition 3p – 5d in
         a hydrogenic ion. The classical elliptic orbits intersect with
         second-order contact at a radius r_i, given by Eq. (3-19), which
         proves to be close to the saddle-point r_s in Eq. (3-18) for
         transitions n, ℓ – n', ℓ + 1.

TABLE 1

Dipole matrix-elements for the Coulomb potential, atomic units

Transition	WKB R(n,1,n',1')	Exact R	percentage difference
1s – 2p	1.2510	1.2903	3.04
1s – 3p	0.4782	0.5167	7.45
1s – 4p	0.2796	0.3046	8.21
1s – 5p	0.1910	0.2087	8.49
2s – 3p	3.1439	3.0648	2.58
2p – 3d	4.6123	4.7480	2.86
2p – 3s	1.0019	0.9384	6.76
3s – 4p	5.7413	5.4693	4.97
3p – 4d	7.6298	7.5654	0.85
3p – 4s	2.6982	2.4435	10.42
3d – 4f	9.9253	10.2303	2.98
3d – 4p	1.4379	1.3023	10.41

$$r_i = \frac{4(\ell + 1)}{\left(\frac{\ell + 3/2}{n}\right)^2 - \left(\frac{\ell + 1/2}{n'}\right)^2} \frac{a_o}{Z} \qquad (3\text{-}19)$$

In examining numerical data, we discovered that the saddle-point is approximately

$$r_s \simeq r_i \, e^{+i\pi/4} \qquad (3\text{-}20)$$

Evidently the radiative transition occurs as a "quantum jump" from initial to final orbit during second-order orbit contact, when the two orbits have nearly the same position and velocity. This is particle language. In terms of wave-mechanics, the transition is a three-wave interaction process. The initial and final electron waves are coupled to the electromagnetic field and satisfy the momentum-matching condition ($q = q'$ + k) over the <u>interaction region</u>, whose size is measured by the factor $[1/g''(r_s)]^{1/2}$ in Eq. (18).

The picture is supported by looking at examples. These show 1.) the matrix-elements are larger for transitions having a larger interaction region, 2.) transitions in which n and ℓ change in the opposite sense have no second-order orbit-contact and much smaller matrix-elements, and 3.) for a series such as $1s \rightarrow np$, the geometry of orbit intersection is constant for n > 3 or 4 and so is the saddle-point. The matrix-element changes only by a factor $1/n^{3/2}$ which comes from the normalization of the upper-state wave-function, Eq. (3-7).

More and Warren (1989a,b) use the restricted interference approximation defined by Eqs. (3-6, 3-16) to calculate matrix-elements for a variety of systems, including dipole and quadrupole line-transitions in hydrogenic ions; photoelectric transitions and sums over series of transitions; Debye-screened Coulomb potentials; bound-electron screening in high-charge ions; spherical harmonic functions and their matrix-elements; and of course the one-dimensional harmonic oscillator. In almost every case the results are within $\simeq$ 10% of the exact quantum matrix-elements.

There is a surprise in all these results. Equation (3-6) does not give wave-functions normalized in the usual sense, i.e., the integral of $|\Psi|^2$ is not unity. This is true because the integrals of $\Psi^{(+)} \Psi^{(+)}$, $\Psi^{(-)} \Psi^{(-)}$ are not small, at least not for low quantum numbers. If we included these terms, in either the matrix-element or the normalization

442

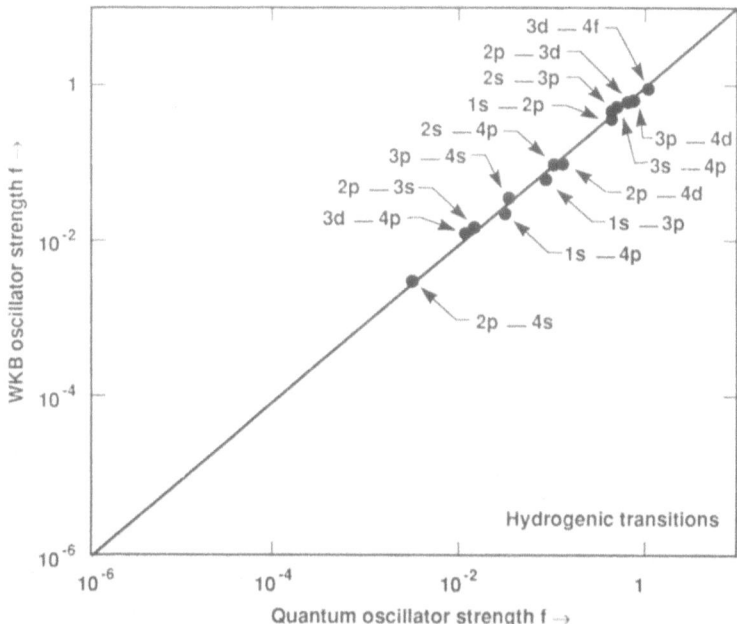

Figure 12   Comparison of WKB and Quantum calculations of the oscillator
strength for the first few transitions of hydrogenic ions.

(or both), the WKB results would not be very good. Why is it better to
omit these terms?

It emerges that there is a coherent semiclassical mechanics which
exists (semi-concealed) in the WKB wave-functions. For example, we have
discovered through a mixture of analytical and numerical calculations
that for each of the systems studied the WKB wave-functions obey a
remarkable one-sided orthonormality relation,

$$\frac{1}{2} R_e \int \Psi_s^{(-)}(r) \; \Psi_{s'}^{(+)}(r) \; dr \quad \simeq \quad \delta_{ss'} \tag{3-21}$$

The quantum numbers are those for which the quantum states would be
orthogonal, i.e., for the radial wave-functions the result refers to
states of the same angular momentum and different principal quantum
number. (It is usually necessary to restrict this equation to s < s').
When Eq. (3-21) is not exactly true, it is a very good approximation
(e.g., for the Coulomb potential the integrals for s ≠ s' are ≤ .01; for
the harmonic oscillator they are zero).

Equation (3-21) is an extraordinary mathematical fact, because the

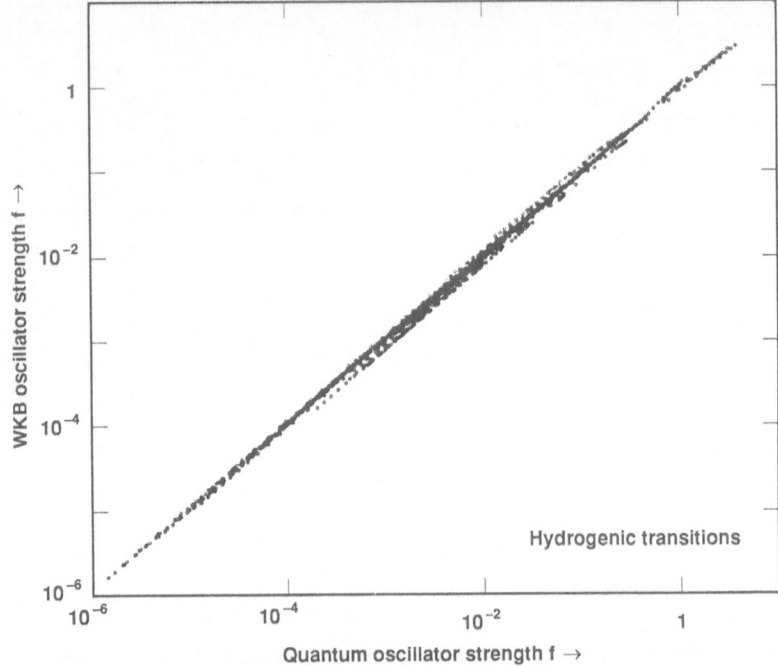

Figure 13   Comparison of WKB and quantum calculations of oscillator
strengths for the first 2500 transitions of hydrogen.   The
results vary over seven orders of magnitude and always agree
to 10–15%.

WKB wave-functions have turning-point singularities in the range of
integration.   It is a surprising physical result, because a fundamental
quantum property is emerging from integrals over classical velocities and
action functions.

Equation (3-21) gives a degree of justification for the use of the
WKB approximation in the calculation of Franck-Condon factors, i.e.,
overlap integrals between wave-functions of different Hamiltonians (for
examples, see Miller, 1974 or Niehaus and Zwakhals, 1983).

For calculation of matrix-elements, there is an additional physical
picture, which helps understand why one should omit the terms $\Psi^{(+)}\ \Psi^{(+)}$ in
Eq. (3-16).   The terms omitted would correspond to a <u>back reflection</u> at
the radiative transition, i.e., a jump from an incoming initial state to
the outgoing part of the final state (Fig. 11).   It is physically evident
that such a back-reflection has small probability because the photon

momentum is small compared to the electron momenta, and therefore it is natural to omit the contribution of these terms.

For transitions with $\Delta n = |n - n'| \ll n, n'$, it is well known that the matrix-element reduces to a Fourier component of the dipole moment taken along a single classical trajectory having the average quantum number. (Kramers, 1923; Percival and Richards, 1975; Naccache, 1972). This relation is one of the original expressions of the correspondence principle; however it is not nearly so general or reliable as Eq. (3–18). For example, the application to the photoelectric effect would involve an extreme extrapolation because there is no meaningful average of bound and free trajectories.

Equation (3–18) is entirely straightforward in this case and continues to give a reasonably accurate answer (Figs. 14–16). The WKB saddle-point is a continuous function of energy through zero energy and the low-energy photoelectric cross-section is an extrapolation of the series-limit line-absorption cross-section.

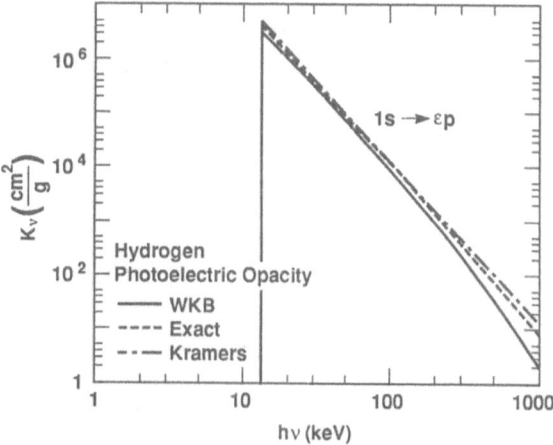

Figure 14   Hydrogen photoelectric cross-section (cm^2/gram) for absorption from the 1s state calculated by WKB, exact nonrelativistic quantum and Kramers methods. The WKB results are within about 15% of the exact over a large range near the threshold, but begin to fail at extreme high energy where the cross-section becomes very small.

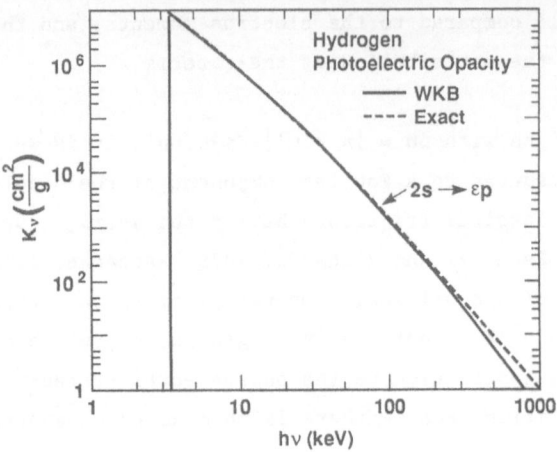

Figure 15   Hydrogen photoelectric cross-section for absorption from the
2s state.

WKB Theory for the Screened-Coulomb Potential

The WKB approximation can also be used to study matrix-elements for
high-charge ions in which the Coulomb potential is screened by either
bound electrons or an external plasma.   Comparison with available
numerical results shows that the WKB method works again for this
application (More and Warren, 1989a,b).

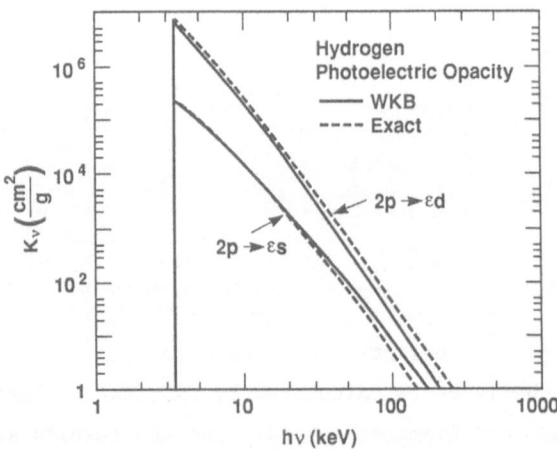

Figure 16   Hydrogen photoelectric cross-section for absorption from the
2p state.   The WKB method is quite accurate for the large
cross-section near the absorption edge.

For the Lyman series (transitions 1s → np) the effect of screening
is to greatly reduce the matrix-element. This was shown for a
Debye-Huckel screening model by Weisheit and Shore (1974) and Hohne and
Zimmerman (1982). Bound-electron screening has the same effect, as shown
in Fig. (17).

The WKB method helps understand exactly why the matrix-element is
reduced by screening. For the cases considered, the screening has little
or no effect on the inner-state (1s) wave-function, but causes the outer
state to relax outward, so that it occupies a larger volume. This change
reduces the normalization constant of the outer-state wave-function and
thereby reduces the matrix-element. In the WKB calculation, this is the
only result of screening, to an accuracy of a few percent.

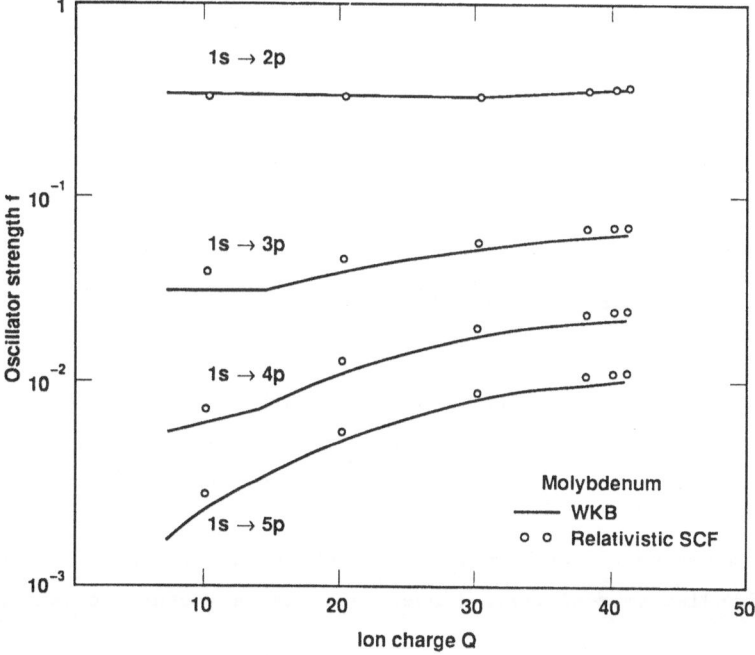

Figure 17 Approximate calculation of oscillator strength for line
           absorption from the K-shell in Molebdenum ions. The WKB
           approximation is compared to relativistic quantum calculations
           performed by D. A. Liberman.

An important job for the future is to extend this analysis to a wider class of transitions and a greater variety of screening conditions, to identify the general behavior of matrix-elements of high-charge ions.

One can visualize what is possible in this approach by looking at the WKB representation of energy-levels in a screened-Coulomb potential. In this development, the essential step is to introduce an average orbit radius $r_n = a_o n^2 / Q(r_n)$ for the state n, ℓ, and to expand the potential in a Taylor series near $r = r_n$ (Sommerfeld, 1934). The series can be rearranged to give

$$V(r) \simeq \frac{Q(r_n)e}{r} + e \; Z'(r_n) + \delta V(r) \qquad\qquad (3\text{-}22)$$

The result is a constant $eZ'(r_n)$ associated with outer screening, a Coulomb potential governed by the core charge $Q(r_n)$, and an additional non-Coulomb potential which vanishes quadratically near r_n.

With this representation of V(r) it is natural to divide the WKB phase integral into a Coulomb part, defined by neglecting $\delta V(r)$, and a correction generated by $\delta V(r)$. This gives an explicit equation for the quantum defect:

$$E_{n\ell} = - \frac{Q^2(r_n) \; e^2}{2 \; a_o (n - \Delta_{n\ell})^2} + e^2 \; Z'(r_n) \qquad\qquad (3\text{-}23)$$

$$\Delta_{n\ell} = \frac{1}{\pi} \left(\int q_{n\ell}(r) \; dr - \int q_{n\ell}^{(o)}(r) \; dr \right) \qquad\qquad (3\text{-}24)$$

Various well-known properties of $\Delta_{n\ell}$ are easily read off from the WKB expression. It is also straightforward to use the quantum defect to parametrize the fine-structure splitting of term values (More, 1982).

This approach to the quantum-defect theory has the advantage that it equally handles core-electron (inner) and plasma (outer) screening phenomena.

If one confines attention to the effects of core-electron screening, there is the well-known method of Bates and Damgaard (1949) which was recently applied to radiative matrix-elements of high-charge ions by Clark and Merts (1987) and Oumarou et al. (1988).

448

4. ELECTRON POPULATION CORRELATIONS AND THE SCREENED-HYDROGENIC MODEL

The most controversial aspect of the self-consistent field theory is the use of average populations which assign fractional occupations to the one-electron states. Can we develope a more precise idea of the consequences of this approximation?

For deeply bound levels or even for excited states of isolated ions, the ion configuration is (presumably) instantaneously defined by integer populations. This means that the Hamiltonian is nearly diagonal in a representation based upon number operators which have integer eigenvalues; the off-diagonal terms are associated with configuration interaction and with environmental perturbations which transfer electrons between states. For high densities and/or loosely bound electrons the Hamiltonian is not diagonal in the number representation because the collision-induced transitions become very rapid. In this limit it is more natural to base the theory upon the average populations which remain well-defined. For these reasons the average-atom model is usually considered to be essentially correct at high densities.

For LTE plasmas, we can show that average populations give adequate results for many purposes. However, for non-LTE plasmas there are certain physical situations in which the approximation clearly disagrees with a more fundamental approach (see below). In these cases the key ingredient turns out to be correlations between the populations of different states.

It is possible to study correlations with a simplified atomic model based on screened hydrogenic energy-levels (Slater, 1951; Mayer, 1947). The screened charges and energy-levels are algebraic functions of the populations, so they can be evaluated rapidly and it becomes practical to calculate and/or average millions of configurations if necessary. With this method we can make precise comparisons of the two theories: the detailed configuration accounting (DCA) theory, based on configurations defined by integer population of one-electron states, and the average-atom (AA) theory, with continuously variable average populations.

Even without an accurate line spectrum, the (screened) hydrogenic model is able to reproduce the ionization potentials to enough accuracy to give a reasonably good idea of plasma ionization balance and the rates for electron-impact and radiative processes. This information is obviously useful for applications.

The screened hydrogenic model is therefore often applied to plasmas in equilibrium and non-equilibrium (non-LTE) conditions. For example, Post, Jensen et al. (1978) gave tables of impurity radiation and cooling rates for low-density plasmas in magnetic fusion reactors, calculated from this theory. Rosen et al. (1979) used the hydrogenic model to compare LTE and non-LTE simulations of laser irradiated disk targets; this work showed the great importance of non-LTE ionization balance in laser-produced high-Z plasmas. The model can be used for low-Z or high-Z elements and even for mixtures of materials, to test the effects of impurities or trace elements in a plasma.

Some recent development and applications are described in papers by Zimmerman and More (1980), More (1982), Yabe et al. (1987), Berthier (1985), Tsakiris and Eidman (1987), and Perlado and Minguez (1986).

Screened Hydrogenic Energy-Levels

A hydrogenic electron configuration is specified by giving the numbers of electrons in states of principal quantum number n. Thus, P_n is the population of the nth shell, restricted to the range $0 \leq P_n \leq D_n = 2n^2$. It is assumed that electrons of the nth shell experience an effective charge Q_n calculated by a linear screening formula,

$$Q_n = Z - \sum_{m<n} \sigma_{nm} P_m - \frac{1}{2} \sigma_{nn} P_n \qquad (4-1)$$

The effective charge Q_n is an approximation to the <u>inner-screened</u> charge, which occurs in the electric field $|\vec{E}(r)| = Q(r) \, e/r^2$. Screening constants σ_{nm} are introduced because the electrons of shell m are not entirely inside the average orbit radius r_n of shell n, even when $r_m <$ r_n. The linear screening formula assumes the overlap between shells m and n is independent of ionization state. This approximation is most accurate for high-charge ions.

As in Eq. (2-13), the hydrogenic average orbit radius is

$$r_n = a_o \, n^2/Q_n \qquad (4-2)$$

In Eq. (2-23) we found a potential $e \, P_m/r_m$ inside a shell of charge $-e \, P_m$ at radius r_m. The <u>outer screening</u> energy of shell n is taken to be $e \, P_n$ times this potential with a screening correction for overlap of shells n, m:

$$E_n^{(0)} = \sum_{m>n} \frac{e^2 \, P_m}{r_m} \sigma_{mn} + \frac{1}{2} \sigma_{nn} \frac{e^2 \, P_n}{r_n} \tag{4-3}$$

In terms of the effective charges of Eq. (2-3) this outer screening energy represents $E_n^{(0)} = -e^2 \, Z'(r_n)$.

The one-electron eigenvalue E_n and total ion energy E_{ion} are taken to be

$$E_n = E_n^{(0)} - \frac{Q_n^2 \, e^2}{2 \, a_o \, n^2} \tag{4-4}$$

$$E_{ion} = - \sum_n \left(\frac{Q_n^2 \, e^2}{2 \, a_o \, n^2} \right) P_n \tag{4-5}$$

Equation (4-4) immediately recalls Eqs. (2-15, 2-20). Equation (4-5) may be understood as an application of the _inner build-up rule_ of Eq. (2-10); as the ion is assembled, shell by shell, each new shell added has a kinetic energy $(mv_n^2/2)P_n$ and a potential energy $-(Q_n^2 \, e^2/a_o \, n^2)P_n$ the interaction with the ion core.

Although Eqs. (4-1 to 4-5) are assumptions here, they closely parallel the results derived in section 2 and thereby provide an algebraic model for the quantum self-consistent field theory. Additional support for the model is given by consistency properties we now describe.

The eigenvalue E_n is the ionization potential for removing electrons from shell n, i.e., there is a _Koopmans' theorem_:

$$E_n = \frac{\partial E_{ion}}{\partial P_n} \tag{4-6}$$

This relation is important in the development of the statistical mechanics.

In the quantum self-consistent field theory it is well known that the sum of eigenvalues doubly-counts the electron-electron interaction energy. The hydrogenic model has an exact analogue of this relation,

$$E_{ion} = \sum_n E_n \, P_n - U_{ee} \tag{4-7}$$

Here the electron-electron energy U_{ee} is determined by the _outer build-up rule_ as

$$U_{ee} = \sum_n E_n^{(o)} P_n \qquad\qquad (4\text{-}8)$$

This equation parallels Eq. (2-9). Together, Eqs. (4-1 to 8) give an algebraic model with the same formal structure as the self-consistent field theory.

Therefore we can employ $E_{ion}(\{P_1, P_2, \ldots\})$ as an effective Hamiltonian for atomic structure. It has the advantage that it is diagonal in the basis of eigenstates of the shell populations P_n.

Now we want to explore the difference between the DCA theory based on integer populations P_n and the AA theory which uses average populations coming from Fermi-Dirac statistics. We can solve the hydrogenic model both ways and compare results; in this way we can identify structural differences which do not depend on the accuracy of the hydrogenic screening model but rather have a general significance for any average-atom theory.

The accuracy of the screening model is limited because it neglects the ℓ-dependence of one-electron eigenvalues and also neglects the term-splitting of many-electron states (an example is given below).

Nevertheless, with an optimized set of screening coefficients (More, 1982) the model is qualitatively correct, and tracks the large changes in one-electron energy-levels produced by variations in the ionization state. There is the possibility of developing a more accurate screening model (Parker, 1986; Perrot, 1988) but for such a model to be equally useful it should retain something like the internal consistency expressed in Eqs. (4-1) through (4-8).

Hydrogenic Configurations

Hydrogenic configurations are defined by assigning integer populations to the principal quantum numbers, i.e., by a set $\{P_n\}$ where $0 \leq P_n \leq D_n = 2n^2$ for each integer P_n. The symbol s is used (here) to denote a <u>class of states</u> having this configuration.

The ion charge Q and number of states in the class s are

$$Q = Z - \sum_n P_n \qquad\qquad (4\text{-}9)$$

$$C_s = \prod_n \left(\frac{P_n!}{D_n! \, (D_n - P_n)!} \right)$$ (4-10)

For example, carbon-like ground-state ions have a hydrogenic configuration $\{P_n\} = \{2, 4, 0, 0, \ldots\}$. The corresponding nonrelativistic quantum states are

** 15 states $1s^2 \, 2s^2 \, 2p^2(^1S, \, ^1D, \, ^3P)$

** 40 states $1s^2 \, 2s \, 2p^3(^1P, \, ^3P, \, ^1D, \, ^3D, \, ^3S, \, ^5S)$

** 15 states $1s^2 \, 2p^4(^1S, \, ^1D, \, ^3P)$

In this case the nonrelativistic Schroedinger theory has three configurations which split to give 12 terms (L.S. coupling) having a total of 70 eigenstates. In the hydrogenic model thëse 70 states are degenerate, forming a single hydrogenic configuration. Equation (4-10) correctly gives $C_s = 70$ for this example.

The high degeneracy of the hydrogenic spectrum obviously leads to a grossly oversimplified line spectrum. By the same token it reduces the number of configurations to a manageable total.

Equilibrium Populations and their Correlations

In thermal equilibrium, the number density N_s of ions in state s is given by the Saha equation, which we write

$$N_s = a \, C_s \, \exp\left(- \frac{E_{ion}(s) + F_{free}(Q, \, p, \, T)}{kT} \right)$$ (4-11)

Here C_s is the number of states in the class s, taken from Eq. (4-10), $E_{ion}(s)$ is the ion energy, taken from Eq. (4-5), and F_{free} is the Helmholtz free energy of Q free electrons in the ion sphere. For ionization states sufficiently near the average, $F_{free}(Q, \, \mu, \, T)$ is a constant plus μQ, where μ is the chemical potential. The coefficient a is fixed by the requirement that N_s add up to the correct total ion density.

The Saha equation can be rearranged, for example, one can divide $E_{ion}(s)$ into a ground-state energy $E_o(Q)$ plus an excitation energy $\Delta E_s = E_{ion}(s) - E_o(Q)$; then one can sum over excited states to obtain an equation for the ratios of populations of ion charge states.

The Saha equation is adapted to moderately high density plasma conditions by adding an extra free energy for the electrostatic interaction of electrons with the ions, which can be taken from Eqs. (2-27) or some more elaborate theoretical picture. (Griem, 1964; More, 1981; Hummer and Mihalas, 1988.)

Noninteracting Electrons

If the electrons did not interact, the eigenvalue E_n would be a constant E_n^c independent of the configuration and the ion energy would be linear in the populations,

$$E_{ion} = \sum_n E_n^c P_n \tag{4-12}$$

In this limit the Saha equation would give ion populations

$$N_s = n_i \prod_n \left[\left(\frac{D_n!}{P_n! (D_n - P_n)!} \right) \frac{e^{-(E_n^c - \mu) P_n/kT}}{\left(1 + e^{-(E_n^c - \mu)/kT}\right)^{D_n}} \right] \tag{4-13}$$

This equation applies to integer populations P_n. However, it is just the Fermi-Dirac distribution written in the detailed configuration (DCA) language, because the averages are

$$<P_n> = \frac{D_n}{1 + \exp(E_n^c - \mu)/kT} \tag{4-14}$$

$$<P_n P_m> = <P_n> <P_m> \tag{4-15}$$

In real high-charge ions, and even in the screened hydrogenic model, the electrons interact and these formulas are no longer exactly valid. This raises an interesting fundamental question: how does the Fermi-Dirac statistics generalize to interacting electrons?

LTE Average-Atom Model

A similar result can be obtained for _interacting_ electrons in the screened hydrogenic model. This is done most simply by writing a free-energy function,

$$F[\{P_n\}] = E_{ion}(\{P_n\}) - TS_{ion} - F_{free}(Q, \mu, T) \tag{4-16}$$

where

454

$$S_{ion} = -k \sum_n \left(P_n \log \frac{P_n}{D_n} + (D_n - P_n) \log \left(\frac{P_n}{D_n} \right) \right) \qquad (4\text{--}17)$$

This is the equation for the entropy of non-interacting Fermions.
Unfortunately, while the electron interaction effect on the energy E_{ion}
is given by Eq. (4-5) we do not have a corresponding simple formula for
the entropy of interacting electrons.

Now the condition of minimum free energy $\delta F = 0$ gives the average-
atom model,

$$\bar{P}_n = \frac{D_n}{1 + \exp (\bar{E}_n - \mu)/kT} \qquad (4\text{--}18)$$

where $\bar{E}_n = \partial E_{ion}/\partial P_n$ is the eigenvalue evaluated with the populations P_n
replaced by their averages.

The result is fundamental to the average-atom theory but the
derivation is not rigorous -- it is based on minimizing an ad hoc free
energy. However Eq. (4-18) can be tested by comparison with numerical
calculations which exactly solve the Saha equation. (Green, 1964; More,
1981; Grimaldi and Grimaldi-Lecourt, 1982; More, Zinamon and Zimmerman,
1988.) The results show that Eq. (4-18) holds to an accuracy of 5% to
10%, and that populations of different shells are independent to similar
accuracy:

$$<P_n P_m> - <P_n> <P_m> \simeq -\frac{1}{10} <P_n> < P_m> \qquad n \neq m \qquad (4\text{--}19)$$

These correlations between populations in different shells are caused by
the electron-electron interaction; when the population of an inner shell n
is large, the outer shell m is less strongly attracted to the nucleus and
is therefore likely to contain fewer bound electrons. However because the
correlation effect is not strong, the results really provide support for
the neglect of correlations in the average-atom theory.

Nonequilibrium Electron Populations and Correlations

The most interesting laser-plasma experiments involve non equilibrium
atomic dynamics (NLTE conditions) and we require an average-atom theory
for these processes which will have practical advantages of giving a good
overall idea of nonequilibrium ionization balance with a minimum of
computation. Such a model is part of many modern laser-plasma computer-
codes (e.g., Rosen et al., 1979).

The question poses itself what are the deficiencies of this simple atomic model. To analyse this question, we must arrange to make systematic comparisons with a more fundamental approach.

At sufficiently low densities, one can readily formulate nonequilibrium atomic rate equations involving the same language as the Saha equation: ions in specific charge and excitation states, and a matrix of rates for transitions between these states. This approach is entirely satisfactory for systems with one or two electrons, where the number of states is limited. (McWhirter, 1965; Duston and Davis, 1980; Hagelstein, 1981.)

For experiments on many-electron ions the number of relevant excited states increases and the direct approach begins to become more difficult. Under these conditions there is a need for some type of hybrid theory, which combines accurate rate equations with for certain levels with a cruder treatment of the other ionization stages or excitation states. (Busquet, 1982; Lee, 1987; Itoh et al., 1987.)

Another useful approach is the formal analysis of the density dependence of systems of atomic rate equations. (Salzmann, 1979; Klapisch, 1988).

In this section we describe an approach based on adding corrections to the average-atom model, specifically those associated with electron population correlations. This method was recently proposed for dielectronic and autoionization processes by More, Zimmerman and Zinamon (1988).

In the detailed configuration theory, nonequilibrium atomic processes are described by a linear rate equation. Each ion is assumed to have a definite ion charge and excitation state, summarized in a formal quantum number s. For the hydrogenic model, s denotes the set of populations $\{P_n\}$ and in the more general theory s is a complete set of quantum numbers. N_s is the number density of ions in the state s, and obeys the DCA rate equation

$$\frac{dN_s}{dt} = \sum_{\substack{\text{process} \\ s' \to s}} N_{s'} \begin{pmatrix} \text{gain} \\ \text{rate} \\ s' \to s \end{pmatrix} - N_s \sum_{\substack{\text{process} \\ s \to s''}} \begin{pmatrix} \text{loss} \\ \text{rate} \\ s \to s'' \end{pmatrix} \qquad (4\text{--}20)$$

Here one sums over the __gain terms__, which are transitions from other states s' to s, and over __loss terms__, transitions from s to other states s". Any

456

specific transition s → s" appears as a loss term in the equation for N_s and as a gain term in the rate equation for s". The processes are excitation, de-excitation, ionization and recombination.

Let f(s) be any well-defined function of the state s: this could be, for example, the ion energy or the population of the L-shell in the ion s. The average of f(s) is

$$<f(s)> = \sum_s f(s) N_s / \sum_s N_s \qquad (4\text{-}21)$$

The sum in the denominator is the total ion density. The average $< f(s) >$ is time-dependent because the populations N_s are time-dependent. One can easily establish the following basic rate equation for $< f(s) >$:

$$\frac{d}{dt} < f(s) > = \sum_{\substack{process \\ s \to s''}} \left\langle [f(s'') - f(s)] \begin{pmatrix} loss \\ rate \\ s \to s'' \end{pmatrix} \right\rangle \qquad (4\text{-}22)$$

Because the population P_n is a unique function of the ionization/excitation state s, this equation can be used to derive a rate equation for the average one-electron population $< P_n >$. That will be the basis for the NLTE average-atom theory.

Line Transitions

We first consider radiative transitions between a pair of levels of principal quantum numbers n and m (e.g., K and L shells). We assume there are P_n electrons in shell n and $(D_m - P_m)$ vacancies in shell m (n < m).

Then the rate of change of the average $< P_m >$ due to these transitions is

$$\frac{d}{dt} <P_m> = + <P_n(D_m - P_m) R_{nm} n_\nu> - <P_m(D_n - P_n) R_{mn}(n_\nu + 1)> \qquad (4\text{-}23)$$

This follows from Eq. (4-22); the first term corresponds to absorptions s → s' where the state s has populations $[P_1, \ldots, P_n, \ldots, P_m, \ldots]$, and during the transition, P_n decreases by one and P_m increases by one. The coefficient is therefore $+ 1 = (P_m + 1) - P_m$. The second term corresponds to spontaneous and stimulated emission, transitions in which P_n increases and P_m decreases by one electron. The coefficient is then $(P_m - 1) - P_m = -1$.

Expression of the n → m rate as $P_n(D_m - P_m) R_{nm} n_\nu$ can be understood two ways, i.) by the kinetic argument that the rate of transitions is

457

jointly proportional to the number of electrons able to make the
transition and to the number of available final states, and ii.) by a
detailed balance argument based on the ratio of equilibrium ion densities
N_s for the initial and final states. Both arguments assume that the
electrons are randomly distributed over subshell states by rapid
collisional transitions n, ℓ, $m \rightarrow n$, ℓ', m'.

The rate coefficient R_{nm} is

$$R_{nm} = \frac{1}{D_m} \frac{\pi e^2}{mc} f_{nm} \qquad (4\text{-}24)$$

R_{mn} is the one-electron radiative rate averaged over states in the
indicated initial and final shells. For the oscillator strength f_{nm} one
can use the Kramers' formula or an improved version which uses effective
charges Q_n to describe the large effects of bound-electron screening
(More, 1981).

For reference we also form the rate equation for $< P_n P_m >$:

$$\frac{d}{dt} <P_n P_m> = \left\langle (P_n - P_m - 1) \left(\begin{matrix} \text{upward} \\ \text{rate} \end{matrix} \right) \right\rangle$$

$$+ \left\langle (P_m - P_n - 1) \left(\begin{matrix} \text{downward} \\ \text{rate} \end{matrix} \right) \right\rangle \qquad (4\text{-}25)$$

In Eq. (4-25) the upward rate is the expression $P_n(D_m - P_m) R_{nm} n_\nu$ as in
Eq. (4-23).

The Average-Atom Radiative Rate

The average-atom approximation results from neglecting correlations
between the populations of upper and lower states in Eq. (4-23), so the
rate is replaced by

$$\frac{d\bar{P}_m}{dt} = \bar{P}_n(D_m - \bar{P}_m) R_{nm} n_\nu - \bar{P}_m(D_n - \bar{P}_n) R_{mn}(n_\nu + 1) \qquad (4\text{-}26)$$

This is the contribution of line radiation to the average-atom rate
equation for $\bar{P}_m$. There are also derivatives $d\bar{P}_m/dt$ from many other
processes.

The average-atom Eq. (4-26) for the radiative rate has two properties
required for a satisfactory description of atomic radiation processes.

458

First, it reduces correctly to the LTE (equilibrium) limit, where the
rate of any process must equal that of its inverse without considering
other processes. For the rate equation of Eq. (4-26) this occurs if the
rate coefficients R_{nm} and R_{mn} are equal,

$$\frac{R_{nm}}{R_{mn}} = \exp\left(\frac{\bar{E}_n + h_\nu - \bar{E}_m}{kT}\right) = 1 \qquad (4-27)$$

To obtain exact cancellation between emission and absorption we require
Eq. (4-27) and also, of course, the electron populations must be the
equilibrium Fermi-Dirac functions and the photon distribution must be a
Bose-Einstein function with the same temperature.

Second, the solution of Eq. (4-26) obeys an important inequality. If
$\bar{P}_m = 0$, the rate equation implies that $d\bar{P}_m/dt \geq 0$. Likewise, if $\bar{P}_m = D_m$,
it is evident that $d\bar{P}_m/dt < 0$. Thus the calculated population is bounded
by

$$0 \leq \bar{P}_m \leq D_m \qquad (4-28)$$

The result $\bar{P}_m$ of averaging integer populations also is obviously
constrained by the same inequality (as is the LTE Fermi function
approximation for $\bar{P}_m$).

Electron-Impact Excitation Rate

The average-atom rate equation for electron-impact excitation is very
similar. In this case we again consider transitions $n \underset{\leftarrow}{\overset{\rightarrow}{}} m$ and have the
decoupled rate equation

$$\frac{d\bar{P}_m}{dt} = n_e \, \bar{P}_n(D_m - \bar{P}_m) \, E_{nm} - n_e \, \bar{P}_m(D_n - \bar{P}_n) \, E_{mn} \qquad (4-29)$$

The rate coefficient E_{nm} is

$$E_{nm} = \frac{1}{D_m} <\sigma_{nm} \, v> \qquad (4-30)$$

where σ_{nm} is the electron-impact cross-section.

The detailed balance relation for this reaction takes the form

$$\frac{E_{nm}}{E_{mn}} = \exp\left(\frac{\bar{E}_n - \bar{E}_m}{kT}\right) \qquad (4-31)$$

Collisional-Radiative Plasma

To see how the NLTE average-atom theory works, we begin with an approximate representation of a low-density plasma, in which we consider an ion with a partially filled valence shell of principal quantum number g and examine the coupling to states in an excited shell n. The rate equation including electron-impact excitation and radiative decay is

$$\frac{d\bar{P}_n}{dt} = n_e \, \bar{P}_g (D_n - \bar{P}_n) \, E_{gn} - \bar{P}_n (D_g - \bar{P}_g) \, R_{ng} \qquad (4-32)$$

As written, Eq. (4-32) omits stimulated emission of the n → g radiative transition and the inverse line absorption process; this omission is reasonable for an optically thin plasma without line trapping, in which the resonant photon population is negligible, but in many real plasmas (especially dense laser plasmas) the approximation would require correction.

Equation (4-32) also omits collisional de-excitation, collisional excitation or impact ionization out of the excited states, and 3-body recombination into the excited state. These processes become less important as the electron density is reduced (the rates are of order n_e^2); however for high quantum numbers the rates are large and these process cannot be neglected; the populations for highly-excited states reach equilibrium with the continuum.

Finally, Eq. (4-32) omits cascade processes, such as impact excitation to a shell m > n followed by radiative decay to shell n, or impact ionization followed by radiative recombination. The cascade rates are comparable to the terms included in Eq. (4-32), at least as far as density dependence is concerned; they are omitted from Eq. (4-32) only for simplicity.

All the processes mentioned are usually included in the computational implementation of the average-atom model.

If the state g corresponds to a partially-filled valence shell, so the typical ion in the plasma has an open-shell configuration, then neither $\bar{P}_g$ nor $(D_g - \bar{P}_g)$ is very small.

Then the steady-state solution of Eq. (4-32) is

$$\frac{\bar{P}_n}{D_n - \bar{P}_n} = n_e \frac{\bar{P}_g}{D_g - \bar{P}_n} \frac{E_{gn}}{R_{ng}}$$

(4-33)

At low densities $\bar{P}_n \ll D_n$ and the left-hand side is approximately $\bar{P}_n / D_n$. Thus the excited state population is proportional to n_e at low densities, which shows why processes like impact ionization out of the excited state are $O(n_e^2)$.

The result in Eq. (4-33) is close to that obtained from the fundamental detailed-configuration approach: the excited state populations are <u>linear</u> in the electron density because (in this model) to have an excitation, one must wait for an electron impact, and the excited state lasts a time determined by the spontaneous radiative decay-rate, so the population is fixed by the ratio of these rates.

Closed-Shell Plasma

Now we ask what happens in the case of a plasma populated by closed-shell ions. It might be difficult to create such a plasma in the laboratory, because there would normally be ions of other charge states. However as a thought experiment we consider a purely neon-like or helium-like plasma and neglect any other charge states. This idealized plasma is of course just what is desired for certain types of x-ray laser.

We assume that the majority of ions are helium-like ground-state ions, which have $P_K = 2$ and $P_L = 0$, but that there are also a small number of helium-like excited ions with $P_K = P_L = 1$. By straightforward averaging we see

$$< P_L (D_K - P_K) > \simeq \bar{P}_L$$

(4-34)

but also

$$(D_K - \bar{P}_K) = \bar{P}_L$$

(4-35)

Thus we have

$$\frac{<P_L (D_K - P_K)>}{\bar{P}_L (D_K - \bar{P}_K)} \gg 1$$

(4-36)

In this case, the average of the product is much larger than the product of the averages, so there is a strong correlation between the populations.

Why is this? What is the meaning of the strong correlation?

The clue comes from examining the (incorrect) average-atom radiative rate; it is proportional to the denominator in Eq. (4-36) and therefore the AA radiative rate is <u>much too small</u>. The physical reason is that omitting the correlation, as in the average-atom description, gives the false impression that the (few) excited electrons have a difficulty to radiate because of a shortage of K-shell holes. In reality, each excited electron is on the same ion as a K-shell hole and is perfectly free to radiate. Keeping track of the correlation restores a reasonable radiative rate.

By giving too small a radiative rate, the average-atom description will exaggerate the excited-state population and therefore might tend to exaggerate the population inversion which can be produced in a helium-like or neon-like plasma.

Because of Eq. (4-36) it is clear that

$$C_{KL} = <P_K \, P_L> - \bar{P}_K \, \bar{P}_L \simeq - \bar{P}_L \; (< 0) \tag{4-37}$$

The population correlation between shells is <u>negative</u> in this case. Intuitively, this is because one active electron is moving between the K and L shells; when it is in the L-shell, the K-shell population is automatically below average.

Ground-State or Coronal Plasma

Next we examine another extreme case which shows a complementary correlation phenomenon. We consider a ground-state plasma, which contains a distribution of charge-states with each ion in its ground-state. This time, the situation is less artificial and occurs normally in any time-independent low-density plasma.

To be definite we assume there are H-like, He-like and Li-like ions, with populations of order unity, and that the excited state populations are negligible. In this case, averaging shows

$$<P_L(D_K - P_K)> \,\tilde{=}\, 0 \tag{4-38}$$

while $\bar{P}_L$ is proportional to the (finite) density of Li-like ions, and $(D_K - \bar{P}_K)$ is proportional to the (finite) density of H-like ions. Thus

462

$$\frac{<P_L(D_K - P_K)>}{\bar{P}_L(D_K - \bar{P}_K)} \ll 1 \qquad\qquad (4\text{--}39)$$

This inequality has the opposite sense from Eq. (4–36), and this time the average–atom radiative rate is <u>much too large</u>. What happened here?

In the average–atom description, the finite population of the L–shell sees a finite density of vacant states in the K–shell, and considers it permitted to radiate. In the actual plasma, ions with an L–shell electron do not have a vacant state in their K–shell; the vacancies occur only on the H–like ions. The correct radiative rate is proportional to $< P_L(D_K - P_K) >$ and this is vanishingly small (proportional to the excited–state population).

In this case, the correlation between shells is <u>positive</u>:

$$C_{KL} = <P_K P_L> - \bar{P}_K \bar{P}_L \simeq \bar{P}_L(D_K - \bar{P}_K) > 0 \qquad\qquad (4\text{--}40)$$

This positive correlation is the result of ionization and recombination; i.e., the way electrons are added to or subtracted from the ground–state ion. When there is a vacancy in the K–shell, the L–shell is already empty, and when the L–shell is occupied, the K–shell is full.

Now we can see why the average–atom model often gets a satisfactory answer. The correlations produced by excitation/de–excitation are <u>negative</u>, while the intershell correlations produced by ionization/ recombination are evidently <u>positive</u>. In the real plasma, both types of process occur and then the resulting correlation is often quite small, as it was in LTE.

Inequalities

The rigorous rate equations are not easily solved, because the equation for $<P_n>$ involves $<P_n P_m>$ and the equation for $<P_n P_m>$ involves higher correlated averages. To get a closed system one must decouple these equations, neglecting some correlations and keeping others. What are the constraints that decoupled equations must satisfy in order to give sensible results? One constraint comes from the requirement to reduce to LTE; this is the <u>detailed balance</u> equation relating rate coefficients. Another constraint comes from examining physical limits where we think we know the answer.

An additional constraint comes from the requirement to keep the variables in their proper ranges. For example, the average $< P_n >$ must obey $0 \leq < P_n > \leq D_n$. For the basic average-atom rate equation, this constraint is built into the structure of the rates. For higher-order rate equations the question is more delicate. Here we examine inequalities which apply to the exact rate equations and give some guidance.

The most important of these constraints are:

i.) $< P_n^2 > \geq \bar{P}_n^2$

ii.) $< P_n^2 > \geq \bar{P}_n$

iii.) If $\bar{P}_n = 0$, then $< P_n P_m > = 0$

iv.)   If $< P_n P_m > = D_n \bar{P}_m$, then $< P_n P_m P_g > = D_n < P_m P_g >$

v.) $< P_n P_m >$ is limited by

$$ < P_n P_m > \quad \geq \quad \text{Max}(0, D_n \bar{P}_m + \bar{P}_n D_m - D_n D_m) \qquad (4\text{-}41) $$

$$ \text{Min}(\bar{P}_n D_m, D_n \bar{P}_m) \quad \geq \quad < P_n P_m > \qquad (4\text{-}42) $$

The first inequality is easily proven by examining the average $<(P_n - \bar{P}_n)^2>$, which is obviously greater than or equal to zero.

For the inequality ii.), it is important that P_n can only take on integer values $0, 1, 2, \ldots, D_n$. Because of this the product $P_n(P_n - 1)$ can only have values $\geq$ zero.

Theorem iii.) is a consequence of the fact that the possible values of P_n are ≥ 0. Therefore $\bar{P}_n$ can only vanish if the probability distribution itself vanishes for all cases with $P_n \neq 0$.

For theorem iv.) the reasoning is similar: if $< (D_n - P_n) P_m > = 0$, that means that only states s for which the product $(D_n - P_n) P_m$ is zero have finite probabilities and therefore that any average $< (D_n - P_n) P_m f(P's) >$ is zero for any function f of the populations.

Finally, v.) is the main inequality for correlations, established by combining the previous results. For example, $< (D_n - P_n) P_m > \quad \geq \quad 0$

translates into one of the upper limits, and $< (D_n - P_n)(D_m - P_m) > \geq 0$ is a lower limit on $< P_n P_m >$.

Of course the rigorous rate equations must preserve the inequalities. To show this requires proof of many small theorems; we give two typical examples.

First consider transitions between a pair of levels n, g, and write the generic rate equations based on Eq. (4-23, 25)

$$\frac{dP_n}{dt} = - <P_n(D_g - P_g)> R_\downarrow + <P_g(D_n - P_n)> R_\uparrow \tag{4-43}$$

$$\frac{d}{dt} <P_n P_g> = <(P_n - P_g - 1) P_n(D_g - P_g)> R_\downarrow +$$

$$<(P_g - P_n - 1)(D_n - P_n) P_g> R_\uparrow \tag{4-44}$$

These are assumed to be exact rate equations (the rate coefficient R has already been decoupled, but this step has no effect on the inequalities). We want to establish that $\bar{P}_n$ is bounded by $0 \leq \bar{P}_n \leq D_n$ and that $< P_n P_g >$ is bounded by Eqs. (4-41, 42), as far as the rates shown in Eqs. (4-43, 44) are concerned.

Using inequalities iii.), iv.) it is clear that the appropriate positive or negative rate in Eq. (4-43) vanish and $\bar{P}_n$ stays in the desired range. Is the same true for $< P_n P_g >$? We examine this rate to show that $< P_n P_g >$ is always ≥ 0. The proof is to assume that $< P_n P_g >$ __is__ zero; if this is the case the derivative becomes

$$\frac{d}{dt} <P_n P_g> = <(P_n - 1) P_n> D_g R_\downarrow + <(P_g - 1) P_g> D_n R_\uparrow$$

Now by inequality ii.), it is clear that the right-hand side is always greater than or equal to zero and so $< P_n P_g >$ cannot further decrease. The remaining inequalities are easily established by similar methods.

SUMMARY

We have found that electron population correlations can give very large corrections to the average-atom model, and that these correlations are largely included when we write rate equations for the averages $< P_n P_m >$. Of course, the correlation effects are already included in the detailed configuration rate equations (i.e., Eq. (4-20) for dN_s/dt), but

465

these equations become somewhat impractical when large number of excited states are involved, as is the case for many-electron ions.

The results obtained are expressed in the notation of the screened hydrogenic model, but apply equally to any other average-atom model and would apply in particular to a self-consistent field non-equilibrium theory.

5. EFFECTS OF HIGH DENSITIES

In this final section we will not attempt an orderly development of the theory of density effects because the requisite theory does not entirely exist, and because too much of the subject involves statistical mechanics rather than atomic physics. Instead we simply raise a number of the questions which must be answered by further research.

Many density effects are relatively well-understood, for example, line-broadening of deeply-bound states for which the plasma effects are only a small perturbation. The theory of line broadening involves a rich mathematical structure (Griem, 1969) capable of describing correlations of perturbations, and probably remains valid for high densities except for the loss of rigor when density effects give large percentage changes in line positions and the perturbation series breaks down.

One central question is how to generalize the "exact" Saha equation which fails at high densities. If one simply solves the Saha equation, one obtains the incorrect prediction of complete recombination at high density. If the Saha equation is modified with a density-dependent reduction of the ionization potential, one can force a degree of pressure ionization, but the results remain rather unsatisfactory: a typical difficulty is the (incorrect) prediction of sharp discontinuities as entire shells are pushed into the continuum, a prediction inconsistent with equation of state experiments on shock-compressed matter (More, 1985).

Probably the physical cause of the difficulties is the fact that the Hamiltonian for one atom (ion) at high densities is no longer diagonal in a basis in which the ion charge state is sharply defined; rather there are inter-atomic hybridization matrix-elements which render the Saha approach invalid. The basic idea here is known in molecular physics as the transition from the Heitler-London limit (equivalent to the Saha equation) to the molecular-orbital picture which applies at small inter-atomic separations.

The average-atom model does not assume quantization of the ion charge and therefore can describe a smooth transition from thermal population fluctuations to quantum population fluctuations. When a level is near the continuum, its individual electronic configurations change rapidly due to electron-electron collisions, but this just means that states away from the average do not persist. As a bound-state rises through zero energy, it becomes a shape resonance, and acquires a quantum lifetime for elastic tunneling as well as a lifetime for interruption by collisions. In principle the population is not very different from the Fermi function; More (1985) derives the formula

$$\frac{P_{n\ell}}{D_{n\ell}} = -\frac{1}{\pi} I_m \int_0^\infty \sqrt{\frac{E}{\tilde{E}_{n\ell}}} \frac{f(E) \ dE}{E - \tilde{E}_{n\ell}} \tag{5-1}$$

for the average population of a resonance state in the self-consistent field theory, where $\tilde{E}_{n\ell} = E_{n\ell} + i \Gamma_{n\ell}$ is the complex eigenvalue for the resonance and $f(E)$ is the Fermi function. All this applies to the self-consistent field theory.

When we think of an atom (ion) in a plasma environment surrounded by other point charges, many new questions come into play.

How close do these other ions come? To answer this we use numerical simulations of the statistical mechanics (Monte Carlo method) and/or the classical mechanics (Molecular Dynamics method) for a system of point-charge ions. The simulations provide a great deal of information but the single quantity most often used is the pair-distribution function $g(r)$, which tells the density of other ions around a given ion. While $g(r)$ is a two-particle correlation function, it reduces to a function of radius r and is therefore sufficiently simple to work with.

Numerical simulations of point charges interacting with Coulomb potentials give the most accurate data for $g(r)$. There is also an elaborate theory which attempts to calculate $g(r)$ from first principles by mathematical rearrangements of the Boltzmann distribution, reviewed by Baus and Hansen (1981).

Can one usefully imply $g(r)$ to construct a more accurate potential? This recurrent idea encounters a severe difficulty because it replaces a point–charge neighbor by a spherical shell of charge. A point charge can bind electrons, but unfortunately a spherically symmetric shell of the same charge cannot. Thus if the neighbor ions carry bound electrons, and if one constructs a potential from their charge density in the form $Ze_n g(r)$, this will predict that no electrons are bound to the neighbor ions.

This problem can only be solved by introducing an effective charge Qe for the neighbor ions but this procedure rapidly becomes ambiguous or dependent on ad hoc assumptions, because the electrons bound to the neighbor ions are treated differently from those attached to the central nucleus.

The fundamental difficulty here is again a question of correlations. It is unsatisfactory to average the potential $V(r)$ before testing for bound states in the potential.

Another example of this difficulty appears in the analysis of bound-states near the continuum limit. In this case, the naive approach is to solve the Schroedinger equation in a relatively long–range Debye-screened Coulomb potential; this predicts a series of excited states extending into the region well beyond the inter–atomic distance. However, electrons occupying such states would be subject to the potentials of neighbor ions, a strong perturbation. One can estimate the strength of the perturbation using the mean–square potential given in Eq. (2-27); unless the binding energy is large compared to the fluctuation in the potential the entire calculation is meaningless.

Recently, this difficulty has been attacked by a new approach, in which exact or partially exact solutions of the two–center problem are combined to provide a tractable model for the strongest density effects (Rose, unpublished; Younger, unpublished; Nguyen et al., unpublished).

ACKNOWLEDGMENT

The author wishes to acknowledge helpful discussions and collaborations with many scientists including D. S. Bailey, J. Green, F. Grimaldi, H. Griem, D. A. Liberman, F. Perrot, K. H. Warren, G. B. Zimmerman, Z. Zinamon. The author is very grateful for the hospitality of the Ecole Polytechnique, GRECO-ILM, Palaiseau, France, where much of this work was performed.

REFERENCES
Bates, D. R. and Damgaard, A., (1949) Phil. Trans. R. Soc. Lond. A242, 101.

Bauche-Arnoult, C., Bauche, J., and Klapisch, M. (1979), Phys. Rev. A 20, 2424; (1982), Phys. Rev. A 25, 2641; (1985), Phys. Rev. A 31, 2248.

Baus, M. and Hansen, J.-P. (1980), Phys. Reports 59, 1.

Berthier, E., Delpech, J.-F., and Vuillemin, M. (1986), J. de Physique 47, C6-327.

Bethe, H. A., and Salpeter, E. E. (1957), Quantum Mechanics of One- and Two-Electron Atoms, Academic Press Inc., New York.

Busquet, M. (1982), Phys. Rev. A 25, 2302.

Cauble, R., Blaha, M., and Davis, J. (1984), Phys. Rev. A 29, 3280.

Charatis, G., Busch, G. E., Shepard, C. L., Campbell, P. M., and Rosen, M. D. (1986), J. de Physique 47, C6-89.

Clark, R. E. H. and Merts, A. K., (1987) J. Quant. Spectrosc. Radiat. Transfer 38, 287.

Condon, E. U., and Shortley, G. H. (1967), Theory of Atomic Spectra, Cambridge University Press.

Duston, D., and Davis, J. (1980), Phys. Rev. A 21, 1664.

Fabbro, R., Faral, R., Virmont, J., Pepin, H., Cottet, F., and Romain, J. P. (1986), Laser and Particle Beams 4, (413).

Feynman, R. P., Metropolis, N., and Teller, E. (1949), Phys. Rev. 75, 1561.

Green, J. (1964), JQSRT 4, 639.

Griem, H. R. (1964), Plasma Spectroscopy, McGraw-Hill, Inc.

Griem, H. R. (1988), J. de Physique 49, C1-293.

Grimaldi, F., and Grimaldi-Lecourt, A. (1982), JQSRT 27, 373.

Hagelstein, P. L. (1981), unpublished report UCRL-53100, Lawrence Livermore National Laboratory.

Hohne, F. E. and Zimmerman, R. (1982), J. Phys. B 15, 2551.

Hummer, D. G., and Mihalas, D. (1988), Ap. J. 331, 794.

Itoh, M., Yabe, T., and Kiyokawa, S. (1987), Phys. Rev. A 35, 233.

Klapisch, M. (1971), Comp. Phys. Comm. 2, 239.

Kramers, H. A. (1923), Phil. Mag. 271, 836.

Lee, Y. T. (1987), JQSRT 38, 131.

Liberman, D. A. (1979), Phys. Rev. B 20, 4981.

Lokke, W. A., and Grasberger, W. H. (1977), unpublished report UCRL-52276, Lawrence Livermore Laboratory.

Mayer, H. (1947), unpublished report LA-647, Los Alamos Scientific Laboratory, Los Alamos, New Mexico.

McWhirter, R. W. P. (1965), Plasma Diagnostic Techniques, p. 201 Ed. by R. Huddlestone and S. Leonard, Academic Press, New York.

Miller, W. H. (1974), Adv. in Chem. Phys., 25, 345.

More, R. M. (1981), unpublished report UCRL-84991, Lawrence Livermore National Laboratory.

469

More, R. M. (1982), J.Q.S.R.T. <u>27</u>, 345.

More, R. M. (1982), JQSRT <u>27</u>, 345.

More, R. M. (1985), Adv. At. and Molec. Phys. <u>21</u>, 305.

More, R. M. and Warren, K. H., (1989a), J. Phys. France <u>50</u>, 35.

More, R. M. and Warren, K. H. (1989b), unpublished.

More, R. M., Warren, K. H., Young, D. A., and Zimmerman, G. B. (1988), Phys. Fluids <u>31</u>, 3059.

More, R. M., Zimmerman, G. B., and Zinamon, Z. (1988), in <u>Atomic Processes in Plasmas</u>, p. 33, Ed. by A. Hauer and A. Merts, AIP Conference Proceedings 168, AIP, New York.

More, R. M., and Warren, K. H. (1989a), J. de Physique <u>50</u>, 35.

Naccache, P. F., J. Phys. <u>B 5</u>, 1308 (1972).

Ng, A., Parfeniuk, D., DaSilva, L., and Celliers, P. (1986), Laser and Particle Beams <u>4</u>, 555.

Niehaus, A. and Zwakhals, C. J., (1983) J. Phys. <u>B 16</u>, L135 (1983).

Oumarou, B., Picart, J., Tran Minh, N., and Chapelle, J., (1988) Phys. Rev. <u>A37</u>, 1885.

Percival, I. C. and Richards, D., (1975), Adv. At. Molec. Phys. <u>11</u>, 1.

Perrot, F. (1982), Phys. Rev. <u>A 26</u>, 1035.

Post, D. E., Jensen, R. V., Tarter, C. B., Grasberger, W. H., and Lokke, W. A. (1978), Atomic Data and Nuclear Tables <u>20</u>, 1.

Rosen, M. D., Phillion, D. W., Rupert, V. C., et al. (1979), Phys. Fluids <u>22</u>, 2020.

Rozsnyai, B. F., (1972), Phys. Rev. <u>A5</u>, 1137.

Salzmann, D. (1979), Phys. Rev. <u>A 20</u>, 1704.

Seely, J. F., Ekberg, J. O., Brown, C. M., Feldman, U., Behring, W. E., Reader, J., and Richardson, M. C., (1986), Phys. Rev. Letters <u>57</u>, 2924.

Slater, J. C. (1951), <u>Quantum Theory of Matter</u>, McGraw-Hill Book Co., New York.

Slater, J. C., and Krutter, H. M. (1935), Phys. Rev. 47, 559.

Sommerfeld, A. (1934), <u>Atomic Structure and Spectral Lines</u>, 3rd Ed., Methuen and Co. LTD, London.

Sommerfeld, A. (1934), <u>Atomic Structure and Spectral Lines</u>, 3rd Edition, Methuen, London.

Tragin, N., Geindre, J.-P., Monier, P., Gauthier, J.-C., Chenais-Popovics, C., Wyart, J.-F., and Bauche-Arnoult, C., (1988), Physica Scripta <u>37</u>, 72.

Tsakiris, G. D., and Eidman, K. (1987), JQSRT 38, 353.

Weisheit, J. C. and Shore, B. W. (1974), Astrophys. J. <u>194</u>, 519.

Zimmerman, G. B., and More, R. M. (1980), JQSRT 23, 517.